Health Properties of Plant Bioactive Compounds: Immune, Antioxidant and Metabolic Effects

Health Properties of Plant Bioactive Compounds: Immune, Antioxidant and Metabolic Effects

Editors

Antonio Carrillo-Vico
Ivan Cruz-Chamorro

MDPI • Basel • Beijing • Wuhan • Barcelona • Belgrade • Manchester • Tokyo • Cluj • Tianjin

Editors

Antonio Carrillo-Vico
Instituto de Biomedicina de Sevilla
IBiS/Hospital Universitario Virgen del Rocío/
CSIC/Universidad de Sevilla
Seville
Spain

Ivan Cruz-Chamorro
Departamento de Bioquímica Médica y Biología Molecular e Inmunología
Facultad de Medicina
Universidad de Sevilla
Seville
Spain

Editorial Office
MDPI
St. Alban-Anlage 66
4052 Basel, Switzerland

This is a reprint of articles from the Special Issue published online in the open access journal *International Journal of Molecular Sciences* (ISSN 1422-0067) (available at: www.mdpi.com/journal/ijms/special_issues/Plant_Bioactive_Peptides).

For citation purposes, cite each article independently as indicated on the article page online and as indicated below:

LastName, A.A.; LastName, B.B.; LastName, C.C. Article Title. *Journal Name* **Year**, *Volume Number*, Page Range.

ISBN 978-3-0365-8013-5 (Hbk)
ISBN 978-3-0365-8012-8 (PDF)

Cover image courtesy of Ivan Cruz-Chamorro

Contents

About the Editors

Antonio Carrillo-Vico

Antonio Carrillo Vico (Ph.D.), B.S. in Biology by the University of Seville in 2005, Ph.D. in Cellular and Molecular Biology by the University of Seville in 2005 (Extraordinary award). He has held several positions as FEBS postdoctoral fellow, lecturer, and associate professor. He completed the postdoctoral period at the Institute of Immunology and Infection Research at the University of Edinburgh. He is full professor of the Department of Medical Biochemistry and Molecular Biology and Immunology at the Faculty of Medicine of the University of Seville and leads the Molecular NeuroInmunoEndocrinology research group at the Institute of Biomedicine of Seville (IBiS). The long-term goals of the Dr. Carrillo Vico research programme is to evaluate melatonin as a therapeutic tool in inflammatory pathologies and to determine the immunoendocrine actions of plant bioactive peptides. He has participated in more than 25 competitive research projects in the fields of biomedicine and science and food technology. He has a track record of publication of 65 JCR articles with 3700 citations and an h-index of 30 (WoS, June-2023). He is also author of books, book chapters, articles of scientific popularisation, and more than 50 communications at national and international congresses. He is also director of 11 Ph.D. theses. He is member of the Editorial Commitees of 3 journals and reviewer of more than 30 journals.

Ivan Cruz-Chamorro

Ivan Cruz-Chamorro (Ph.D.) is a post-doctoral researcher at the University of Seville, Spain. Since 2012, he has been a component of the Molecular Neuroimmunoendocrinology Laboratory of the Institute of Biomedicine of Seville (IBiS). In 2018, he discussed his doctoral thesis, "Assessment of the health effects of plant bioactive peptides: immunomodulatory, antioxidant, and metabolic effects", obtaining an outstanding *cum laude* rating. Ivan is involved in studying the biological effects (immunomodulators, antioxidants, lipid-lowering, etc.) of protein hydrolysates from several plant sources, such as lupine, olive, and hemp. He has directed a doctoral thesis in which the capacity of protein hydrolysate from lupine has been explored in the metabolic syndrome (MetS) and in the preclinical mouse model of hypercholesterolemia. Iván collaborates with different international groups and he is exploring the effects of different natural matrices on some diseases, such as MetS, metabolic-associated fatty liver disease, and multiple sclerosis. At last, his future objective is to explore the biological effects of these matrices on the cancer models.

Preface to "Health Properties of Plant Bioactive Compounds: Immune, Antioxidant and Metabolic Effects"

Given that the best way to improve the quality of life of a human being is prevention of illness, the study of functional compounds, natural food compounds that, in addition to meeting nutritional needs, have beneficial properties on health, has become an emerging field in the area of food technology and health due to the increasing popularity of functional foods among consumers who continually demand healthier products that improve their quality of life and thus prevent or treat diseases. Therefore, this reprint includes a collection of research findings published in the Special Issue of the *International Journal of Molecular Sciences*, titled "Health Properties of Plant Bioactive Compounds: Immune, Antioxidant and Metabolic Effects". This collection provides an updated overview of studies focusing on the positive multifunctional effects of vegetable-derived compounds, such as peptides, extracts, essential oils, phytosterols, and phenolic compounds, on the main components of chronic diseases that confer their potential as novel functional foods.

Antonio Carrillo-Vico and Ivan Cruz-Chamorro
Editors

International Journal of
Molecular Sciences

MDPI

Editorial

Health Properties of Plant Bioactive Compounds: Immune, Antioxidant, and Metabolic Effects

Ivan Cruz-Chamorro [1,2,*] and Antonio Carrillo-Vico [1,2,*]

1 Departamento de Bioquímica Médica y Biología Molecular e Inmunología, Facultad de Medicina, Universidad de Sevilla, 41009 Seville, Spain
2 Instituto de Biomedicina de Sevilla, IBiS/Hospital Universitario Virgen del Rocío/CSIC/Universidad de Sevi-lla, 41013 Seville, Spain
* Correspondence: icruz-ibis@us.es (I.C.-C.); vico@us.es (A.C.-V.)

In recent decades, people in the industrialized world have increased the demand for meat-free foods motivated by health, environmental, and animal welfare reasons. Thus, plant products and their derivatives have gained great popularity in nutrition [1]. Meanwhile, the World Health Organization (WHO) recommends the frequent consumption of plant foods instead of foods of animal origin, which contain a considerable amount of saturated fat and cholesterol [2].

The high consumption of plant products has generated an increase in the study of the beneficial properties of their components beyond their basic macro- and micro-nutrients [3,4]. In this context, several studies have shown that vegetable-derived peptides have multifunctional effects related to the main components of chronic diseases, which have attracted interest from the food, nutraceutical, and pharmaceutical industries.

In this Special Issue, several studies in which different vegetable extracts were tested have been compiled. The enzyme-assisted extraction of plant compounds is a widely used method and can be a strategy for sustainable and functional applications [5]. This method was used to produce an extract of *Castana sativa* (chestnut) that was shown to have antimicrobial activity against Streptococcus and Staphylococcus, among others [6]. These microorganisms are present in the oral cavity during oral mucositis, a common side effect of oncological treatment. Moreover, the high phenolic content of this extract has been shown to possess antioxidant activity. Other extracts of *Chenopodium quinoa* (quinoa), *Amaranthus retroflexus* (amaranth), and *Fagopyrum esculentum* (buckwheat), obtained by simulated gastrointestinal digestion, showed protective effects against IL-1-induced inflammation in vitro [7]. Additionally, an extract of *Protium heptaphyllum* gum resin has been shown to reduce cholesterol production in human liver cells and regulate the gene expression of several proteins involved in cholesterol metabolism [8].

Methyl p-coumarate, an esterified derivative of p-coumaric acid and a naturally occurring compound in plants, has been shown to be an effective agent for reducing inflammation in an experimental in vivo model of allergic asthma, reducing the influx of immune cells and mucus secretion in the lung, among other activities [9].

Another natural compound was also tested in a mouse model of anxiety. In that study, a protein hydrolysate from *Lupins angustifolius* (lupin) was able to reduce anxiety in mice evaluated by the elevated plus maze and Morris water maze behavior tests [10].

The clinical application of a vegetable extract of *Rosmarinus officinalis* (rosemary) was studied in four patients with osteogenesis imperfecta, a genetic connective tissue disease [11]. In this study, the extract reduced collagen accumulation in the fibroblasts of patients, due to increased autophagy of fibroblasts. Furthermore, the rosemary extract attenuated pro-apoptotic markers, such as cleaved caspase 3.

Finally, the biological effects of *Lycium barbarum* berries [12], the essential oils of *Annonaceae* species [13], antioxidant compounds involved in several disorders [14], as well

Citation: Cruz-Chamorro, I.; Carrillo-Vico, A. Health Properties of Plant Bioactive Compounds: Immune, Antioxidant, and Metabolic Effects. *Int. J. Mol. Sci.* **2023**, *24*, 7916. https://doi.org/10.3390/ijms24097916

Received: 2 February 2023
Accepted: 17 March 2023
Published: 26 April 2023

as phytosterols in neurodegenerative diseases [15], and different natural compounds in the prevention and treatment of oral mucositis [16] were reviewed.

This Special Issue provides an overview of the current research being carried out with vegetable-derived bioactive compounds. All of these findings confirm and point out that these compounds may be possible new nutraceuticals capable of preventing or treating several diseases.

Author Contributions: Conceptualization, I.C.-C. and A.C.-V.; writing—original draft preparation, I.C.-C.; writing—review and editing, A.C.-V. All authors have read and agreed to the published version of the manuscript.

Funding: Funding was provided by the Spanish Government, Ministerio de Economía y Competitividad (AGL2012-40247-C02-01 and AGL2012-40247-C02-02), the Andalusian Government Ministry of Health (PC-0111-2016-0111, PEMP-0085-2020 co-financed with FEDER funds, from the Resolution Call of 7 July 2021 of the General Secretary for Research, Development and Innovation in Health, which called for grants to finance research, development and innovation in biomedicine and health sciences in Andalusia by 2021), and the PAIDI Program from the Andalusian Government [CT-S160]. I.C.-C. was supported by a postdoctoral fellowship from the Andalusian Government Ministry of Economy, Knowledge, Business, and University (DOC_00587/2020).

Conflicts of Interest: The authors declare no conflict of interest.

References

1. Christopher, A.; Bartkowski, J.P.; Haverda, T. Portraits of veganism: A comparative discourse analysis of a second-order subculture. *Societies* **2018**, *8*, 55. [CrossRef]
2. Martínez-Villaluenga, C.; Zieliński, H.; Frias, J.; Piskuła, M.K.; Kozłowska, H.; Vidal-Valverde, C. Antioxidant capacity and polyphenolic content of high-protein lupin products. *Food Chem.* **2009**, *112*, 84–88. [CrossRef]
3. García, M.; Puchalska, P.; Esteve, C.; Marina, M. Vegetable foods: A cheap source of proteins and peptides with antihypertensive, antioxidant, and other less occurrence bioactivities. *Talanta* **2013**, *106*, 328–349. [CrossRef] [PubMed]
4. Rivero-Pino, F.; Espejo-Carpio, F.J.; Guadix, E.M. Identification of dipeptidyl peptidase IV (DPP-IV) inhibitory peptides from vegetable protein sources. *Food Chem.* **2021**, *354*, 129473. [CrossRef] [PubMed]
5. Streimikyte, P.; Viskelis, P.; Viskelis, J. Enzymes-Assisted Extraction of Plants for Sustainable and Functional Applications. *Int. J. Mol. Sci.* **2022**, *23*, 2359. [CrossRef] [PubMed]
6. Ferreira, A.S.; Silva, A.M.; Pinto, D.; Moreira, M.M.; Ferraz, R.; Švarc-Gajić, J.; Costa, P.C.; Delerue-Matos, C.; Rodrigues, F. New Perspectives on the Sustainable Employment of Chestnut Shells as Active Ingredient against Oral Mucositis: A First Screening. *Int. J. Mol. Sci.* **2022**, *23*, 14956. [CrossRef] [PubMed]
7. Capraro, J.; Benedetti, S.D.; Heinzl, G.C.; Scarafoni, A.; Magni, C. Bioactivities of pseudocereal fractionated seed proteins and derived peptides relevant for maintaining human well-being. *Int. J. Mol. Sci.* **2021**, *22*, 3543. [CrossRef] [PubMed]
8. Mannino, G.; Iovino, P.; Lauria, A.; Genova, T.; Asteggiano, A.; Notarbartolo, M.; Porcu, A.; Serio, G.; Chinigò, G.; Occhipinti, A. Bioactive triterpenes of protium heptaphyllum gum resin extract display cholesterol-lowering potential. *Int. J. Mol. Sci.* **2021**, *22*, 2664. [CrossRef] [PubMed]
9. Park, J.-W.; Choi, J.; Lee, J.; Park, J.-M.; Kim, S.-M.; Min, J.-H.; Seo, D.-Y.; Goo, S.-H.; Kim, J.-H.; Kwon, O.-K. Methyl P-Coumarate Ameliorates the Inflammatory Response in Activated-Airway Epithelial Cells and Mice with Allergic Asthma. *Int. J. Mol. Sci.* **2022**, *23*, 14909. [CrossRef] [PubMed]
10. Santos-Sánchez, G.; Ponce-España, E.; López, J.C.; Álvarez-Sánchez, N.; Álvarez-López, A.I.; Pedroche, J.; Millán, F.; Millán-Linares, M.C.; Lardone, P.J.; Bejarano, I. A lupin (*Lupinus angustifolius*) protein hydrolysate exerts anxiolytic-like effects in Western diet-fed ApoE$^{-/-}$ mice. *Int. J. Mol. Sci.* **2022**, *23*, 9828. [CrossRef] [PubMed]
11. Sutkowska-Skolimowska, J.; Brańska-Januszewska, J.; Strawa, J.W.; Ostrowska, H.; Botor, M.; Gawron, K.; Galicka, A. Rosemary Extract-Induced Autophagy and Decrease in Accumulation of Collagen Type I in Osteogenesis Imperfecta Skin Fibroblasts. *Int. J. Mol. Sci.* **2022**, *23*, 10341. [CrossRef] [PubMed]
12. Teixeira, F.; Silva, A.M.; Delerue-Matos, C.; Rodrigues, F. Lycium barbarum Berries (Solanaceae) as Source of Bioactive Compounds for Healthy Purposes: A Review. *Int. J. Mol. Sci.* **2023**, *24*, 4777. [CrossRef] [PubMed]
13. Cascaes, M.M.; Carneiro, O.D.S.; Nascimento, L.D.D.; de Moraes, Â.A.B.; de Oliveira, M.S.; Cruz, J.N.; Guilhon, G.M.S.P.; Andrade, E.H.D.A. Essential oils from annonaceae species from Brazil: A systematic review of their phytochemistry, and biological activities. *Int. J. Mol. Sci.* **2021**, *22*, 12140. [CrossRef] [PubMed]
14. Khutami, C.; Sumiwi, S.A.; Khairul Ikram, N.K.; Muchtaridi, M. The Effects of Antioxidants from Natural Products on Obesity, Dyslipidemia, Diabetes and Their Molecular Signaling Mechanism. *Int. J. Mol. Sci.* **2022**, *23*, 2056. [CrossRef] [PubMed]

15. Sharma, N.; Tan, M.A.; An, S.S.A. Phytosterols: Potential metabolic modulators in neurodegenerative diseases. *Int. J. Mol. Sci.* **2021**, *22*, 12255. [CrossRef] [PubMed]
16. Ferreira, A.S.; Macedo, C.; Silva, A.M.; Delerue-Matos, C.; Costa, P.; Rodrigues, F. Natural Products for the Prevention and Treatment of Oral Mucositis—A Review. *Int. J. Mol. Sci.* **2022**, *23*, 4385. [CrossRef] [PubMed]

International Journal of
Molecular Sciences

MDPI

Review

Enzymes-Assisted Extraction of Plants for Sustainable and Functional Applications

Paulina Streimikyte *, Pranas Viskelis * and Jonas Viskelis *

Lithuanian Research Centre for Agriculture and Forestry, Institute of Horticulture, 54333 Babtai, Lithuania
* Correspondence: paulina.streimikyte@lammc.lt (P.S.); pranas.viskelis@lammc.lt (P.V.); jonas.viskelis@lammc.lt (J.V.)

Abstract: The scientific community and industrial companies have discovered significant enzyme applications to plant material. This rise imparts to changing consumers' demands while searching for 'clean label' food products, boosting the immune system, uprising resistance to bacterial and fungal diseases, and climate change challenges. First, enzymes were used for enhancing production yield with mild and not hazardous applications. However, enzyme specificity, activity, plant origin and characteristics, ratio, and extraction conditions differ depending on the goal. As a result, researchers have gained interest in enzymes' ability to cleave specific bonds of macroelements and release bioactive compounds by enhancing value and creating novel derivatives in plant extracts. The extract is enriched with reducing sugars, phenolic content, and peptides by disrupting lignocellulose and releasing compounds from the cell wall and cytosolic. Nonetheless, depolymerizing carbohydrates and using specific enzymes form and release various saccharides lengths. The latest studies show that oligosaccharides released and formed by enzymes have a high potential to be slowly digestible starches (SDS) and possibly be labeled as prebiotics. Additionally, they excel in new technological, organoleptic, and physicochemical properties. Released novel derivatives and phenolic compounds have a significant role in human and animal health and gut-microbiota interactions, affecting many metabolic pathways. The latest studies have contributed to enzyme-modified extracts and products used for functional, fermented products development and sustainable processes: in particular, nanocellulose, nanocrystals, nanoparticles green synthesis with drug delivery, wound healing, and antimicrobial properties. Even so, enzymes' incorporation into processes has limitations and is regulated by national and international levels.

Keywords: enzyme-assisted extraction; plant material; phenolic compounds; oligosaccharides; prebiotic; nanocellulose; nanofibers; fermentation; sustainability

Citation: Streimikyte, P.; Viskelis, P.; Viskelis, J. Enzymes-Assisted Extraction of Plants for Sustainable and Functional Applications. *Int. J. Mol. Sci.* **2022**, *23*, 2359. https://doi.org/10.3390/ijms23042359

Academic Editors: Antonio Carrillo Vico and Ivan Cruz-Chamorro

Received: 12 January 2022
Accepted: 17 February 2022
Published: 21 February 2022

Publisher's Note: MDPI stays neutral with regard to jurisdictional claims in published maps and institutional affiliations.

1. Introduction

The demand for new and natural compounds, 'clean label' trend, rising drug resistance, holistic wellbeing approach of the post-pandemic period, and sustainable living has intensified the development of plant-derived compounds called biologically active components [1–3]. Biologically active substances bind by interaction or binding to specific receptors in stem cells, improving a particular physiological function of the body. Unfortunately, many such compounds are present in cytosolic cell spaces and plant cell walls [4]. Many extraction methods cannot achieve these compounds and thus obtain the highest components yields. That is why enzymes incorporation in various extractions is currently one of the few methods to provide this result. Enzymes with specific hydrolytic properties are used to degrade this matrix to gain access to biologically active components from cytosolic spaces and cell walls [5]. Nevertheless, the global market for industrial enzymes is expected to grow up to $9.2 billion by 2027 [6]. One of the advantages of the usage of enzymes is that they can be added to hydrophilic and multi-step lipophilic extractions, especially for by-product valorization [7]. For example, in Europe, grain, fruit, and vegetable food loss from post-harvest to distribution varies from 20, 41, and 46%, respectively [7,8]. However, enzyme-assisted incorporation increases phenolic content in lipophilic extracts, which is potently applicable for nutraceuticals or pharmaceuticals. However, limitations in the safety of by-products have risen, and greater attention is given to this topic [8]. In comparison, for hydrophilic extracts, enzymes efficiently increase the water-soluble content of novel derivatives applicable in food industries [9].

Due to the high demand for diverse health outcomes, functional food categories arise during the post-pandemic period, especially probiotic and prebiotic categories [10]. Incorporating these foods into human diets may reduce obesity. According to WHO, from 2018 to 2030, obese children will reach from 150 billion to 250 billion, respectively [11]. Gut microbiota modulates lipogenesis and cholesterol synthesis. Dysbiosis initiates higher absorption of sugars in the small intestines by modulating membrane transport [11,12]. Moreover, acetate, the metabolite made by the gut microbiome in the proper amount, can boost immune responses by promoting B10 cells, and in higher amounts can lead to adiposity [13–15]. These challenges invite scientists to search for sustainable and functional food development worldwide. One scope is enzymes, usually used for plant-based drinks production and syrups for saccharification, decreased viscosity, higher yield, and low toxicity in the food industry. However, lately, studies suggest that controlled enzyme-assisted extraction could lead to a higher and broader density of nutrients [16,17]. For example, dietary fibers with three or more monomeric units, phenolic compounds, and complexes can be suggested prebiotics and used for functional food development [18]. Chen et al. [19] investigated amylolytic and cellulolytic enzymes impact of releasing phenolic compounds and the correlation in solid-state fermentation with significant results for increased phenolic compounds content and antioxidant activities [19].

Because the plant material is complex, with varied compositions and matrices, enzymes are used in mixtures or cocktails. Besides releasing secondary metabolites and small peptides, they cleave long-chain molecules into shorter ones. Likewise, these substances are soluble in the solvent and can enhance organoleptic, technological, and functional properties. Moreover, enzymatic extraction methods are characterized by mild reaction conditions, substrate specificity, industrial applicability, and many other advantages [20]. These extracts may be used continuously in many fields and, surprisingly, in green synthesis development [16,21,22].

Green nanoparticle synthesis in aqueous plant extracts has increased over the last decade. Scientific discussion and research indicate the appropriate size of nanoparticles with high potential antimicrobial properties, involving the most common pathogenic bacteria like *Escherichia coli*, *Staphylococcus aureus*, and widely spread, highly resistant *Candida albicans* [23,24]. Studies identify that phenolic compounds and sugars play an essential capping and stabilizing role in green nanoparticles synthesis, and enzymes incorporation could increase the synthesized media yield with economically friendly conditions [25,26].

This review briefly suggests scientific approaches of commercially used hydrolases and carbohydrases for various plant materials to extract functional ingredients, products development, and possible applications.

2. Carbohydrases and Phenolic Compounds in Plants

In this review, carbohydrases and hydrolases get a more profound overview due to the plant cell wall mainly consisting of various carbohydrates, trapping active biological components. The cell contains various linear heterogeneous polymeric carbohydrates homologous to cellulose, such as xyloglucans and mannan, and hemicellulose is covalently linked to cellulose microfibrils and lignin to form complex structural branches. This multi-component structure in the plant cell wall is called lignocellulose. However, plant polymeric substances are usually categorized to waste. The global agricultural sector is estimated to produce 5×10^9 tons of plant-derived biomass each year, where the total amount of lignocellulosic waste is about 2×10^{11} tons per year [27,28]. The structure of lignocellulose gives the stability and resistance of the cell to the extraction of internal cellular components

(Figure 1), where various enzymatic activities are required to degrade all the different forms of hemicellulose.

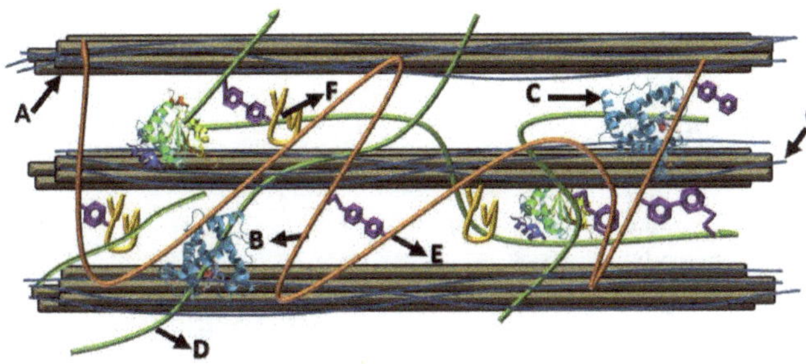

Figure 1. Plant cell wall graphical scheme, describing cross-linked phenolic compounds, peptides, and polysaccharides network adapted from Acosta et al., 2014 and Carpita et al., 2020 [29,30]. A—cellulose from cellulose microfibrils, B—hemicelluloses consisting of xyloglucans, glucuronoarabinoxylan, (1–3) (1–4) β glucans and glucomannan. C—structural proteins, D—pectin consisting of homogalacturonan, xylogalacturonan, and rhamnogalacturonans I and II; E—phenolic compounds, F—lignin; G—xylan and mannan coating of cellulose microfibrils.

Enzymes are derived from bacteria, fungi, yeasts, archaea, animal organs, or plant extracts. However, microbial enzymes are more stable compared to ones having a plant or animal origin. Moreover, the production of the enzyme during microbial fermentation is cost-effective and easily adapted to modifications and high purity [31]. Carbohydrases can be categorized in starch-degrading enzymes: amylases and glucoamylases; and non-starch polysaccharides (NSP) catalyzing enzymes with cellulolytic, pectinolytic xylanolytic activities [31–33]. In general, NSP enzymes also can be named xylases, cellulases, and pectinases due to being composed of glycoside hydrolases, carbohydrate esterases, xylanases, etc. (Table 1).

Table 1. Structure of commonly found carbohydrates in plant material and enzymes usage of their cleavage [30–33].

	Non-Starch Polysaccharides		
	Hemicellulose	**Cellulose**	**Pectin**
Consist of	xyloglucans glucuronoarabinoxylan β–glucan glucomannan	cellulose nanofibrils: (a) xylan (b) mannan	homogalacturonan rhamnogalacturonan I and II xylogalacturonan
Enzymes used in processes	Xylanases: exoxylanases β–xylosidases, xylan–1,4–β-xylosidase endoxylanases	Cellulases: endo–(1,4)–β–d–glucanase (EC 3.2.1.4) exo–(1,4)–β–d–glucanase (EC 3.2.1.91) β–glucosidases (EC 3.3.1.21) β–glucosidases (EC 3.3.1.21)	Pectinases: polygalacturonases pectin esterases pectate lyase
	Starches		
Consist of	amylose amylopectin		
Enzymes used in processes	α–Amylases (EC 3.2.1.1) β–amylase (EC 3.2.1.2) glucoamylase (EC 3.2.1.3) α-glucosidase (EC 3.2.1.20) pullulanase or amylopullulanase (EC 3.2.1.41) cyclodextrin glycosyltransferase (EC 2.4.1.19)		

NSP enzymes are preferred as a part of commercial enzyme mixture, thus ensuring complete lysis of cell walls while contributing to a cost-effective means [5,34]. Various fungi, including *Trichoderma* sp. and *Aspergillus* sp., produce carbohydrate-hydrolyzing enzymes. For many years, *Trichoderma* sp. has been extensively studied for high cellulase production [35]. However, most strains of *Trichoderma* are known to have low β-glucosidase activity, which causes cellobiose accumulation. Although much effort has been made to obtain *T. reesei* mutants by classical mutagenicity, such as RUT-C30, the relatively low activity of β-glucosidase remains one of the significant barriers to efficient cellulose hydrolysis [36]. *Aspergillus* sp. is important in xylanase production, and the latest studies showed UV-irradiated *Aspergillus* mutants for a higher yield of enzymes [37,38]. Endoxynalases, specifically endo-β-1,4-xynalases, are the most important, depolymerizing xylan polymer into small branches. Xylooligosaccharides later are converted to xylotriose, xylobiose, and xylose [33]. Another essential component of the cell wall is pectin. It is a polymer of α-D-galacturonate and L-rhamnose units linked to α-1,4 or 1,2 to form so-called pectic elbows. Pectin, associated with cellulose, imparts stiffness and cohesion to cell walls [5,34]. Galactose, mannose, fucose, arabinose, xylose, and L-methyl, O-acetyl groups all these components make four main regions of pectin structure: rhamnogalacturonan I (RG-I), rhamnogalacturonan II (RG-II), homogalacturonan (HG), and xylogalacturonan (XG), which are involved in reducing inflammatory processes in human [39]. Pectinolytic enzymes or pectinases were the first enzymes commercially available in wine and fruit juices, although the cell wall structure was only determined later [31]. They consist of three main classes: protopectinases, esterases, and depolymerases. Protopectinases occur naturally and are responsible for dissolving otherwise undissolved protopectins from immature fruit during maturation. Esterases or pectin methylesterases remove esterified units to remove methoxy esters. Depolymerases are represented by lyses and hydrolases that catalyze the fragmentation of glycosidic bonds. Today, pectinases are used in the fruit juice industry because of their effectiveness: higher juice yield, filterability, lower viscosity, and increased transparency. Pectinolytic enzymes have been found to be particularly effective in the extraction of polyphenols, particularly in the release of anthocyanidins from glycosides. Many pectinolytic mixtures sold today contain all three types of the above classes and a mixture with cellulases and hemicellulases to achieve an overall synergistic effect [5,40,41].

In general, biological raw material systems range from 5000 to 25,000 individual phytochemicals that can have biological activity. Biologically active substances are metabolites synthesized in plants that perform plant protection and other functions. There is growing evidence that biologically active substances can help maintain optimal health and reduce the risk of chronic diseases such as cancer, cardiovascular disease, stroke, and Alzheimer's disease (AD) [42,43]. For example, quercetin found in pines, buckwheats, and many other plants has a high binding link through hydrogen bonds to the cyclin-dependent kinase 6 (CDK6) and inhibits its activity, which plays an essential role in the progression of different types of cancer [44]. However, phenolic compounds are usually soluble conjugates (glycosides) or insoluble forms (phenolic acids) covalently linked to carbohydrate radicals or cell wall structural components such as cellulose, hemicellulose, lignin, and pectin. In the insoluble form, the phenolic compounds are linked to the structural cell wall components by covalent bonds. Phenolic acids bind to lignin through the hydroxyl groups of their aromatic rings and various polysaccharides and other proteins through ester linkages. They perform a protective function. Significant interests are given to flavonoids which are found in the form of glycosides bound to sugar residues via –OH groups (O-glycosides) or via –C–C– bonds (C-glycosides) [44,45]. The latest article discovers the promising neuroprotective and memory-enhancer characteristics of flavonoid-rich food and plant extracts by increasing the functions of neurons and cell proliferation [46]. Flavonoids are involved in regulating kinase-signaling cascades, including PI3K/Akt, PKC, and MAPK pathways [46,47]. In the food industry, functional food components are used to replace preservatives due to their antioxidant and antimicrobial properties [48–50]. Unfortunately, solvent-based extraction of biologically active substances often suffers from low extraction efficiency, requires a long extraction time, and often leaves traces of organic solvents in the final product, which reduces the quality of the product [51,52]. Therefore, there is a need to develop optimized, step-by-step extraction methods for beneficial substances for each type of raw material. Therefore, detailed protocols are required to produce bioactive compounds, especially from plants where the cell wall may inhibit the extraction efficiency [7,49]. Enzymatic extraction of biologically active compounds from plants is a potential alternative to conventional solvent-based extraction methods. Enzymes are ideal catalysts for extracting, modifying, or synthesizing complex biologically active compounds of natural origin. Enzyme-assisted extraction is based on the inherent ability of enzymes to catalyze reactions with exceptional specificity and the ability to function under mild processing conditions in aqueous solutions [53]. However, parameters such as enzyme specificity, activity, botanical origin, pH, enzyme and substrate, liquid to solid ratios, temperature, and time are essential for obtaining the highest value results [54,55].

Carbohydrases-active enzymes database (CAZy; https://www.cazy.org/, accessed on 10 February 2022) classified five main classes: glycoside hydrolases, glycosyltransferases, polysaccharides lyases, carbohydrate esterases, and auxiliary activities [56]. A mixture of these enzymes has to be implemented into reactions on the results maintained to achieve. Ideal selected enzymes have high activity and regio-/stereo-selectivity [57]. Moreover, raw materials of plant origin consist of a complex system of various macro-/micro-components [29]. Plant origin, morphological part, cultivar, and growing conditions impact the storage of secondary metabolites. The approach of Woo et al. [58] of fifteen cultivars in oats showed the variation by cultivars and harvesting days of phenolic content from 3748.4 to 5700.0 mg/100 g [58]. Enzyme usage usually requires low temperatures, usually 50–60 °C; short extraction periods of up to several hours, waste-free production possibilities, reduced substrate specificity, allowing extraction of many bioactive compounds, otherwise, were made inaccessible. [45,59]. Examples of various plant material extraction are presented in Table 2. Additionally, the yield of the extract is often characterized by high quality and bioavailability. Gonzalez et al. [60] compared ultrasound-assisted extraction with enzyme-assisted extraction for anthocyanins from blackcurrants, where a significant difference was not evaluated. Major two Cyanidin 3–O–rutinoside (C3R) and Cyanidin 3–O–glucoside were extracted at similar quantities and accounted for up to 90%. C3R stimulates the mechanism of insulin secretion through INS-pancreatic β-cells by promoting calcium channels and activating the PLC-IP3 pathway [61].

Table 2. Examples of various substrates technological parameters for enzyme-assisted extraction.

Enzymes	Producent	Substrate	Enzymes Quantity	Liquid to Substrate Ratio	pH	Temp. °C	Time	Ref.
Xylanase cocktail	*A. niger*	Citrus fiber	0.45%	1:20	4.5–6.5	50	120 min	Song et al. [62]
Cellulase	*A. niger*	Coffee by-products	5–15 U	1:25	5.0–6.0	50	30–20 min	Belmiro et al. [63]
Cellulase from Celluclast 1.5 L	*T. reesei*	Banana peel	5 FPU/ml	1:20	6.0–7.0	50	120 h	Phirom-on et al. [64]
Pectinase	*A. niger*	Guava pulp	0.10%	2:5	2.97–3.97	45	3–90 min	Ninga et al. [65]
Pectinase	*A. niger*	Blackcurrant	108 U/g	0.1:15 and 0.2:15	5–6	60	10–90 min	Gonzalez et al. [60]
Heat stable alpha-amylase	*Bacillus sp.*	Oat flours	0.01%	1:5	5.0–9.0	100	15–75 min	Chen et al. [66]

Furthermore, the method also allows applying greener chemistry in the food industry and pharmaceutical companies to optimize purer ways to extract new compounds. Enzymes can disrupt specific bonds and interactions in cell walls and membranes, resulting in higher extraction yields of bioactive substances [40,67,68].

3. Enzymatically Treated Polysaccharides as Possible Functional Components

Lignocellulose and starch content differs depending on plant material. Grain materials usually consist of more starches because of the endosperm in the seed, whereas other plant parts, like brans, leaves, and others, consist of non-starch polysaccharides. Shah et al. [62] described three categories of starches: rapidly digestible starch (RDS), resistant starch (RS), and slowly digestible starch (SDS), which relate glucose release during digestion. Mainly, starches are found in cereals and pseudocereals where β-glucans are the most widely investigated non-starch polysaccharide and have enormous health-promoting properties [69]. However, β-glucans properties may vary depending on molar mass, which can be from 209 up to 2500 (kg mol^{-1}) depending on the cultivar, variety, and the location of growth [70,71]. Nonetheless, β-glucans are found in fungi, yeasts, bacteria, and seaweed. The main difference is the ratio, length, and linkage between D-glucose of β-(1,3) and β-(1,4) for cereal grains β-(1,3) and β-(1,6) for fungi and yeasts [72]. For oats, novel oligosaccharides were produced by hydrolyzing β-glucans with dual enzyme-assisted hydrolysis using α-amylases and transglucosidase in result shorter α-(1,6)-branch linkage glucans, which stabilize the glucose release during the intestinal phase in vitro and may be prebiotic [73]. Another study in vitro also resulted in SDS increasing with corn starch using branching enzymes and maltogenic alpha-amylase [74]. Moreover, as a major non-starch polysaccharide in cereal grains, arabinoxylans are getting attention for their prebiotic properties, where enzymatically hydrolyzed fiber releases ferulic acid or combines with arabinose [75,76]. In general, dietary fibers, with more than three monomeric units known as prebiotics, are a growing trend for immune support through the microbiome and may prevent the uprising resistance to drugs for humans and animals [10]. Moreover, because antibiotic-free products are getting more attention from consumers, feed enhancement with dietary fibers may support the immune system of animals [19,75]. However, the clinical study by Wastyk et al. [77] for identifying humans' health marker variation while having the intervention of 10 weeks of high-fiber diet showed an increase in short-chain fatty acid (SCFAs), the density of microbes for higher protein content, and quantity but no diversity of microbiome [77]. It suggests that microbiota was insufficient to process increased fiber intake, even though glycan-degrading carbohydrate-active enzymes increased [77,78]. These studies implement the essential role of carbohydrate lengths for human diets. Animal studies with rats showed the xylan-oligosaccharides (XOS) as a potent dietary supplement of obesity prevalence. XOS promotes growth of *Bifidobacterium* strains involved in the development of obesity and insulin resistance [14,79]. In between, galacto-oligosaccharides (GOS) in studies with rats implement the potency of gut recoveries after alcohol withdrawal by a significant increase of butyric and propionic acid and the proliferation of *Lactobacillus* and *Bifidobacterium* strains [80]. Another animal study with GOS extracted from mulberry treated with β-mannanases showed that 200 mg/kg/day can initiate expression of crucial glycolysis enzymes such as GK, PK, and PCB and proteins p110 and Akt of key signaling intermediates, which results in the prevalence of diabetes and obesity [81]. A graphical scheme of summarized polysaccharides and enzymes combination results is presented in Figure 2.

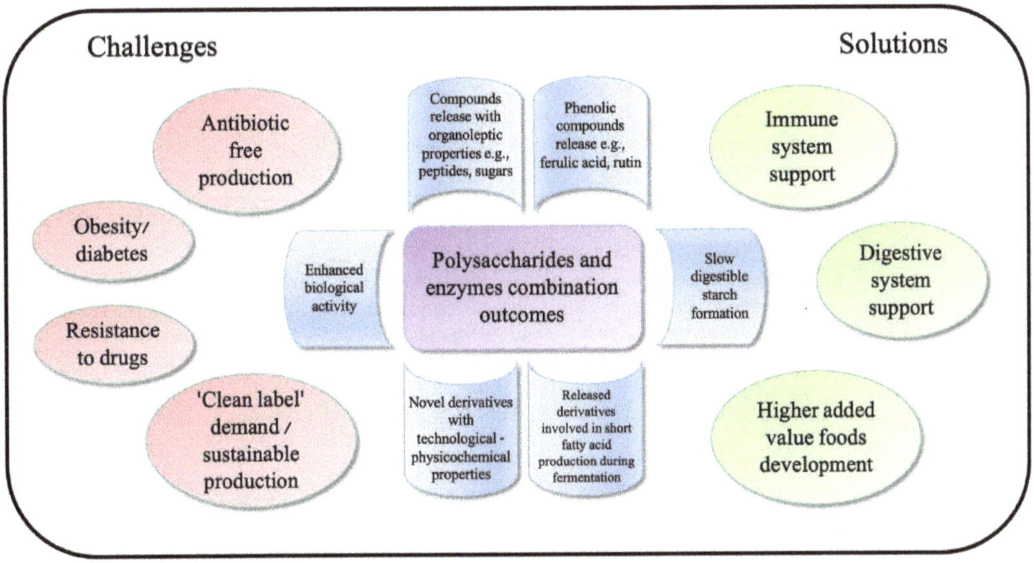

Figure 2. Summary graphic of polysaccharides and enzyme combination outcomes connected to solutions for uprising challenges.

4. Enzymes-Assisted Processes for Plant Materials

4.1. Bioactives Extraction from By-Products

In order to use the enzyme or their mixtures efficiently in extraction methods, it is essential to understand their catalytic mechanism of action and the optimal activity conditions for the recovery of individual biological raw materials and substances: e.g., a mixture of cellulose, pectin, and hemicellulose enzymes in a grapefruit peel during hydrolysis releases sugar into monomeric compounds that microorganisms can use later to produce ethanol and other fermentation products [82]. Another example is in tomatoes: lycopene is found mainly in the peel, giving it a red color. Carotenoids, especially lycopene, are one of the most potent antioxidants of plant origin, with a role in more than twenty different induced-signaling pathways and cell cycles described by Qi et al. [83]. According to scientific knowledge, lycopene is better absorbed from processed products than fresh tomatoes [84,85]. The digestive enzyme pancreatin is recommended before the solvent extraction of lycopene. Its use increased the yield of lycopene 2.5-fold compared to that obtained using the traditional extraction method [54]. Using Pectinex Ultra SP–L (P, pectinolytic enzyme), Celluclast (C, cellulolytic enzyme), and Viscozyme L commercial enzymes for solvent extractions, the yield of lycopene was increased as well as antioxidant activity compared with samples without enzyme mixtures [84]. Moreover, the scientific literature shows that in the stepwise extraction of buckwheat husk using xylanase commercial preparations, the yield of soluble fractions increased 4–5 times compared to the control extraction [7]. The use of enzymes reduces the amount of solvent required for extraction and increases the yield of extractable compounds. For example, tannase, pectase, cellulase, and hemicellulase are widely used in juices to increase product yield and improve quality [40,86]. However, when it is enzymatic extraction alone, compare with other extraction methods. Zheng et al. [87] described different extraction methods of palm kernel expeller. Dried extracts prepared with enzymes, hydrochloric acid, carboxymethylation, and hydroxipropylation were compared among chemical composition and physicochemical and functional properties. All the extract methods increased dietary content, phenolic content, and physicochemical properties. However, the goals of extract application have to be implemented because all extract methods were suitable, although enzyme assistance required fewer organic solvents, which implements in sustainability coverage [87].

Enzyme-assisted extraction (EAE) presents applicability to extract pectins from wastes and by–products by increasing plant cell wall permeability [88,89]. Enzymes are applicable to extract many phenolic compounds, including flavonoids and anthocyanidins [90]. Enzyme activity, treatment time, substrate ratio, and particle size are essential to get the highest efficiency during enzymatic treatment. Optimized conditions were discussed for pistachio green hull extraction by Yazdi et al. [84]. This research used cellulases, pectinases, tannases, and their mixtures for the extraction. Results indicated that all of the three enzymes at the same time used to extract phenolics were giving the highest score [91]. Moreover, by-products may be applied as enzymes production during solid-state fermentation (SSF). Some of the highest by-products of cacao are bean shell, brewers' spent grain, and wheat bran, which have an estimated production of 140 million tons per year. These were implemented for SSF with *Aspergillus awamori*, *Aspergillus niger*, and *Aspergillus oryzae*, which produce feruloyl-esterases and amylases. These enzymes in bread enhanced ferulic acid quantity, total phenolic content, and antioxidant activity [92,93].

4.2. Plant-Based Drinks from Grains and Fermented Drinks Production

There is also no surprise that dairy milk substitute from grains production often requires enzymatic assistance for increasing extraction yields, proteins, and total solids content [22]. Annually, global plant-based dairy substitutes were marked to be grown by 10% and by 2019 had reached US 1.8 billion dollars [94]. Moreover, created derivatives and released sugars create sensory-acceptant organoleptic properties [16]. Amylolytic enzymes are required due to the amylose and amylolytic ratio of the starches, resulting in different rheological and textural properties. However, disrupting amylose and amylopectin molecules increases liquefying properties of grain beverages. Many starch-modifying enzymes including α-amylase (EC 3.2.1.1), β-amylase (EC 3.2.1.2), glucoamylase (EC 3.2.1.3), α-glucosidase (EC 3.2.1.20), pullulanase or amylopullulanase (EC 3.2.1.41), and cyclodextrin glycosyltransferase (EC 2.4.1.19) are used in industries that cleave the α-1,4- or α-1,6-glycosidic linkages, leading to the release of reducing sugars and oligosaccharides which are possibly prebiotic [32,95]. Interestingly, a quantitative prebiotic score of plant-based dairy substitutes can be identified, describing which prebiotics foster probiotic strains' selective growth. Phenolic compounds, reduced sugars, and oligosaccharides are also great nutrients for bacterial growth, and the fermented food sector is rising due to their possible probiotic functions [66,96]. However, different plant materials are different sources for particular strains [76,97]. Enzymes are widely applicable and potent as precursors of nutrient production and initiate the growth of fermentation starter cultures: e.g., *Lactobacillus* strains *L. reuteri* L45, *L. plantarum* L47, and *L. johnsonii* L63 growth were initiated by cellulase- and pectinase-degraded rapeseed fibers [98]. Nissen et al. [99] identified prebiotic scores with different selectivity of hemp, soy, rice, and their mixtures, where hemp and soy drinks initiated the highest growth of *L. plantarum* 98b, and for growth *B. bifidum* B700795 mixture of hemp-rice and hemp-soy drinks, showed the potent scores [99,100]. Five main steps are incorporated in the flow chart for manufacturing enzymatically treated and by-the-step fermented plant-based drinks from plant materials: plant disruption, extraction, formulation, fermentation, and packaging [94]. However, it is also important not to forget the cultivar's selectivity to get the highest value product [101]. Figure 3 represents the summarized flow chart of plant-based drinks production, where additional importance is given for plant material selection due to European Commission strategy incorporation 'from farm to fork', which is the heart of the European Green Deal priorities from 2019–2024. Continuing on, the flow chart represents the prebiotic plant-based drink production, which further on leads to probiotic drink production which can be mono- or multi-microbial [94]. Shori et al. [102] reviewed the plant-based dairy substitutes fermented with probiotic strains functionality, shelf-life expansions, and nutrition value enhancement [102].

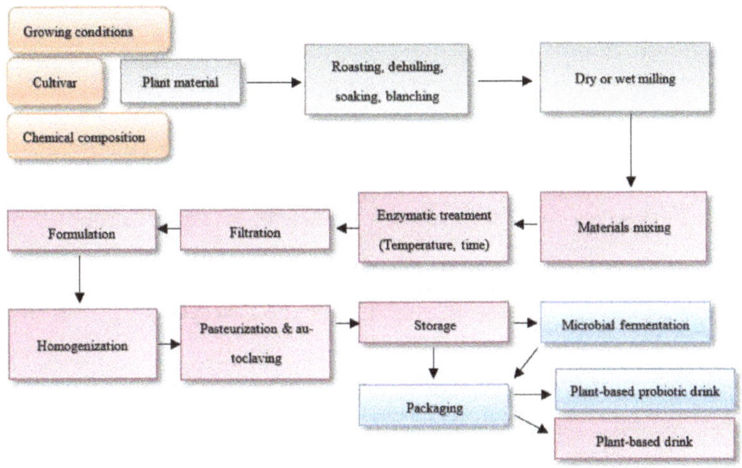

Figure 3. Chart flow of plant-based prebiotic and probiotic drinks production.

Moreover, plant-based fermented products have properties against pathogenic bacteria, and pH is lower than regular plant-based drinks, affecting product stability [103]. However, another aspect of ingested probiotic food pathway and viability may be investigated. The texture and nutrient density and the complexity of the food matrix are responsible for releasing the nutrients delivering specific metabolites and strains to the intestines [104,105]. Moreover, the density and variety of live cultures may be the essential indicators for increasing the variety of gut microbiome [77].

4.3. Nanocrystals, Nanofibers, and Nanocellulose

The latest studies indicate that phenolic content increased in fermented products [96,106,107]. As a sidestream nanocellulosic material, it is usually produced by *Komagataeibacater, Acetobacter, Gluconacetobacter* strains which might be used as, e.g., wound healing biofilm [108]. Specific enzymes release, cleave, transport, and form derivatives from different plant origins by opening the ability to discover green synthesis applications for nanofibers, nanocrystals, and nanoparticles. Aqueous different plant extracts are the new scientific approach for synthesizing nanoparticles by changing environmentally disruptive chemical and physical methods. Enzyme-assistance by disrupting plant cell wall microfibrils and amorphous zones is visible through Transmission Electron Microscopy (TEM), Atomic Force Microscopy (AFM), or Scanning Electron Microscopy (SEM), which also implies an increase of extract yield [109,110]. Plant extracts contain high phenolic content and reducing sugars and reducing or stabilizing agents [26,111]. Additionally, as mentioned before, increasing drug resistance and nanofibers formation is getting attention for possible antibacterial and drug delivery properties and nanocellulose formation for biodegradability, non-toxicity, and potential physicochemical properties [112]. Yarbrough et at. [113] described cellulolytic enzymes performance of nanofibrils and nanocrystals formation by depolymerizing carbohydrates into smaller units [113]. Depending on the plant origin, charges and forces change using specific cellulolytic activity characterized enzymes [68].

Summarized application of enzymes-assisted extraction and their products is described in Figure 4.

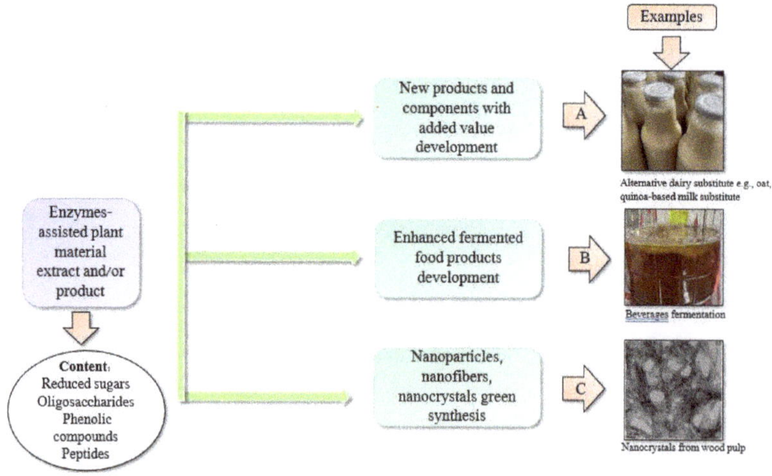

Figure 4. Enzyme-assisted extract or possible product applications, where A presents alternative dairy substitutes development [22], B—fermented beverages production [96], C—nanocrystals formation from enzymatically treated wood pulp [114]. A and B photographs were taken by the authors at the Lithuanian Research Centre for Agriculture and Forestry.

5. Existing Rules on the Use of Enzymes and Their Products

At present, in the European Union, food enzymes usage is described under Regulation (EC) No 1332/2008 of the European Parliament and of the Council of 16 December 2008 on food enzymes. Enzymes are considered food additives because they are found in the final products only in inactivated form and in smaller amounts. The definition of food enzymes are described in Regulation (EC) No 1169/2011 [34]. Specifically, directive 2000/13/EC describes labeling requirements for food enzymes. However, some limitations occur while extracting by-products: first, the lack of specific regulations to assure the safety of valorized products. For human consumption, some substances of food by-products are categorized as food additives and described in EC Regulation No. 1881/2006. However, it is categorized as nutraceutical; these products are covered by food supplement Directive 200/46/EC. Moreover, the contamination of food by-products must be implemented, which is described in Council Regulation 315/93/EEC. Secondly, if by-products are enzymatically modified, EC regulation No 258/97 (1997) could apply to food and food ingredients with a new or intentionally modified primary molecule structure if not used before May 1997. This regulation ensures the safety assessment before entering the EU market. While entering the USA market, Food and Drug Administration (FDA) regulates food additives (Sec. 348), new dietary ingredients (Sec. 350b), pesticide chemical residues (Sec. 346a), and others through Federal Food, Drug, and Cosmetic Act (Chapter 9, Subchapter IV) [8]. Moreover, for worldwide fair food trade, international food standard CODEX Alimentarius, which is covered by the World Health Organization (WHO) and Food and Agriculture Organization of the United Nations (FAO), also describes food enzymes as food additives. Depending on how a product is made, the usage is characterized by enzyme name, the producent, and the highest dosage. In general, for the standard of food additives CODEX STAN 192-1995 for flours and starch products, two enzymes are described. In detail, for α-amylase and glucoamylase, whose producers are *Bacillus subtilis* (INS 1100(iii)) and *Aspergillus oryzae* var. (INS 1100(iii)), dosage is not identified. However, all food additives, including food enzymes such as xylanases, pectinases, and others, are collected in the list of CODEX Specification for Food Additives CXN 6–2019 [115]. However, for enzymes producers, the recommended purity specifications for food-grade enzymes are given by the

Joint FAO/WHO Expert Committee on Food Additives (JECFA) and the Food Chemical Codex (FCC).

6. Conclusions

The use of enzymes in extracting biological raw material compounds is an up-and-coming area from small-scale, laboratory optimization studies to large-scale, industrial applications. It implies food processing, functional components, and medical devices development for high antioxidant, anti-inflammatory, and antimicrobial characteristics. However, success in this area requires interdisciplinary research from various life sciences disciplines. An important area of research is investigating the stability of enzymes and their interaction with other food and plant ingredients during processing and storage; repeatability is also questionable because the plant material differs from the origin, cultivars, and growing, harvesting, and storage conditions. Additionally, limitations occur in the form of worldwide regulations of enzymes usage and dosages due to the novel components that are produced during these processes. However, enzyme-assisted processes are reaching for more sustainable development of innovations in a broad spectrum of industries.

Author Contributions: Conceptualization, P.S.; software, P.S.; validation, P.S., J.V. and P.V.; formal analysis, P.S.; investigation, P.S.; resources, P.S.; data curation, P.S.; writing—original draft preparation, P.S.; writing—review and editing, P.S., J.V., P.V.; visualization, P.S.; supervision, J.V. and P.V.; project administration, J.V.; funding acquisition, P.V. All authors have read and agreed to the published version of the manuscript.

Funding: This research received no external funding.

Institutional Review Board Statement: Not applicable.

Informed Consent Statement: Not applicable.

Data Availability Statement: The data presented in this study are available in the article.

Acknowledgments: The authors wish to thank the Lithuanian Research Centre for Agriculture and Forestry for the support of this study. The work is partly attributed to the project "Innovative application of biologically active substances for the prevention of cardiovascular insufficiency and sarcopenia" (Nr. 01.2.1-LVPA-K-856-01-0065) under grant agreement with the Lithuanian Business Support Agency (LVPA).

Conflicts of Interest: The authors declare no conflict of interest. The funders had no role in the design of the study; in the collection, analyses, or interpretation of data; in the writing of the manuscript, or in the decision to publish the results.

References

1. Rose, N.; Reynolds, T.; Kolodinsky, J. P90 Consumer Use of Food Labels Increases as "Clean Label" Trend Continues. *J. Nutr. Educ. Behav.* **2020**, *52*, S58–S59. [CrossRef]
2. Chawla, G.K.; Garg, K.; Kaur, K.; Chopra, V.; Suri, R. Pattern of drug resistance among patients of pulmonary tuberculosis. *Indian J. Tuberc.* **2021**. [CrossRef]
3. Grunert, K.G.; De Bauw, M.; Dean, M.; Lähteenmäki, L.; Maison, D.; Pennanen, K.; Sandell, M.A.; Stasiuk, K.; Stickel, L.; Tarrega, A.; et al. No lockdown in the kitchen: How the COVID-19 pandemic has affected food-related behaviours. *Food Res. Int.* **2021**, *150*, 110752. [CrossRef] [PubMed]
4. Gligor, O.; Mocan, A.; Moldovan, C.; Locatelli, M.; Crișan, G.; Ferreira, I.C. Enzyme-assisted extractions of polyphenols—A comprehensive review. *Trends Food Sci. Technol.* **2019**, *88*, 302–315. [CrossRef]
5. Noda-Garcia, L.; Liebermeister, W.; Tawfik, D.S. Metabolite–Enzyme Coevolution: From Single Enzymes to Metabolic Pathways and Networks. *Annu. Rev. Biochem.* **2018**, *87*, 187–216. [CrossRef] [PubMed]
6. BlueWeave Consulting and Research Pvt Ltd. Global industrial enzymes market projected to reach worth $9.2 bn by 2027. *Focus Catal.* **2021**, *2021*, 2. [CrossRef]
7. Mackėla, I.; Andriekus, T.; Venskutonis, P.R. Biorefining of buckwheat (Fagopyrum esculentum) hulls by using supercritical fluid, Soxhlet, pressurized liquid and enzyme-assisted extraction methods. *J. Food Eng.* **2017**, *213*, 38–46. [CrossRef]
8. Socas-Rodríguez, B.; Álvarez-Rivera, G.; Valdés, A.; Ibáñez, E.; Cifuentes, A. Food by-products and food wastes: Are they safe enough for their valorization? *Trends Food Sci. Technol.* **2021**, *114*, 133–147. [CrossRef]

9. Lao, Y.; Zhang, M.; Li, Z.; Bhandari, B. A novel combination of enzymatic hydrolysis and fermentation: Effects on the flavor and nutritional quality of fermented Cordyceps militaris beverage. *LWT* **2019**, *120*, 108934. [CrossRef]
10. Cunningham, M.; Vinderola, G.; Charalampopoulos, D.; Lebeer, S.; Sanders, M.E.; Grimaldi, R. Applying probiotics and prebiotics in new delivery formats—Is the clinical evidence transferable? *Trends Food Sci. Technol.* **2021**, *112*, 495–506. [CrossRef]
11. Włodarczyk, M.; Śliżewska, K. Obesity as the 21st Century's major disease: The role of probiotics and prebiotics in prevention and treatment. *Food Biosci.* **2021**, *42*, 101115. [CrossRef]
12. Alou, M.T.; Lagier, J.-C.; Raoult, D. Diet influence on the gut microbiota and dysbiosis related to nutritional disorders. *Hum. Microbiome J.* **2016**, *1*, 3–11. [CrossRef]
13. Janssen, A.W.F.; Kersten, S. The role of the gut microbiota in metabolic health. *FASEB J.* **2015**, *29*, 3111–3123. [CrossRef] [PubMed]
14. Delgado, G.T.C.; Tamashiro, W.M.D.S.C. Role of prebiotics in regulation of microbiota and prevention of obesity. *Food Res. Int.* **2018**, *113*, 183–188. [CrossRef] [PubMed]
15. Daïen, C.I.; Tan, J.; Audo, R.; Mielle, J.; Quek, L.-E.; Krycer, J.R.; Angelatos, A.S.; Duraes, M.; Pinget, G.V.; Ni, D.; et al. Gut-derived acetate promotes B10 cells with antiinflammatory effects. *JCI Insight* **2021**, *6*, 144156. [CrossRef]
16. Mäkinen, O.E.; Wanhalinna, V.; Zannini, E.; Arendt, E.K. Foods for Special Dietary Needs: Non-dairy Plant-based Milk Substitutes and Fermented Dairy-type Products. *Crit. Rev. Food Sci. Nutr.* **2016**, *56*, 339–349. [CrossRef]
17. Penha, C.B.; Santos, V.D.P.; Speranza, P.; Kurozawa, L.E. Plant-based beverages: Ecofriendly technologies in the production process. *Innov. Food Sci. Emerg. Technol.* **2021**, *72*, 102760. [CrossRef]
18. Cantu-Jungles, T.M.; Zhang, X.; Kazem, A.E.; Iacomini, M.; Hamaker, B.R.; Cordeiro, L.M. Microwave treatment enhances human gut microbiota fermentability of isolated insoluble dietary fibers. *Food Res. Int.* **2021**, *143*, 110293. [CrossRef]
19. Zhang, Y.-G.; Kan, H.; Chen, S.-X.; Thakur, K.; Wang, S.; Zhang, J.-G.; Shang, Y.-F.; Wei, Z.-J. Comparison of phenolic compounds extracted from Diaphragma juglandis fructus, walnut pellicle, and flowers of Juglans regia using methanol, ultrasonic wave, and enzyme assisted-extraction. *Food Chem.* **2020**, *321*, 126672. [CrossRef]
20. Puri, M.; Sharma, D.; Barrow, C.J. Enzyme-assisted extraction of bioactives from plants. *Trends Biotechnol.* **2012**, *30*, 37–44. [CrossRef]
21. Islam, S.M.; Ju, L.-K. Enzymatic soybean flour processing: Modeling for insights into optimal carbohydrases composition and carbohydrate monomerization from complex biomass. *Catal. Commun.* **2021**, *149*, 106244. [CrossRef]
22. Jeske, S.; Zannini, E.; Cronin, M.F.; Arendt, E.K. Impact of protease and amylase treatment on proteins and the product quality of a quinoa-based milk substitute. *Food Funct.* **2018**, *9*, 3500–3508. [CrossRef] [PubMed]
23. Dhasarathan, P.; AlSalhi, M.S.; Devanesan, S.; Subbiah, J.; Ranjitsingh, A.; Binsalah, M.; Alfuraydi, A.A. Drug resistance in Candida albicans isolates and related changes in the structural domain of Mdr1 protein. *J. Infect. Public Health* **2021**, *14*, 1848–1853. [CrossRef] [PubMed]
24. Balciunaitiene, A.; Viskelis, P.; Viskelis, J.; Streimikyte, P.; Liaudanskas, M.; Bartkiene, E.; Zavistanaviciute, P.; Zokaityte, E.; Starkute, V.; Ruzauskas, M.; et al. Green Synthesis of Silver Nanoparticles Using Extract of *Artemisia absinthium* L., *Humulus lupulus* L. and *Thymus vulgaris* L., Physico-Chemical Characterization, Antimicrobial and Antioxidant Activity. *Process* **2021**, *9*, 1304. [CrossRef]
25. Chandhirasekar, K.; Thendralmanikandan, A.; Thangavelu, P.; Nguyen, B.-S.; Nguyen, T.-A.; Sivashanmugan, K.; Nareshkumar, A.; Nguyen, V.-H. Plant-extract-assisted green synthesis and its larvicidal activities of silver nanoparticles using leaf extract of Citrus medica, Tagetes lemmonii, and Tarenna asiatica. *Mater. Lett.* **2021**, *287*, 129265. [CrossRef]
26. Jalab, J.; Abdelwahed, W.; Kitaz, A.; Al-Kayali, R. Green synthesis of silver nanoparticles using aqueous extract of Acacia cyanophylla and its antibacterial activity. *Heliyon* **2021**, *7*, 08033. [CrossRef]
27. Naidu, D.; Hlangothi, S.P.; John, M.J. Bio-based products from xylan: A review. *Carbohydr. Polym.* **2018**, *179*, 28–41. [CrossRef]
28. De Bhowmick, G.; Sarmah, A.K.; Sen, R. Lignocellulosic biorefinery as a model for sustainable development of biofuels and value added products. *Bioresour. Technol.* **2018**, *247*, 1144–1154. [CrossRef]
29. Acosta-Estrada, B.A.; Gutiérrez-Uribe, J.A.; Serna-Saldívar, S.O. Bound phenolics in foods, a review. *Food Chem.* **2014**, *152*, 46–55. [CrossRef]
30. Carpita, N.C.; McCann, M.C. Redesigning plant cell walls for the biomass-based bioeconomy. *J. Biol. Chem.* **2020**, *295*, 15144–15157. [CrossRef]
31. Raveendran, S.; Parameswaran, B.; Ummalyma, S.B.; Abraham, A.; Mathew, A.K.; Madhavan, A.; Rebello, S.; Pandey, A. Applications of Microbial Enzymes in Food Industry. *Food Technol. Biotechnol.* **2018**, *56*, 16–30. [CrossRef] [PubMed]
32. Miao, M.; Jiang, B.; Jin, Z.; BeMiller, J.N. Microbial Starch-Converting Enzymes: Recent Insights and Perspectives. *Compr. Rev. Food Sci. Food Saf.* **2018**, *17*, 1238–1260. [CrossRef] [PubMed]
33. Puchart, V.; Šuchová, K.; Biely, P. Xylanases of glycoside hydrolase family 30—An overview. *Biotechnol. Adv.* **2021**, *47*, 107704. [CrossRef] [PubMed]
34. Danalache, F.; Mata, P.; Alves, V.D.; Moldão-Martins, M. Chapter 10—Enzyme-Assisted Extraction of Fruit Juices. In *Fruit Juices*; Rajauria, G., Brijesh, K., Eds.; Academic Press: San Diego, CA, USA, 2018; pp. 183–200. ISBN 978-0-12-802230-6.
35. Kawamori, M.; Morikawa, Y.; Takasawa, S. Induction and production of cellulases by l-sorbose in Trichoderma reesei. *Appl. Microbiol. Biotechnol.* **1986**, *24*, 449–453. [CrossRef]
36. Juturu, V.; Wu, J.C. Microbial cellulases: Engineering, production and applications. *Renew. Sustain. Energy Rev.* **2014**, *33*, 188–203. [CrossRef]

37. Pedersen, N.R.; Ravn, J.L.; Pettersson, D. A multienzyme NSP product solubilises and degrades NSP structures in canola and mediates protein solubilisation and degradation in vitro. *Anim. Feed Sci. Technol.* **2017**, *234*, 244–252. [CrossRef]
38. Ire, F.S.; Chima, I.J.; Ezebuiro, V. Enhanced xylanase production from UV-mutated Aspergillus niger grown on corn cob and sawdust. *Biocatal. Agric. Biotechnol.* **2021**, *31*, 101869. [CrossRef]
39. Tan, H.; Nie, S. Deciphering diet-gut microbiota-host interplay: Investigations of pectin. *Trends Food Sci. Technol.* **2020**, *106*, 171–181. [CrossRef]
40. Muley, A.B.; Thorat, A.S.; Singhal, R.S.; Babu, K.H. A tri-enzyme co-immobilized magnetic complex: Process details, kinetics, thermodynamics and applications. *Int. J. Biol. Macromol.* **2018**, *118*, 1781–1795. [CrossRef]
41. Ladole, M.R.; Pokale, P.B.; Varude, V.R.; Belokar, P.G.; Pandit, A.B. One pot clarification and debittering of grapefruit juice using co-immobilized enzymes@chitosanMNPs. *Int. J. Biol. Macromol.* **2021**, *167*, 1297–1307. [CrossRef]
42. Biesalski, H.-K.; Dragsted, L.O.; Elmadfa, I.; Grossklaus, R.; Müller, M.; Schrenk, D.; Walter, P.; Weber, P. Bioactive compounds: Definition and assessment of activity. *Nutrition* **2009**, *25*, 1202–1205. [CrossRef] [PubMed]
43. Ashraf, M.A.; Iqbal, M.; Rasheed, R.; Hussain, I.; Riaz, M.; Arif, M.S. Chapter 8—Environmental Stress and Secondary Metabolites in Plants: An Overview. In *Plant Metabolites and Regulation under Environmental Stress*; Ahmad, P., Ahanger, M.A., Singh, V.P., Tripathi, D.K., Alam, P., Nasser, M., Eds.; Academic Press: Cambridge, MA, USA, 2018; pp. 153–167. ISBN 978-0-12-812689-9.
44. Shamsi, A.; Shahwan, M.; Khan, M.S.; Husain, F.M.; Alhumaydhi, F.A.; Aljohani, A.S.M.; Rehman, T.; Hassan, I.; Islam, A. Elucidating the Interaction of Human Ferritin with Quercetin and Naringenin: Implication of Natural Products in Neurodegenerative Diseases: Molecular Docking and Dynamics Simulation Insight. *ACS Omega* **2021**, *6*, 7922–7930. [CrossRef] [PubMed]
45. Paz, A.; Outeiriño, D.; Guerra, N.P.; Domínguez, J.M. Enzymatic hydrolysis of brewer's spent grain to obtain fermentable sugars. *Bioresour. Technol.* **2019**, *275*, 402–409. [CrossRef]
46. Uddin, S.; Kabir, T.; Niaz, K.; Jeandet, P.; Clément, C.; Mathew, B.; Rauf, A.; Rengasamy, K.R.; Sobarzo-Sánchez, E.; Ashraf, G.M.; et al. Molecular Insight into the Therapeutic Promise of Flavonoids against Alzheimer's Disease. *Molecules* **2020**, *25*, 1267. [CrossRef] [PubMed]
47. Spencer, J.P.E. Flavonoids: Modulators of brain function? *Br. J. Nutr.* **2008**, *99*, ES60–ES77. [CrossRef]
48. Gulpinar, A.R.; Orhan, I.E.; Kan, A.; Senol, F.S.; Celik, S.A.; Kartal, M. Estimation of in vitro neuroprotective properties and quantification of rutin and fatty acids in buckwheat (Fagopyrum esculentum Moench) cultivated in Turkey. *Food Res. Int.* **2012**, *46*, 536–543. [CrossRef]
49. Urbanavičiūtė, I.; Liaudanskas, M.; Bobinas, Č.; Šarkinas, A.; Rezgienė, A.; Viskelis, P. Japanese Quince (*Chaenomeles japonica*) as a Potential Source of Phenols: Optimization of the Extraction Parameters and Assessment of Antiradical and Antimicrobial Activities. *Foods* **2020**, *9*, 1132. [CrossRef]
50. Raudonė, L.; Liaudanskas, M.; Vilkickytė, G.; Kviklys, D.; Žvikas, V.; Viškelis, J.; Viškelis, P. Phenolic Profiles, Antioxidant Activity and Phenotypic Characterization of *Lonicera caerulea* L. Berries, Cultivated in Lithuania. *Antioxidants* **2021**, *10*, 115. [CrossRef]
51. Yang, B.; Jiang, Y.; Shi, J.; Chen, F.; Ashraf, M. Extraction and pharmacological properties of bioactive compounds from longan (Dimocarpus longan Lour.) fruit—A review. *Food Res. Int.* **2011**, *44*, 1837–1842. [CrossRef]
52. Teo, C.C.; Tan, S.N.; Yong, J.W.H.; Hew, C.S.; Ong, E.S. Pressurized hot water extraction (PHWE). *J. Chromatogr. A* **2010**, *1217*, 2484–2494. [CrossRef]
53. Gardossi, L.; Poulsen, P.B.; Ballesteros, A.; Hult, K.; Švedas, V.K.; Vasić-Rački, Đ.; Carrea, G.; Magnusson, A.; Schmid, A.; Wohlgemuth, R.; et al. Guidelines for reporting of biocatalytic reactions. *Trends Biotechnol.* **2010**, *28*, 171–180. [CrossRef] [PubMed]
54. Chen, H.; Zhou, X.; Zhang, J. Optimization of enzyme assisted extraction of polysaccharides from Astragalus membranaceus. *Carbohydr. Polym.* **2014**, *111*, 567–575. [CrossRef] [PubMed]
55. Zuorro, A.; Lavecchia, R.; González-Delgado, Á.D.; García-Martinez, J.B.; L'Abbate, P. Optimization of Enzyme-Assisted Extraction of Flavonoids from Corn Husks. *Process* **2019**, *7*, 804. [CrossRef]
56. Drula, E.; Garron, M.-L.; Dogan, S.; Lombard, V.; Henrissat, B.; Terrapon, N. The carbohydrate-active enzyme database: Functions and literature. *Nucleic Acids Res.* **2021**, *50*, D571–D577. [CrossRef] [PubMed]
57. Bulmer, G.S.; de Andrade, P.; Field, R.A.; van Munster, J.M. Recent advances in enzymatic synthesis of β-glucan and cellulose. *Carbohydr. Res.* **2021**, *508*, 108411. [CrossRef]
58. Woo, S.-Y.; Yang, J.Y.; Lee, H.; Ahn, H.J.; Lee, Y.B.; Do, S.H.; Kim, J.Y.; Seo, W.D. Changes in metabolites with harvest times of seedlings of various Korean oat (*Avena sativa* L.) cultivars and their neuraminidase inhibitory effects. *Food Chem.* **2021**, *373*, 131429. [CrossRef]
59. Cole, M.R.; Eggleston, G.; Gaines, D.K.; Heckemeyer, M. Development of an enzyme cocktail to bioconvert untapped starch in sweet sorghum processing by-products: Part I. *Ind. Crop. Prod.* **2019**, *133*, 142–150. [CrossRef]
60. González, M.J.A.; Carrera, C.; Barbero, G.F.; Palma, M. A comparison study between ultrasound–assisted and enzyme–assisted extraction of anthocyanins from blackcurrant (Ribes nigrum L.). *Food Chem. X* **2021**, *13*, 100192. [CrossRef]
61. Kongthitilerd, P.; Thilavech, T.; Marnpae, M.; Rong, W.; Yao, S.; Adisakwattana, S.; Cheng, H.; Suantawee, T. Cyanidin-3-rutinoside stimulated insulin secretion through activation of L-type voltage-dependent Ca2+ channels and the PLC-IP3 pathway in pancreatic β-cells. *Biomed. Pharmacother.* **2021**, *146*, 112494. [CrossRef]
62. Song, L.-W.; Qi, J.-R.; Liao, J.-S.; Yang, X.-Q. Enzymatic and enzyme-physical modification of citrus fiber by xylanase and planetary ball milling treatment. *Food Hydrocoll.* **2021**, *121*, 107015. [CrossRef]

63. Belmiro, R.H.; Oliveira, L.D.C.; Geraldi, M.V.; Junior, M.R.M.; Cristianini, M. Modification of coffee coproducts by-products by dynamic high pressure, acetylation and hydrolysis by cellulase: A potential functional and sustainable food ingredient. *Innov. Food Sci. Emerg. Technol.* **2021**, *68*, 102608. [CrossRef]
64. Phirom-On, K.; Apiraksakorn, J. Development of cellulose-based prebiotic fiber from banana peel by enzymatic hydrolysis. *Food Biosci.* **2021**, *41*, 101083. [CrossRef]
65. Ninga, K.A.; Desobgo, Z.S.C.; De, S.; Nso, E.J. Pectinase hydrolysis of guava pulp: Effect on the physicochemical characteristics of its juice. *Heliyon* **2021**, *7*, 08141. [CrossRef] [PubMed]
66. Chen, G.; Liu, Y.; Zeng, J.; Tian, X.; Bei, Q.; Wu, Z. Enhancing three phenolic fractions of oats (Avena sativa L.) and their antioxidant activities by solid-state fermentation with Monascus anka and Bacillus subtilis. *J. Cereal Sci.* **2020**, *93*, 102940. [CrossRef]
67. Pinelo, M.; Arnous, A.; Meyer, A.S. Upgrading of grape skins: Significance of plant cell-wall structural components and extraction techniques for phenol release. *Trends Food Sci. Technol.* **2006**, *17*, 579–590. [CrossRef]
68. Zhou, Z.; Ju, X.; Chen, J.; Wang, R.; Zhong, Y.; Li, L. Charge-oriented strategies of tunable substrate affinity based on cellulase and biomass for improving in situ saccharification: A review. *Bioresour. Technol.* **2021**, *319*, 124159. [CrossRef]
69. HozoVá, B.; Kuniak, L.; Moravcikova, P.; Gajdosova, A. Determination of Water-Insoluble Beta-D-Glucan in the Whole-Grain Cereals and Pseudocereals. *Czech J. Food Sci.* **2007**, *25*, 316. [CrossRef]
70. Schmidt, M. Cereal beta-glucans: An underutilized health endorsing food ingredient. *Crit. Rev. Food Sci. Nutr.* **2020**, 1–20. [CrossRef]
71. Nishantha, M.D.L.C.; Zhao, X.; Jeewani, D.C.; Bian, J.; Nie, X.; Weining, S. Direct comparison of β-glucan content in wild and cultivated barley. *Int. J. Food Prop.* **2018**, *21*, 2218–2228. [CrossRef]
72. Du, B.; Meenu, M.; Liu, H.; Xu, B. A Concise Review on the Molecular Structure and Function Relationship of β-Glucan. *Int. J. Mol. Sci.* **2019**, *20*, 4032. [CrossRef]
73. Shah, A.; Masoodi, F.; Gani, A.; Ashwar, B. Dual enzyme modified oat starch: Structural characterisation, rheological properties, and digestibility in simulated GI tract. *Int. J. Biol. Macromol.* **2018**, *106*, 140–147. [CrossRef] [PubMed]
74. Zhong, Y.; Herburger, K.; Kirkensgaard, J.J.K.; Khakimov, B.; Hansen, A.R.; Blennow, A. Sequential maltogenic α-amylase and branching enzyme treatment to modify granular corn starch. *Food Hydrocoll.* **2021**, *120*, 106904. [CrossRef]
75. Zhang, Z.; Yang, P.; Zhao, J. Ferulic acid mediates prebiotic responses of cereal-derived arabinoxylans on host health. *Anim. Nutr.* **2021**. [CrossRef]
76. Pereira, G.V.; Abdel-Hamid, A.M.; Dutta, S.; D'Alessandro-Gabazza, C.N.; Wefers, D.; Farris, J.A.; Bajaj, S.; Wawrzak, Z.; Atomi, H.; Mackie, R.I.; et al. Degradation of complex arabinoxylans by human colonic Bacteroidetes. *Nat. Commun.* **2021**, *12*, 1–21. [CrossRef] [PubMed]
77. Wastyk, H.C.; Fragiadakis, G.K.; Perelman, D.; Dahan, D.; Merrill, B.D.; Yu, F.B.; Topf, M.; Gonzalez, C.G.; Van Treuren, W.; Han, S.; et al. Gut-microbiota-targeted diets modulate human immune status. *Cell* **2021**, *184*, 4137–4153.e14. [CrossRef]
78. Vangay, P.; Johnson, A.; Ward, T.L.; Al-Ghalith, G.A.; Shields-Cutler, R.R.; Hillmann, B.M.; Lucas, S.K.; Beura, L.K.; Thompson, E.A.; Till, L.M.; et al. US Immigration Westernizes the Human Gut Microbiome. *Cell* **2018**, *175*, 962–972.e10. [CrossRef]
79. Zhang, C.; Abdo, A.A.A.; Kaddour, B.; Wu, Q.; Xin, L.; Li, X.; Fan, G.; Teng, C. Xylan-oligosaccharides ameliorate high fat diet induced obesity and glucose intolerance and modulate plasma lipid profile and gut microbiota in mice. *J. Funct. Foods* **2020**, *64*, 103622. [CrossRef]
80. Yang, F.; Wei, J.-D.; Lu, Y.-F.; Sun, Y.-L.; Wang, Q.; Zhang, R.-L. Galacto-oligosaccharides modulate gut microbiota dysbiosis and intestinal permeability in rats with alcohol withdrawal syndrome. *J. Funct. Foods* **2019**, *60*, 103423. [CrossRef]
81. Li, E.; Long, X.; Liao, S.; Pang, D.; Li, Q.; Zou, Y. Effect of mulberry galacto-oligosaccharide isolated from mulberry on glucose metabolism and gut microbiota in a type 2 diabetic mice. *J. Funct. Foods* **2021**, *87*, 104836. [CrossRef]
82. Wilkins, M.R.; Widmer, W.W.; Grohmann, K.; Cameron, R. Hydrolysis of grapefruit peel waste with cellulase and pectinase enzymes. *Bioresour. Technol.* **2007**, *98*, 1596–1601. [CrossRef]
83. Qi, W.J.; Sheng, W.S.; Peng, C.; Xiaodong, M.; Yao, T.Z. Investigating into anti-cancer potential of lycopene: Molecular targets. *Biomed. Pharmacother.* **2021**, *138*, 111546. [CrossRef] [PubMed]
84. Catalkaya, G.; Kahveci, D. Optimization of enzyme assisted extraction of lycopene from industrial tomato waste. *Sep. Purif. Technol.* **2019**, *219*, 55–63. [CrossRef]
85. Urbonaviciene, D.; Viskelis, P. The cis-lycopene isomers composition in supercritical CO2 extracted tomato by-products. *LWT* **2017**, *85*, 517–523. [CrossRef]
86. Kundu, D.; Karmakar, S.; Banerjee, R. Simultaneous debittering and clarification of enzyme mediated mixed citrus juice production. *Appl. Food Res.* **2021**, *2*, 100031. [CrossRef]
87. Zheng, Y.; Li, Y.; Tian, H. Effects of carboxymethylation, acidic treatment, hydroxypropylation and heating combined with enzymatic hydrolysis on structural and physicochemical properties of palm kernel expeller dietary fiber. *LWT* **2020**, *133*, 109909. [CrossRef]
88. Marić, M.; Grassino, A.N.; Zhu, Z.; Barba, F.J.; Brnčić, M.; Brnčić, S.R. An overview of the traditional and innovative approaches for pectin extraction from plant food wastes and by-products: Ultrasound-, microwaves-, and enzyme-assisted extraction. *Trends Food Sci. Technol.* **2018**, *76*, 28–37. [CrossRef]
89. Yuliarti, O.; Goh, K.K.; Matia-Merino, L.; Mawson, J.; Brennan, C. Extraction and characterisation of pomace pectin from gold kiwifruit (Actinidia chinensis). *Food Chem.* **2015**, *187*, 290–296. [CrossRef]

90. Rodríguez De Luna, S.L.; Ramírez-Garza, R.E.; Serna-Saldívar, S.O. Environmentally friendly methods for flavonoid extraction from plant material: Impact of their operating conditions on yield and antioxidant properties. *Sci. World J.* **2020**, *2020*, 6792069. [CrossRef]
91. Yazdi, A.P.G.; Barzegar, M.; Sahari, M.A.; Gavlighi, H.A. Optimization of the enzyme-assisted aqueous extraction of phenolic compounds from pistachio green hull. *Food Sci. Nutr.* **2018**, *7*, 356–366. [CrossRef]
92. Costa, R.D.S.; de Almeida, S.S.; Cavalcanti, E.D.C.; Freire, D.M.G.; Moura-Nunes, N.; Monteiro, M.; Perrone, D. Enzymes produced by solid state fermentation of agro-industrial by-products release ferulic acid in bioprocessed whole-wheat breads. *Food Res. Int.* **2021**, *140*, 109843. [CrossRef]
93. Yin, Z.; Wu, W.; Sun, C.; Lei, Z.; Chen, H.; Liu, H.; Chen, W.; Ma, J.; Min, T.; Zhang, M.; et al. Comparison of releasing bound phenolic acids from wheat bran by fermentation of three Aspergillus species. *Int. J. Food Sci. Technol.* **2018**, *53*, 1120–1130. [CrossRef]
94. Tangyu, M.; Muller, J.; Bolten, C.J.; Wittmann, C. Fermentation of plant-based milk alternatives for improved flavour and nutritional value. *Appl. Microbiol. Biotechnol.* **2019**, *103*, 9263–9275. [CrossRef] [PubMed]
95. Rempel, A.; Machado, T.; Treichel, H.; Colla, E.; Margarites, A.C.; Colla, L. Saccharification of Spirulina platensis biomass using free and immobilized amylolytic enzymes. *Bioresour. Technol.* **2018**, *263*, 163–171. [CrossRef] [PubMed]
96. Silva, K.A.; Uekane, T.M.; de Miranda, J.F.; Ruiz, L.F.; da Motta, J.C.B.; Silva, C.B.; Pitangui, N.D.S.; Gonzalez, A.G.M.; Fernandes, F.F.; Lima, A.R. Kombucha beverage from non-conventional edible plant infusion and green tea: Characterization, toxicity, antioxidant activities and antimicrobial properties. *Biocatal. Agric. Biotechnol.* **2021**, *34*, 102032. [CrossRef]
97. Bai, J.; Li, Y.; Zhang, W.; Fan, M.; Qian, H.; Zhang, H.; Qi, X.; Wang, L. Source of gut microbiota determines oat β-glucan degradation and short chain fatty acid-producing pathway. *Food Biosci.* **2021**, *41*, 101010. [CrossRef]
98. Zhu, X.; Wang, L.; Zhang, Z.; Ding, L.; Hang, S. Combination of fiber-degrading enzymatic hydrolysis and lactobacilli fermentation enhances utilization of fiber and protein in rapeseed meal as revealed in simulated pig digestion and fermentation in vitro. *Anim. Feed Sci. Technol.* **2021**, *278*, 115001. [CrossRef]
99. Nissen, L.; di Carlo, E.; Gianotti, A. Prebiotic potential of hemp blended drinks fermented by probiotics. *Food Res. Int.* **2020**, *131*, 109029. [CrossRef]
100. Huebner, J.; Wehling, R.; Parkhurst, A.; Hutkins, R. Effect of processing conditions on the prebiotic activity of commercial prebiotics. *Int. Dairy J.* **2008**, *18*, 287–293. [CrossRef]
101. Du, J.; Pan, R.; Obadi, M.; Li, H.; Shao, F.; Sun, J.; Wang, Y.; Qi, Y.; Xu, B. In vitro starch digestibility of buckwheat cultivars in comparison to wheat: The key role of starch molecular structure. *Food Chem.* **2021**, *368*, 130806. [CrossRef]
102. Shori, A.B.; Aljohani, G.S.; Al-Zahrani, A.J.; Al-Sulbi, O.S.; Baba, A.S. Viability of probiotics and antioxidant activity of cashew milk-based yogurt fermented with selected strains of probiotic *Lactobacillus* spp. *LWT* **2021**, *153*, 112482. [CrossRef]
103. Panghal, A.; Janghu, S.; Virkar, K.; Gat, Y.; Kumar, V.; Chhikara, N. Potential non-dairy probiotic products—A healthy approach. *Food Biosci.* **2018**, *21*, 80–89. [CrossRef]
104. Guimarães, J.T.; Balthazar, C.F.; Silva, R.; Rocha, R.S.; Graça, J.S.; A Esmerino, E.; Silva, M.C.; Sant'Ana, A.S.; Duarte, M.C.K.H.; Freitas, M.Q.; et al. Impact of probiotics and prebiotics on food texture. *Curr. Opin. Food Sci.* **2020**, *33*, 38–44. [CrossRef]
105. Shori, A.B.; AL Zahrani, A.J. Non-dairy plant-based milk products as alternatives to conventional dairy products for delivering probiotics. *Food Sci. Technol.* **2021**. [CrossRef]
106. Zhang, Y.; Liu, W.; Wei, Z.; Yin, B.; Man, C.; Jiang, Y. Enhancement of functional characteristics of blueberry juice fermented by Lactobacillus plantarum. *LWT* **2021**, *139*, 110590. [CrossRef]
107. Gao, B.; Wang, J.; Wang, Y.; Xu, Z.; Li, B.; Meng, X.; Sun, X.; Zhu, J. Influence of fermentation by lactic acid bacteria and in vitro digestion on the biotransformations of blueberry juice phenolics. *Food Control* **2021**, *133*, 108603. [CrossRef]
108. Ahmed, J.; Gultekinoglu, M.; Edirisinghe, M. Bacterial cellulose micro-nano fibres for wound healing applications. *Biotechnol. Adv.* **2020**, *41*, 107549. [CrossRef]
109. Rossi, B.R.; Pellegrini, V.O.; Cortez, A.A.; Chiromito, E.M.; Carvalho, A.J.; Pinto, L.O.; Rezende, C.A.; Mastelaro, V.R.; Polikarpov, I. Cellulose nanofibers production using a set of recombinant enzymes. *Carbohydr. Polym.* **2021**, *256*, 117510. [CrossRef]
110. Meftahi, A.; Samyn, P.; Geravand, S.A.; Khajavi, R.; Alibkhshi, S.; Bechelany, M.; Barhoum, A. Nanocelluloses as skin biocompatible materials for skincare, cosmetics, and healthcare: Formulations, regulations, and emerging applications. *Carbohydr. Polym.* **2021**, *278*, 118956. [CrossRef]
111. Ceylan, R.; Demirbas, A.; Ocsoy, I.; Aktumsek, A. Green synthesis of silver nanoparticles using aqueous extracts of three Sideritis species from Turkey and evaluations bioactivity potentials. *Sustain. Chem. Pharm.* **2021**, *21*, 100426. [CrossRef]
112. Liu, S.; Qamar, S.A.; Qamar, M.; Basharat, K.; Bilal, M. Engineered nanocellulose-based hydrogels for smart drug delivery applications. *Int. J. Biol. Macromol.* **2021**, *181*, 275–290. [CrossRef]
113. Yarbrough, J.M.; Zhang, R.; Mittal, A.; Wall, T.V.; Bomble, Y.J.; Decker, S.R.; Himmel, M.E.; Ciesielski, P.N. Multifunctional Cellulolytic Enzymes Outperform Processive Fungal Cellulases for Coproduction of Nanocellulose and Biofuels. *ACS Nano* **2017**, *11*, 3101–3109. [CrossRef] [PubMed]
114. Dai, J.; Chae, M.; Beyene, D.; Danumah, C.; Tosto, F.; Bressler, D.C. Co-Production of Cellulose Nanocrystals and Fermentable Sugars Assisted by Endoglucanase Treatment of Wood Pulp. *Materials* **2018**, *11*, 1645. [CrossRef] [PubMed]
115. Ladics, G.S.; Fan, L.; Sewalt, V.J.; Spök, A. Chapter Eight—Safety Assessment and Regulation of Food Enzymes. In *Enzymes*; Kermasha, S., Michael, N.A., Eds.; Academic Press: San Diego, CA, USA, 2021; pp. 203–258. ISBN 978-0-12-800217-9.

International Journal of *Molecular Sciences*

MDPI

Article

New Perspectives on the Sustainable Employment of Chestnut Shells as Active Ingredient against Oral Mucositis: A First Screening

Ana Sofia Ferreira [1], Ana Margarida Silva [1], Diana Pinto [1], Manuela M. Moreira [1], Ricardo Ferraz [2], Jaroslava Švarc-Gajić [3], Paulo C. Costa [4], Cristina Delerue-Matos [1] and Francisca Rodrigues [1,*]

1 REQUIMTE/LAQV—Instituto Superior de Engenharia do Porto, Rua Dr. António Bernardino de Almeida, 431, 4249-015 Porto, Portugal
2 Ciências Químicas e Das Biomoléculas (CQB) e Centro de Investigação Em Saúde e Ambiente (CISA), Escola Superior de Saúde do Instituto Politécnico do Porto, 4400-330 Porto, Portugal
3 Faculty of Technology, University of Novi Sad, Bulevar cara Lazara, 1, 21000 Novi Sad, Serbia
4 REQUIMTE/UCIBIO—MedTech-Laboratory of Pharmaceutical Technology, Department of Drug Sciences, Faculty of Pharmacy, University of Porto, Rua de Jorge Viterbo Ferreira, 228, 4050-313 Porto, Portugal
* Correspondence: francisca.rodrigues@graq.isep.ipp.pt or franciscapintolisboa@gmail.com; Tel.: +351-22-83-40-500; Fax: +351-22-83-21-159

Citation: Ferreira, A.S.; Silva, A.M.; Pinto, D.; Moreira, M.M.; Ferraz, R.; Švarc-Gajić, J.; Costa, P.C.; Delerue-Matos, C.; Rodrigues, F. New Perspectives on the Sustainable Employment of Chestnut Shells as Active Ingredient against Oral Mucositis: A First Screening. *Int. J. Mol. Sci.* **2022**, *23*, 14956. https://doi.org/10.3390/ijms232314956

Academic Editors: Antonio Carrillo-Vico and Ivan Cruz-Chamorro

Received: 22 September 2022
Accepted: 26 November 2022
Published: 29 November 2022

Abstract: Oral mucositis (OM), a common side effect of oncological treatment, is an oral mucosal disorder characterized by painful ulcerations and increased risk of infection. The use of natural antioxidants to suppress the redox imbalance responsible for the OM condition has emerged as an interesting approach to prevent/treat OM. This study aims to explore the chestnut (*Castana sativa*) shells as potential active ingredient against OM. Therefore, chestnut shells were extracted at different temperatures (110–180 °C) by Subcritical Water Extraction (SWE), aiming to recover antioxidants. The extracts were also evaluated against microorganisms present in the oral cavity as well as on human oral cell lines (TR146 and HSC3). The highest phenolic content was obtained with the extraction temperature of 110 °C, exhibiting the best antioxidant/antiradical activities and scavenging efficiencies against HOCl (IC_{50} = 4.47 µg/mL) and ROO• (0.73 µmol TE/mg DW). High concentrations of phenolic acids (e.g., gallic and protocatechuic acids) and flavanoids (catechin, epicatechin and rutin) characterized the phenolic profile. The antimicrobial activity against several oral microorganisms present in the oral cavity during OM, such as *Streptococcus*, *Staphylococcus*, *Enterococcus*, and *Escherichia*, was demonstrated. Finally, the effects on HSC3 and TR146 cell lines revealed that the extract prepared at 110 °C had the lowest IC_{50} (1325.03 and 468.15 µg/mL, respectively). This study highlights the potential effects of chestnut shells on OM.

Keywords: *Castanea sativa* shells; phenolic compounds; Subcritical Water Extraction (SWE); antimicrobial activity; oral mucositis

1. Introduction

Oral mucositis (OM) is one of the most common side effects of antineoplastic treatments, namely chemo- and/or radiotherapy, and is defined as an acute inflammatory, ulcerative, and painful disorder that severely degrades the quality of life of patients [1–3]. Ulceration of buccal mucosa compromises the natural defensive barrier, allowing oral bacteria and inflammatory cytokines to enter underlying tissues and the systemic blood supply [2]. During the antineoplastic treatments, reactive oxygen species (ROS) are generated, leading to the development of oxidative stress, one of the principal phenomena in the basis of OM development [2,3]. An imbalance between the body's synthesis of oxidants and antioxidant defenses of human body is known as oxidative stress, which may result in a wide range of diseases, including diabetes, cardiovascular or neurodegenerative diseases, and, at the last step, cancer [4,5]. Therefore, the oxidative stress pathway could be a therapeutic target against OM, particularly in what concerns natural antioxidants that have a strong potential to prevent and control the generation of free radicals and restore cellular homeostasis [6]. A recent analysis by our research team highlighted the benefits of using natural substances to treat this oral condition [1]. As the pharma-, nutra-, and cosmeceutical industries shift their focus on natural bioactive chemicals, natural product research is increasing due to their unique properties as source of therapeutic phytochemicals, commonly associated with efficacy, safety, and minimal side effects [7,8]. Among natural products, food by-products arise as a unique opportunity, considering sustainable principles and circular economy questions, such as low cost, abundance, and accessibility. Agro-industries generate large amounts of by-products in the form of shells, skins, flowers, leaves, and pulps, that may be used as raw resources to establish added-value products, a key strategy for a sustainable production [9,10]. This is the case of chestnut shells. European or sweet chestnut (*Castanea sativa*, Mill.) production has, as its primary purpose, the natural fruit consumption, or the production of frozen chestnut [11,12]. Portugal is one of the major European producers, with its main distribution coming from the northeast part of the country, covering 35,000 ha, with an estimated fruit production of 1800 kg/ha/year by 2022, expected to rise to about 3000 kg/ha/year in 2030 [11]. The inner and outer shells are generated by chestnut peeling processes, representing almost 10–15% of the whole chestnut weight [12,13]. Shells are mostly used as fuel and as source of tannins for wine aging, also being a major origin of fermentable sugars for the development of biofuels, while the extracts are known for their high polyphenol content (with emphasis on phenolic acids, flavonoids, and tannins) [12–14].

Extraction is the most vital phase for the recovery of bioactive compounds from natural matrices. A great challenge to start implementing eco-friendly and effective extraction methods, rather than the use of standard solvent extraction, has arisen in industry [15,16]. Greener alternative technologies, such as Subcritical Water Extraction (SWE), are being described as more ecological and efficient to recover valued compounds and produce high-quality extracts [17,18]. SWE presents a major advantage by using water as solvent, which is heated at a temperature range of 100–374 °C and a pressure state between 10–100 bar to maintain in its liquid form [19,20]. Under these parameters, water's dielectric constant decreases, exhibiting a behavior like organic solvents and enabling a faster extraction, higher extraction yield, and efficient recovery of both polar and nonpolar compounds [19]. The short extraction time, energy savings, selectivity, discontinuation of organic solvents, as well as the low environmental impact and high cost-effectiveness ratio are positive aspects that should be considered [21,22]. Nevertheless, some phenolic compounds can be completely or partially degraded at high temperatures, producing new ones. Therefore, temperature is a key parameter in the SWE [19,21,23]. In 2019, our research team published the first study that evaluated the subcritical water extraction of chestnut shells, but the influence of temperature was not deeply explored, posing a challenge. Determining the ideal SWE temperature (110–180 °C) to extract bioactive compounds from C. *sativa* shells was the main aim of this study, to produce an optimal extract with promising effects against OM. While the *in vitro* effects on several buccal cell lines and the antibacterial activities against oral pathogens ensure the final OM effects, the phenolic profiles were evaluated to understand the specific compounds responsible for the antioxidant/antiradical activities. To the best of our knowledge, this is the first research that provides insights on how OM can be treated using a food by-product.

2. Results and Discussion

2.1. Extraction Yield

Extraction is a critical step in the phytochemical processing pathway to obtain the highest quantity of bioactive compounds from plant sources. To standardize plant products, it is crucial to use an appropriate extraction method [24,25]. Furthermore, for upscaling reasons, it is crucial to select an appropriate extraction technique and optimize various parameters, such as temperature, sample-to-solvent ratio, pH, and extraction period [22]. Various extraction techniques such as maceration, infusion and Soxhlet extraction, and alternative and greener methods, such as microwave-assisted extraction (MAE), ultrasound-assisted extraction (UAE), supercritical fluid extraction (SFE), and SWE, are available in the market [18,24,26–28]. Green extraction techniques are faster and more environmentally friendly in terms of solvent and energy use, leading to a continuously increasing application in recent years. In some circumstances, the green extraction yield is considerably higher than that of the conventional approaches [26,28–30]. Table 1 summarizes the extraction yields of the different extracts of chestnut shells obtained at different temperatures (110 °C, 120 °C, 140 °C, 160 °C, and 180 °C).

Table 1. Extraction yields, total phenolic content (TPC), and antioxidant/antiradical activities of *C. sativa* extracts prepared by SWE were evaluated by DPPH, ABTS, and FRAP assays. Results are expressed as mean ± standard deviation ($n = 3$). Different letters (a, b, c, d, e) in the same column indicate significant differences between extracts ($p < 0.05$).

SWE Extracts	Extraction Yield (%)	TPC (mg GAE/g DW)	DPPH IC_{50} (μg/mL)	ABTS IC_{50} (μg/mL)	FRAP (μmol FSE/g)
110 °C	20.88 ± 0.78	239.53 ± 23.17 ^{a}	426.88 ± 13.42	148.68 ± 13.20 ^{a}	4240.38 ± 10.04 ^{a}
120 °C	21.00 ± 1.49	201.75 ± 13.02 ^{b}	489.49 ± 5.96	228.32 ± 16.48 ^{b}	4092.98 ± 92.82 ^{b}
140 °C	20.83 ± 0.62	162.52 ± 7.14 ^{c}	460.21 ± 14.81	232.97 ± 4.23 ^{b}	3398.98 ± 57.23 ^{c}
160 °C	20.97 ± 1.74	122.31 ± 5.89 ^{d}	496.15 ± 8.75	220.40 ± 7.83 ^{b}	2728.06 ± 35.04 ^{d}
180 °C	20.29 ± 0.93	126.19 ± 6.33 ^{d}	583.47 ± 14.68	256.59 ± 5.55 ^{b}	2464.32 ± 68.44 ^{e}

IC_{50} = In vitro concentration required to decrease the reactivity of the studied radical species in the tested media by 50%. GAE = gallic acid equivalent; FSE = ferrous sulphate equivalent.

As it is possible to observe, no significant extraction differences were observed between the extract's yields, with the extract prepared at 180 °C achieving the lowest value (20.29%), oppositely to the extract obtained at 120 °C (21.00%). Cacciola et al. (2019) evaluated three different methods (conventional liquid extraction, UAE, and MAE) using water as a solvent to recover bioactive compounds from chestnut shells. According to the authors, the extraction yields obtained for conventional liquid extraction, UAE, and MAE were, respectively, 5.2%, 2.2%, and 3.8% [31], at least four times lower than those shown in Table 1. In another study, Fernández-Agulló et al. (2014) reported higher extraction yields for chestnut shells extracted using 50% of organic solvents, namely methanol (12.09%) and ethanol (13.27%), when compared to conventional extraction at 75 °C with water (8.72%) and MAE (10.75%, 10.38% and 5.64%, respectively, for methanol, ethanol, and water) [32]. The authors demonstrated the influence of temperature on the extraction efficiency, as the extraction yields increased with the augmented extraction temperatures tested [32]. In addition, Pinto et al. (2021) determined the efficiency of MAE in extracting *C. sativa* shells and reported extraction yields ranging from 6.52% (25% aqueous extract) to 11.13% (100% ethanol) [33]. The same research group tested different SWE times and stated yields between 6.70 and 9.19% [19].

The influence of sample-to-solvent ratio and mass transport enhancement on chestnut shells extraction was also evaluated in Microwave-Assisted Subcritical Water Extraction (MASWE) using a range of temperatures between 100 °C and 220 °C [34]. According to the authors, it was evidenced an increase of the extract production proportional to the increase of temperature, but minor deviations were observed for the sample-to-solvent ration (1:20–1:30), with exception of the 150 °C extract (322.37 mg/g DW and 416.56 mg/g DW, respectively) [34].

The SWE extraction yields reported in Table 1 were consistent with those determined by Wang et al. (2014) for citrus peel [35], Silva et al. (2022) for kiwiberry leaves [36], and Rodrigues et al. (2019) for papaya seed extraction [37].

2.2. Extraction Yield

Table 1 summarizes the TPC, and antioxidant/antiradical activities of SWE extracts obtained at different temperatures. As shown, the TPC ranged between 126.19 and 239.53 mg GAE/g DW. The highest results were obtained for the extracts prepared at 120 °C (201.75 mg GAE/g DW) and 110 °C (239.53 mg GAE/g DW), and significant differences were noted between the extracts obtained at 120 °C and 140 °C ($p = 0.008$). Thermal degradation is probably responsible for the decrease in polyphenols content at higher temperatures. Nonetheless, this phenomenon alone cannot explain polyphenol behavior, as in conventional extraction methods, and the polyphenols breakdown can also occur at 80 °C. In fact, the increase in TPC observed at high SWE temperatures could also be attributed to the disruption of lignin–phenolic acid linkages or to the breakdown of lignin itself, resulting in large amounts of phenolic acids [23].

Pinto et al. (2021) assessed the phenolic profile, bioactivity, and cell viability of *C. sativa* shells extracted by MAE using different solvents (water and ethanol). The research team concluded that the highest TPC was achieved by the aqueous extract (173.89 mg GAE/g DW) [33]. The same authors tested this matrix by SWE extraction using response surface methodology (RSM) and reported higher TPC values for the extract prepared at 220 °C (417.30 mg GAE/g DW); however, the phenolic profile was different from that obtained in the present study. According to the authors, the response surface 3D plots of the TPC showed that the temperature of 220 °C was the optimal temperature and the phenolic content at 80 °C is higher than that at 150 °C, probably due to the degradation of tannin [19].

IC_{50} values for the DPPH assay ranged from 426.88 g/mL (110 °C) to 583.47 μg/mL (180 °C). Statistical analysis demonstrated significant differences between the extracts ($p = 0.002$), which aligns with the TPC results. At lower temperatures the extracts presented higher phenolic content, resulting in higher antioxidant powers.

Regarding the ABTS assay, the IC_{50} results varied between 148.68 μg/mL and 256.59 μg/mL for the extracts obtained at 110 °C and 180 °C, respectively. Significant differences were noted between the extracts obtained at 110 °C, 120 °C, and 140 °C ($p < 0.016$), as well as between the extracts obtained at 120 °C, 160 °C, and 180 °C ($p < 0.05$).

According to Table 1, the extract's ability to decrease ferric ions increased with temperature was 180 °C < 160 °C < 140 °C < 120 °C < 110 °C. The extract processed at 110 °C had the highest value (4240.38 μmol FSE/g DW), while the extract produced at 180 °C had the lowest activity (2464.32 μmol FSE/g DW), an almost two-fold difference. Significant differences ($p = 0.031$) were observed between the extracts obtained at 110 °C and 160 °C.

Fernández-Agulló et al. (2014) also used DPPH and FRAP assays to evaluate the antiradical and antioxidant properties of chestnut shells and found that the aqueous extract (75 °C) obtained by conventional extraction produced the best results ($IC_{50} = 0.031$ mg/mL for DPPH; $IC_{50} = 0.284$ mg/mL for ABTS). The same extract obtained the highest FRAP (56.23 g GAE/100 g extract), in the accordance with TPC results [32]. In the chestnut shells SWE study, the extract prepared at 220 °C was assessed through DPPH, FRAP, and ABTS assays and showed good results, respectively, 901.16 mg AAE/g DW, 7994.26 mg FSE/g DW and 815.66 mg TE/g DW [19]. Regarding the UAE extraction of chestnut shells, the extract (70 °C/40 min) also had good antiradical and antioxidant results in the DPPH ($IC_{50} = 44.1$ μg/mL), ABTS ($IC_{50} = 65.4$ μg/mL) and FRAP ($IC_{50} = 32.0$ μg/mL) assays [38]. Analyzing the results from Table 1 with the authors mentioned above, it is evident that the extraction conditions are crucial. When the plant matrix was the same, the different extraction methods used were subjected to several variations. A detailed understanding of the bioactive composition, particularly the phenolic profile, is, therefore, essential, since the spectrophotometric assays have limitations such as the sensitivity to pH, reaction time, and presence of interferents (e.g., sugars and amino acids) [39].

2.3. Phenolic Profile Identification and Quantification by HPLC-PDA

The phenolic profiles of the *C. sativa* shells extracts prepared using SWE are detailed in Table 2.

Table 2. Identification and quantification (mg/g DW) of phenolic compounds present in *C. sativa* shells SWE extracts by HPLC-PDA analysis. Results are expressed as mean ± standard deviation (n = 3).

Phenolic Compound	110 °C	120 °C	140 °C	160 °C	180 °C
			mg/g DW		
Alkaloids					
Caffeine	1.18 ± 0.06	1.14 ± 0.06	1.01 ± 0.05	0.33 ± 0.02	0.44 ± 0.02
Chalconoids					
Phloridzin	0.83 ± 0.04	0.50 ± 0.03	ND	ND	ND
Flavanols					
Catechin	4.10 ± 0.20	3.65 ± 0.18	1.60 ± 0.08	0.74 ± 0.04	0.69 ± 0.03
Epicatechin	1.68 ± 0.08	1.14 ± 0.06	0.75 ± 0.04	<LOQ	<LOQ
Flavonols					
Rutin	1.04 ± 0.05	0.64 ± 0.03	0.69 ± 0.03	0.103 ± 0.005	0.136 ± 0.007
Flavonones					
Naringin	<LOQ	<LOQ	ND	ND	ND
Phenolic acids					
3,5-di-caffeolquinic acid	<LOQ	<LOQ	<LOQ	ND	ND
4-O-caffeyolquinic acid	0.49 ± 0.02	0.46 ± 0.02	0.40 ± 0.02	0.34 ± 0.02	0.21 ± 0.01
4,5-DI-O-caffeoylquinic acid	0.126 ± 0.006	0.22 ± 0.01	0.34 ± 0.02	0.026 ± 0.001	0.28 ± 0.01
Caftaric acid	0.41 ± 0.02	0.26 ± 0.01	0.53 ± 0.03	0.16 ± 0.01	0.58 ± 0.03
Caffeic acid	0.17 ± 0.01	0.16 ± 0.01	0.12 ± 0.01	0.069 ± 0.003	0.12 ± 0.01
Clorogenic acid	0.43 ± 0.02	0.27 ± 0.01	0.18 ± 0.01	0.081 ± 0.004	0.13 ± 0.01
p-Coumaric acid	<LOD	<LOD	<LOD	<LOD	<LOD
Ellagic acid	0.136 ± 0.007	0.157 ± 0.008	0.181 ± 0.009	0.064 ± 0.003	0.108 ± 0.005
Gallic acid	5.99 ± 0.30	5.08 ± 0.25	4.64 ± 0.23	3.48 ± 0.17	2.40 ± 0.12
Ferulic acid	<LOQ	<LOQ	<LOQ	0.011 ± 0.001	0.056 ± 0.003
Neochlorogenic acid	0.71 ± 0.04	0.59 ± 0.03	0.49 ± 0.02	0.21 ± 0.01	0.28 ± 0.01
Protocatechuic acid	1.68 ± 0.08	1.37 ± 0.07	1.15 ± 0.06	1.31 ± 0.07	1.59 ± 0.08
Sinapic acid	ND	ND	ND	<LOQ	<LOQ
Syringic acid	ND	ND	ND	0.019 ± 0.001	0.063 ± 0.003
Vanillic acid	1.02 ± 0.05	0.91 ± 0.05	0.55 ± 0.03	0.084 ± 0.004	0.12 ± 0.01
Quercetin-3-O-galactoside	0.115 ± 0.006	0.099 ± 0.005	0.136 ± 0.007	0.012 ± 0.001	0.120 ± 0.006
Quercetin-3-O-glucopyranoside	0.057 ± 0.003	0.069 ± 0.003	0.081 ± 0.004	0.043 ± 0.002	0.036 ± 0.002
Stilbenoids					
Resveratrol	<LOD	<LOD	<LOD	ND	ND
Trans-polydatin	<LOQ	<LOD	<LOD	<LOD	<LOD
Total	20.20	16.71	12.85	7.07	7.37

ND, not detected; LOD, limit of detection; LOQ, limit of quantitation.

As shown in Table 2, health-promoting compounds were grouped into seven different classes. Overall, 25 phenolic compounds were identified and quantified, with phenolic acids being the main class of compounds. The extract obtained at 110 °C presented the highest amount of phenolic acids (20.20 mg phenolic acids/g DW), though higher temperatures led to a significant decrease, such as the extract prepared at 160 °C (7.07 mg phenolic acids/g DW). This may be due to the diminution of the dielectric constant and polarity of water as temperature increases, which leads to the dissolution of non-polar compounds such as polyphenols [40]. However, high temperatures have a negative effect on polyphenols above 120 °C, resulting in the breakdown of thermally unstable polyphenols [41].

As it is possible to observe, the major phenolic acid present in all extracts is gallic acid, particularly at 110 °C, 120 °C and 140 °C (5.99, 5.08 and 4.64 mg/g DW, respectively). This compound is extremely important for human health, and several studies have reported its biological and pharmacological activities, including anti-inflammatory, antioxidant, anticancer, and antimicrobial activities [42]. Moreover, the presence of protocatechuic acid may indicate the thermal degradation of catechin [19]. Flavonoids were also identified in considerable amounts, corresponding to more than 30% of the total phenolic content. Particularly, catechin ranged from 3.65 to 4.10 mg/g DW for the extracts obtained at 120 °C and 110 °C, respectively. The epicatechin amounts in the extracts obtained at 110 °C and 120 °C were, respectively, 1.68 and 1.14 mg/g DW, while for rutin, the quantities were 1.04 (110 °C) and 0.64 mg/g DW (120 °C). However, for most of the compounds mentioned, the results indicate a high temperature sensitivity that should be considered.

Oxidative stress is the main factor in the basis development of OM, which induces to the activation of the apoptotic process, leading to the clinical manifestation of mucosal damage. Pro-inflammatory cytokines are in a positive feedback mechanism, conducting to ulceration, which compromises the mucosal integrity, allowing several microorganisms to penetrate tissues and increasing the sepsis risk [2,43]. The five phenolic compounds mentioned above were found at higher concentrations in all the extracts. Because of their capacity to control crucial cellular enzyme activities that lead to anti-inflammatory, antioxidant, antimicrobial, and anticancer activities, all of them have been associated with a wide range of beneficial properties and are expected to be potential treatments and prevention options for OM [44].

Previous studies have reported discrepancies in the phenolic composition of chestnut shells extracts, which can be explained by the use of diverse extraction techniques and also varied extraction settings. For example, Vásquez et al. (2012) analyzed chestnut shells extracted by organic solvents using conventional methods. The authors reported 11 compounds, with catechin/epicatechin and gallocatechin/epigallocatechin being the principal ones, while hydrolysable tannins, such as galloyl-glucoses and ellagic acid, were also present and identified [45]. Gallic acid esters of glucose were particularly present in chestnut shells extracted with water at 75 °C (conventional extraction) as well as gallocatechin and catechin [32]. In the chestnut shells extract prepared by UAE, ellagic acid (40.4 μg/mg DW), caffeic acid derivative (15.4 μg/mg DW) and epigallocatechin (15.3 μg/mg DW) were identified in high quantities [38], while by MAE the gallic acid and protocatechuic acid were also detected (117.58 and 16.8 mg/g DW, respectively) [31].

2.4. ROS Scavenging Capacity

A major global issue is the development of disorders related to oxidative stress caused by an imbalance between reactive species and antioxidants commonly present in the human body, being a key factor for the OM pathogenesis. However, its mechanism remains unclear. Nonetheless, it is known that chemo and/or radiotherapy triggers an increase in ROS generation [1]. Reactive species, such as hypochlorous acid (HOCl), superoxide ($O_2^{\bullet-}$), peroxyl radical ($ROO^{\bullet}$), and peroxynitrite ($ONOO^-$), when produced in excess in living cells are responsible, directly, or indirectly, for a number of diseases, including OM and, at last stage, oral cancer [46,47]. One of the most significant classes of natural antioxidants that can prevent or reduce the harm caused by oxidative stress is the polyphenol family, which has a positive effect on OM [43]. Table 3 shows the ROS scavenging activity results.

The extract prepared at 140 °C (IC_{50} = 31.14 μg/mL) showed the highest activity in terms of $O_2^{\bullet-}$ scavenging capacity, whereas the extract prepared at 180 °C showed the lowest activity (IC_{50} = 73.18 μg/mL). Gallic acid was used as a positive control for this test, and it produced the best results (IC_{50} = 23.82 μg/mL). No significant differences (p = 0.123) were observed between the extracts. However, HOCl was effectively scavenged by all extracts. While the extract obtained at 160 °C had the lowest free radical scavenging capacity (IC_{50} = 22.85 °C), the extract prepared at 110 °C had the best effectiveness (IC_{50} = 4.47 °C). Significant differences were noted between extracts obtained at 110 °C and 140 °C (p = 0.026), and 160 °C and 180 °C (p = 0.027). Regarding the $ROO^{\bullet}$ quenching, the efficacy increased in the following order: gallic acid >110 °C > 140 °C > 120 °C >160 °C >180 °C. No significant difference (p = 0.379) was observed between extracts.

Table 3. Superoxide anion radical ($O_2^{\bullet-}$), hypochlorous acid (HOCl), peroxyl radical ($ROO^{\bullet}$) and peroxynitrite ($ONOO^-$) scavenging capacities of *C. sativa* shells SWE extracts. Values are expressed as mean ± standard deviations (n = 3). Different letters (a, b, c, d) in the same column indicate significant differences between extracts (p < 0.05).

	Reactive Oxygen Species			Reactive Nitrogen Species	
SWE Extracts	$O_2^{\bullet-}$	**HOCl**	$ROO^{\bullet}$	**ONOO⁻ in Presence of** HCO_3^-	**ONOO⁻ in Absence of** HCO_3^-
	IC_{50} (µg/mL)	IC_{50} (µg/mL)	(µmol TE/mg DW)	IC_{50} (µg/mL)	
110 °C	44.36 ± 0.34 [cd]	4.47 ± 0.29 [b]	0.73 ± 0.024 [b]	2.52 ± 0.19 [c]	2.53 ± 0.22 [bc]
120 °C	46.31 ± 1.37 [d]	6.49 ± 0.18 [b]	0.54 ± 0.016 [cd]	1.41 ± 0.11 [b]	3.32 ± 0.19 [c]
140 °C	31.14 ± 2.05 [b]	8.27 ± 0.77 [b]	0.61 ± 0.037 [bc]	2.77 ± 0.29 [c]	2.09 ± 0.21 [b]
160 °C	35.34 ± 3.14 [bc]	22.85 ± 1.67 [c]	0.50 ± 0.032 [cd]	2.79 ±0.26 [c]	1.89 ± 0.09 [b]
180 °C	73.18 ±1.86 [e]	20.74 ± 1.27 [c]	0.49 ± 0.009 [d]	2.93 ± 0.03 [c]	5.94 ± 0.30 [d]
Positive control					
Gallic acid	23.82 ± 0.82 [a]	1.68 ± 0.16 [a]	4.23 ± 0.25 [a]	0.30 ± 0.03 [a]	0.05 ± 0.00 [a]

IC_{50} = In vitro concentration required to decrease the reactivity of the studied reactive species in the tested media by 50%. TE = Trolox Equivalents.

Comparing with other studies, the SWE extracts prepared in the present study showed lower ROS scavenging activity. Lameirão et al. (2021) related that the chestnut shells extract obtained by UAE (at 70 °C during 40 min) achieved an IC_{50} of 14.1 µg/mL, 0.7 µg/mL and 0.3 µg/mL regarding $O_2^{\bullet-}$, HOCl and $ROO^{\bullet}$, respectively [38]. Likewise, the optimal SWE extract (obtained at 220 °C) of chestnut shells prepared by Pinto et al. (2021) exhibited high scavenging activity regarding HOCl (IC_{50} = 0.79 µg/mL) and $O_2^{\bullet-}$ (IC_{50} = 12.92 µg/mL). However, as summarized in Table 3, the extracts in the present study had a better ability to quench $ROO^{\bullet}$, as Pinto et al. reported an Ssample/STrolox = 0.084 for the optimal extract [19].

Regarding RNS, the peroxynitrite ($ONOO^-$) scavenging activity was evaluated in the absence and in presence of hydrogen carbonate (HCO_3^-) to mimic the in vivo environment. Although the results presented in Table 3 do not attest to a temperature-dependent correlation, a significant ability of the extracts to act as effective reducers of peroxynitrite-mediated damage can be observed. The best IC_{50} calculated was 1.41 µg/mL (for the extract prepared at 120 °C) and 1.89 µg/mL (for the extract obtained at 160 °C) in the presence and absence of HCO_3^-, respectively. However, the gallic acid achieved best results (IC_{50} = 0.30 and 0.05 µg/mL, respectively, in presence and absence of HCO_3^-). To the best of our knowledge, this is the first peroxynitrite assay performed on chestnut shells extracts. The phenolic profile, and thus the potential for compounds to degrade, as previously mentioned, could play a role in the observed discrepancies. No significant differences between the extracts were observed in the presence (p = 0.285) or absence (p = 0.357) of HCO_3^-. However, the radical scavenging capacity calculations, particularly for HOCl and $ROO^{\bullet}$, are generally consistent with the spectrophotometric results (Table 1), indicating that the extract prepared at 180 °C had the lowest potential, while the extract obtained at 110 °C had the most promising characteristics for OM treatment.

2.5. Antimicrobial Activity

The oral cavity contains complex microbiota compromising more than 700 different bacterial species. The oral cavity is an ideal environment for microbial growth due to its pH (6.5–7.5), temperature (37 °C), soft tissues (palate, tongue, and buccal mucosa), and hard tissues (teeth) [48]. The oral microbial population is generally balanced however, pathologies may appear when the ecological balance is disturbed. Chemotherapy-induced myelosuppression and radiation-induced xerostomia are known to cause microbiome dysbiosis, which compromise the saliva flow, a major barrier against potential pathogens [49]. Owing to its ability to affect the innate immune response, which is an activator of the mucositis pathway, the oral microbiota can aggravate or sustain oral mucositis [50]. Examples of the main species of bacteria commonly isolated from the oral cavity are *Porphyromonas, Lactobacillus, Veillonella, Actinobacillus, Streptococcus, Staphylococcus, Enterococcus*, and *Escherichia* [51,52].

The antimicrobial capacity of SWE *C. sativa* shells extracts was analyzed against several Gram-negative and Gram-positive bacteria, the majority of which are present in the oral cavity in case of OM [52,53]. MIC and MBC were determined for each microorganism at different extract concentrations (2–64 mg/mL) using a microplate dilution assay. Table 4 summarizes the results for all strains.

Table 4. Minimum Inhibition Concentration (MIC) and Minimum Bactericidal Concentration (MBC) of *C. sativa* shells SWE extracts on the bacterial strains tested, at concentrations range of 2–64 mg/mL. Values are expressed as mean (n = 3).

SWE Extracts		Gram-Negative			Gram-Positive		
		E. coli ATCC 25922	*E. coli* ATCC 8739	*E. coli* CTX M2 *	*S. aureus* ATCC 25913	*S. aureus* MRSA *	*E. faecalis*
110 °C	MIC	8	4	4	-	4	64
	MBC	64	-	-	-	32	-
120 °C	MIC	8	4	8	64	8	2
	MBC	64	-	-	-	16	-
140 °C	MIC	4	16	4	4	32	64
	MBC	-	-	-	-	64	-
160 °C	MIC	4	4	4	4	4	8
	MBC	32	-	-	64	32	64
180 °C	MIC	8	8	-	4	4	16
	MBC	16	4	-	64	16	64

* drug-resistant bacteria; - absence of MIC/MBC.

As can be observed, the chestnut shells extracts obtained at different temperatures exhibited diverse effects on the growth of microorganisms, which is probably due to the phenolic content of the extracts.

The extract prepared at 110 °C did not present MIC and MBC values for *S. aureus* (Gram-positive) at the concentrations tested. However, the extract showed a significant antimicrobial activity against two drug-resistant bacteria, *S. aureus* MRSA and *E. coli* CTX M2 (MIC of 4 mg/mL and MBC of 32 mg/mL in *S. aureus* MRSA). This extract also exhibited MIC in *E. coli* ATCC 8739, *E. coli* ATCC 25922, and *E. faecalis* (8, 4, and 64 mg/mL, respectively) and MBC of 64 mg/mL in *E. coli* ATCC 25922. *E. coli* CTX M2 was resistant to the extract obtained at 180 °C. Compared to the 110 °C extract, the extract prepared at 180 °C presented higher MIC values for the two non-resistant strains of *E. coli* (8 mg/mL), but a lower MIC for *E. faecalis* (16 mg/mL). At this temperature, the extract exhibited MBC values for all strains that had growth inhibition. As shown in Table 4, the extract prepared at 160 °C showed better antimicrobial activity, as it showed lower MIC values for most of the strains tested and displayed MBC values for all the strains, except for the drug-resistant *E. coli* CTX M2.

Phenolic compounds may affect bacterial cells by damaging their cellular membranes, binding to the cell wall, and leading to enzyme inactivation and DNA damage [54]. Flavanols (e.g., catechin and epicatechin) are known to inhibit the in vitro growth of several strains, including *S. mutans* and *E. coli* [55]. Flavonols (e.g., rutin) also exhibit a significant activity against *S. aureus* (Gram-positive), possibly through an aggregation mechanism. Nevertheless, phenolic acids, such as gallic acid, also showed antibacterial activity against

S. aureus and *E. coli*. Although these compounds present weaker efficiency in bacterial growth inhibition when compared to flavonoids, they still prove to be more efficient than standard antibiotics, such as gentamicin and streptomycin [55,56]. Kang et al. (2008), studied the antimicrobial effects of gallic acid, one of the main compounds present in *Rhus chinensis* L., and reported MIC values of 8 mg/mL for *S. mutans*. The authors also reported in vitro inhibition of *S. mutans* biofilms, demonstrating the antimicrobial activity of gallic acid [57]. Likewise, Passos et al. (2021) reported the antimicrobial and anti-adherence activity of gallic acid, an isolated compound of the ethanolic extract of *L. ferrea*, which exhibited significant antimicrobial activity and inhibition of adhesion of *S. mutans* biofilms [58].

Previous research has demonstrated a correlation between phenolic compounds and the antimicrobial activity, as the quantity of phenolic compounds has a proportional inhibitory effect on the microorganism growth [59]. Though, the present results do not demonstrate a correlation as the extract prepared at 110 °C has a higher amount of phenolic compounds and the extract obtained at 160 °C showed better antimicrobial capacity. One of the possible explanations for such data is the antagonistic effect of the combination of polyphenolic compounds present in the SWE extracts, which may decrease the antimicrobial ability.

2.6. Cells Viability Assays

Considering the potential use of SWE *C. sativa* extracts in OM treatment, it is particularly important to assess their effects in in vitro models. In this study, SWE *C. sativa* shells extracts were evaluated on HSC3 and TR146 cell lines by MTT assay. Both cell lines mimic the normal human buccal epithelium and are tumorigenic, making them suitable as oral mucosal models for testing medication absorption. In particular, the TR146 cell line was established from a neck metastasis of a buccal carcinomic tumor, and the HSC3 cell line was derived from tumors of metastatic lymph nodes originating from tongue squamous cell carcinoma [60–62]. Figure 1 summarizes the obtained results for both cell lines using various extract concentrations (125–2000 μg/mL), while Table 5 presents the IC_{50} values.

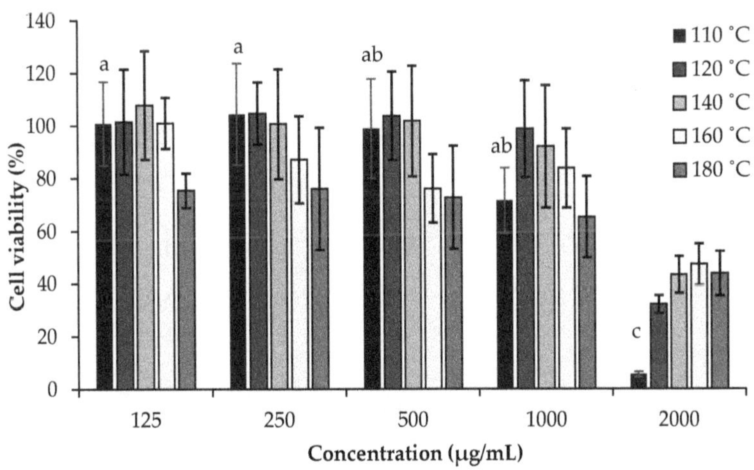

Figure 1. Effects of *C. sativa* shells SWE extracts on the viability of HSC3 cells at a range of concentrations of 125–2000 μg/mL, measured by MTT assay, at 24 h. Values are expressed as mean ± standard deviation (n = 3). Different letters indicate significant differences between concentrations of the same sample ($p < 0.005$) according to Tukey's HSD test.

Table 5. IC_{50} values of oral model cellular lines treated with *C. sativa* shells SWE extracts, measured by MTT assay. Values are expressed as mean (n = 3). IC_{50} = In vitro concentration required to decrease cell viability by 50%.

	Cell Lines IC_{50} (μg/mL)	
SWE Extracts	**HSC3**	**TR146**
110 °C	1325.03	468.15
120 °C	1730.39	496.45
140 °C	1862.17	663.83
160 °C	1923.94	553.79
180 °C	1707.33	572.18

Regarding the HSC3 results, the extracts obtained at 160 °C and 180 °C presented viabilities of 87.16% and 76.03%, respectively, after exposure with the low concentrations tested (250 μg/mL). The extracts obtained at 110 °C, 120 °C, and 140 °C showed viability reductions at concentrations of 500, 1000, and 2000 μg/mL, respectively. Significant differences were noted for the different concentrations of the extracts obtained at 110 °C (500–2000 μg/mL; $p = 0.044$) and 180 °C (125–250 μg/mL and 125–2000 μg/mL; $p < 0.03$). These results might be explained by the fact that the extract prepared at 110 °C contained the highest concentration of phenolic acids, which was in accordance with Table 1. As shown in Figure 2, except for the extract prepared at 110 °C, none of the extracts obtained at different temperatures resulted in a reduction in the viability of TR146 cells at a concentration of 125 μg/mL. However, with increasing concentrations, all extracts significantly decreased cellular viability, leading to viability values below 57% for all extracts at 500 μg/mL. Significant differences were observed for the extract prepared at 140 °C (125–250 μg/mL; $p = 0.022$) and 180 °C (25–1000 μg/mL; $p = 0.016$). These results proved that TR146 cells were the most sensitive to the SWE chestnut shells extracts (Table 5). Moreover, it can be concluded that the extracts exhibited antiproliferative effects for both cell lines, with the extract prepared at 110 °C achieving the lowest IC_{50} value in TR146 (468.15 μg/mL) and in HSC3 (468.15 μg/mL). This may be due to the higher content of phenolic compounds, as shown in Table 1.

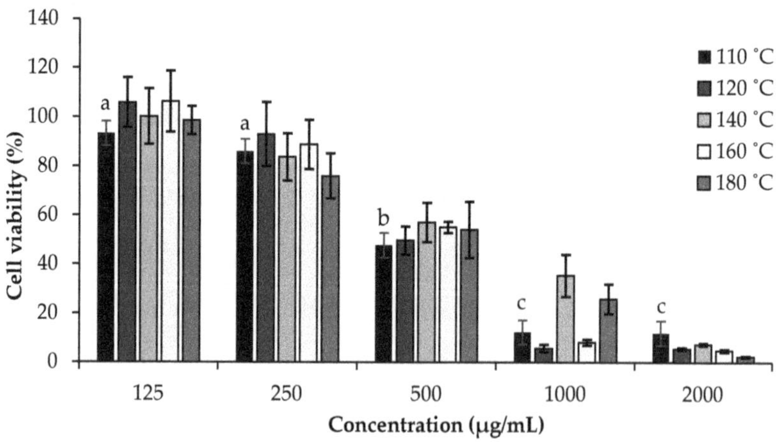

Figure 2. Effects of *C. sativa* shells SWE extracts on the viability of TR146 cells at concentrations range of 125–2000 μg/mL, measured by MTT assay, at 24 h. Values are expressed as mean ± standard deviation (n = 3). Different letters indicate significant differences between concentrations of the same sample ($p < 0.005$), according to Tukey's HSD test.

Lambert et al. reported that polyphenols can inhibit different phases of carcinogenesis, including initiation, promotion, and progression of tumor cells, demonstrating anticancer properties [63]. In addition, polyphenols have been reported to prevent oral cancer, as they come into direct contact with tissues prior to being absorbed, suppressing the proliferation of oral cancer cells on the surface of epithelial cells [64]. Catechins, for example, have been demonstrated the capacity to decrease the activity of oral squamous cell carcinoma (OSCC) in OC2 cells by inhibiting the synthesis of matrix metalloproteinases (MMPs). Hence, invasion and migration of cancer cells can be regulated [65,66]. To the best of our knowledge, this is the first study that assessed the effects of SWE chestnut shells extract on oral model cell lines. This highlights its potential as an ingredient against OM condition.

3. Materials and Methods

3.1. Chemicals

α,α′-azodiisobutyramidine dihydrochloride (AAPH), β-nicotinamide adenine dinucleotide (NADH), dihydrorhodamine 123 (DHR), 2,2-diphenyl-1-picryl-hydrazyl (DPPH), sodium hypochlorite solution with 4% available chlorine, phenazine methosulphate (PMS), nitroblue tetrazolium chloride (NBT), fluorescein sodium salt and Trolox were obtained from Sigma-Aldrich (Steinheim, Germany). Catechin, gallic acid, Folin–Ciocalteu's reagent and Triton X-100 were purchased from Sigma Chemical Co. (St. Louis, MO, USA).

For microbial culture, Mueller Hinton Broth (MHB) medium was purchase to Liofilmchem (Teramo, Italy), Tryptic Soy Broth (TSB), and Tryptic Soy Agar (TSA) medium was obtained from Biolife (Milan, Italy). HPLC solvents were delivered by Sigma-Aldrich (Milan, Italy). Individual's standards of phenolic compounds used for the identification and quantification in extracts were purchased from Sigma-Aldrich (Steiheim, Germany). HPLC solvents were delivered by Sigma-Aldrich (Milan, Italy). Individual's standards of phenolic compounds, namely gallic acid (≥99%), protocatechuic acid (99.63%), neochlorogenic acid (≥98%), caftaric acid (≥97%), chlorogenic acid (>95%), 4-*O*-caffeyolquinic acid (≥98%), vanillic acid (≥97%), caffeic acid (≥98%), syringic acid (≥98%), *p*-coumaric acid (≥98%), *trans*-ferulic acid (≥99%), sinapic acid (≥99%), 3,5-di-*O*-caffeyolquinic acid (≥95%), ellagic acid (≥95%), 4,5-di-*O*-caffeyolquinic acid (≥90%), (+)-catechin (≥98%), (-)-epicatechin (≥90%), naringin (≥95%), quecetin-3-*O*-galactoside (≥97%), quercetin-3-*O*-glucopyranoside (≥99%), rutin (≥94%), phloridzin (99%), *trans*-polydatin (≥98%), resveratrol (≥99%) and caffeine (≥95%), used for the identification and quantification in extracts were purchased from Sigma-Aldrich (Steiheim, Germany). Dulbecco's Modified Eagle Medium (DMEM), penicillin, trypsin-EDTA and streptomycin were provided by Invitrogen Corporation (Life Technologies, S.A., Madrid, Spain). Dimethyl sulfoxide (DMSO) was purchased by AppliChem (Darmstadt, Germany).

TR146 and HSC3 cell lines were obtained from the American Type Culture Collection (ATCC, Manassas, VA, USA). The following reference bacterial strains were obtained from the American Culture Collection (ATCC): *Escherichia coli* ATCC 25922, *E. coli* ATCC 8739, and *Staphylococcus aureus* ATCC 25913. All other strains, including resistant bacteria, were clinical isolates: *Enterococcus faecalis* and the resistant bacteria, *Escherichia coli CTX M2* and *Staphylococcus aureus MRSA*.

3.2. Sample

Castanea sativa shells were generously provided by Sortegel, located in Sortes, Bragança, Portugal. Dehydration of shells occurred at 41 °C for 24 h (Excalibur Food Dehydrator, Pennsylvania, USA) and then were pulverized in a miller to 1 mm particles (Ultra Centrifugal Mill ZM 200, Retsch, Germany). Prior to extraction, samples were carefully blended and kept at room temperature (20 °C), in the dark.

3.3. Subcritical Water Extraction (SWE) of C. sativa Shells

Extraction of *C. sativa* shells was performed in a homemade subcritical batch-type extractor of 1.7 L [67]. For it, a built-in valve and 99.99% nitrogen (Messer) pressurization were used. The samples were then extracted for 30 min at various temperatures between 110 °C and 180 °C using 20 bars of pressure and a 1:30 sample-to-solvent ratio. The vibrating platform (3 Hz) housing the extraction vessel agitated the sample/water mixture during the extraction. The extraction vessel was then cooled using a flow-through water bath (20 ± 2 °C), and the pressure was released by opening the valve. Filtration was used to separate the extracts, and the solutions were subsequently lyophilized (Telstar, model Cryodos −80 °C, Spain) and stored at room temperature (20 °C) until further analysis. The ratio between the total weight of the lyophilized extract and the total weight of the extract obtained following SWE was used to calculate the extraction yield as follows:

$$\text{Extraction yield } (\%) = \frac{\text{weight of lyophilized extract}}{\text{weight of liquid extract}} \times 100$$

3.4. *Determination of Total Phenolic Content*

The Folin–Ciocalteu method was used to measure the total phenolic content (TPC) using spectrophotometry, with minor changes [19]. A calibration curve (linearity range = 5–100 μg/mL; $R^2 > 0.996$) was created using gallic acid solution. Results were expressed as mg of Gallic Acid Equivalents (GAE) per g of dry weight (mg GAE/g DW).

3.5. *Antioxidant and Antiradical Activities*

3.5.1. Ferric Reducing Antioxidant Power (FRAP) Assay

FRAP assay was carried out as described by Benzie and Strain with a few minor changes [36]. A calibration curve (linearity range: 25–500 μM; $R^2 > 0.993$) was established using a standard ferrous sulfate ($FeSO_4 \cdot 7H_2O$) solution at a concentration of 1 mM. Results were expressed in μmol of ferrous sulfate equivalents (FSE) per gram of DW (μmol FSE/g DW).

3.5.2. 2,2-Diphenyl-1-picrylhydrazyl (DPPH) Free Radical Scavenging Assay

DPPH free radical scavenging assay was performed according to the method described by Pinto et al. [68]. Trolox was used as reference for the calibration curve (linearity range: 5–125 μg/mL; $R^2 > 0.989$). Results were presented as the radical scavenging activity's half-maximal inhibitory concentration (IC_{50}, μg/mL).

3.5.3. 2,2′-Azino-bis-3-ethylbenzothiazoline-6-sulfonic Acid (ABTS) Radical Scavenging Activity Assay

Extracts were tested for ABTS radical scavenging activity according to Re et al. [69], with slight adjustments. The calibration curve standard (linearity range: 5–100 μg/mL; $R^2 > 0.998$) was ascorbic acid and the results were expressed as IC_{50} (μg/mL).

3.6. *HPLC-PDA Analysis*

HPLC was used to identify and quantify polyphenols using photodiode array (PDA) detection [70]. Separation was performed at 25 °C using a Gemini C18 column (250 mm 4.6 mm, 5 m, Phenomenex, Alcobendas, Spain). Results on DW (mg/g DW) were represented as mg of each phenolic component per gram of extract. The calibration data used for the quantification of individual phenolic compounds in the chestnut shells extracts are presented in Table S1.

3.7. *Reactive Oxygen Species and Reactive Nitrogen Species Scavenging Capacity*

3.7.1. Superoxide Radical Scavenging Assay

Superoxide radical ($O_2^{\bullet-}$) scavenging assay was performed as described by Gomes et al. [71]. Results were expressed as IC_{50} of the NBT reduction to diformazan.

3.7.2. Hypochlorous Acid Scavenging Assay

Hypochlorous acid (HOCl) scavenging assay was performed accordingly to Gomes et al. [71]. Gallic acid was used as a positive control. The results were presented as an inhibition of the HOCl-induced oxidation of DHR, measured in IC_{50}.

3.7.3. Peroxyl Radical Scavenging Assay

Peroxyl radical (ROO$^•$) assay was carried out according to the method described by Ou et al. [72]. Trolox was used as standard control. The ratio of the slopes obtained for the extracts or positive controls to the slope of Trolox (Sample/STrolox), which represents the ROO$^•$ induced oxidation of fluorescein, was used to express the results.

3.7.4. Peroxynitrite Scavenging Assay

The effects of the studied extracts on the of inhibition of $ONOO^-$ induced oxidation of non-fluorescent DHR to fluorescent rhodamine 123 were assessed by peroxynitrite ($ONOO^-$) scavenging activity. $ONOO^-$ was synthesized as described by Whiteman et al. [73]. To simulate physiological CO_2 values, parallel tests were run in the presence of 187.5 mM $NaHCO_3$, and the results were represented as IC_{50}.

3.8. Antimicrobial Activity

3.8.1. Minimum Inhibitory Concentration (MIC) and Minimum Bactericidal Concentration (MBC)

MIC values were determined by the broth microdilution method using a 96-well plate with MHB medium with minor modifications [74]. Microorganisms were exposed to extracts (0.05 mL of dilutions + 0.05 mL of MHB medium) in the following concentration range: 2–64 mg/mL. The extracts were dissolved in deionized water and the results were compared with those bacteria that had grown in MHB in the presence of water as a control. The MIC of each extract was determined as the lowest concentration that showed no turbidity after 24 h of incubation at 37 °C. All the 96-well plates and petri dish plates were sealed.

3.8.2. Growth Rate (GR) Studies

The GR parameter was determined using the broth microdilution method in a 96-well plate using TSB. All different strains of bacteria were exposed to extracts concentrations superior, equal, or inferior to the determined MIC, performing 1 h interval readings of the optical density (OD) at 625 nm, 37 °C, for 24 h [74]. Growth rate was determined by fitting a linear function to an exponential (log) phase curve.

3.8.3. Minimum Bactericidal Concentration (MBC) Determination

MBC assays were evaluated by the broth dilution method, where MBC was established as the lowest extract concentration that inhibited 99.9% of the bacterial inocula after 24 h of incubation at 37 °C. Briefly, 5 µL was taken from the well obtained from the MIC experiment from each extract and its dilutions and added to a petri dish plate with TSA medium. All the petri dish plates were sealed. The plate was divided into six equal areas, and for all extracts, the area of the lower dilution with no visible colonies was determined as MBC [75].

3.9. Cell Viability Assay

A cell viability assay was performed using two human oral cancer cell lines, namely TR146 and HSC3. Passages 28 and 6 were used for TR146 and HSC3, respectively. The cells were cultured in a CellCulture® CO_2 Incubator (ESCO GB Ltd., Barnsley, UK) for 24 h at 37 °C and exposed to various extract concentrations (125–2000 g/mL). This assay was performed according to Pinto et al. [19]. Extracts were dissolved in cell culture medium or absent and then incubated (2.5×10^4 cells/mL). MTT reagent was added, and the cells were incubated for 3 h at 37 °C, the number of viable cells was estimated, and the crystals were solubilized using DMSO. The positive and negative control were DMEM and 1% (*w/v*) Triton X-100. The background absorbance (630 nm) was subtracted at the 590 nm reading. Cell viability results were expressed as percentages.

3.10. Statistical Analysis

Results are expressed as mean ± standard deviation (n = 3). IBM SPSS Statistics 27.0 software (SPSS Inc., Chicago, IL, USA) was used to conduct statistical analysis. Tukey's HSD test was used for post hoc comparisons of the means, and one-way ANOVA was used to analyze sample differences. p values lower than 0.05 ($p < 0.05$) were recognized as significant.

4. Conclusions

The scientific data presented in this study support the valorization of *C. sativa* shells extracts generated by SWE, an innovative and environmentally safe extraction method that may be applied as a therapeutic agent for the treatment of OM. The extract that produced the best results in terms of antioxidant and antiradical activities as well as the ability to scavenge free radicals was that produced at 110 °C, which was the optimal temperature for extraction. This extract also had a higher variety of polyphenols (25 phenolic compounds), with a particular focus on gallic acid, protocatechuic acid, catechin, epicatechin, caffeine, and rutin. Furthermore, antimicrobial activity assays demonstrated that, in general, all extracts significantly inhibited bacterial growth against strains commonly found in OM patients. The chestnut shells extracts induced cytotoxic effects in HSC3 and TR146 cells, with a particular focus on TR146 cell line, and the extract obtained at 110 °C exhibited the lowest IC_{50} values for both cell lines. All the biological properties exhibited by the SWE *C. sativa* shells extract, particularly the extract obtained at 110 °C, may target the oxidative stress caused by anticancer treatments through the antioxidant activity and radical species scavenging capacity and, consequently, the inflammatory state. In addition, in advanced conditions (ulceration), it can manage bacterial proliferation and exert antiproliferative effects in head and neck squamous cell carcinoma. More research is required to reach a conclusion on the effects on the oral mucosa, particularly on the absorption of phenolic compounds via the buccal route. These may include in vitro oral epithelial permeation assays and oxidative stress experiments in cell lines. Furthermore, the sustainable use of chestnut shells will improve the applicability of food by-products, improve their economic value, and contribute to environmental sustainability.

Supplementary Materials: The following supporting information can be downloaded at: https://www.mdpi.com/article/10.3390/ijms232314956/s1.

Author Contributions: A.S.F. formal analysis, investigation, methodology, validation, writing—original draft; A.M.S. investigation, methodology; D.P. investigation, methodology; M.M.M. investigation, methodology; R.F. investigation, methodology, writing—review and editing; J.Š.-G. investigation, methodology, writing—review and editing; P.C.C. investigation, methodology, resources, supervision, writing—review and editing; C.D.-M. resources, supervision, writing—review and editing; F.R. conceptualization, funding acquisition, investigation, methodology, project administration, resources, supervision, writing—review and editing. All authors have read and agreed to the published version of the manuscript.

Funding: This work received financial support from project PTDC/ASP-AGR/29277/2017—*Castanea sativa* shells as a new source of active ingredients for Functional Food and Cosmetic applications: a sustainable approach, and project 5537 DRI, Sérvia 2020/21 from Portuguese-Serbia Bilateral Cooperation - Development of functional foods incorporating a chestnut shells extract obtained by subcritical water. This work was also financially supported by Portuguese national funds through projects UIDB/50006/2020, UIDP/50006/2020, and LA/P/0008/2020, from the Fundação para a Ciência e a Tecnologia (FCT)/Ministério da Ciência, Tecnologia e Ensino Superior (MCTES). This work was also financed by national funds from FCT—Fundação para a Ciência e a Tecnologia, I.P., in the scope of the project UIDP/04378/2020 and UIDB/04378/2020 of the Research Unit on Applied Molecular Biosciences—UCIBIO and the project LA/P/0140/2020 of the Associate Laboratory Institute for Health and Bioeconomy—i4HB.

Institutional Review Board Statement: Not applicable.

Informed Consent Statement: Not applicable.

Data Availability Statement: Not applicable.

Acknowledgments: Ana Sofia Ferreira (SFRH/BD/7519/2020), Diana Pinto (SFRH/BD/144534/2019) and Ana Margarida Silva (SFRH/BD/144994/2019) are thankful for their Ph.D. grants financed by POPH-QREN and subsidized by the European Science Foundation and Ministério da Ciência, Tecnologia e Ensino Superior. Francisca Rodrigues (CEECIND/01886/2020) and Manuela Moreira (CEECIND/02702/2017) are thankful for their contracts financed by FCT/MCTES—CEEC Individual Program Contract. Jaroslava Švarc-Gajić is grateful to the Science Fund of the Republic of Serbia (Grant No. 7747845, In situ pollutants removal from waters by sustainable green nano-technologies—CleanNanoCatalyze), and to the Ministry of education, science and technological development of the Republic of Serbia (Grant No. 451-03-68/2020–14/200134).

Conflicts of Interest: The authors declare no conflict of interest.

References

1. Ferreira, A.S.; Macedo, C.; Silva, A.M.; Delerue-Matos, C.; Costa, P.; Rodrigues, F. Natural Products for the Prevention and Treatment of Oral Mucositis-A Review. *Int. J. Mol. Sci.* **2022**, *23*, 4385. [CrossRef] [PubMed]
2. Nguyen, H.; Sangha, S.; Pan, M.; Shin, D.H.; Park, H.; Mohammed, A.I.; Cirillo, N. Oxidative Stress and Chemoradiation-Induced Oral Mucositis: A Scoping Review of In Vitro, In Vivo and Clinical Studies. *Int. J. Mol. Sci.* **2022**, *23*, 4863. [CrossRef] [PubMed]
3. Soutome, S.; Yanamoto, S.; Nishii, M.; Kojima, Y.; Hasegawa, T.; Funahara, M.; Akashi, M.; Saito, T.; Umeda, M. Risk factors for severe radiation-induced oral mucositis in patients with oral cancer. *J. Dent. Sci.* **2021**, *16*, 1241–1246. [CrossRef] [PubMed]
4. Liguori, I.; Russo, G.; Curcio, F.; Bulli, G.; Aran, L.; Della-Morte, D.; Gargiulo, G.; Testa, G.; Cacciatore, F.; Bonaduce, D.; et al. Oxidative stress, aging, and diseases. *Clin. Interv. Aging* **2018**, *13*, 757–772. [CrossRef]
5. Forman, H.J.; Zhang, H. Targeting oxidative stress in disease: Promise and limitations of antioxidant therapy. *Nat. Rev. Drug Discov.* **2021**, *20*, 689–709. [CrossRef]
6. Ramana, K.V.; Reddy, A.B.M.; Majeti, N.; Singhal, S.S. Therapeutic Potential of Natural Antioxidants. *Oxid. Med. Cell. Longev.* **2018**, *2018*, 9471051. [CrossRef] [PubMed]
7. Drasar, P.B.; Khripach, V.A. Growing Importance of Natural Products Research. *Molecules* **2019**, *25*, 6. [CrossRef] [PubMed]
8. Osorio, L.; Florez-Lopez, E.; Grande-Tovar, C.D. The Potential of Selected Agri-Food Loss and Waste to Contribute to a Circular Economy: Applications in the Food, Cosmetic and Pharmaceutical Industries. *Molecules* **2021**, *26*, 515. [CrossRef] [PubMed]
9. Freitas, L.C.; Barbosa, J.R.; da Costa, A.L.C.; Bezerra, F.W.F.; Pinto, R.H.H.; de Carvalho Junior, R.N. From waste to sustainable industry: How can agro-industrial wastes help in the development of new products? *Resour. Conserv. Recycl.* **2021**, *169*, 105466. [CrossRef]
10. Ravindran, R.; Hassan, S.S.; Williams, G.A.; Jaiswal, A.K. A Review on Bioconversion of Agro-Industrial Wastes to Industrially Important Enzymes. *Bioengineering* **2018**, *5*, 93. [CrossRef] [PubMed]
11. Rodrigues, A.; Gonçalves, A.B.; Costa, R.L.; Gomes, A.A. GIS-Based Assessment of the Chestnut Expansion Potential: A Case-Study on the Marvão Productive Area, Portugal. *Agriculture* **2021**, *11*, 1260. [CrossRef]
12. Pinto, D.; Cadiz-Gurrea, M.L.; Vallverdu-Queralt, A.; Delerue-Matos, C.; Rodrigues, F. Castanea sativa shells: A review on phytochemical composition, bioactivity and waste management approaches for industrial valorization. *Food Res. Int.* **2021**, *144*, 110364. [CrossRef] [PubMed]
13. Vella, F.M.; Laratta, B.; La Cara, F.; Morana, A. Recovery of bioactive molecules from chestnut (*Castanea sativa* Mill.) by-products through extraction by different solvents. *Nat. Prod. Res.* **2018**, *32*, 1022–1032. [CrossRef] [PubMed]
14. Braga, N.; Rodrigues, F.; Oliveira, M.B. *Castanea sativa* by-products: A review on added value and sustainable application. *Nat. Prod. Res.* **2015**, *29*, 1–18. [CrossRef] [PubMed]
15. Lefebvre, T.; Destandau, E.; Lesellier, E. Selective extraction of bioactive compounds from plants using recent extraction techniques: A review. *J. Chromatogr. A* **2021**, *1635*, 461770. [CrossRef] [PubMed]
16. Chuo, S.C.; Nasir, H.M.; Mohd-Setapar, S.H.; Mohamed, S.F.; Ahmad, A.; Wani, W.A.; Muddassir, M.; Alarifi, A. A Glimpse into the Extraction Methods of Active Compounds from Plants. *Crit. Rev. Anal. Chem.* **2022**, *52*, 667–696. [CrossRef]
17. Alara, O.R.; Abdurahman, N.H.; Ukaegbu, C.I. Extraction of phenolic compounds: A review. *Curr. Res. Food Sci.* **2021**, *4*, 200–214. [CrossRef]
18. El Maaiden, E.; Bouzroud, S.; Nasser, B.; Moustaid, K.; El Mouttaqi, A.; Ibourki, M.; Boukcim, H.; Hirich, A.; Kouisni, L.; El Kharrassi, Y. A Comparative Study between Conventional and Advanced Extraction Techniques: Pharmaceutical and Cosmetic Properties of Plant Extracts. *Molecules* **2022**, *27*, 2074. [CrossRef] [PubMed]
19. Pinto, D.; Vieira, E.F.; Peixoto, A.F.; Freire, C.; Freitas, V.; Costa, P.; Delerue-Matos, C.; Rodrigues, F. Optimizing the extraction of phenolic antioxidants from chestnut shells by subcritical water extraction using response surface methodology. *Food Chem.* **2021**, *334*, 127521. [CrossRef] [PubMed]
20. Mikucka, W.; Zielinska, M.; Bulkowska, K.; Witonska, I. Subcritical water extraction of bioactive phenolic compounds from distillery stillage. *J. Environ. Manag.* **2022**, *318*, 115548. [CrossRef]

21. Dorosh, O.; Moreira, M.M.; Pinto, D.; Peixoto, A.F.; Freire, C.; Costa, P.; Rodrigues, F.; Delerue-Matos, C. Evaluation of the Extraction Temperature Influence on Polyphenolic Profiles of Vine-Canes (Vitis vinifera) Subcritical Water Extracts. *Foods* **2020**, *9*, 872. [CrossRef] [PubMed]
22. Machmudah, S.; Wahyu Fitriana, M.; Fatbamayani, N.; Wahyudiono; Kanda, H.; Winardi, S.; Goto, M. Phytochemical compounds extraction from medicinal plants by subcritical water and its encapsulation via electrospraying. *Alex. Eng. J.* **2022**, *61*, 2116–2128. [CrossRef]
23. Antony, A.; Farid, M. Effect of Temperatures on Polyphenols during Extraction. *Appl. Sci.* **2022**, *12*, 2107. [CrossRef]
24. Muhamad, I.I.; Hassan, N.D.; Mamat, S.N.H.; Nawi, N.M.; Rashid, W.A.; Tan, N.A. Extraction Technologies and Solvents of Phytocompounds From Plant Materials: Physicochemical Characterization and Identification of Ingredients and Bioactive Compounds From Plant Extract Using Various Instrumentations. In *Ingredients Extraction by Physicochemical Methods in Food*; Academic Press: Cambridge, MA, USA, 2017; pp. 523–560. [CrossRef]
25. Azmir, J.; Zaidul, I.S.M.; Rahman, M.M.; Sharif, K.M.; Mohamed, A.; Sahena, F.; Jahurul, M.H.A.; Ghafoor, K.; Norulaini, N.A.N.; Omar, A.K.M. Techniques for extraction of bioactive compounds from plant materials: A review. *J. Food Eng.* **2013**, *117*, 426–436. [CrossRef]
26. Chaves, J.O.; de Souza, M.C.; da Silva, L.C.; Lachos-Perez, D.; Torres-Mayanga, P.C.; Machado, A.; Forster-Carneiro, T.; Vazquez-Espinosa, M.; Gonzalez-de-Peredo, A.V.; Barbero, G.F.; et al. Extraction of Flavonoids From Natural Sources Using Modern Techniques. *Front. Chem.* **2020**, *8*, 507887. [CrossRef] [PubMed]
27. Yahya, N.A.; Attan, N.; Wahab, R.A. An overview of cosmeceutically relevant plant extracts and strategies for extraction of plant-based bioactive compounds. *Food Bioprod. Process.* **2018**, *112*, 69–85. [CrossRef]
28. Picot-Allain, C.; Mahomoodally, M.F.; Ak, G.; Zengin, G. Conventional versus green extraction techniques—A comparative perspective. *Curr. Opin. Food Sci.* **2021**, *40*, 144–156. [CrossRef]
29. Pagano, I.; Campone, L.; Celano, R.; Piccinelli, A.L.; Rastrelli, L. Green non-conventional techniques for the extraction of polyphenols from agricultural food by-products: A review. *J. Chromatogr. A* **2021**, *1651*, 462295. [CrossRef] [PubMed]
30. Barba, F.J.; Zhu, Z.; Koubaa, M.; Sant'Ana, A.S.; Orlien, V. Green alternative methods for the extraction of antioxidant bioactive compounds from winery wastes and by-products: A review. *Trends Food Sci. Technol.* **2016**, *49*, 96–109. [CrossRef]
31. Cacciola, N.A.; Squillaci, G.; D'Apolito, M.; Petillo, O.; Veraldi, F.; La Cara, F.; Peluso, G.; Margarucci, S.; Morana, A. *Castanea sativa* Mill. Shells Aqueous Extract Exhibits Anticancer Properties Inducing Cytotoxic and Pro-Apoptotic Effects. *Molecules* **2019**, *24*, 3401. [CrossRef]
32. Fernández-Agulló, A.; Freire, M.S.; Antorrena, G.; Pereira, J.A.; González-Álvarez, J. Effect of the Extraction Technique and Operational Conditions on the Recovery of Bioactive Compounds from Chestnut (*Castanea sativa*) Bur and Shell. *Sep. Sci. Technol.* **2014**, *49*, 267–277. [CrossRef]
33. Pinto, D.; Silva, A.M.; Freitas, V.; Vallverdú-Queralt, A.; Delerue-Matos, C.; Rodrigues, F. Microwave-Assisted Extraction as a Green Technology Approach to Recover Polyphenols from *Castanea sativa* Shells. *ACS Food Sci. Technol.* **2021**, *1*, 229–241. [CrossRef]
34. Cravotto, C.; Grillo, G.; Binello, A.; Gallina, L.; Olivares-Vicente, M.; Herranz-Lopez, M.; Micol, V.; Barrajon-Catalan, E.; Cravotto, G. Bioactive Antioxidant Compounds from Chestnut Peels through Semi-Industrial Subcritical Water Extraction. *Antioxidants* **2022**, *11*, 988. [CrossRef] [PubMed]
35. Wang, X.; Chen, Q.; Lü, X. Pectin extracted from apple pomace and citrus peel by subcritical water. *Food Hydrocoll.* **2014**, *38*, 129–137. [CrossRef]
36. Silva, A.M.; Luís, A.S.; Moreira, M.M.; Ferraz, R.; Brezo-Borjan, T.; Švarc-Gajić, J.; Costa, P.C.; Delerue-Matos, C.; Rodrigues, F. Influence of temperature on the subcritical water extraction of Actinidia arguta leaves: A screening of pro-healthy compounds. *Sustain. Chem. Pharm.* **2022**, *25*, 100593. [CrossRef]
37. Gonçalves Rodrigues, L.G.; Mazzutti, S.; Vitali, L.; Micke, G.A.; Ferreira, S.R.S. Recovery of bioactive phenolic compounds from papaya seeds agroindustrial residue using subcritical water extraction. *Biocatal. Agric. Biotechnol.* **2019**, *22*, 101367. [CrossRef]
38. Lameirao, F.; Pinto, D.; E, F.V.; Peixoto, A.F.; Freire, C.; Sut, S.; Dall'Acqua, S.; Costa, P.; Delerue-Matos, C.; Rodrigues, F. Green-Sustainable Recovery of Phenolic and Antioxidant Compounds from Industrial Chestnut Shells Using Ultrasound-Assisted Extraction: Optimization and Evaluation of Biological Activities In Vitro. *Antioxidants* **2020**, *9*, 267. [CrossRef]
39. Munteanu, I.G.; Apetrei, C. Analytical Methods Used in Determining Antioxidant Activity: A Review. *Int. J. Mol. Sci.* **2021**, *22*, 3380. [CrossRef]
40. Lachos-Perez, D.; Baseggio, A.M.; Mayanga-Torres, P.C.; Maróstica, M.R.; Rostagno, M.A.; Martínez, J.; Forster-Carneiro, T. Subcritical water extraction of flavanones from defatted orange peel. *J. Supercrit. Fluids* **2018**, *138*, 7–16. [CrossRef]
41. Luo, X.; Cui, J.; Zhang, H.; Duan, Y. Subcritical water extraction of polyphenolic compounds from sorghum (*Sorghum bicolor* L.) bran and their biological activities. *Food Chem.* **2018**, *262*, 14–20. [CrossRef]
42. Kahkeshani, N.; Farzaei, F.; Fotouhi, M.; Alavi, S.S.; Bahramsoltani, R.; Naseri, R.; Momtaz, S.; Abbasabadi, Z.; Rahimi, R.; Farzaei, M.H.; et al. Pharmacological effects of gallic acid in health and diseases: A mechanistic review. *Iran. J. Basic Med. Sci.* **2019**, *22*, 225–237. [CrossRef] [PubMed]
43. Zhang, H.; Tsao, R. Dietary polyphenols, oxidative stress and antioxidant and anti-inflammatory effects. *Curr. Opin. Food Sci.* **2016**, *8*, 33–42. [CrossRef]
44. Panche, A.N.; Diwan, A.D.; Chandra, S.R. Flavonoids: An overview. *J. Nutr. Sci.* **2016**, *5*, e47. [CrossRef] [PubMed]

45. Vázquez, G.; Pizzi, A.; Freire, M.S.; Santos, J.; Antorrena, G.; González-Álvarez, J. MALDI-TOF, HPLC-ESI-TOF and 13C-NMR characterization of chestnut (*Castanea sativa*) shell tannins for wood adhesives. *Wood Sci. Technol.* **2012**, *47*, 523–535. [CrossRef]
46. More, G.K.; Makola, R.T. In-Vitro analysis of free radical scavenging activities and suppression of LPS-induced ROS production in macrophage cells by *Solanum sisymbriifolium* extracts. *Sci. Rep.* **2020**, *10*, 6493. [CrossRef] [PubMed]
47. Di Meo, S.; Reed, T.T.; Venditti, P.; Victor, V.M. Role of ROS and RNS Sources in Physiological and Pathological Conditions. *Oxid. Med. Cell. Longev.* **2016**, *2016*, 1245049. [CrossRef]
48. Orlandi, E.; Iacovelli, N.A.; Tombolini, V.; Rancati, T.; Polimeni, A.; De Cecco, L.; Valdagni, R.; De Felice, F. Potential role of microbiome in oncogenesis, outcome prediction and therapeutic targeting for head and neck cancer. *Oral Oncol.* **2019**, *99*, 104453. [CrossRef]
49. Vasconcelos, R.M.; Sanfilippo, N.; Paster, B.J.; Kerr, A.R.; Li, Y.; Ramalho, L.; Queiroz, E.L.; Smith, B.; Sonis, S.T.; Corby, P.M. Host-Microbiome Cross-talk in Oral Mucositis. *J. Dent. Res.* **2016**, *95*, 725–733. [CrossRef]
50. Hong, B.Y.; Sobue, T.; Choquette, L.; Dupuy, A.K.; Thompson, A.; Burleson, J.A.; Salner, A.L.; Schauer, P.K.; Joshi, P.; Fox, E.; et al. Chemotherapy-induced oral mucositis is associated with detrimental bacterial dysbiosis. *Microbiome* **2019**, *7*, 66. [CrossRef]
51. Gendron, R.; Grenier, D.a.; Maheu-Robert, L.-F. The oral cavity as a reservoir of bacterial pathogens for focal infections. *Microbes Infect.* **2000**, *2*, 897–906. [CrossRef]
52. Vanhoecke, B.; De Ryck, T.; Stringer, A.; Van de Wiele, T.; Keefe, D. Microbiota and their role in the pathogenesis of oral mucositis. *Oral Dis.* **2015**, *21*, 17–30. [CrossRef] [PubMed]
53. Gupta, N.; Quah, S.Y.; Yeo, J.F.; Ferreira, J.; Tan, K.S.; Hong, C.H.L. Role of oral flora in chemotherapy-induced oral mucositis in vivo. *Arch. Oral Biol.* **2019**, *101*, 51–56. [CrossRef] [PubMed]
54. Duda-Chodak, A.; Tarko, T.; Satora, P.; Sroka, P. Interaction of dietary compounds, especially polyphenols, with the intestinal microbiota: A review. *Eur. J. Nutr.* **2015**, *54*, 325–341. [CrossRef] [PubMed]
55. Daglia, M. Polyphenols as antimicrobial agents. *Curr. Opin. Biotechnol.* **2012**, *23*, 174–181. [CrossRef] [PubMed]
56. Sarjit, A.; Wang, Y.; Dykes, G.A. Antimicrobial activity of gallic acid against thermophilic Campylobacter is strain specific and associated with a loss of calcium ions. *Food Microbiol.* **2015**, *46*, 227–233. [CrossRef]
57. Kang, M.S.; Oh, J.S.; Kang, I.C.; Hong, S.J.; Choi, C.H. Inhibitory effect of methyl gallate and gallic acid on oral bacteria. *J. Microbiol.* **2008**, *46*, 744–750. [CrossRef]
58. Passos, M.R.; Almeida, R.S.; Lima, B.O.; Rodrigues, J.Z.S.; Macedo Neres, N.S.; Pita, L.S.; Marinho, P.D.F.; Santos, I.A.; da Silva, J.P.; Oliveira, M.C.; et al. Anticariogenic activities of *Libidibia ferrea*, gallic acid and ethyl gallate against *Streptococcus mutans* in biofilm model. *J. Ethnopharmacol.* **2021**, *274*, 114059. [CrossRef]
59. Smullen, J.; Koutsou, G.A.; Foster, H.A.; Zumbe, A.; Storey, D.M. The antibacterial activity of plant extracts containing polyphenols against *Streptococcus mutans*. *Caries Res.* **2007**, *41*, 342–349. [CrossRef]
60. Michi, Y.; Morita, I.; Amagasa, T.; Murota, S. Human oral squamous cell carcinoma cell lines promote angiogenesis via expression of vascular endothelial growth factor and upregulation of KDR/¯k-1 expression in endothelial cells. *Oral Oncol.* **2000**, *36*, 81–88. [CrossRef]
61. Lin, G.C.; Leitgeb, T.; Vladetic, A.; Friedl, H.P.; Rhodes, N.; Rossi, A.; Roblegg, E.; Neuhaus, W. Optimization of an oral mucosa in vitro model based on cell line TR146. *Tissue Barriers* **2020**, *8*, 1748459. [CrossRef]
62. Soares, S.; Soares, S.; Brandao, E.; Guerreiro, C.; Mateus, N.; de Freitas, V. Oral interactions between a green tea flavanol extract and red wine anthocyanin extract using a new cell-based model: Insights on the effect of different oral epithelia. *Sci. Rep.* **2020**, *10*, 12638. [CrossRef] [PubMed]
63. Lambert, J.D.; Hong, J.; Yang, G.-y.; Liao, J.; Yang, C.S. Inhibition of carcinogenesis by polyphenols: Evidence from laboratory investigations. *Am. J. Clin. Nutr.* **2005**, *81*, 284S–291S. [CrossRef] [PubMed]
64. Walle, T.; Browning, A.M.; Steed, L.L.; Reed, S.G.; Walle, U.K. Flavonoid Glucosides Are Hydrolyzed and Thus Activated in he Oral Cavity in Humans. *J. Nutr.* **2005**, *135*, 48–52. [CrossRef]
65. Ho, Y.C.; Yang, S.F.; Peng, C.Y.; Chou, M.Y.; Chang, Y.C. Epigallocatechin-3-gallate inhibits the invasion of human oral cancer cells and decreases the productions of matrix metalloproteinases and urokinase-plasminogen activator. *J. Oral Pathol. Med.* **2007**, *36*, 588–593. [CrossRef] [PubMed]
66. Balappanavar, A.Y. Tea and Oral Health. In *Tea–Chemistry and Pharmacology*; Justino, G., Ed.; IntechOpen: London, UK, 2020. [CrossRef]
67. Švarc-Gajić, J.; Cvetanović, A.; Segura-Carretero, A.; Linares, I.B.; Mašković, P. Characterisation of ginger extracts obtained by subcritical water. *J. Supercrit. Fluids* **2017**, *123*, 92–100. [CrossRef]
68. Pinto, D.; Braga, N.; Rodrigues, F.; Oliveira, M. *Castanea sativa* Bur: An Undervalued By-Product but a Promising Cosmetic Ingredient. *Cosmetics* **2017**, *4*, 50. [CrossRef]
69. Re, R.; Pellegrini, N.; Proteggente, A.; Pannala, A.; Yang, M.; Rice-Evans, C. Antioxidant activity applying an improved ABTS radical cation decolorization assay. *Free. Radic. Biol. Med.* **1999**, *26*, 1231–1237. [CrossRef]
70. Moreira, M.M.; Barroso, M.F.; Boeykens, A.; Withouck, H.; Morais, S.; Delerue-Matos, C. Valorization of apple tree wood residues by polyphenols extraction: Comparison between conventional and microwave-assisted extraction. *Ind. Crops Prod.* **2017**, *104*, 210–220. [CrossRef]
71. Gomes, A.; Fernandes, E.; Silva, A.M.; Santos, C.M.; Pinto, D.C.; Cavaleiro, J.A.; Lima, J.L. 2-Styrylchromones: Novel strong scavengers of reactive oxygen and nitrogen species. *Bioorg. Med. Chem.* **2007**, *15*, 6027–6036. [CrossRef]

72. Ou, B.; Hampsch-Woodill, M.; Prior, R.L. Development and Validation of an Improved Oxygen Radical Absorbance Capacity Assay Using Fluorescein as the Fluorescent Probe. *J. Agric. Food Chem.* **2001**, *49*, 4619–4626. [CrossRef]
73. Whiteman, M.; Ketsawatsakul, U.; Halliwell, B. A Reassessment of the Peroxynitrite Scavenging Activity of Uric Acid. *Ann. N. Y. Acad. Sci.* **2002**, *962*, 242–259. [CrossRef] [PubMed]
74. CLSI. *Methods for Dilution Antimicrobial Susceptibility Tests for Bacteria That Grow Aerobically; Approved Standard*, 9th ed.; CLSI document. M07-A9; Clinical and Laboratory Standards Institute: Wayne, PA, USA, 2012.
75. Ozturk, S.; Ercisli, S. Chemical composition and in vitro antibacterial activity of Seseli libanotis. *World J. Microbiol. Biotechnol.* **2006**, *22*, 261–265. [CrossRef]

International Journal of
Molecular Sciences

MDPI

Article

Bioactivities of Pseudocereal Fractionated Seed Proteins and Derived Peptides Relevant for Maintaining Human Well-Being

Jessica Capraro , Stefano De Benedetti , Giuditta Carlotta Heinzl , Alessio Scarafoni * and Chiara Magni

Department of Food, Environmental and Nutritional Sciences, Università degli Studi di Milano, 20133 Milano, Italy; jessica.capraro@unimi.it (J.C.); stefano.debenedetti@unimi.it (S.D.B.); giuditta.heinzl@unimi.it (G.C.H.); chiara.magni@unimi.it (C.M.)
* Correspondence: alessio.scarafoni@unimi.it; Tel.: +39-0250326820

Abstract: Food proteins and peptides are able to exert a variety of well-known bioactivities, some of which are related to well-being and disease prevention in humans and animals. Currently, an active trend in research focuses on chronic inflammation and oxidative stress, delineating their major pathogenetic role in age-related diseases and in some forms of cancer. The present study aims to investigate the potential effects of pseudocereal proteins and their derived peptides on chronic inflammation and oxidative stress. After purification and attribution to protein classes according to classic Osborne's classification, the immune-modulating, antioxidant, and trypsin inhibitor activities of proteins from quinoa (*Chenopodium quinoa* Willd.), amaranth (*Amaranthus retroflexus* L.), and buckwheat (*Fagopyrum esculentum* Moench) seeds have been assessed in vitro. The peptides generated by simulated gastro-intestinal digestion of each fraction have been also investigated for the selected bioactivities. None of the proteins or peptides elicited inflammation in Caco-2 cells; furthermore, *all protein fractions* showed different degrees of protection of cells from IL-1β-induced inflammation. Immune-modulating and antioxidant activities were, in general, higher for the albumin fraction. Overall, seed proteins can express these bioactivities mainly after hydrolysis. On the contrary, higher trypsin inhibitor activity was expressed by globulins in their intact form. These findings lay the foundations for the exploitation of these pseudocereal seeds as source of anti-inflammatory molecules.

Keywords: seed storage proteins; peptides; anti-inflammatory; antioxidant; trypsin inhibitors; quinoa; amaranth; buckwheat; in vitro gastro-intestinal digestion

Citation: Capraro, J.; Benedetti, S.D.; Heinzl, G.C.; Scarafoni, A.; Magni, C. Bioactivities of Pseudocereal Fractionated Seed Proteins and Derived Peptides Relevant for Maintaining Human Well-Being. *Int. J. Mol. Sci.* **2021**, *22*, 3543. https://doi.org/10.3390/ijms22073543

Academic Editors: Antonio Carrillo Vico and Ivan Cruz-Chamorro

Received: 9 February 2021
Accepted: 24 March 2021
Published: 29 March 2021

1. Introduction

Modern trends in dietary approaches and ethical and environmental sustainability concerns catalyze researchers' interest in plant proteins as an alternative to animal proteins [1]. In this frame, the consumption of pseudocereals has increased significantly over the past few years, which is also due to their integration into vegan/vegetarian and gluten-free diets [2]. The protein content of their seeds varies from 9.1 to 16.7% in quinoa, 13.1 to 21.5% in amaranth, and 5.7 to 14.2% in buckwheat [3]. Their amino acid composition is generally well balanced, with higher contents of lysine, methionine, and cysteine than common cereals and legumes [2]. The distribution of the three aforementioned pseudocereal seed proteins according to their solubility reveals a closer relationship to legumes than to cereal proteins, since albumins and globulins are the most abundant protein species [4].

It has been well established that foods may contain molecules able to exert a variety of bioactivities, some of which are related to well-being and disease prevention in humans and animals [5,6]. Beyond their nutritional value, recent research identified the potential health benefits of food proteins and peptides. They can exert antioxidant, antihypertensive, antilipidemic, hypocholesterolemic, antimicrobial, and immune-modulating effects [7,8]. Among these peptides, many have been identified in legume and cereal seeds, including soybean (*Glycine max*), lupin (*Lupinus* ssp.), corn (*Zea mays*), horse gram (*Macrotyloma uniflorum*), cowpea (*Vigna unguiculata*), common bean (*Phaseolus vulgaris*), and lentil (*Lens culinaris*) [9–14]. However, there is still a general lack of information about the potential effects of pseudocereal proteins and their derived peptides on chronic inflammation and the various forms of oxidative stress, these being considered as major causes of age-related diseases and of some forms of cancer [15].

The present study aims to advance the knowledge of selected activities potentially exerted by pseudocereal proteins. The immune-modulating, antioxidant, and trypsin inhibitor activities of proteins from quinoa (*Chenopodium quinoa* Willd.), amaranth (*Amaranthus retroflexus* L.), and buckwheat (*Fagopyrum esculentum* Moench) seeds have been assessed in vitro, after purification and separation in different fractions. The three biological activities considered in this work are intimately linked to each other with regards to their implications on human health.

Indeed, inflammation plays an important role in the ability of the immune system to fend off pathogens and harmful agents. However, an unregulated inflammatory response can lead to tissue damage and the development of chronic inflammatory diseases [16]. Several food-derived compounds are able to modulate the immune response in humans [17]. For example, many compounds may mediate inflammation by altering the DNA-binding capacities of NF-κB, the major effector of immune response pathways, and other transcription factors [18]. NF-κB acts as a central inflammatory mediator by regulating a vast array of genes involved in the immune and inflammatory responses. It responds to a large variety of molecules, including cytokine IL-1β, and its activation induces the expression of inflammatory cytokines, chemokines, and adhesion molecules [16]. Hence, the control of the NF-κB pathway represents a potential strategy for preventing inflammation-associated diseases [17]. In the present work, the effects on cell inflammation of proteins and their peptides obtained by simulated gastro-intestinal digestion have been studied using cultivated intestinal Caco-2 cells, whose immune response was triggered by IL-1β.

The dampening of oxidative processes is of great importance to human well-being [19]. When free radicals are overproduced or the cellular defenses are impaired, biomolecules such as lipids, proteins, and DNA may be damaged by oxidative stress [20], ultimately leading to pathological conditions. Plant foods are rich in antioxidant molecules, especially phenolic compounds [6]. However, an increasing body of evidence suggests that proteins and peptides can also exert this protective effect [11,21–29]. The capacity of inhibition on the oxidation of cellular components can be exerted through multiple mechanisms of action, including free radical scavenging, metal ion chelation, and hydroperoxides and reactive oxygen species reduction [30]. In addition, the typical amphipathicity of most peptides allows them to act both in aqueous and lipidic systems [31].

Although protease inhibitors (PIs) have long been considered anti-nutritional compounds because of their negative effects on protein digestibility, several recent studies have shown that they may play important roles in the treatment or prevention of inflammation-associated diseases, such as some types of cancers [32,33], autoimmune diseases [34], coagulation diseases [35], metabolic syndrome, and obesity [36]. These studies focused mainly on PIs from leguminous plants, and information about PIs from pseudocereal seeds continues to remain limited [37–41]. It is known that serine proteases act as modulators of the immune system and inflammatory response by regulating cytokine and chemokine production. Aberrant functioning of serine proteases may contribute to the development of disorders derived from inflammatory cell activation that lead to immunological problems and excessive activation of inflammation [34]. Thus, the inhibition of serine proteases by PIs may play a role in the prevention of these diseases [42].

2. Results and Discussion

2.1. Purification of Pseudocereal Protein Fractions and Their In Vitro Digestion

The isolation procedure we adopted allowed us, as a first step, to obtain a water-soluble fraction, namely albumin, and a salt-soluble fraction, corresponding to globulins. Albumins and globulins are the most abundant seed proteins of pseudocereals. Amaranth, buckwheat, and quinoa contain different proportions of each [4]. Albumins include many enzymes involved in cotyledon cell metabolism and plant defense, whereas globulin proteins essentially play a storage role. Due to the low selective pressure, seed storage proteins (SSPs) show common characteristics among species. Globulins may be classified according to the sedimentation coefficient as 2S, 7–8S, and 11–13S, also known as vicilin-like and legumin-like globulin, respectively [43]. In order to visualize the distribution of the proteins in the obtained fractions, these latter were analyzed by SDS-PAGE (Figure 1). The albumin electrophoretic patterns are clearly different from those of the respective globulin fractions, confirming that the procedure allowed the differential solubilization of the two protein classes (Figure 1).

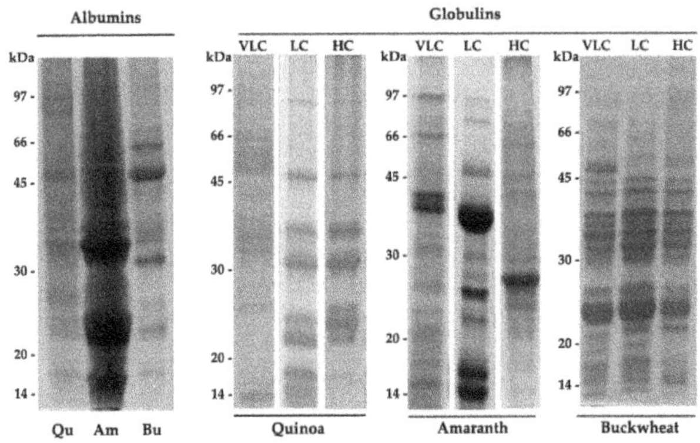

Figure 1. SDS-PAGE analysis of albumins and globulin fractions from quinoa (Qu), amaranth (Am), and buckwheat (Bu) seed proteins. VLC: very low charge; LC: low charge; HC: high charge.

Due to the combination of different factors, including the and the post-translational modifications, globulin polypeptide chains show size and charge heterogeneity [7]. For this reason, we further fractionated globulin by ion-exchange chromatography, according to their net surface charge at pH 8.0, in three fractions named VLC (very low charge), LC (low charge) and HC (high charge). The quinoa VLC fraction consists of a large range of different polypeptides. Notably, the SDS-PAGE pattern was very similar to that of the salt-soluble fraction obtained following isoelectric precipitation of chenopodins, as shown by Brinegar et al. [44], and very different from the two retained LC and HC fractions. These two fractions include isoforms of the 11–13S globulin chenopodin, the main quinoa SSPs [45]. The bands with molecular weights (MWs) of 35–37 and 22–25 kDa correspond to the α and β chains of chenopodin, respectively. Thus, the purification procedure we set up allowed the separation of the quinoa proteins using only a single chromatographic step.

Buckwheat VLC, LC, and HC globulin separation produced very similar patterns, with the only difference being one extra polypeptide of about 50 kDa in the VLC fraction. Again, this result is due to the charge heterogeneity of constituent polypeptides of buckwheat globulins [46,47]. The two visible main groups of polypeptides with MWs of 30–47 and 23–25 kDa likely correspond to the α and β chains of 13S globulin, the most abundant buckwheat globulin [4].

Amaranth separated globulins showed more complex polypeptide profiles than either the quinoa or the buckwheat samples. The LC fraction included bands likely attributable to 11S amaranthin. Amaranthin subunits consists of acidic (34–36 kDa) and basic (22–24 kDa) polypeptides. A polypeptide of about 55 kDa is also present in this fraction. In all probability, it is the uncleaved precursor (52–59 kDa) [48]. Vicilin-like 7S globulins are represented by several polypeptides with MWs between 15 and 90 kDa, also present in amaranth seeds [4]. Due to their more acidic pI, in our case, they are mainly distributed in the VLC fraction. A typical band of about 38 kDa is visible [49].

Bioactive cryptic peptides are specific protein fragments inactive in the whole intact protein but active when released by proteolysis. Dietary proteins may supply active peptides able to promote human health and well-being [7]. The most studied cryptic peptides are those released from milk, chicken eggs, and meat protein digestion, whereas plant proteins are much less studied as source of food-derived active peptides. To obtain peptides as similar as possible to those that would be obtained following ingestion, a simulated gastro-intestinal digestion was performed using well-established procedures [50]. The proteolytic treatment caused the breakdown of proteins into fragments with molecular weights equal to or smaller than 14 kDa, as revealed by SDS-PAGE (Figure 2). In all samples, only a small number of bands with MWs higher than 30 kDa remained faintly visible.

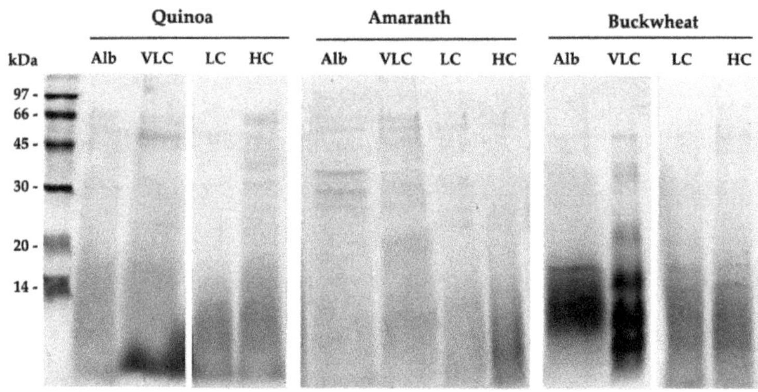

Figure 2. SDS-PAGE analysis of albumins and globulin fractions after simulated gastro-intestinal digestion. Alb: albumin fraction; VLC: very low-charge globulins; LC: low-charge globulins; HC: high-charge globulins.

2.2. The Immune-Modulating Capacity of Protein Fractions and Peptides in Caco-2 Cells

To study the immune-modulating activity of intact protein fractions, we evaluated NF-κB activation by transiently transfecting epithelial intestinal Caco-2 cells with an NF-κB luciferase reporter construct that includes five NF-κB binding sites. The presence of NF-κB-activating molecules stimulated the expression of the luciferase gene. The results are shown in Figure 3. After incubation with the protein fractions, NF-κB activation was comparable with cells incubated with the culturing media to suggest that amaranth, quinoa, and buckwheat proteins did not exert pro-inflammatory effects by themselves. Under stimulation with the pro-inflammatory cytokine IL-1β, the presence of pseudocereal protein fractions induced the decrease of inflammatory response in all cases but to different extents.

Overall, albumin and VLC globulin showed higher anti-inflammatory activity (a drop of inflammation from 31 to 74%) than LC and HC globulin fractions (from 20 to 39%). Quinoa albumin was the most active fraction, showing a 74% inflammation reduction. The immune-modulating effects of quinoa LC and HC globulin fractions are similar to those previously described [45] and are slightly higher than those of amaranth and buckwheat globulins. In the present study, we demonstrated that amaranth, buckwheat, and quinoa seed proteins potentially modulate an inflammation response by decreasing NF-κB activation. To date, very few plant proteins in the intact form are known to possess anti-inflammatory properties [45,51–53]. The most studied protein presently is lunasin, a 43-amino-acid peptide isolated from soybean which acts on macrophage cells by inhibiting the NF-κB pathway [54–56]. Recently, lunasin was detected in quinoa seeds as well and its bioactive effects were confirmed [55]. Since lunasin extraction efficiency in water is high [57], it is likely that, in our case, it is mainly present in the albumin fraction. This supports the observed high protective effect of this fraction (Figure 3). The binding of lunasin with αVβ3 integrin through an Arg-Gly-Asp (RGD) motif has been associated with inhibition of the NF-κB pathways [54]. Interestingly, the RGD/RGE motif was found in three sequences of chenopodin [45]. Searching protein sequence databases, an RGE motif was found only in sequence ABO93593.1 of buckwheat BW10KD allergen protein, corresponding to 2S albumin [58].

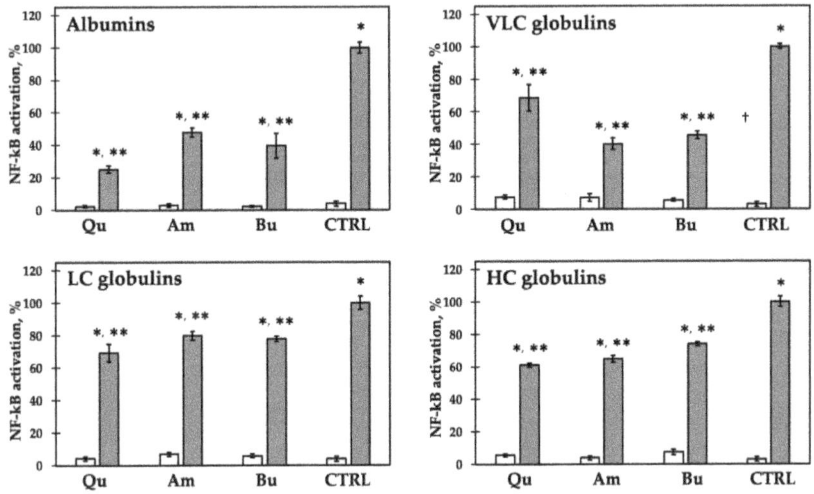

Figure 3. Inflammatory response of Caco-2 cells incubated with 1.0 mg/mL of each protein class without (white bars) or with (gray bars) IL-1β. Response to IL-1β alone was set as 100% (CTRL). Qu: quinoa; Am: amaranth; Bu: buckwheat. * $p < 0.05$ vs. CTRL without IL-1β; ** $p < 0.05$ vs. CTRL with IL-1β. Each experiment was performed in triplicate.

Finally, we assessed the potential anti-inflammatory capacity of the pool of peptides originated from the single fractionated proteins by simulated gastro-intestinal digestion. In a previous work [45], we showed that the expression of chemokine IL-8 mirrors the expression of NF-κB in undifferentiated Caco-2 cells stimulated with IL-1β and treated with quinoa chenopodin isoforms. Since the levels of NF-κB and IL-8 can be considered to be modulated in the same way, we decided to directly quantify the expression of IL-8 by qRT-PCR to evaluate the anti-inflammatory effect of the peptides (Figure 4).

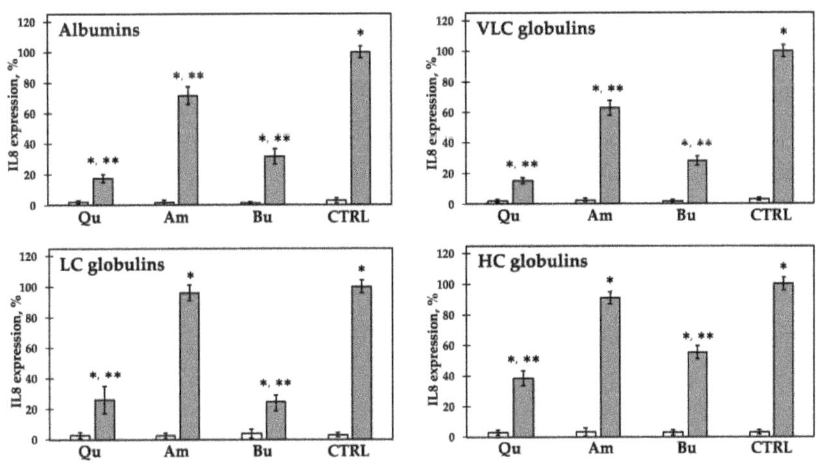

Figure 4. Inflammatory response of Caco-2 cells incubated with 1.0 mg/mL of peptides originated by in vitro simulated gastro-intestinal digestion of each protein class, without (white bars) or with (gray bars) IL-1β. Response to IL-1β alone was set as 100% (CTRL). Qu: quinoa; Am: amaranth; Bu: buckwheat. * $p < 0.05$ vs. CTRL without IL-1β; ** $p < 0.05$ vs. CTRL with IL-1β. Each experiment was performed in triplicate.

Again, the peptides alone, originating from the different samples, exerted no inflammatory activity in Caco-2 cells. Peptides from quinoa and buckwheat protein fractions were effective in lowering the inflammatory stimulus due to the presence of IL-1β with a higher extent if compared to the effect exerted by the undigested proteins. On the contrary, amaranth peptides showed a minor protective shielding. LC globulin peptides from amaranth appeared absolutely ineffective.

2.3. The In Vitro Antioxidant Activity of Protein Fractions and Peptides

The antioxidant activity of peptides obtained through generic enzymatic treatments of total protein extracts from amaranth, buckwheat, and quinoa has been reported [29,59–64]. This finding, although interesting, does not provide information on the most active protein classes, on their functional implications, or on what might actually happen when proteins are ingested. Therefore, in order to obtain indications on the specifics of each fraction, we focused our attention on the radical scavenging capacity of every single pseudocereal protein fraction and of the peptides generated by their simulated gastro-intestinal digestion.

Figure 5 shows the DPPH radical inhibition percentage at two different protein concentrations (0.5 mg/mL and 1 mg/mL). In general, the different globulin fractions showed low antioxidant activity when tested in the undigested form (amaranth VLC, quinoa and amaranth LC, and all HC fractions).

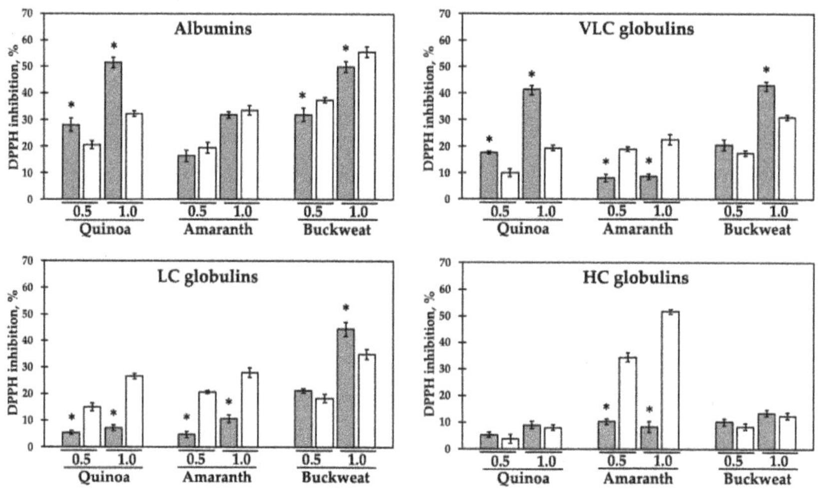

Figure 5. DPPH (2,2-diphenyl-1-picrylhydrazyl) radical inhibition percentage at two different protein concentrations (0.5 mg/mL and 1.0 mg/mL) of each separated protein class, in the native form (gray bars) and after simulated gastro-intestinal digestion (white bars). * $p < 0.05$ vs. digested form. Each experiment was performed in triplicate.

As described in Materials and Methods, the radical scavenging activity was measured via use of the DPPH (1,1-diphenyl-2-picrylhydrazyl) assay, a well-established method to evaluate the antioxidant potential of a compound [65]. Many in vitro methods for the assessment of antioxidant activity in foods of plant origin are available and currently used in a number of studies [66]. Although the principle of each group of antioxidant assay methods are similar, their applicability depends on various factors, including the presence of lipophilic compounds [66]. Gallego et al. [67] reported that the antioxidant activity of peptides is related more to their amino acid composition, structure, and hydrophobicity rather than their size. Indeed, studies indicate that a high radical scavenging activity of protein hydrolysates or peptides is usually associated with a high hydrophobic, aromatic, or sulphur amino acid content [28,63,68]. For all these reasons, the DPPH method was used in this study. Different protein fractions, especially from buckwheat and quinoa, showed antioxidant activity when tested in the intact form (all albumin, buckwheat VLC and LC, and quinoa VLC fractions).

The results indicate that the effect of proteolysis on the antioxidant activity varies according to the protein species. Digested albumins from amaranth and buckwheat possess scavenging activity similar to, or slightly higher than, the undigested forms. Globulin fragmented forms showed increased activity in all the cases for amaranth and in quinoa LC, whereas in the case of buckwheat, the digested globulins scavenging potential was similar or slightly less when compared to the undigested form. Thus, digestion of these protein fractions potentially leads to the release of encrypted antioxidant peptides. Indeed, the most striking result was the observed substantial increase of antioxidant potential in all the digested globulin fractions from amaranth. An increase in the scavenging capacity as a result of the degree of proteolysis was reported for amaranth, buckwheat, quinoa, and chickpea total protein hydrolysates [25,29,59–61,63]. Our results allowed us to quantify the antioxidant activity of each separated protein class and to identify the protein and derived peptides fractions showing the highest activities. For example, our data indicate that the antioxidant peptides were released by LC chenopodin. These data support the findings of Vilcacundo et al. [64], who identified 17 potential antioxidant peptides from quinoa proteins, 13 of which derived from acidic and basic subunits of chenopodin.

It has been demonstrated that the fractions of hydrolyzed buckwheat proteins contained di-, tri-, and tetrameric peptides. When tryptophan, proline, valine, leucine, and phenylalanine were present, these fractions exhibited the strongest radical scavenging activity [29]. Based on these observations, we investigated the amino acid composition of albumin and globulin protein classes for each pseudocereal. Table 1 shows the sum of the percentage content of the amino acids which may be involved in redox reactions (phenylalanine, tyrosine, tryptophan, histidine, and cysteine).

Table 1. Content of amino acids (AA) whose side chains may be involved in redox reactions.

	Quinoa		Amaranth		Buckwheat	
AA	Alb [a]	Glob [b]	Alb [c]	Glob [c]	Alb [d]	Glob [e]
C	6.1 ± 0.6	1.1 ± 0.3	1.9	1.5	7.9	0.9 ± 0.1
F+Y	2.9 ± 0.9	6.1 ± 0.9	5.9	4.8	4	5.9 ± 0.2
W	0.7 ± 0.1	0.8 ± 0.1	/	/	0.8	1.0 ± 0.3
H	1.4 ± 1.1	3.0 ± 0.4	2.3	2.3	0.8	2.3 ± 1.0
Tot	11.1 ± 1.5	11.0 ± 0.9	10.1	8.6	13.5	10.1 ± 1.1
n	2	4	/	/	1	3

Data are expressed as a percentage of the total amino acid number of the proteins (mean ± SD, where applicable). n = available sequence number (NCBI Proteins and UniprotKB databases accessed on 10 January 2020). [a] NCBI: XP_021758596.1, XP_021758543.1; [b] Capraro et al. [45]; [c] Segura-Nieto et al. [48]; [d] UniProt: Q2PS07; [e] UniProt: O23878, O23880, Q9XFM4.

This in silico analysis supports our experimental findings. Indeed, buckwheat albumin, the richest fraction in antioxidant amino acids, is the most active both in the intact and digested form. The redox amino acid content of buckwheat albumin is higher than the globulin fraction and this fits well with the observed difference in the DPPH scavenging activities of these fractions, especially in the digested forms. However, the experimentally determined high activity of digested amaranth HC globulin does not match with the calculated low content of antioxidant amino acids. Possibly, this could be due to the presence of co-purified polypeptides rich in antioxidant amino acids.

2.4. The Trypsin Inhibitor Activity of Protein Fractions and Peptides

The inhibitory activity was measured in vitro on the L-BAPA substrate by comparison with the activity of standard trypsin, set to 100%. Table 2 shows the inhibition percentages of the pseudocereal protein fractions.

Table 2. Trypsin inhibition percentage of protein fractions as such (native) and following simulated gastro-intestinal digestion (hydrolyzed).

	Fraction	Native	Hydrolyzed
Quinoa	Albumins	74.6 ± 6.2	2.1 ± 0.3
	VLC	65.5 ± 4.1	41.4 ± 2.8
	LC	51.8 ± 4.3	32.3 ± 3.2
	HC	42.3 ± 3.6	40.8 ± 2.4
Amaranth	Albumins	n.d. [a]	46.3 ± 5.1
	VLC	41.5 ± 2.9	33.2 ±3.2
	LC	50.2 ± 5.7	31.9 ± 2.7
	HC	52.7 ± 3.5	21.7 ± 1.6
Buckwheat	Albumins	51.7 ± 6.8	101.5 ± 7.7
	VLC	79.6 ± 5.1	18.6 ± 1.8
	LC	83.6 ± 6.2	6.1 ± 3.9
	HC	0.4 ± 0.1	1.5 ± 0.2

[a] n.d.: not determined.

Buckwheat VLC and LC globulins were the most active, with about 79% and 83% of trypsin inhibition, respectively. The HC fraction showed no inhibitory activity. A remarkable inhibitory capacity was carried out also by albumin and VLC globulin of quinoa (74% and 65% inhibition, respectively). Among amaranth proteins, only LC and HC globulin fractions showed little inhibitory activity (about 50%); on the other hand, a low content of trypsin inhibitors (TI) has already been described for this species [69]. Different isoforms of buckwheat trypsin inhibitors with molecular weights between 5 to 7 KDa have been identified [70]. Subsequently, we evaluated the residual trypsin inhibitor activity of proteins after proteolysis. The results showed that TI activity is lower in all cases except, interestingly, in buckwheat albumin fraction, where it increased to up 100% after proteolysis, suggesting the possible release of strongly active peptides. To our knowledge, no previous report has been published describing the release of trypsin inhibitor activity by proteolysis of seed proteins. However, a peptide from a soybean Bowman–Birk inhibitor, containing 16 amino acid residues, with strong anti-tryptic activity has been described [71]. No activity was observed for all other protein hydrolyzed fractions tested.

Serine proteinases have been demonstrated to be involved in immunity [72]: they can inhibit the proteases secreted at the time of injury or infection, which are the prime initiators of inflammation, and the release of neutrophils from lysosomes [73]. Our results indicate that the protein fractions with higher TI activity also possess radical scavenging and immune-modulating properties, supporting the hypothesis of the involvement of TI in protection against oxidation and inflammation. The Bowman–Birk protease inhibitor (BBI) from soybean is one of the most studied protease inhibitors. Different mechanisms were proposed to explain the BBI anti-proliferative and anti-inflammatory effects: the capacity to inhibit the proteolytic proteasome 20S activity [74], whose inhibition may result in the induction of apoptosis [75]; the inhibition of cathepsin G, elastase, and mast cell chymase, proteases involved in inflammatory processes [42]; and the reduction of the neutrophil infiltration and TNF-α during inflammation [76] and the increased production of IL-10, an anti-inflammatory cytokine that plays an important role in suppressing autoimmune diseases [34]. Intriguingly, seed protease inhibitors with antioxidant properties have been observed [73,77,78].

Overall, we analyzed, in vitro, the different biological effects of pseudocereal-derived proteins both in intact and in digested form. The evaluation of their bioavailabilities and metabolic fate was outside the scope of this work. However, previous studies have shown that some plant proteins introduced with the diet arrive in the intestinal tract in intact form and may be absorbed by intestinal cells [79]. The susceptibility of proteins and peptides to digestion and their bioavailability is influenced by diverse conditions, including extreme pH values and plasma membranes phospholipids interaction [80]. Considering this, the bioactivity of intact proteins may be improved by oral administration through gastro-protected tablets.

3. Materials and Methods

3.1. Protein Extraction and Purification

Sequential extraction of albumin and globulin protein classes was performed according to their differential solubilization. Seeds were ground to a meal until sifted through a 60-mesh sieve. Each of the following steps were performed twice. The flours were suspended in distilled water (1:10, *w*/*v*) and stirred for 4 h at 4 °C. The suspension was centrifuged at 10,000× *g* at 4 °C for 30 min to separate soluble albumin fraction. Pellets were resuspended (1:20, *w*/*v*) in a 50 mM sodium phosphate buffer, pH 7.5, containing 500 mM NaCl. The salt-soluble globulins were extracted for 4 h under stirring at 4 °C. The suspension was centrifuged at 10,000× *g* for 30 min at 4 °C and the supernatant was recovered. Albumins were dialyzed against water and freeze-dried. Globulins were gel-filtered using a Sephadex G-50 column, equilibrated in 50 mM phosphate buffer, pH 7.5, containing 100 mM NaCl. The desalted globulins were immediately loaded on a DEAE-cellulose column (20 × 180 mm) equilibrated with the same buffer (2 mg protein/mL resin). The unretained fractions were called VLC proteins, whereas the bound proteins were fractionated using the same buffer with stepwise addition of 150 and 250 mM NaCl (fractions LC and HC, respectively). All three fractions were then dialyzed, freeze-dried, and kept in sealed tubes at 4 °C until used.

3.2. SDS-PAGE

SDS-PAGE was carried out according to [81], on 12% polyacrylamide gel. For peptides separations, 16% polyacrylamide gels were used. Gels were stained by Coomassie Blue G-250 (BioRad, Milan, Italy.) Low-range SDS PAGE Standards (Bio-Rad, Hercules, CA, USA) were used for *Mr* calculations

3.3. Protein Hydrolysis

Simulated gastro-intestinal digestion was carried out as previously described [50], focusing the action only on the proteolytic phase. Briefly, freeze-dried protein samples were dissolved at a concentration of 2 mg/mL in 0.01 M HCl. Pepsin solution was added to protein samples in a ratio 1:100 (*w*/*w*) and the samples were incubated at 37 °C for 10 min. The pH of the solution was then adjusted to 8.0 with 0.5 M NaOH and pancreatin was added to samples in a ratio 1:100 (*w*/*w*). The samples were incubated under shaking at 37 °C for 10 min. At the end, the reaction was stopped with a protease inhibitor cocktail (Merck Life Science, Milan, Italy) as recommended by the manufacturer.

3.4. Caco-2 Cells Cultivation and Immuno-Modulation Assay

All methods and protocols were essentially performed as previously described [45,82].

Cells were cultured in 75 cm^2 flasks at 37 °C under 5% CO_2, using DMEM supplemented with 10% inactivated fetal bovine serum, 2 mM L-glutamine, 100 U/mL penicillin, and 0.1 mg/mL streptomycin as growing medium.

Caco-2 cells were transfected with the plasmid pNiFty2-Luc (InvivoGen, Rho, Italy), after 24 h they were seeded in 24-multiwell plates at a density of 2×10^5 cells/cm^2. The immuno-modulation assay was then performed 24 h after transfection. The medium was replaced with fresh DMEM containing protein samples at the concentration of 1.0 mg/mL and incubated for 4 h at 37 °C. When appropriate, interleukin 1β (IL-1β) was added to the cell medium (10 ng/mL) for stimulating cell immune response. After incubation, plates were chilled on ice for 15 min and cells were scraped from the wells, transferred into 1.5 mL tubes and sonicated three times for 10 s on an ice bath, using a Soniprep 150 device (MSE, Heathfield, East Sussex, UK). After spinning, 100 µL of each supernatant was transferred in a 96-well microtitre plate (PerkinElmer, Waltham, MA, USA) and added with 25 µL of a solution containing ATP and D-luciferin (final concentrations: 1 mM and 10 μM, respectively. Luminescence was measured every 2 min using a Victor3 1420 Multilabel Counter (Perkin-Elmer, Waltham, MA, USA). The maximum recorded signal was considered for comparing the results.

For IL-8 expression quantifications, Caco-2 cells were seeded in 12-multiwell plates using complete DMEM and incubated as described until they reached confluence. Subsequently, cells were treated with 1 mg/mL of proteins or peptide samples for 1 h in the presence or absence of IL-1β (20 ng/mL), in complete DMEM. Total RNA was purified using the Aurum Total RNA Tissue Kit (Bio-Rad, Hercules, CA, USA) according to the manufacturer's instructions. One μg total RNA was reverse transcribed into cDNA (20 μL final volume) using iScript Reverse Transcription Supermix for RT-qPCR kit (Bio-Rad, Hercules, CA, USA) and a Mastercycler Personal 5332 Thermal Cycler (Eppendorf Italia, Milano, Italy). Reaction conditions were: 5 min at 25 °C, 20 min at 46 °C, and 1 min at 95 °C. The cDNAs were then diluted 1:100 with sterile water and 2 μl used as templates in quantitative PCR, using SsoAdvanced Universal SYBR Green Supermix and a CFX Connect Real-Time PCR detection system (Bio-Rad, Hercules, CA, USA). IL-8 specific primers were: 5′-ATGACTTCCAAGCTGGCCGTGGCT-3′ and 5′-TCTCAGCCCTCTTCAAAAACTTCTC-3′ [83]. The GAPDH reference gene was amplified with the following primers: 5′-GGAAGGTGAAGGTCGGAGTC-3′ and 5′-CACAAGCTTCCCGTTCTCAG-3′ [84]. Cycling conditions were: 3 min at 95 °C, then 40 cycles of denaturation (20 s at 95 °C), annealing (30 s at 55 °C), and extension (30 s at 72 °C). For each experiment, cDNA from unstimulated Caco-2 cells was used as the calibrator. Negative controls were performed without cDNA. Relative amounts of target genes compared to the GAPDH reference gene were calculated according to Livak [85]. Effects of the different molecules on inflammation were expressed as fold changes in target genes expression relative to the untreated control sample. Each individual treatment was performed in triplicate.

3.5. Antioxidant Activity Determinations

Antioxidant activities were assessed using the 2,2-diphenyl-1-picrylhydrazyl (DPPH) radical scavenging method, as previously described [86]. Briefly, 0.5 mL of DPPH (Merck Life Science, Milan, Italy) solution in methanol (0.03 mg/mL) were added to 0.5 mL of sample solutions at different concentrations and incubated for 15 min in the dark. Subsequently, the absorbance at 515 nm was measured by a spectrophotometer (Lambda 2, Perkin-Elmer, Waltham, MA, USA). The results were expressed as IC50, namely, the extract concentration that scavenged 50% of DPPH [87].

3.6. Trypsin Inhibitor Activity Assay

Trypsin inhibitor activity (TIA) was measured according to ISO 14902 standard method with slight modifications [88]. Trypsin activity was quantitatively determined by using the synthetic substrate Nα-Benzoyl-L-arginine 4-nitroanilide hydrochloride (BAPA, Merck Life Science, Milan, Italy). Trypsin stock solution was prepared dissolving 27 mg of trypsin (Merck Life Science, Milan, Italy) in 100 mL of 1 mM HCl with 5 mM $CaCl_2$. This solution was then diluted 1:20 prior to the test execution. BAPA working solution was obtained diluting 1:100 of the stock solution (1.5 mM in DMSO) in 50 mM Tris-HCl, pH 8.2, and 5 mM $CaCl_2$. The assay was performed by mixing 0.1 mL of protein extract (0,4 mg/mL) with 0.1 mL of BAPA working solution and 0.2 mL of water. After 10 min of incubation at 37 °C, 0.1 mL of trypsin solution was added and the sample was incubated for 10 min at 37 °C. The reaction was stopped by adding 0.1 mL of 30% acetic acid. The absorbance for the sample reading at 410 nm was a measure of the trypsin activity in the presence of the sample inhibitors (As). The reaction was also run in the absence of inhibitors by replacing the sample extract with an equal amount of water (standard). The corresponding absorbance was the reference reading (Ar). In addition, reagent blanks for the sample readings (Abs) and a reagent blank for the reference readings (Abr) were also prepared by adding the acetic acid solution before the trypsin solution.

TIA was calculated as $i = \left[\frac{(Ar - Abr) - (As - Abs)}{Ar - Abr}\right] \times 100$

Each experiment was performed in triplicate.

3.7. Statistical Analysis

Data reported in the histograms are expressed as the mean ± SE. Data from RT-*q*PCR were analyzed by using the CFX Maestro software (Bio-Rad, Hercules, CA, USA). Data were analyzed by a *t*-test; p values < 0.05 were considered statistically significant.

4. Conclusions

In this work, we explored three different biological effects exerted by each class of proteins and peptides derived from different pseudocereals, using biochemical methodologies and in vitro model cell systems. We designed our experiments so that the results complement what was previously reported. We also describe a fairly realistic depiction of what the gastro-intestinal proteolysis might produce.

On the whole, the obtained results, thanks to the separations in single protein classes, alongside increasing knowledge about their individual effect, open new perspectives for their possible application, for example, in nutraceutical formulations.

Author Contributions: Conceptualization, A.S., J.C., and C.M.; methodology, A.S. and J.C.; validation, A.S., J.C., S.D.B., and C.M.; investigation, J.C., S.D.B., and G.C.H.; writing—original draft preparation, J.C., C.M., and A.S.; writing—review and editing, A.S., J.C., S.D.B., and C.M.; funding acquisition, A.S. All authors have read and agreed to the published version of the manuscript.

Funding: J.C. was supported by Università degli Studi di Milano (Assegno di Ricerca tipo A, 2018-RPDF-0048). The APC was funded by the University of Milan.

Institutional Review Board Statement: Not applicable.

Informed Consent Statement: Not applicable.

Data Availability Statement: The data that support the findings of this study are available from the corresponding author upon reasonable request.

Acknowledgments: The authors are in debt with Stefania Arioli for the helpful assistance with luminometer determinations, and Annalisa Bellucci for assistance during purifications. All experiments involving cultured Caco-2 cells were carried out at the DeFENS Cell Culture Laboratory, a core facility of the Department of Food, Environmental and Nutritional Sciences (University of Milan, Italy).

Conflicts of Interest: The authors declare no conflict of interest.

References

1. Clark, M.; Hill, J.; Tilman, D. The Diet, health, and environment trilemma. *Ann. Rev. Environ. Res.* **2018**, *43*, 109–134. [CrossRef]
2. Schoenlechner, R.; Siebenhandl, S.; Berghofer, E. Chapter 7–Pseudocereals. In *Food Science and Technology, Gluten-Free Cereal Products and Beverages*; Elke, K., Arendt, F.D.B., Eds.; Academic Press: Cambridge, MA, USA, 2008; pp. 149–190. [CrossRef]
3. Martínez-Villaluenga, C.; Peñas, E.; Hernández-Ledesma, B. Pseudocereal grains: Nutritional value, health benefits and current applications for the development of gluten-free foods. *Food Chem. Toxicol.* **2020**, *137*, 111178. [CrossRef]
4. Janssen, F.; Pauly, A.; Rombouts, I.; Jansens, K.J.; Deleu, L.J.; Delcour, J.A. Proteins of amaranth (*Amaranthus* spp.), buckwheat (*Fagopyrum* spp.), and quinoa (*Chenopodium* spp.): A food science and technology perspective. *Compr. Rev. Food Sci. Food Saf.* **2017**, *16*, 39–58. [CrossRef]
5. Liu, R.H. Dietary bioactive compounds and their health implications. *J. Food Sci.* **2013**, *78*, A18–A25. [CrossRef] [PubMed]
6. Ribeiro, P.V.M.; Andrade, P.A.; Hermsdorff, H.H.M.; Dos Santos, C.A.; Cotta, R.M.M.; Estanislau, J.A.S.G.; Campos, A.A.O.; Rosa, C.O.B. Dietary non-nutrients in the prevention of non-communicable diseases: Potentially related mechanisms. *Nutrition* **2019**, *66*, 22–28. [CrossRef] [PubMed]
7. Hernández-Ledesma, B.; Hsieh, C.C. *Bioactive Food Peptides in Health and Disease*; IntechOpen: London, UK, 2013. [CrossRef]
8. Majumder, K.; Mine, Y.; Wu, J. The potential of food protein-derived anti-inflammatory peptides against various chronic inflammatory diseases. *J. Sci. Food Agric.* **2016**, *96*, 2303–2311. [CrossRef] [PubMed]
9. Duranti, M. Grain legume proteins and nutraceutical properties. *Fitoterapia* **2006**, *77*, 67–82. [CrossRef]
10. Roy, F.; Boye, J.I.; Simpson, B.K. Bioactive proteins and peptides in pulse crops: Pea, chickpea and lentil. *Food. Res. Int.* **2010**, *43*, 432–442. [CrossRef]

11. Pyo, Y.H.; Lee, T.C. The potential antioxidant capacity and angiotensin I-converting enzyme inhibitory activity of Monascus-fermented soybean extracts: Evaluation of Monascus-fermented soybean extracts as multifunctional food additives. *J. Food Sci.* **2007**, *72*, S218–S223. [CrossRef]
12. Wang, W.; De Mejia, E.G. A new frontier in soy bioactive peptides that may prevent age-related chronic diseases. *Compr. Rev. Food Sci. Food Saf.* **2005**, *4*, 63–78. [CrossRef]
13. Reyes-Díaz, A.; Del-Toro-Sánchez, C.L.; Rodríguez-Figueroa, J.C.; Valdéz-Hurtado, S.; Wong-Corral, F.J.; Borboa-Flores, J.; González-Osuna, M.F.; Perez-Perez, L.M.; González-Vega, R.I. Legume proteins as a promising source of anti-inflammatory peptides. *Curr. Protein Pept. Sci.* **2019**, *20*, 1204–1217. [CrossRef]
14. Petchiammal, C.; Hopper, W. Antioxidant activity of proteins from fifteen varieties of legume seeds commonly consumed in india. *Int. J. Pharm. Pharm. Sci.* **2014**, *6*, 476–479.
15. Khansari, N.; Shakiba, Y.; Mahmoudi, M. Chronic inflammation and oxidative stress as a major cause of age-related diseases and cancer. *Recent Pat. Inflamm. Allergy Drug Discov.* **2009**, *3*, 73–80. [CrossRef] [PubMed]
16. Liu, T.; Zhang, L.; Joo, D.; Sun, S.C. NF-κB signaling in inflammation. *Sig. Transduct. Target Ther.* **2017**, *2*, 17023. [CrossRef] [PubMed]
17. Zhu, W.; Ren, L.; Zhang, L.; Qiao, Q.; Farooq, M.Z.; Xu, Q. The potential of food protein-derived bioactive peptides against chronic intestinal inflammation. *Mediat. Inflamm.* **2020**, 6817156. [CrossRef]
18. Zhu, F.; Du, B.; Xu, B. Anti-inflammatory effects of phytochemicals from fruits, vegetables, and food legumes: A review. *Crit. Rev. Food Sci. Nutr.* **2018**, *58*, 1260–1270. [CrossRef] [PubMed]
19. Pizzino, G.; Irrera, N.; Cucinotta, M.; Pallio, G.; Mannino, F.; Arcoraci, V.; Squadrito, F.; Altavilla, D.; Bitto, A. Oxidative Stress: Harms and Benefits for Human Health. *Oxid. Med. Cell. Longev.* **2017**, 8416763. [CrossRef]
20. Mendis, E.; Kim, M.; Rajapakse, N.; Kim, S. An in vitro cellular analysis of the radical scavenging efficacy of chitooligosaccharides. *Life Sci.* **2007**, *80*, 2118–2127. [CrossRef]
21. Chen, H.M.; Muramoto, K.; Yamauchi, F.; Nokihara, K. Antioxidant activity of designed peptides based on the antioxidative peptide isolated from digests of a soybean protein. *J. Agric. Food Chem.* **1996**, *44*, 2619–2623. [CrossRef]
22. Peña-Ramos, E.A.; Xiong, Y.L. Whey and soy protein hydrolysates inhibit lipid oxidation in cooked pork patties. *Meat Sci.* **2003**, *64*, 259–263. [CrossRef]
23. Chiue, H.; Kusano, T.; Iwami, K. Antioxidative activity of barley hordein and its loss by deamidation. *J. Nutr. Sci. Vitaminol.* **1997**, *43*, 45–54. [CrossRef]
24. Kong, B.; Xiong, Y.L. Antioxidant activity of zein hydrolysates in a liposome system and the possible mode of action. *J. Agric. Food Chem.* **2006**, *54*, 6059–6068. [CrossRef] [PubMed]
25. Li, Y.; Jiang, B.; Zhang, T.; Mu, W.; Liu, J. Antioxidant and free radical-scavenging activities of chickpea protein hydrolysate (CPH). *Food Chem.* **2008**, *106*, 444–450. [CrossRef]
26. Kong, X.; Zhou, H.; Hua, Y. Preparation and antioxidant activity of wheat gluten hydrolysates (WGHs) using ultrafiltration membranes. *J. Sci. Food Agric.* **2008**, *88*, 920–926. [CrossRef]
27. Cumby, N.; Zhong, Y.; Naczk, M.; Shahidi, F. Antioxidant activity and water-holding capacity of canola protein hydrolysates. *Food Chem.* **2008**, *109*, 144–148. [CrossRef] [PubMed]
28. Pihlanto, A.; Akkanen, S.; Korhonen, H.J. ACE-inhibitory and antioxidant properties of potato (Solanum tuberosum). *Food Chem.* **2008**, *109*, 104–112. [CrossRef]
29. Ma, Y.; Xiong, Y.L.; Zhai, J.; Zhu, H.; Dziubla, T. Fractionation and evaluation of radical-scavenging peptides from in vitro digests of buckwheat protein. *Food Chem.* **2010**, *118*, 582–588. [CrossRef]
30. Esfandi, R.; Walters, M.E.; Tsopmo, A. Antioxidant properties and potential mechanisms of hydrolyzed proteins and peptides from cereals. *Heliyon* **2019**, *5*. [CrossRef]
31. Wu, H.C.; Chen, H.M.; Shiau, C.Y. Free amino acids and peptides as related to antioxidant properties in protein hydrolysates of mackerel (*Scomber austriasicus*). *Food Res. Int.* **2003**, *36*, 949–957. [CrossRef]
32. Kennedy, A.R. Chemopreventive agents: Protease inhibitors. *Pharmacol. Ther.* **1998**, *78*, 167–209. [CrossRef]
33. Clemente, A.; Moreno, F.J.; Marín-Manzano, M.C.; Jiménez, E.; Domoney, C. The cytotoxic effect of Bowman-Birk isoinhibitors, IBB1 and IBBD2, from soybean (Glycine max) on HT29 human colorectal cancer cells is related to their intrinsic ability to inhibit serine proteases. *Mol. Nutr. Food Res.* **2010**, *54*, 396–405. [CrossRef] [PubMed]
34. Safavi, F.; Rostami, A. Role of serine proteases in inflammation: Bowman-Birk protease inhibitor (BBI) as a potential therapy for autoimmune diseases. *Exp. Mol. Pathol.* **2012**, *93*, 428–433. [CrossRef] [PubMed]
35. Machado, R.J.A.; Monteiro, N.K.V.; Migliolo, L.; Silva, O.N.; Pinto, M.F.S.; Oliveira, A.S.; Franco, O.L.; Kiyota, S.; Bemquerer, M.P.; Uchoa, A.F.; et al. Characterization and Pharmacological Properties of a Novel Multifunctional Kunitz Inhibitor from Erythrina velutina Seeds. *PLoS ONE* **2013**, *8*, e63571. [CrossRef] [PubMed]
36. de Lima, V.C.O.; Piuvezam, G.; Maciel, B.L.L.; de Araújo Morais, A.H. Trypsin inhibitors: Promising candidate satietogenic proteins as complementary treatment for obesity and metabolic disorders? *J. Enzyme Inhib. Med. Chem.* **2019**, *34*, 405–419. [CrossRef]
37. Tamir, S.; Bell, J.; Finlay, T.H.; Sakal, E.; Smirnoff, P.; Gaur, S.; Birk, Y. Isolation, characterization, and properties of a trypsin-chymotrypsin inhibitor from amaranth seeds. *J. Protein Chem.* **1996**, *15*, 219–229. [CrossRef]

38. Valdes-Rodriguez, S.; Segura-Nieto, M.; Chagolla-Lopez, A.; Verver y Vargas-Cortina, A.; Martinez-Gallardo, N.; Blanco-Labra, A. Purification, characterization, and complete amino acid sequence of a trypsin inhibitor from amaranth (*Amaranthus hypochondriacus*) seeds. *Plant Physiol.* **1993**, *103*, 1407–1412. [CrossRef]
39. Tsybina, T.A.; Dunaevsky, Y.E.; Musolyamov, A.K.; Egorov, T.A.; Belozersky, M.A. Cationic inhibitors of serine proteinases from buckwheat seeds. *Biochemistry* **2001**, *66*, 941–947. [CrossRef]
40. Park, S.S.; Abe, K.; Kimura, M.; Urisu, A.; Yamasaki, N. Primary structure and allergenic activity of trypsin inhibitors from the seeds of buckwheat (*Fagopyrum esculentum* Moench). *FEBS Lett.* **1997**, *400*, 103–107. [CrossRef]
41. Pesoti, A.R.; de Oliveira, B.M.; de Oliverira, A.C.; Pompeu, D.G.; Gonçalves, D.B.; Marangoni, S.; da Silva, J.A.; Granjeiro, P.A. Extraction, purification and characterization of inhibitor of trypsin from Chenopodium quinoa seeds. *Food Sci. Technol.* **2015**, *35*. [CrossRef]
42. Clemente, A.; del Carmen Arques, M. Bowman-Birk inhibitors from legumes as colorectal chemopreventive agents. *World J. Gastroenterol.* **2014**, *20*, 10305–10315. [CrossRef]
43. Shewry, P.R.; Napier, J.A.; Tatham, A.S. Seed storage proteins: Structures and biosynthesis. *Plant Cell.* **1995**, *7*, 945–956. [CrossRef]
44. Brinegar, C.; Goundan, S. Isolation and characterization of chenopodin, the 11S seed storage protein of quinoa (Chenopodium quinoa). *J. Agric. Food Chem.* **1993**, *41*, 182–185. [CrossRef]
45. Capraro, J.; De Benedetti, S.; Di Dio, M.; Bona, E.; Abate, A.; Corsetto, P.A.; Scarafoni, A. Characterization of Chenopodin Isoforms from Quinoa Seeds and Assessment of Their Potential Anti-Inflammatory Activity in Caco-2 Cells. *Biomolecules* **2020**, *10*, 795. [CrossRef] [PubMed]
46. Radović, S.R.; Maksimović, V.R.; Varkonji-Gašić, E.I. Characterization of buckwheat seed storage proteins. *J. Agric. Food Chem.* **1996**, *44*, 972–974. [CrossRef]
47. Capraro, J.; Magni, C.; Giorgi, A.; Scarafoni, A.; Duranti, M. Comparative 1d- and 2d-electrophoretic protein profiles of ancestral and modern buckwheat seeds grown in the Italian alpine region. *Ital. J. Food Sci.* **2018**, *30*, 497–503. [CrossRef]
48. Segura-Nieto, M.; Vazquez-Sanchez, N.; Rubio-Velazquez, H.; Olguin-Martinez, L.E.; Rodriguez-Nester, C.E.; Herrera-Estrella, L. Characterization of amaranth (*Amaranthus hypochondriacus* L.) seed proteins. *J. Agric. Food Chem.* **1992**, *40*, 1553–1558. [CrossRef]
49. Barba de la Rosa, A.P.; Fomsgaard, I.S.; Laursen, B.; Mortensen, A.G.; Olvera-Martínez, L.; Silva-Sánchez, C.; Mendoza-Herrera, A.; González-Castañeda, J.; De León-Rodríguez, A. Amaranth (*Amaranthus hypochondriacus*) as an alternative crop for sustainable food production: Phenolic acids and flavonoids with potential impact on its nutraceutical quality. *J. Cereal Sci.* **2009**, *49*, 117–121. [CrossRef]
50. Brodkorb, A.; Egger, L.; Alminger, M.; Alvito, P.; Assunção, R.; Balance, S.; Bohn, T.; Bourlieu-Lacanal, C.; Boutrou, R.; Carrière, F.; et al. INFOGEST static in vitro simulation of gastrointestinal food digestion. *Nat. Protoc.* **2019**, *14*, 991–1014. [CrossRef]
51. Lima-Cabello, E.; Morales-Santana, S.; Foley, R.C.; Melser, S.; Alché, V.; Siddique, K.H.M.; Singh, K.B.; Alché, J.D.; Jimenez-Lopez, J.C. Ex vivo and in vitro assessment of anti-inflammatory activity of seed β-conglutin proteins from *Lupinus angustifolius*. *J. Funct. Foods* **2018**, *40*, 510–519. [CrossRef]
52. Lima-Cabello, E.; Alché, J.D.; Morales-Santana, S.; Clemente, A.; Jimenez-Lopez, J.C. Narrow-Leafed Lupin (Lupinus angustifolius L.) Seeds Gamma-Conglutin is an Anti-Inflammatory Protein Promoting Insulin Resistance Improvement and Oxidative Stress Amelioration in PANC-1 Pancreatic Cell-Line. *Antioxidants* **2020**, *9*, 12. [CrossRef]
53. Sullivan, A.C.; Pangloli, P.; Dia, V.P. Kafirin from Sorghum bicolor inhibition of inflammation in THP-1 human macrophages is associated with reduction of intracellular reactive oxygen species. *Food Chem. Toxicol.* **2018**, *111*, 503–510. [CrossRef] [PubMed]
54. Cam, A.; de Mejia, E.G. RGD-peptide lunasin inhibits Akt-mediated NF-κB activation in human macrophages through interaction with the αVβ3 integrin. *Mol. Nutr. Food Res.* **2012**, *56*, 1569–1581. [CrossRef]
55. Ren, G.; Zhu, Y.; Shi, Z.; Li, J. Detection of lunasin in quinoa (*Chenopodium quinoa* Willd.) and the in vitro evaluation of its antioxidant and anti-inflammatory activities. *J. Sci. Food Agric.* **2017**, *97*, 4110–4116. [CrossRef] [PubMed]
56. Hernández-Ledesma, B.; Hsieh, C.C.; de Lumen, B.O. Antioxidant and anti-inflammatory properties of cancer preventive peptide lunasin in RAW 264.7 macrophages. *Biochem. Biophys. Res. Commun.* **2009**, *390*, 803–808. [CrossRef]
57. Seber, L.E.; Barnett, B.W.; McConnell, E.J.; Hume, S.D.; Cai, J.; Boles, K.; Davis, K.R. Scalable purification and characterization of the anticancer lunasin peptide from soybean. *PLoS ONE* **2012**, *7*, e35409. [CrossRef] [PubMed]
58. Matsumoto, R.; Fujino, K.; Nagata, Y.; Hashiguchi, S.; Ito, Y.; Aihara, Y.; Takahashi, Y.; Maeda, K.; Sugimura, K. Molecular characterization of a 10-kDa buckwheat molecule reactive to allergic patients' IgE. *Allergy* **2004**, *59*, 533–538. [CrossRef] [PubMed]
59. Orsini Delgado, M.; Tironi, V.; Añón, M. Antioxidant activity of amaranth proteins or their hydrolysates under simulated gastrointestinal digestion. *LWT Food Sci. Technol.* **2011**, *44*, 1752–1760. [CrossRef]
60. Tironi, V.; Añón, M. Amaranth as a source of antioxidant peptides. Effect of proteolysis. *Food Res. Int.* **2010**, *43*, 315–322. [CrossRef]
61. Aluko, R.; Monu, E. Functional and Bioactive Properties of Quinoa Seed Protein Hydrolysates. *J. Food Sci.* **2003**, *68*, 1254–1258. [CrossRef]
62. Tang, C.H.; Peng, J.; Zhen, D.W.; Chen, Z. Physicochemical and antioxidant properties of buckwheat (Fagopyrum esculentum Moench) protein hydrolysates. *Food Chem.* **2009**, *115*, 672–678. [CrossRef]
63. Nongonierma, A.B.; Le Maux, S.; Dubrulle, C.; Barre, C.; FitzGerald, R.J. Quinoa (Chenopodium quinoa Willd.) protein hydrolysates with in vitro dipeptidyl peptidase IV (DPP-IV) inhibitory and antioxidant properties. *J. Cereal Sci.* **2015**, *65*, 112–118. [CrossRef]

64. Vilcacundo, R.; Miralles, B.; Carrillo, W.; Hernández-Ledesma, B. In vitro chemopreventive properties of peptides released from quinoa (Chenopodium quinoa Willd.) protein under simulated gastrointestinal digestion. *Food Res. Int.* **2018**, *105*, 403–411. [CrossRef]
65. Nicklisch, S.C.T.; Waite, J.H. Optimized DPPH assay in a detergent-based buffer system for measuring antioxidant activity of proteins. *MethodsX* **2014**, *1*, 233–238. [CrossRef] [PubMed]
66. Huang, D.; Ou, B.; Prior, R.L. The chemistry behind antioxidant capacity assays. *J. Agric. Food Chem.* **2005**, *53*, 1841–1856. [CrossRef] [PubMed]
67. Gallego, M.; Mora, L.; Hayes, M.; Reig, M.; Toldrá, F. Effect of cooking and in vitro digestion on the antioxidant activity of dry-cured ham by-products. *Food Res. Int.* **2017**, *97*, 296–306. [CrossRef]
68. Rajapakse, N.; Mendis, E.; Byun, H.G.; Kim, S.K. Purification and in vitro antioxidative effects of giant squid muscle peptides on free radical-mediated oxidative systems. *J. Nutr. Biochem.* **2005**, *16*, 562–569. [CrossRef] [PubMed]
69. Correa, A.D.; Jokl, L.; Carlsson, R. Chemical constituents, in vitro protein digestibility, and presence of antinutritional substances in amaranth grains. *Arch. Latinoam. Nutr.* **1986**, *36*, 319–326.
70. Tsybina, T.; Dunaevsky, Y.; Musolyamov, A.; Egorov, T.; Larionova, N.; Popykina, N.; Belozersky, M. New protease inhibitors from buckwheat seeds: Properties, partial amino acid sequences and possible biological role. *Biol. Chem.* **2004**, *385*, 429–434. [CrossRef]
71. Nishino, N.; Izumiya, N. Anti-tryptic activity of a synthetic bicyclic fragment of soybean bowman-birk inhibitor. *Biochim. Biophys. Acta* **1982**, *708*, 233–235. [CrossRef]
72. Heutinck, K.M.; ten Berge, I.J.; Hack, C.E.; Hamann, J.; Rowshani, A.T. Serine proteases of the human immune system in health and disease. *Mol. Immunol.* **2010**, *47*, 1943–1955. [CrossRef] [PubMed]
73. Shamsi, T.N.; Parveen, R.; Afreen, S.; Azam, M.; Sen, P.; Sharma, Y.; Haque, Q.M.R.; Fatma, T.; Manzoor, N.; Fatima, S. Trypsin Inhibitors from Cajanus cajan and Phaseolus limensis Possess Antioxidant, Anti-Inflammatory, and Antibacterial Activity. *J. Diet. Suppl.* **2018**, *15*, 939–950. [CrossRef]
74. Souza Lda, C.; Camargo, R.; Demasi, M.; Santana, J.M.; de Sá, C.M.; de Freitas, S.M. Effects of an anticarcinogenic Bowman-Birk protease inhibitor on purified 20S proteasome and MCF-7 breast cancer cells. *PLoS ONE* **2014**, *9*, e86600. [CrossRef] [PubMed]
75. An, B.; Goldfarb, R.H.; Siman, R.; Doul, Q.P. Novel dipeptidyl proteasome inhibitors overcome Bcl-2 protective function and selectively accumulate the cyclin-dependent kinase inhibitor p27 and induce apoptosis in transformed, but not normal, human fibroblasts. *Cell. Death Differ.* **1998**, *5*, 1062–1075. [CrossRef] [PubMed]
76. Gitlin-Domagalska, A.; Maciejewska, A.; Dębowski, D. Bowman-Birk Inhibitors: Insights into Family of Multifunctional Proteins and Peptides with Potential Therapeutical Applications. *Pharmaceuticals* **2020**, *13*, 421. [CrossRef] [PubMed]
77. Hou, W.C.; Chen, Y.C.; Chen, H.J.; Lin, Y.H.; Yang, L.L.; Lee, M.H. Antioxidant activities of trypsin inhibitor, a 33 KDa root storage protein of sweet potato (*Ipomoea batatas* (L.) Lam cv. Tainong 57). *J. Agric. Food Chem.* **2001**, *49*, 2978–2981. [CrossRef]
78. Shamsi, T.N.; Parveen, R.; Fatima, S. Trypsin inhibitors demonstrate antioxidant activities, inhibit A549 cell proliferation, and increase activities of reactive oxygen species scavenging enzymes. *Indian J. Pharmacol.* **2017**, *49*, 155–160. [CrossRef]
79. Capraro, J.; Clemente, A.; Rubio, L.A.; Magni, C.; Scarafoni, A.; Duranti, M. Assessment of the lupin seed glucose-lowering protein intestinal absorption by using in vitro and ex vivo models. *Food Chem.* **2011**, *125*, 1279–1283. [CrossRef]
80. Capraro, J.; Spotti, P.; Magni, C.; Scarafoni, A.; Duranti, M. Spectroscopic studies on the pH-dependent structural dynamics of γ-conglutin, the blood glucose-lowering protein of lupin seeds. *Int. J. Biol. Macromol.* **2010**, *47*, 502–507. [CrossRef]
81. Laemmli, U.K. Cleavage of structural proteins during the assembly of the head of bacteriophage T4. *Nature* **1970**, *227*, 680–685. [CrossRef]
82. Parizad, P.A.; Capraro, J.; Scarafoni, A.; Bonomi, F.; Blandino, M.; Marengo, M.; Giordano, D.; Carpen, A.; Iametti, S. The bio-functional properties of pigmented cereals may involve synergies among different bioactive species. *Plant Foods Hum. Nutr.* **2019**, *74*, 128–134. [CrossRef]
83. Li, J.; Moran, T.; Swanson, E.; Julian, C.; Harris, J.; Bonen, D.K.; Hedl, M.; Nicolae, D.L.; Abraham, C.; Cho, J.H. Regulation of IL-8 and IL-1 expression in Crohn's disease associated NOD2/CARD15 mutations. *Hum. Mol. Genet.* **2004**, *13*, 1715–1725. [CrossRef] [PubMed]
84. Bottero, V.; Imbert, V.; Frelin, C.; Formento, J.L.; Peyron, J.F. Monitoring NF-κB transactivation potential via Real-Time PCR quantification of IκB-α Gene Expression. *Mol. Diagn.* **2003**, *7*, 187–194. [CrossRef] [PubMed]
85. Livak, K.J.; Schmittgen, T.D. Analysis of relative gene expression data using real-time quantitative PCR and the 2(-Delta Delta C(T)) method. *Methods* **2001**, *25*, 402–408. [CrossRef] [PubMed]
86. Parizad, P.A.; De Nisi, P.; Adani, F.; Sciarria, T.P.; Squillace, P.; Scarafoni, A.; Iametti, S.; Scaglia, B. Antioxidant and anti-Inflammatory activities of the crude extracts of raw and fermented tomato pomace and their correlations with aglycate-polyphenols. *Antioxidants* **2020**, *9*, 179. [CrossRef] [PubMed]
87. Mishra, K.; Ojha, H.; Chaudhury, N.K. Estimation of antiradical properties of antioxidants using DPPH assay: A critical review and results. *Food Chem.* **2012**, *130*, 1036–1043. [CrossRef]
88. International Organization for Standardization. *Standard 14902: Animal Feeding Stuffs-Determination of Trypsin Inhibitor Activity of Soya Products*; International Organization for Standardization: Geneva, Switzerland, Approved October 2001, Reapproved August 2017.

International Journal of *Molecular Sciences*

MDPI

Article

Bioactive Triterpenes of *Protium heptaphyllum* Gum Resin Extract Display Cholesterol-Lowering Potential

Giuseppe Mannino [1], Piera Iovino [2], Antonino Lauria [1], Tullio Genova [3], Alberto Asteggiano [2,4], Monica Notarbartolo [1], Alessandra Porcu [5], Graziella Serio [1], Giorgia Chinigò [3], Andrea Occhipinti [5], Andrea Capuzzo [5], Claudio Medana [4], Luca Munaron [3] and Carla Gentile [1,*]

1 Department of Biological, Chemical and Pharmaceutical Sciences and Technologies (STEBICEF), University of Palermo, 90128 Palermo, Italy; giuseppe.mannino@unipa.it (G.M.); antonino.lauria@unipa.it (A.L.); monica.notarbartolo@unipa.it (M.N.); graziella.serio01@unipa.it (G.S.)
2 Biosfered S.R.L., 10148 Turin, Italy; p.iovino@biosfered.com (P.I.); a.asteggiano@biosfered.com (A.A.)
3 Department of Life Sciences and Systems Biology, University of Turin, 10123 Turin, Italy; tullio.genova@unito.it (T.G.); giorgia.chinigo@unito.it (G.C.); luca.munaron@unito.it (L.M.)
4 Department of Molecular Biotechnology and Health Sciences, University of Torino, 10125 Torino, Italy; alberto.asteggiano@unito.it (A.A.); claudio.medana@unito.it (C.M.)
5 Abel Nutraceuticals S.R.L., 10148 Turin, Italy; a.porcu@abelnutraceuticals.com (A.P.); a.occhipinti@abelnutraceuticals.com (A.O.); a.capuzzo@abelnutraceuticals.com (A.C.)
* Correspondence: carla.gentile@unipa.it; Tel.: +39-091-2388-6472

Citation: Mannino, G.; Iovino, P.; Lauria, A.; Genova, T.; Asteggiano, A.; Notarbartolo, M.; Porcu, A.; Serio, G.; Chinigò, G.; Occhipinti, A.; et al. Bioactive Triterpenes of *Protium heptaphyllum* Gum Resin Extract Display Cholesterol-Lowering Potential. *Int. J. Mol. Sci.* **2021**, *22*, 2664. https://doi.org/10.3390/ijms22052664

Academic Editors: Antonio Carrillo Vico and Ivan Cruz-Chamorro

Received: 3 February 2021
Accepted: 2 March 2021
Published: 6 March 2021

Abstract: Hypercholesterolemia is one of the major causes of cardiovascular disease, the risk of which is further increased if other forms of dyslipidemia occur. Current therapeutic strategies include changes in lifestyle coupled with drug administration. Statins represent the most common therapeutic approach, but they may be insufficient due to the onset of resistance mechanisms and side effects. Consequently, patients with mild hypercholesterolemia prefer the use of food supplements since these are perceived to be safer. Here, we investigate the phytochemical profile and cholesterol-lowering potential of *Protium heptaphyllum* gum resin extract (PHE). Chemical characterization via HPLC-APCI-HRMS^2 and GC-FID/MS identified 13 compounds mainly belonging to ursane, oleanane, and tirucallane groups. Studies on human hepatocytes have revealed how PHE is able to reduce cholesterol production and regulate the expression of proteins involved in its metabolism. (*HMGCR*, *PCSK9*, *LDLR*, *FXR*, *IDOL*, and *PPAR*). Moreover, measuring the inhibitory activity of PHE against HMGR, moderate inhibition was recorded. Finally, molecular docking studies identified acidic tetra- and pentacyclic triterpenoids as the main compounds responsible for this action. In conclusion, our study demonstrates how PHE may be a useful alternative to contrast hypercholesterolemia, highlighting its potential as a sustainable multitarget natural extract for the nutraceutical industry that is rapidly gaining acceptance as a source of health-promoting compounds.

Keywords: hypercholesterolemia; gene expression; HMGCR; PCSK9; PPARα; enzymatic activity; molecular docking; statin; monacolin; breu branco

1. Introduction

Hypercholesterolemia is a form of hyperlipidemia that is characterized by the presence of high levels of circulating low-density lipoprotein cholesterol (LDL-C). It is the major cause of arteriosclerosis [1], which may lead to serious cardiovascular diseases (CVDs) such as hypertension, cerebrovascular disease, peripheral arterial disease, coronary disease, deep vein thrombosis, and pulmonary embolism [2]. CVDs are the leading cause of death globally and were responsible for 31% of all global deaths in 2016, as estimated by WHO (World Health Organization) [3]. According to the different hypercholesterolemia risk levels, a number of therapeutic strategies are actually available [1]. As a first step, changes in lifestyle and diet behaviors are strongly recommended [1,4]. In particular, the consumption of foods with high carbohydrates, saturated fats, and high cholesterol levels should be avoided, while increasing the intake of foods rich in fiber, potassium, unsaturated fats, and saponins [5,6]. Statin therapy represents the most common pharmacological approach for the management of hypercholesterolemia, and the current guidelines recommend a treatment with the maximally tolerated dose of statins [1]. This class of drugs exerts cholesterol-lowering effects by affecting the enzymatic activity of 3-hydroxy-3-methyl-glutaryl coenzyme A reductase (HMGCR) and also increasing the hepatic expression of the LDL receptor (LDLR) and reducing low density lipoprotein (LDL) and very low density lipoprotein (VLDL) synthesis [7]. However, in most cases, statin monotherapy was insufficient to achieve the recommended LDL-C level [2,8], also because of the onset of drug-resistance mechanisms [7]. In these cases, statin therapy is replaced or accompanied by the administration of other drugs, such as fibrate, niacin, and ezetimibe [1,8]. Furthermore, recent epidemiological evidence has shown that patients under treatment with statin, alone or in combination with other medicines, might present muscle-related symptoms or other pharmacological side effects, which lead to the discontinuation of treatments [9,10]. For these reasons, patients with a clinical picture of low hypercholesteremic severity prefer applying nonpharmacological approaches, such as phytotherapeutic ones, because they are perceived as natural, safer, and free of side effects [11]. Among the different phytochemical approaches, red yeast rice (RYR) has been used for centuries as herbal medicine in China to alleviate symptoms derived from hypercholesterolemia conditions [12–14]. Currently, RYR is also present in the global market as a dietary supplement (DS), and, in 2008, it was estimated that USD 20 million was spent by American consumers exclusively on this DS category [12]. The antihypercholesterolemic activity of RYR is attributed to the presence of monacolin K [7,13] and citrinin [15]. While monacolin K efficacy is linked to its structural equality to lovastatin [16,17], citrinin is a mycotoxin with documented antihypercholesterolemic properties [15,18]. However, recent clinical studies attributed significant side effects to both compounds. In particular, equally to other statins, monacolin K displayed anaphylaxis, hepatotoxicity, central nervous system complaints, rhabdomyolysis effects, and a high possibility of developing diabetes mellitus [7,14,19]. On the other hand, the nephrotoxicity, carcinogenicity, cytotoxicity, and teratogenicity of citrinin are well known even at a very low dosage [20]. For these reasons, in the last few decades, the research of new medicinal plants as an alternative approach for the treatment of hypercholesterolemia is on-going [21–23]. In particular, attention is focused on the identification of new phytochemicals that may exert an antihypercholesterolemic action through a different mechanism of action from what has been reported for statins [24].

Burseraceae is a pantropical-distributed family consisting of 21 genera and more than 700 species [25], of which 135 belong to the *Protium genus* [25,26]. *Protium heptaphyllum* is a traditional medicinal plant distributed in the Amazon region, especially in Brazil [27]. Equally to other species belonging to the *Burseraceae* family, *P. heptaphyllum* also yields large amounts of resin from the trunk wood [26]. The resin, popularly known as "almécega" or "breu branco", is collected and used locally in folk medicine practices for its analgesic, anti-inflammatory, and expectorant effects; some recent works have demonstrated several of those activities. In particular, concerning central actions, Aragão and colleagues showed in a murine model system both anticonvulsant effects related to the GABAergic system [28,29], and anxiolytic and antidepressant effects involving benzodiazepine-type receptors and noradrenergic mechanisms [26]. In addition, antinociceptive potential was also demonstrated [30,31]. The traditional use of the oily resin in inflammatory syndromes is justified by data from several in vivo studies. In particular anti-inflammatory effects of *P. heptaphyllium* gum resin extracts were showed in pleuritis [32], gastric ulcer [30,33], carrageenan induced-edema [34] acute periodontitis [35], and pancreatitis [35]. However, the mechanisms involved in the observed anti-inflammatory action were not clarified. More recently, antibacterial activity [36–38] and effects on blood sugar level and lipid profile have also been highlighted for the essential oil obtained from *P. heptaphyllum* gum resin [39–41]. The observed effects on lipid metabolism involved a decrease of total cholesterol, LDL-C, serum triglycerides, and VLDL, with an elevation of high-density lipoprotein cholesterol (HDL-C). In addition, antilipidemic effects were associated with modulatory actions on the secretion of hormones related to fat and carbohydrate metabolism [40]. In a more recent study, in order to investigate the mechanism involved in the hypolipidemic action of the resin, De Melo et al. evaluated the antiadipogenic activity of an α- and β-amyrin mixture in 3T3-L1 murine adipocytes [42]. They showed significant suppression of adipocyte differentiation by downregulation of adipogenesis-related transcription factors, including PPARγ (Peroxisome Proliferator-Activated Receptor α9 and C/EBPα (CCAAT-enhancer binding protein α) [42].

In the past twenty years, the chemical composition of the essential oil obtained from the resin of *P. heptaphyllum* has been largely investigated. In particular, De Lima [38] and Rüdigera [26] showed as this resin was rich in terpenoid compounds, especially monoterpenes (α- and γ-terpinene, p-cymenol, terpinolene, α-pinene, p-cymene, 3-carene, limonene), sesquiterpenes (g- and D-cadinene), and pentacylic triterpenes belonging the classes of ursane, oleanane, tirucalane, lupane, and taraxtane. Among them, α-amyrin and β-amyrin have always been identified as the main phytochemicals responsible for the observed in vitro and in vivo biological activities.

In this work, we investigate the hypocholesterolemic potential of a new chemically characterized *P. heptaphyllum* resin extract (PHE). With this purpose, we measured the in vitro effects of PHE on cholesterol biosynthesis in comparison with monacolin K [43]. Moreover, we also evaluated if PHE affects HMGCR, evaluating both changes in enzymatic activity and the binding ability of the molecules identified in PHE. Finally, studies on the expression of genes related to cholesterol metabolism were performed.

2. Results and Discussion

2.1. HPLC APCI-HRMS^2(High Performance Liquid Chromatography coupled with Atmospheric Pressure Chemical Ionization and High Resolution Mass Spectrometry) and GC-FID/MS (Gas Chromatography coupled with Flame Ionization Detector and Mass Spectrometry) Analyses identified Tetra- and Pentacyclic Triterpenoic compounds as the main constituent of PHE

The composition of PHE was assessed by gas and liquid chromatographic analyses coupled to mass spectrometry. Eleven compounds were identified (Figure 1) and quantified (Table 1).

Among them, two were small-sized volatile organic compounds (Δ3-carene (#1); p-cymene (#2)), six were nonacidic pentacyclic triterpenes (α-amyrin (#3); β-amyrin (#4); α-amyron (#5); β-amyron (#6); brein (#7); maniladiol (#8)), two were acid pentacyclic triterpenois (oleanolic acid (#9); ursolic acid (#10)), and three were tetracyclic triterpenoids (elemonic acid (#11); α-elemolic acid (#12); β-elemolic acid (#13)).

The fraction containing the volatile organic compounds (VOCs) was the least representative in PHE, reaching only about 1.5% of total composition. Meanwhile, Compound #1 was a bicyclic monoterpene consisting of fused cyclohexene and cyclopropane rings, and #2 was the alkylbenzenic form of #1, which originated after pyrolysis under oxidative conditions [44]. Both compounds are naturally distributed in several aromatic plants and gum resins [45], and they are widely used as ingredients for the realization of cosmetics, such as perfumes [45].

Nonacidic triterpenoids (NATs) and acidic triterpenoids (ATs) were the most abundant molecules in PHE, representing more than 28% and 39% of total composition, respectively. Among the identified NATs, the compounds having a ursane skeleton (#3, #5, and #7) were more predominant with respect to those composed by an oleanane skeleton (#4, #6, and #8). Compounds #3 and #4 contributed to more than 20% of the PHE composition. They are compounds well-known in the literature to exert various pharmacological actions, including analgesic, anti-inflammatory, gastroprotective, hepatoprotective, and hypolipidemic properties [46]. #5, #6, #7, and #8 are compounds originated by the structural changes of #3 and #4. In particular, #5 and #6 are the oxidized forms, respectively, of #3 and #4; meanwhile, #7 and #8 are their 5-hydroxy forms. However, #5, #6, #7, and #8 contribute, together, just 20% of the identified NATs and only 6% of total identified and quantified compounds in PHE. Concerning ATs, HPLC analysis revealed the presence of two pentacycles (#9 and #10) and three tetracycles (#11, #12, and #13) having at least one acid function. Regarding the ATS with a tetracyclic core, the identified compounds had a very similar structure, belonging to the tirucallane groups and exclusively differing, for the substituent group, in the C_3 position (Figure 1). Among the ATs having pentacyclic scaffolds, #9 belongs to the oleanane group and #10 has a ursane skeleton. ATs are widely distributed in the plant kingdom, and they are most abundant in particular plant gum resins [47–49]. Growing evidence has shown a greater attitude for ATs to bind the active site of different enzymes, thanks to the presence of one or more carboxylic substituents in the chemical scaffolds that are able to accept noncovalent bonds easily [50–53].

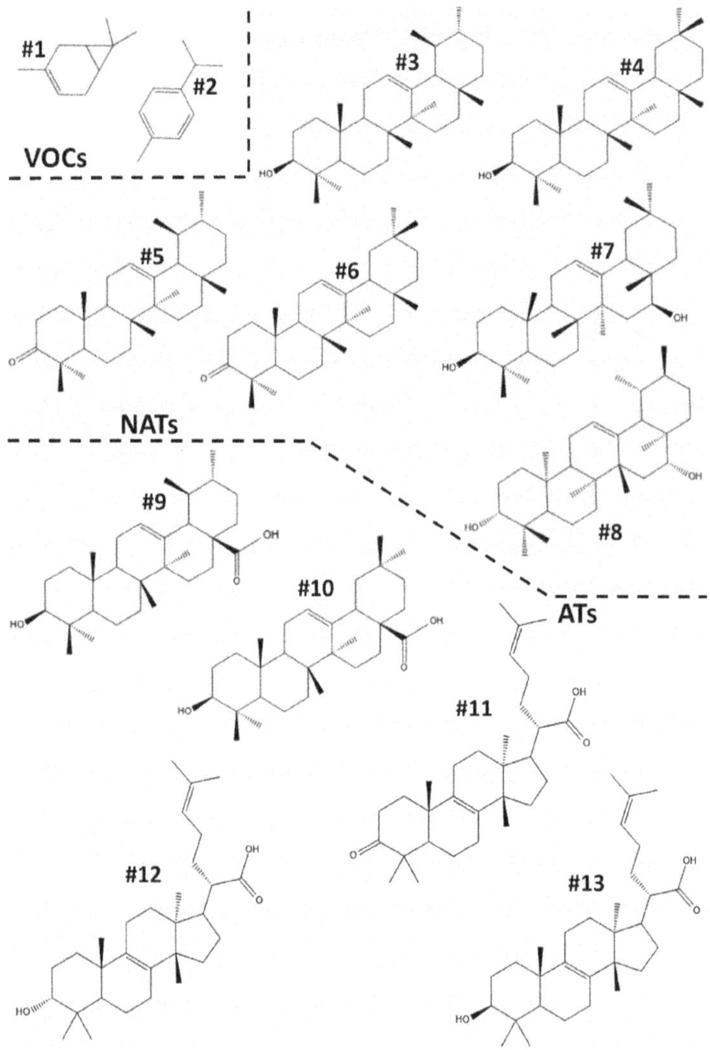

Figure 1. Chemical structures of VOC, NAT, and AT compounds, characterized and quantified in *Protium heptaphyllum* ethanolic extract (PHE). VOCs: volatile organic compounds (mono- and sesquiterpenes); NATs: nonacidic triterpenes; ATs: acidic triterpenes containing a carboxyl group.

Table 1. Chemical composition of *P. heptaphyllum* gum resin extract. For each compound, CAS ID, MW, chemical formula, name, and percentage composition are reported in the table.

#	CAS ID	MW	Chemical Formula	Class	Name	Composition (%)
1	13466-78-9	136.23	$C_{10}H_{16}$	VOCs	Δ3-carene	0.70 ± 0.08%
2	99-87-6	134.22	$C_{10}H_{14}$		p-cymene	0.90 ± 0.02%
3	638-95-9	556.9	$C_{39}H_{56}O_2$	NATs	α-amyrin	12.40 ± 1.39%
4	559-70-6	426.7	$C_{30}H_{50}O$		β-amyrin	9.50 ± 1.17%
5	638-96-0	424.7	$C_{30}H_{48}O$		α-amyron	2.20 ± 0.08%
6	638-97-1	424.7	$C_{30}H_{48}O$		β-amyron	1.70 ± 0.61%
7	465-08-7	442.7	$C_{30}H_{50}O_2$		Brein	1.30 ± 0.08%
8	595-17-5	442.7	$C_{30}H_{50}O$		Maniladiol	0.90 ± 0.06%
9	508-02-1	456.7	$C_{30}H_{48}O_3$	ATs	Oleanolic acid	0.05 ± 0.01%
10	77-52-1	456.7	$C_{30}H_{48}O_3$		Ursolic acid	0.06 ± 0.01%
11	28282-25-9	454.7	$C_{30}H_{46}O_3$		Elemonic acid	15.60 ± 0.06%
12	28282-27-1	456.7	$C_{30}H_{48}O_3$		α-elemolic acid	10.57 ± 0.76%
13	28282-54-4	456.7	$C_{30}H_{48}O_3$		β-elemolic acid	12.80 ± 1.38%

CAS ID: Chemical Abstracts Service identification number; MW: molecular weight; VOCs: volatile organic compounds (mono- and sesquiterpenes); NATs: nonacidic triterpenes; ATs: acidic triterpenes containing a carboxyl group.

2.2. PHE Decreases Total Cholesterol in Hepatocytes

In order to evaluate cholesterol production, hepatocytes were treated for 6 or 12 h with PHE (25–200 μg mL^{-1}). The same experimental conditions were used to assay LHA (0.01–10 μg mL^{-1}) as positive control (Figure 2). LHA is a molecule originated by in vivo metabolism of lovastatin [54,55]. Indeed, lovastatin is a prodrug presenting a γ-lactone closed ring in the form that it is administered. The closed ring strongly limits its inhibitory action on HMGCR, but it can be easily hydrolyzed in vivo into different active metabolites, which appear in plasma after 24 h after oral administration [55,56]. Among them, the most effective cholesterol-lowering agent is LHA, in which the γ-lactone closed ring is hydrolyzed, forming a β-hydroxy acid function [54,55].

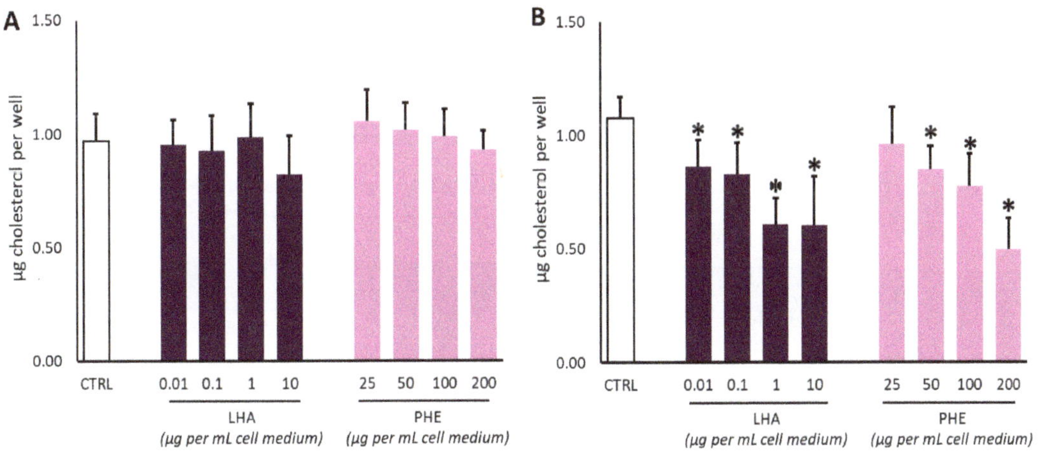

Figure 2. Effect of lovastatin hydroxy acid (LHA) and *Protium heptaphyllum* resin extract (PHE) on cholesterol levels in THLE-3 cells after 6 h (**A**) or 12 h (**B**) of the treatment. Data are expressed as mean ± SD of three different replicates. The symbol "*", when present, indicates significant ($p < 0.05$) differences between treated and control samples, as calculated by the Mann–Whitney test.

No significant decreases in cholesterol production were observed at the tested concentrations after 6 h of treatment with LHA or PHE (Figure 2A). On the other hand, both LHA and PHE were able to negatively affect the cholesterol amount in a dose-dependent manner upon 12 h incubation (Figure 2B). In particular, LHA lowered the cholesterol levels at all the tested dosages, with a decrement of about 50% at 1 μg mL^{-1} or higher concentrations. Our data are in good agreement with the literature for LHA [55–57]. Regarding PHE, treatment with 200 μg of the gum resin extract was able to reduce cholesterol levels in a similar way to 10 μg mL^{-1} of the pure LHA molecule.

The observed effect on cholesterol after treatment with PHE may be correlated to the peculiar phytochemical composition of the gum resin extract. In particular, chromatographic analysis has shown how PHE is a rich source of triterpenes and triterpenoic acids mainly belonging to ursane (**#3**, **#5**, **#8**, and **#10**), oleanane (**#4**, **#6**, **#7**, and **#9**), and tirucallane (**#11**, **#12**, and **#13**) classes. Previous studies have evidenced how cyclic triterpenes may exert several beneficial effects in metabolic disorders, modulating factors involved in glucose, lipid, and cholesterol metabolism. For example, Tang and coworkers demonstrated the decrease of cholesterol and fatty acid biosynthesis by the inhibition of SREBP (sterol regulatory element-binding protein) after interaction of SCAP (SREBP cleavage activating protein) with betulin, a cyclic triterpene having a lupane skeleton very similar to the compounds identified in PHE [58].

Concerning the phytochemicals identified and quantified in PHE, #3 and #4 are two pentacyclic triterpenoids whose cholesterol-lowering activity is well known and already reported in the literature [46]. In particular, a recent in vivo study highlighted the preventive role of #3 in attenuating high fructose diet-induced metabolic syndrome [59]. Additionally, #9 and #10 have been reported to exert in vivo cardiovascular, antihyperlipidemic, and antioxidant effects in a Dahl salt-sensitive rat model [60]. Moreover, a potential mechanism of action involving PPAR-α activation, a key role receptor involved in glucose and lipid homeostasis, has been proposed [61].

2.3. *Triterpenoic Acids of PHE Modulate HMGCR Activity Assuming Similar Poses to LHE in the Active Site of the Enzyme*

Statins regulate cholesterol levels by acting as reversible and competitive inhibitors of the enzyme HMGCR, which is involved in the conversion of acetyl-CoA into mevalonate, a key step that triggers the biosynthetic process of cholesterol metabolism [62]. This action is due to the structural similarity of activated statins, which have a β-hydroxy-acid function that is mistakenly recognized by the enzyme as its natural ligand (β-hydroxy β-methylglutaryl-CoA) [63–66]. The experimental evidence on in vitro biosynthesis showed a decrement of total cholesterol after 12 h of hepatocyte treatment with 200 μg mL^{-1} PHE, similar to 10 μg mL^{-1} LHA (Figure 2B). Therefore, we investigated the possibility that phytochemical content in PHE could exert a cholesterol-lowering activity through a statin-like action mechanism and the modulation of HMGCR enzymatic activity. Consequently, HMGR activity was monitored after treatment with LHA (0.01–100 μg mL^{-1}) or PHE (10–200 μg mL^{-1}) (Figure 3). LHA dampened HMGCR activity in a dose dependent manner. In particular, a complete enzymatic abolition was observed at 100 μg mL^{-1} LHA, with a 50% inhibition between 1–10 μg mL^{-1} LHA. On the other hand, the treatments with PHE exerted a weaker effect, although a dose-dependent inhibition of enzymatic activity was detectable (about 26% reduction at the highest concentration). Kashyap and colleagues, reviewing different patents related to extracts composed of two compounds that we identified in PHE (**#9** and **#10**), have reported inhibitory activity against HMGCR [67], although scientific evidence supporting this hypothesis has not been currently reported in the literature.

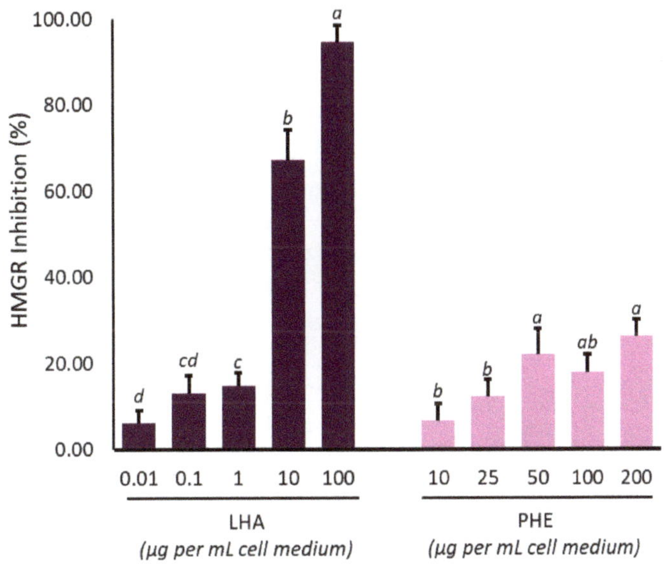

Figure 3. Effect of lovastatin hydroxy acid (LHA) and *Protium heptaphyllum* resin extract (PHE) on HMG-CoA reductase activity. Data are expressed as mean ± SD of three different replicates. Among the two conditions (LHA or PHE), different lowercase letters indicate significant ($p < 0.05$) changes among treatment concentrations, as calculated by one-way ANOVA analysis followed by Tukey's test.

Molecular docking studies on HMGCR were performed with the aim of analyzing the binding properties of the identified compounds in PHE (Table 1, Figure 1) and hypothesizing the binding site with the AAs involved. We started in silico studies by selecting both lovastatin and its activated form (LHA) as known HMGCR inhibitors (Table 2). In the past, similar studies have also been performed on phytochemical compounds present in extracts of *Withania coagulans* fruits [68,69].

Table 2. Molecular docking results of the identified compounds in *Protium heptaphyllum* resin extract (PHE).

#	Compound Name	Docking Score	Prime Energy	IFD Score
	Lovastatin	−6.247	−29124.9	−1462.494
	Lovastatin Hydroxy Acid	−14.667	−29097.0	−1469.500
1	Δ3-carene	−3.220	−29015.2	−1453.979
2	p-cymene	−4.009	−29030.2	−1455.521
3	α-amyrin	−4.637	−29073.5	−1458.313
4	β-amyrin	−4.405	−29079.4	−1458.376
5	α-amyron	−5.196	−29066.6	−1458.528
6	β-amyron	−4.245	−29030.2	−1455.753
7	Brein	−5.307	−29066.2	−1458.666
8	Maniladiol	−5.356	−29151.0	−1462.869
9	Oleanolic acid	−9.835	−29059.2	−1462.795
10	Ursolic acid	−8.008	−29074.3	−1461.721
11	Elemonic acid	−6.786	−29212.4	−1467.408
12	α-elemolic acid	−4.724	−29220.5	−1465.750
13	β-elemolic acid	−6.828	−29166.3	−1465.142

IFD Score: induced fit docking score.

The position of lovastatin in the binding site of HMGCR is shown as the AAs involved are Val^{663}, Ser^{684}, Ser^{661}, Leu^{857}, Cys^{561}, Ala^{856}, Leu^{853}, Asn^{755}, Hid^{752}, and Lys^{692}. However, only Ala^{751}, Lys^{735}, and Asn^{755} played an important role in binding (Figure 4A). On the other hand, when the AAs involved in LHA–HMGCR complex stabilization were analyzed, significant differences were observed with respect to the lovastatin–HMGCR complex (Figure 4B). In particular, Asp^{690}, Ser^{664}, Lys^{735}, Ala^{751}, Lys^{692}, Hid^{752}, Lys^{691}, Glu^{559}, Arg^{590}, Asn^{755}, Leu^{857}, Ser^{661}, and Ser^{684} surrounded the docked structure, and the complex was further stabilized by the formation of six bonds with some AAs residues (Asp^{690}, Ser^{684}, Lys^{735}, Glu^{559}, and Arg^{590}). To be precise, these bonds exclusively involved the β-hydroxy-acid moiety that is present only after in vivo activation of lovastatin.

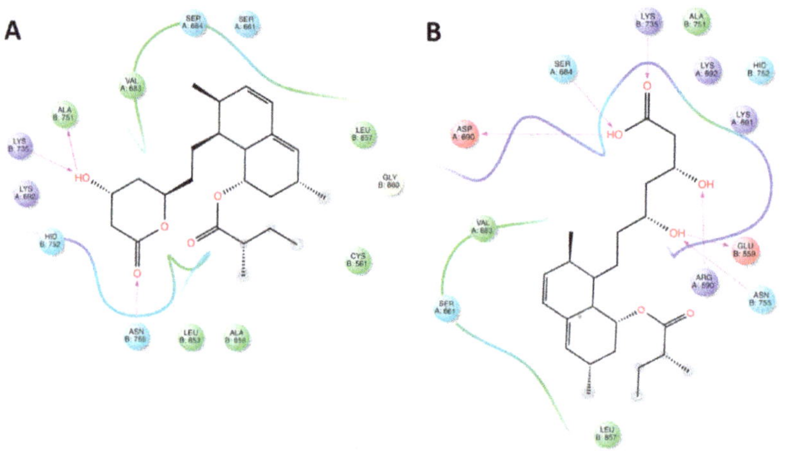

Figure 4. Maps of amino acid (AA) residues involved in the binding of lovastatin (**A**) and lovastatin hydroxy acid (LHA; (**B**)) into 3-hydroxy-3-methyl-glutaryl-coenzyme A reductase (HMGR).

Table 2 also reports the molecular docking results with regard to interactions between HMGCR and several compounds previously identified in PHE (Table 1, Figure 1). Most of the molecules identified in the extract showed low docking score values (IFD score < −1460), indicating a weak interaction with the active site of HMGCR. In particular, #1 and #2 showed low binding affinity to the receptor site. Regarding the compounds having docking scores ranging between −4.5 and −6.0 (#3, #4, #5, #6, #7, #8, and #9), a limited surrounding by AA residues was observed. However, despite #8 recording a docking score equal to −5.356 because of the surrounding of only eight AA residues (Leu^{857}, Val^{683}, Leu^{853}, Arg^{590}, Ala^{856}, Asp^{690}, Lys^{691}, and Asn^{755}), it had an IFD score comparable to lovastatin, thanks to the significative interactions established with Lys^{691} and Asn^{755} (Figure 5).

On the other hand, among the docked structures, only ATs showed docking scores and IFD values indicative of good interaction with the enzyme (Table 2). Among them, the pentacyclic triterpenoids showed the highest docking scores but the lower IFD values with respect to the tetracycles. In particular, the docked structures of #9 and #10 were surrounded by a large number of AA residues (11 surrounded #9 and 13 surrounded #10). Ten of the observed AA residues surrounded both the molecules (Hid^{752}, Lys^{691}, Lys^{692}, Ser^{684}, Asp^{690}, Lys^{735}, Leu^{857}, Arg^{590}, Val^{683}, Leu^{853}), suggesting their important contribution in the stabilizing of both compounds in the active site of HMGCR (Figure 6). Interestingly, nine of them were also found to be involved in the stabilization of LHA inside the binding site (Figure 4B). Concerning the bonds directly involved in the stabilization of the complex, the carboxyl group of the two ATs seems to be as important as in LHA. However, while the acidic function of #9 and #10 made only three bonds (Lys^{692}, Ser^{684}, and Lys^{735}; Figure 6A,C), LHA was further stabilized by three additional bonds (Lys^{692}, Ser^{684}, Asp^{690}, Lys^{735}, Leu^{857}, and Arg^{590}) involving β- and δ-hydroxyl groups (Figure 4B). The differences in binding properties between ATs and LHA may explain the lower docking score values of #9 and #10 with respect to LHA.

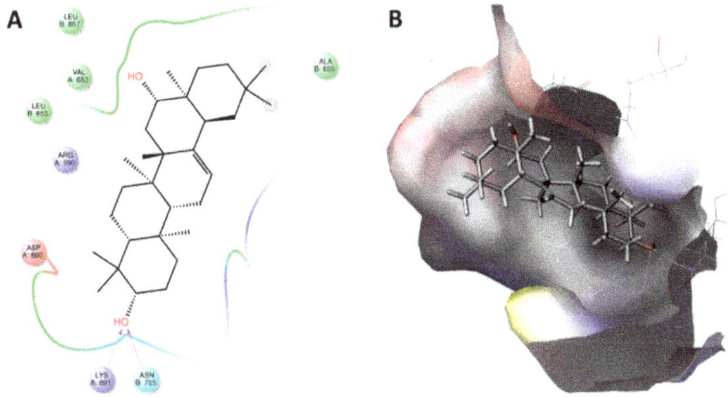

Figure 5. Molecular docking of maniladiol (#8) with the active site of 3-hydroxy-3-methyl-glutaryl-coenzyme A reductase (HMGR). Panel (**A**) shows the amino acid (AA) residues map of the binding. Panel (**B**) shows the 3D docking pose of the molecule into the hydrophobic cavity of HMGCR.

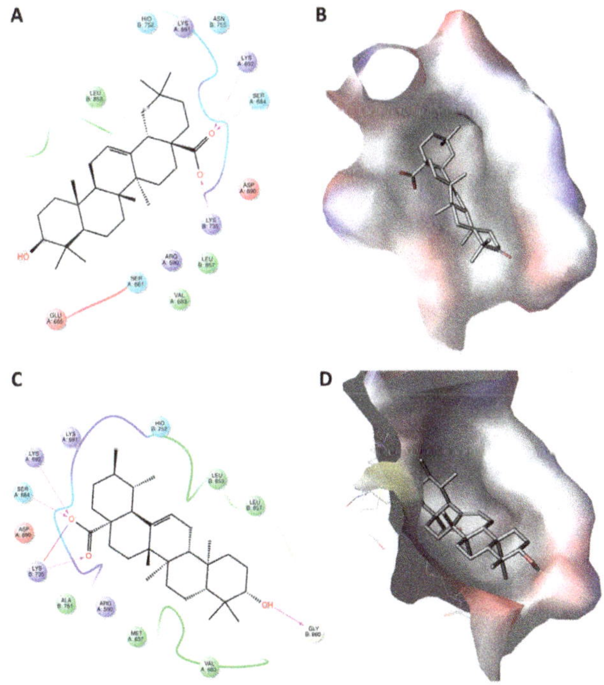

Figure 6. Molecular docking of oleanolic acid (#9) and ursolic acid (#10) with the active site of 3-hydroxy-3-methyl-glutaryl-coenzyme A reductase (HMGR). Panels (**A**,**C**) show the interactions between the amino acid (AA) residues in the active site of HMGR and oleanolic acid and ursolic acid, respectively. Panels (**B**,**D**) show the 3D docking structures of oleanolic acid and ursolic acid, respectively, inserted into the hydrophobic cavity of HMGCR.

Concerning the acidic tetracyclic terpenoids (#11, #12, and #13), despite having low–medium docking scores, they reported the highest IFD scores among the identified compounds in PHE. Moreover, even though they did not reach the IFD values measured for LHA, our data suggest that the three tetracycles were able to better interact with the AA residues present in the HMGR receptor site and, consequently, to exert a comparable in silico inhibitory activity. Differently from the docked structures of #9 and #10, the AAs surrounding #11, #12, and #13 were quite different. In particular, #11 was surrounded by 18 AA residues (Met^{657}, Asp^{767}, Glu^{559}, Gly^{560}, Leu^{562}, Cys^{561}, Ser^{852}, Ser^{565}, Arg^{568}, Ala^{856}, Leu^{857}, Leu^{853}, Arg^{590}, Hid^{752}, Lys^{691}, Asp^{690}, Asn^{755}, and Met^{655}); meanwhile, #13 was surrounded by 16 AAs residues (Ala^{751}, Asp^{690}, Lys^{692}, Ser^{684}, Lys^{735}, Ser^{661}, Val^{683}, Leu^{857}, Glu^{665}, Arg^{590}, Met^{657}, Lys^{691}, Asn^{755}, Hid^{752}, Leu^{853}, and Leu^{562}) (Figure 7). Among them, the same AA residues that seem to be important for the stabilization of LHA in the HMGR binding site through interaction with its β-hydroxy acid function (Figure 4B) were also observed in the docked structures of #11 (Arg^{590}, Hid^{752}, Lys^{691}, Asp^{690}, Asn^{755}, and Met^{655}) and #13 (Asp^{690}, Ala^{751}, Lys^{692}, Ser^{684}, and Lys^{735}). This last aspect may explain the higher values recorded for tetracycles with respect to pentacycles.

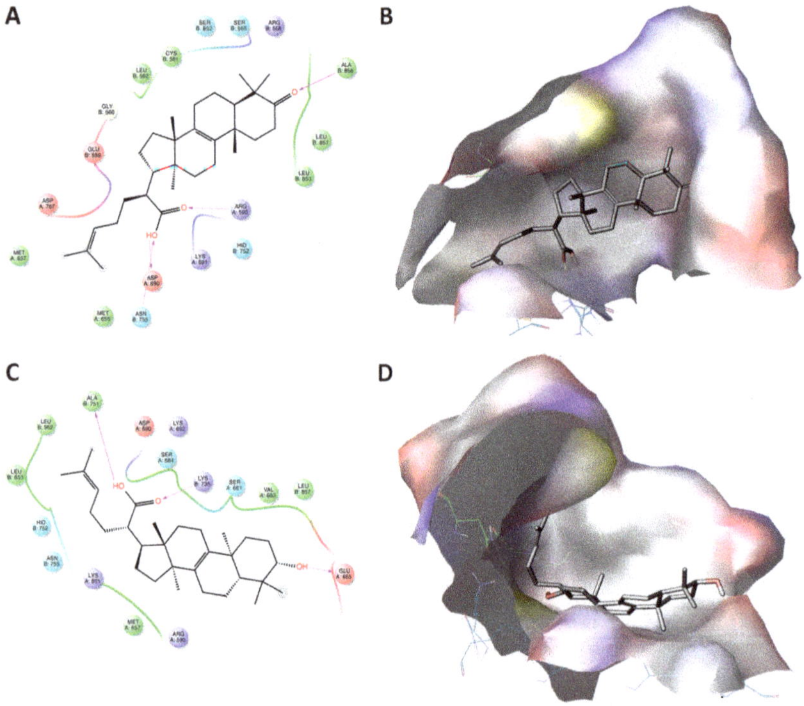

Figure 7. Molecular docking of elemonic acid (#11) and β-elemolic acid (#13) with the active site of 3-hydroxy-3-methyl-glutaryl-coenzyme A reductase (HMGR). Panels (**A**,**C**) show the interactions between the amino acid (AA) residues in the active site of HMGR and elemonic acid and β-elemolic acid, respectively. Panels (**B**,**D**) show the 3D docking structures of elemonic acid and β-elemolic acid, respectively, inserted into the hydrophobic cavity of HMGCR.

2.4. PHE Modulates the Expression of Cholesterol-Related Genes

We demonstrated that cell exposure to 200 μg mL^{-1} PHE for 12 h was able to reduce cholesterol production in a comparable way to 10 μg mL^{-1} lovastatin (Figure 2). On the other hand, results from enzymatic assay showed that 200 μg mL^{-1} of PHE produced about 20% inhibition of HMGCR activity; meanwhile, the same enzyme was inhibited by more than 90% by lovastatin at 10 μg mL^{-1} (Figure 3), as previously reported for statins [70,71]. These results suggested differences in the action mechanism of PHE in reducing cholesterol synthesis with respect to lovastatin. In particular, the effects on cholesterol production may involve mechanisms other than post-translational ones. This hypothesis is also supported by molecular docking studies, in which weaker interactions between the identified compounds in PHE (Table 1, Figure 1) and HMGCR were highlighted when compared with the interaction established with statins (Table 2).

In the past, the ability of plant bioactive compounds to regulate the expression of target genes has been largely demonstrated [72,73]. In particular, it has been reported that gene expression may be modulated by phytochemicals via both genetic and epigenetic mechanisms [74,75]. In particular, phytochemical compounds may affect gene expression via the modulation of several biological targets, such as DNA, lipid rafts, and transcriptional factors, directly binding them or indirectly affecting the cellular redox state [75]. In this work, we investigated whereas the observed effect related to the decrease of cholesterol level after PHE treatment could be linked to changes in gene expression of *HMGCR* via qRT-PCR (quantitative Reverse Transcription Polymerase Chain Reaction) analysis. Cells exposed to PHE (1–50 μg mL^{-1}) displayed a dose-dependent downregulation of *HMGCR*. In particular, 10 μg mL^{-1} PHE was able to inhibit 50% *HMGCR* gene expression (Figures 8 and 9). This result is in contrast with the effect reported for statins. Indeed, it is well known in the literature that statins, including lovastatin, interfere with the negative feedback of cholesterol on the transcription of the *HMGCR* gene, resulting in its upregulation [76]. Upregulation of the *HMGCR* gene is also a strong limiting factor for the pharmacological effect of statins [7,76].

In addition to de novo synthesis, cholesterol homeostasis depends on other processes, including cellular uptake, intestinal absorption, and metabolic turnover in biliary salts [77]. With the aim of evaluating if PHE may influence other important processes for cholesterol homeostasis, we evaluated the effects derived by cell exposure to PHE on the gene expression of cholesterol-metabolism-related genes.

LDLRs (low-density lipoprotein receptor) play a key role in regulating LDL-C levels in the blood by influencing the clearance of LDL-C from circulation. The gene expression of the receptor is regulated at both transcriptional and post-translational levels [78]. Meanwhile, transcriptional regulation is dependent on the availability of intracellular cholesterol; post-translational mechanisms depend on the action of PCSK9 (Proprotein Convertase Subtilisin/Kexin type 9) and IDOL (Inducible Degrader of Low-density Lipoprotein Receptor) [78,79]. These two proteins are synthesized in hepatocytes and secreted into plasma. Here, PCSK9 binds LDLR in an extracellular site, interfering with receptor in-membrane recycling, while IDOL marks LDLR with ubiquitin, leading to its degradation by lysosomes [80]. Consequently, high plasmatic levels of PCSK9 and IDOL reduce the amount of LDLR in the hepatocyte membrane, resulting in higher levels of plasmatic LDL-C. Therefore, the discovery of new agents that may inhibit *PCSK9* and *IDOL* expression is a useful tool for the treatment of hypercholesterolemia. This finding is also supported by the role of new anti-PCSK9 monoclonal antibody drugs such as alirocumab and evolocumab [1]. With the aim of evaluating whether PHE could contribute to the regulation of LDLR levels through transcriptional mechanisms, we checked the effects of the treatment on the expression of *LDLR*, *PCSK9*, and *IDOL* genes. Surprisingly, the *LDLR* gene was downregulated (Figure 8). On the other hand, under our experimental conditions, we recorded a reduction in the expression of both *PCSK9* and *IDOL* after treatment with PHE in the tested concentration range (Figure 8). In particular, 80% of *IDOL* and *PCSK9* inhibition was obtained at 50 and 10 μg mL^{-1} PHE, respectively. These results suggest a stabilization of LDLR through post-translational mechanisms.

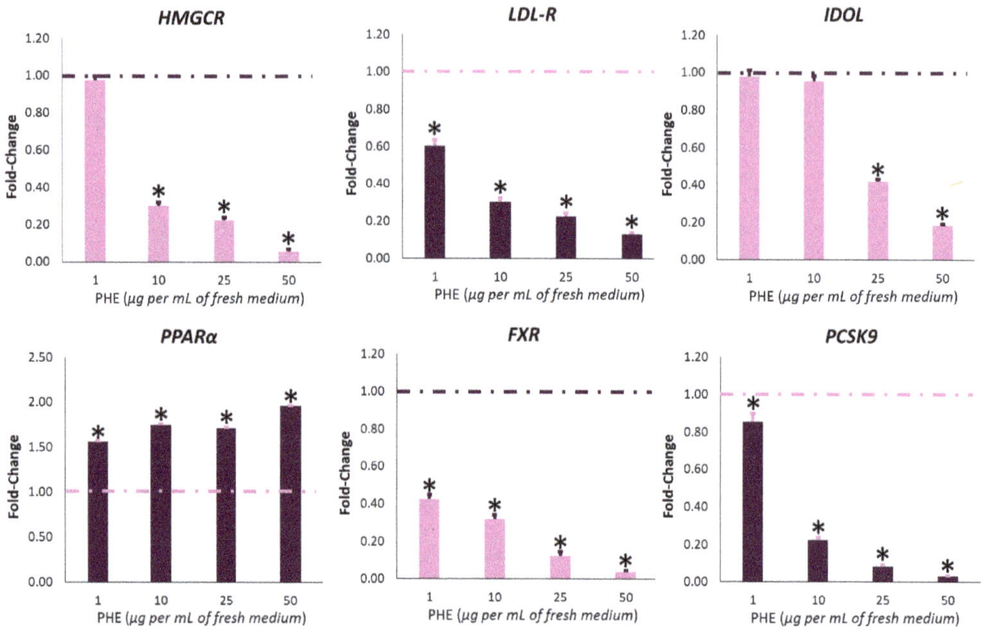

Figure 8. Effect of *Protium heptaphyllum* resin extract (PHE) on gene expression of *HMGCR, LDL-R, IDOL, PPARα, FXR*, and *PCSK9* on HepG2 cells. After the seeding, cells were treated for 6 h with PHE. Untreated cells were used as control (dashed line). Bars represent the mean ± SD of three qRT-PCR analyses. Values are expressed as fold change with respect to the gene expression of control cells. The symbol "*", when present, indicates significant ($p < 0.05$) differences between treated and control samples, as calculated by the Mann–Whitney test. Representative heat map analysis coupled with hierarchical cluster analysis is reported in Figure 9. *HMGCR*: 3-hydroxy-3-methyl-glutaryl-coenzyme A reductase; *FXR*: farnesoid X receptor; *LDLR*: low-density lipoprotein receptor; *IDOL*: inducible degrader of low-density lipoprotein receptor; *PCSK9*: proprotein convertase subtilisin/kexin type 9; *PPARα*: peroxisome proliferator-activated receptor α.

Since high levels of intracellular cholesterol induce the inhibition of the *LDLR* gene, it has been reported that statins indirectly induce its gene expression [81–83]. *PCSK9* is also regulated by intracellular sterol levels. However, despite low intracellular cholesterol levels inhibiting *PCSK9* gene expression, statins have been shown to induce its expression [81–83]. This process causes an attenuation of the lipid-lowering effects of statins. Our results, with respect to the observed statin's gene expression modulation, demonstrate that the phytocomponents present in PHE have an opposite effect on transcriptional and post-transductional regulation of membrane *LDLR*.

To compensate for the fraction that escapes enterohepatic circulation and is removed by feces, bile salts are de novo synthesized from cholesterol in hepatocytes [84]. This conversion represents the major route for the removal of cholesterol from the body, and it is well known that an increased turnover of bile salts is able to reduce LDL-C [84]. Bile salts regulate their own synthesis in a negative feedback loop by activating *FXR* (Farnesoid X Receptor). As a consequence of the activation, *FXR* reduces the expression of cholesterol-7α-hydroxylase (*CYP7A1*), the first and main rate-controlling enzyme in the pathway of bile salt synthesis [85]. Cell exposure to PHE determined a strong downregulation of *FXR* under all tested concentration ranges (Figure 8). In particular, 60% of its inhibition was already observed at the lowest PHE concentration (1 μg mL^{-1}). Since FXR is considered the major regulator of bile salt synthesis, our result suggests that PHE may strongly increase cholesterol turnover.

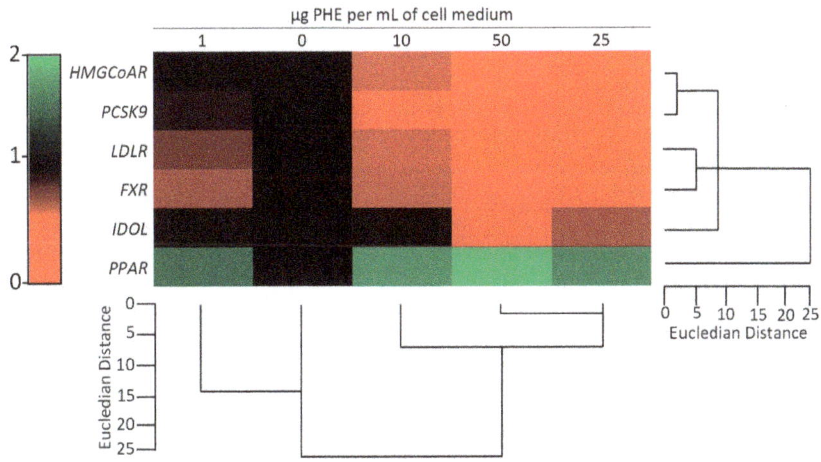

Figure 9. Hierarchical clustering analysis and heatmap visualization of the gene expression of *HMGCR*, *LDL-R*, *IDOL*, *PPARα*, *FXR*, and *PCSK9* on HepG2 cells evaluated via qRT-PCR analysis. For each row, diverse colors indicate differences in gene expression for each gene among the different treatments. *HMGCR*: 3-hydroxy-3-methyl-glutaryl-coenzyme A reductase; *FXR*: farnesoid X receptor; *LDLR*: low-density lipoprotein receptor; *IDOL*: inducible degrader of low-density lipoprotein receptor; *PCSK9*: proprotein convertase subtilisin/kexin type 9; *PPARα*: peroxisome proliferator-activated receptor α.

Increased activity of *CYP7A1* is also associated with the activation of nuclear receptor PPARα (Peroxisome Proliferator-Activated Receptor α), which is principally involved in lipid metabolism regulation. Concerning cholesterol metabolism, PPARα activation intensely downregulates the expression of *HMGCR* [86], promotes cholesterol conversion into bile acids [87], and influences the absorption of cholesterol in the intestinal tract [88]. Consequently, the clinical use of PPARα agonists improves the overall plasmatic lipid profile and reduces cardiovascular risks [89]. In our experimental conditions, cell exposure to PHE increased the gene expression of PPARα. Surprisingly, already at the lowest PHE concentration, an upregulation of the transcript was observed (Figure 8).

3. Materials and Methods

3.1. Plant Material and Extract Preparation

Protium heptaphyllum oleum-resin was collected in Brazil in April 2019. The bioactive compounds present in the raw material were extracted by Abel Nutraceuticals (Turin, Italy; Extract Batch #P75-2-0) through a patent-pending hydroalcoholic extraction process targeted to increase the concentration of acidic triterpenes present in the resin. The form of the extract provided for this study (which is branded as Hepamyr® by the company in its final commercial powder form) was the solid crude extract of the resin obtained after extraction, purification, and complete solvent removal. For both high-pressure liquid chromatography (HPLC) analysis and biochemical assays, the dried extract was completely resuspended in ethanol (Sigma-Aldrich, Berlin, Germany, EU); meanwhile, diethyl ether (Sigma-Aldrich, Germany) was used for gas-chromatographic (GC) analysis sample preparation.

3.2. Chemical Characterization Via HPLC-APCI-HRMS^2

Semi-untargeted qualitative and quantitative analyses were performed by an HPLC system (Dionex ultimate 3000 HPLC, ThermoFisher Scientific, Waltham, MA, USA) coupled via atmospheric-pressure chemical ionization (APCI) to a high-resolution tandem mass spectrometry instrument (HRMS^2; LTQ Orbitrap, ThermoFisher Scientific, Waltham, MA, USA). Separation was carried out with Luna C18(2) 150 × 2 mm, 100 Å, 3 µm (Phenomenex,

Torrance, CA, USA) using 0.1% (*v*/*v*) formic acid (solvent A) and 0.1% (*v*/*v*) acetonitrile (solvent B) as mobile phases. Chromatographic separation consisted of a solvent ramp from 5% to 100% solvent B for 30 min, followed by column reconditioning of 15 min. The flow rate was set to 0.2 mL min^{-1} and the injection volume to 10 µL. Detection parameters were the following: negative ionization mode; capillary temperature: 250 °C; APCI vaporizer temperature: 450 °C; sheath gas: 35 Arb; auxiliary gas: 15 Arb; discharge needle: 5 kV; the acquisition was carried out in Dependent Scan mode with a mass range from 220 to 1000 *m*/*z*, normalized CE: 35V. Raw data obtained were analyzed with Thermo Xcalibur software (ThermoFisher Scientific, Waltham, MA, USA); molecules were identified using the MetFrag online tool [90]. Quantification of identified acidic triterpenes was performed with a calibration curve of moronic acid (TCI-Europe, Bruxelles, Belgium, EU).

3.3. Chemical Characterization via GC-MS and GC-FID

Volatile compounds (VOC) were profiled via GC coupled with a mass spectrometer (MS) and quantified by GC coupled with a flame ionization detector (FID; GCMS-QP2010 SE, Shimadzu, Japan). Qualitative and quantitative analyses were performed using an Rtx-5MS column (30 m; 0.25 mm ID; 0.25 µm film thickness; Restek, Milan, Italy, EU). The injection port was in split mode (split ratio 1:5) for GC-FID analysis or in splitless mode for GC-MS analysis. The temperature of the injection port was kept at 280 °C. The carrier gas was helium, with a constant flow of 1 mL min^{-1}. The temperature gradient was as follows: initial temperature 50 °C, then a 3 °C min^{-1} ramp-up to 140 °C and a 12 °C min^{-1} ramp-up to 320 °C. The final temperature was held for 10 min. MS detector conditions were as follows: ionization energy 70 eV, ion source 200 °C, and quadrupole 150 °C; the acquisition was in scan mode (scan range 50–650 *m*/*z*). VOC-targeted identification and quantitation were achieved with a VOC terpene analytical-standard mix (Cannabis Terpene Mix B; CRM40937; Sigma-Aldrich, USA). Meanwhile, the identification was carried out with the NIST database and the FFNSC3 Shimadzu mass spectra library. Identification and quantification of not acid triterpenes (NATs) were carried by GC-MS (TRACE 1310 coupled to TSQ Quantum Ultra, Thermo Scientific, USA). The detection was led in full-mass mode (50–450 *m*/*z* scan range), and the identification was performed by the NIST database. The injection volume was 0.5 µL, and the injector was a PTV in splitless mode at constant temperature (280 °C). The carrier gas was He (1.2 mL min^{-1}), and a DB-1 column (30 m × 0.53 mm ID × 5 µm film thickness) was assembled on the GC instrument. The temperature gradient was 150 °C initial temperature, 150 °C at 4 min, 320 °C at 12.5 min, 320 °C at 20 min. For the quantification, the extracted ions were the following: 218, 203, 426, 189 *m*/*z*. Quantification of identified triterpenes was performed with a calibration curve of α- and β-amyrin (TCI-Europe, Belgium).

3.4. Determination of Cholesterol Levels

Cholesterol production was evaluated from THLE-3 ATCC® CRL-11233 cells (American Type Culture Collection ATCC, USA). THLE-3 cells were derived from primary normal liver cells by infection with SV40 large T-antigens. Cells were cultured in bronchial epithelial cell growth medium (BEGM; Lonza/Clonetics Corporation, Walkersville, MD, USA) in 5% CO_2 at 37 °C. The culture flasks used for the experimentation were precoated with a mixture containing 0.01 mg mL^{-1} fibronectin, 0.03 mg mL^{-1} bovine collagen type I, and 0.01 mg mL^{-1} bovine serum albumin dissolved in BEBM medium.

To evaluate cholesterol production by THLE-3 cells, a cholesterol/cholesteryl ester assay kit (Abcam, Cambridge, United Kingdom, EU) was employed. Fluorescent determination was performed following the instructions provided by the manufacturer. Briefly, cells were seeded on 6-well coated plate, and 48 h after seeding, they were treated with the active carboxylate form of lovastatin (CAS 75225-50-2, Enzo Life Sciences, Inc., Farmingdale, NY, USA) at concentrations ranging 0.01–10 µg mL^{-1} in cell medium, PHE (25–200 µg mL^{-1} in cell medium), or simply with BEGM. Cells were treated for 6 and 12 h. After incubation time, the cell monolayer was washed 5 times with phosphate-buffered saline (PBS). Lipid extraction was performed by using 200 μL of 7:11:0.1 (*v*/*v*/*v*) chloroform:isopropanol:NP40 solution. The extracts were then centrifuged for 10 min at 15,000× *g*. Samples were desiccated in a vacuum chamber, and then dried lipids were dissolved in assay buffer, treated with cholesterol reaction mix, and finally incubated for 60 min at 37 °C. Fluorescence was recorded using an F5 FilterMax microplate reader (Molecular Devices, San Jose, CA, USA) set at 535 nm for excitement and at 587 nm for emission.

3.5. Evaluation of HMGCR Activity

In order to evaluate the effect of PHE on HMGCR activity, the Colorimetric HMG-CoA Reductase Activity Assay Kit (Abcam, UK) was employed. This assay is based on the consumption of NADPH by the enzyme, which can be consequently measured by the decrease of the absorbance at 340 nm. The assay was performed following the instructions provided by the manufacturer, testing different concentrations of LHA (100–0.01 μg mL^{-1}) as a positive control and PHE (10–200 μg mL^{-1}). The absorbance was measured using an F5 FilterMax microplate reader (Molecular Devices, San Jose, CA, USA) in Kinect mode for 30 min every 2 min at 37 °C.

3.6. Molecular Docking Studies and Validation on HMGCR

3.6.1. Ligand and Protein Preparation

Before in silico studies, both ligands and the protein–ligand complex were prepared as described below. The ligands were optimized for molecular docking using the default setting of the LigPrep Tool implemented in Schrödinger's software (v2017-1) [91]. All the potential tautomer and stereoisomer combinations were also generated for the biologically relevant pH (pH = 7.0 ± 0.4) using the Epik ionization method [92]. Moreover, energy minimization was done using the integrated OPLS 2005 force field [93]. Regarding the protein, the high-resolution crystal structure of the complex related to the catalytic portion of human HMGCR and lovastatin was downloaded from the Protein Databank (PDB ID: 1HWK) [62,94]. The Protein Preparation Wizard of Schrödinger's software was employed for protein structure preparation using the condition previously described [95]. Bond orders, hydrogen atoms, and the protonation of the heteroatom states were assigned using the Epik-tool set at biologically relevant pH values (pH = 7.0 ± 2.0). Additionally, the H-bond network was adjusted and optimized. Finally, a restrained energy minimization step (RMSD of the atom displacement for terminating the minimization set at 0.3 Å) was done using the Optimized Potentials for Liquid Simulations (OPLS)-2005 force field on the obtained structure [93].

3.6.2. Molecular Docking

Molecular docking studies were performed using the Glide program [96–98]. The receptor grid was prepared by assigning lovastatin as the centroid of the grid box. The 3D structures of the conformers previously generated were then docked into the receptor model using the Standard Precision mode as the scoring function. Five different poses for each ligand conformer were included in the postdocking minimization step, and at least two docking poses for each ligand conformer were generated. The proposed docking procedure was able to redock the crystallized lovastatin within the receptor-binding pocket with RMSD < 0.51 Å. An IFD application is an accurate and robust method aimed at accounting for both ligand and receptor flexibility [99], and it was performed with the aim of inducing fit docking simulations [91,100] using the Schrödinger software suite [101,102]. The atomic coordinates for c-Kit were downloaded from the Protein Databank (PDB id 1HWK) and submitted to the Protein Preparation Wizard module in Schrödinger in order to add hydrogen, assign partial charges (using the OPLS-2001 force field), confer protonation states, and remove crystal waters. The IFD protocol was carried out as previously reported [103,104]. Briefly, the ligands were docked into the rigid receptor model with scaled-down van der Waals (vdW) radii. The Glide Standard Precision (SP) mode was used for the docking, and the different ligand poses were retained for protein structural refinements [97,98]. In order to include all amino acid (AA) residues within the dimensions of 25 × 25 × 25 Å, the docking boxes were defined from the center of the original ligands [105,106]. The induced-fit protein–ligand complexes were produced using Prime software [105,106]. The previously determined ligand poses, including all residues with at least one atom located within 5.0 Å, were submitted to both side chain and backbone refinements. All poses generated were then hierarchically classified, refined, and further minimized into the active site grid before being finally scored using the proprietary GlideScore function, defined as Equation (1):

$$GScore = 0.065\ x\ vdW + 0.30\ x\ Coul + Lipo + Hbond + Metal + BuryP + RotB + Site \quad (1)$$

where *vdW* is the energy term for van der Waals interactions; *Coul* is the Coulomb energy; *Lipo* is the lipophilic contact term that refers to potential favorable hydrophobic interactions; *Hbond* is the H-bonding term; *Metal* is the metal-binding term; *BuryP* is a penalty term assigned to buried polar groups; *RotB* is a penalty factor assigned for freezing rotatable bonds; *Site* is a term used to describe favorable polar interactions in the active site.

Finally, in order to account for both protein–ligand interaction energy and total energy of the system, the *IFD score* was calculated according to Equation (2):

$$IFD\ score = 1.0\ Glide\_Gscore + 0.05\ Prime\_Energy \quad (2)$$

As a result of this equation, the compounds were ranked for its *IFD score*, considering that the most negative *IFD score* is conferred to the most favorable binding.

3.7. Gene Expression via qRT-PCR

Gene expression was evaluated on a HepG2 cancer cell line (American Type Culture Collection ATCC, Manassas, VA, USA). The cells were cultured in RPMI supplemented with 5% (v/v) FBS, 2 mM L-glutamine, 50 IU mL^{-1} penicillin, and 50 µg mL^{-1} streptomycin, and they were grown in 75 cm^2 culture flasks under humidity (85–95% H_2O), temperature (36.5–37.3 °C), and CO_2 (4.9–5.1%) controlled conditions [22].

When the cells reached about 80% confluence, they were collected using a gamma-irradiated solution containing 0.12% trypsin and 0.02% EDTA in Dulbecco's phosphate-buffered saline (133.00 mg L^{-1} $CaCl_2$, 100.00 mg L^{-1} $MgCl_2$, 200.00 mg L^{-1} KCl, 200.00 mg L^{-1} KH_2PO_4, 8000 mg L^{-1} NaCl, and 1143.56 mg L^{-1} Na_2HPO_4) without phenol red [22]. The cells were seeded in 24-multiwell plates at a density equal to 5 × 10^5 cells/well. After 24 h, they were treated for 8 h with PHE (1–50 µg mL^{-1} cell medium) in fresh FBS-free RPMI, and total cellular RNA was isolated after their lysis using a commercial kit (GenElute™ Direct mRNA Miniprep Kits, Sigma-Aldrich, USA) [72]. For each sample, the same amount of total RNA (one microgram) was reverse-transcribed using a High-Capacity cDNA Reverse Transcription Kit (Applied Biosystems, USA) and oligo (dT), following the protocol provided by the manufacturer [107]. qMAXSen™ Green qPCR MasterMix (Low ROX™; Canvan, Australia) was employed to analyze the obtained cDNA for quantitative real-time PCR (qRT-PCR) using a QuantStudio™ 3 Real-Time PCR System (Thermofisher, USA) [108]. The qRT-PCR was performed as previously described [72], using the primers reported in Table 3. Finally, relative expression was calculated according to the Pfaffl method [109] using Equation (3):

$$Fold\ Change = \frac{(E_{GOI})^{\Delta Ct\ GOI}}{(E_{HKG})^{\Delta Ct\ HKG}} \quad (3)$$

where *E* is the primer efficiency; *GOI* is the gene of interest; *HKG* is the housekeeping gene; *Ct* is the cycle threshold; Δ*Ct* is the difference between the control average and average *Ct*.

Table 3. Genes employed for qRT-PCR analysis. For each gene, the name, acronym, and nucleotide sequences for both Forward (F) and Reverse (R) primers are reported in the table.

Gene Acronym		Sequence (5′−3′)	Accession Number
HMGCR	F R	TGATTGACCTTTCCAGAGCAAG CTAAAATTGCCATTCCACGAGC	NM_000859.2
FXR	F R	TCTCCTGGGTCGCCTGACT ACTGCACGTCCCAGATTTCAC	NM_005123
LDLR	F R	AGTTGGCTGCGTTAATGTGA TGATGGGTTCATCTGACCAGT	NM_000527.4
IDOL	F R	AAGTTCTTCGTGGAGCCTCA ACTGAGTTCCACTGCCTGCT	NM_013262
PCSK9	F R	GCTGAGCTGCTCCAGTTTCT AATGGCGTAGACACCCTCAC	NM_174936.3
PPARα	F R	CTGGAAGCTTTGGCTTTACG GTTGTGTGACATCCCGACAG	NM_005036.4
βACT	F R	CGGGAAATCGTGCGTGACAT GGACTCCATGCCCAGGAAGG	NM_001101.3

HMGCR: 3-hydroxy-3-methyl-glutaryl-coenzyme A reductase; *FXR*: farnesoid X receptor; *LDLR*: low-density lipoprotein receptor; *IDOL*: inducible degrader of low-density lipoprotein receptor; *PCSK9*: proprotein convertase subtilisin/kexin type 9; *PPARα*: peroxisome proliferator-activated receptor α; *βACT*: β-actin.

3.8. Statistical Analysis

For each experiment, at least three different replicates were performed. The results are expressed as mean values ± standard deviation (SD). Significative differences among the sample were evaluated by the Mann–Whitney test or one-way ANOVA, followed by Tukey's test. The Mann–Whitney test was employed to analyze data related to cholesterol release and gene expression experiments, in which a direct comparison between control and treatment was required. One-way ANOVA, followed by Tukey's test, was employed to evaluate what different concentrations of treatment (LHE or PHE) could give in terms of HMGCR activity. Hierarchical cluster analysis, coupled with heat map visualization, was generated using SPPSS software (v24).

4. Conclusions

Our results provided information on the functional properties of *Protium heptaphyllum* gum resin extract. Specifically, we have shown that PHE is a rich source of triterpenoic compounds, mainly belonging to the ursane, oleanane, and tirucallane classes, which are limitedly distributed in nature. Moreover, our observations suggest that *Protium heptaphyllum* gum resin has a specific effect on cholesterol metabolism. Therefore, PHE may be a useful alternative to improve dyslipidemia and its clinical complication, such as atherosclerosis and CVDs. For this reason, *Protium heptaphyllum* may become an interesting raw material for the nutraceutical industry in addition to gaining acceptance as a source of health-promoting compounds. Finally, the obtained results may boost the production demand of this resin, improving the income of local rainforest communities. This, together with the application of innovative and sustainable models in the production chain, based on the use of vegetable nontimber forest products, are the base concept of the circular economy [110].

5. Patents

A patent related to a part of this work is pending (application number 102020000015598).

Author Contributions: Conceptualization, P.I., A.O., A.C., and C.G.; data curation, G.M., T.G., A.L., A.A., M.N., C.M., L.M., and C.G.; formal analysis, G.M., T.G., A.L., A.A., G.C., C.G., G.S., and M.N.; funding acquisition, C.M., L.M., and C.G.; investigation, G.M., T.G., A.L., and C.G.; methodology, G.M., A.L., A.A., M.N., C.M., L.M., and C.G.; project administration, P.I., A.P., A.C., and C.G.; resources, P.I., A.L., A.P., M.N., A.O., A.C., C.M., L.M., and C.G.; software, G.M., A.L., and A.A.; supervision, A.O., A.C., C.M., L.M., and C.G.; validation, G.M., T.G., A.A., M.N., C.M., L.M., and C.G.; visualization, G.M., P.I., and A.L.; writing—original draft, G.M., T.G., A.L., A.A., and C.G.; writing—review and editing, G.M., P.I., A.L., A.P., A.O., A.C., C.M., L.M., and C.G. All authors have read and agreed to the published version of the manuscript.

Funding: This research was supported partly by the Ph.D Program in Chemical and Material Sciences of the University of Turin (XXXVI cycle) and partly by the companies Abel Nutraceuticals S.R.L. (Turin, Italy) and Biosfered S.R.L. (Turin, Italy).

Institutional Review Board Statement: Not applicable.

Informed Consent Statement: Not applicable.

Data Availability Statement: The data presented in this study are available on request from the corresponding author. The data are not publicly available because a patent related to a part of this work is pending (application number 102020000015598).

Acknowledgments: The authors are grateful to Regione Piemonte for supporting the activity of the Ph.D. Program in the apprenticeship of A.A. Moreover, the University of Turin and the University of Palermo are thankful to Abel Nutraceuticals S.R.L. for providing sample materials free of charge.

Conflicts of Interest: P.I. was employed by Biosfered S.R.L.; A.P., A.O., and A.C. were employed by Abel Nutraceuticals S.R.L.; A.A. was a Ph.D student of the Ph.D Program in Chemical and Material Sciences of the University of Turin (XXXVI cycle) and was an apprentice at Biosfered S.R.L. The remaining authors declare that the research was conducted in the absence of any commercial or financial relationships that could be construed as a potential conflict of interest. The funders had no role in the design of the study, in the collection, analyses, or interpretation of data, in the writing of the manuscript, or in the decision to publish the results.

References

1. Raal, F.J.; Hovingh, G.K.; Catapano, A.L. Familial hypercholesterolemia treatments: Guidelines and new therapies. *Atherosclerosis* **2018**, *277*, 483–492. [CrossRef]
2. Santos, R.D.; Gidding, S.S.; Hegele, R.A.; Cuchel, M.A.; Barter, P.J.; Watts, G.F.; Baum, S.J.; Catapano, A.L.; Chapman, M.J.; Defesche, J.C. Defining severe familial hypercholesterolaemia and the implications for clinical management: A consensus statement from the International Atherosclerosis Society Severe Familial Hypercholesterolemia Panel. *Lancet Diabetes Endocrinol.* **2016**, *4*, 850–861. [CrossRef]
3. Rosso, A.; Pitini, E.; D'Andrea, E.; Massimi, A.; De Vito, C.; Marzuillo, C.; Villari, P. The cost-effectiveness of genetic screening for familial hypercholesterolemia: A systematic review. *Ann. Ig.* **2017**, *29*, 464–480. [PubMed]
4. Jiang, J.; Wu, C.; Zhang, C.; Zhao, J.; Yu, L.; Zhang, H.; Narbad, A.; Chen, W.; Zhai, Q. Effects of probiotic supplementation on cardiovascular risk factors in hypercholesterolemia: A systematic review and meta-analysis of randomized clinical trial. *J. Funct. Foods* **2020**, *74*, 104177. [CrossRef]
5. Yuningrum, H.; Rahmuniyati, M.E.; Sumiratsi, N.N.R. Consumption of Fried Foods as A Risk Factor for Hypercholesterolemia: Study of Eating Habits in Public Health Students. *JHE J. Health Educ.* **2020**, *5*, 78–85. [CrossRef]
6. Al-Muzafar, H.M.; Amin, K.A. Efficacy of functional foods mixture in improving hypercholesterolemia, inflammatory and endothelial dysfunction biomarkers-induced by high cholesterol diet. *Lipids Health Dis.* **2017**, *16*, 194. [CrossRef]
7. Stancu, C.; Sima, A. Statins: Mechanism of action and effects. *J. Cell. Mol. Med.* **2001**, *5*, 378–387. [CrossRef]
8. Reiner, Ž. Management of patients with familial hypercholesterolaemia. *Nat. Rev. Cardiol.* **2015**, *12*, 565. [CrossRef]
9. Joy, T.R.; Hegele, R.A. Narrative review: Statin-related myopathy. *Ann. Intern. Med.* **2009**, *150*, 858–868. [CrossRef]
10. Joy, T.R.; Hegele, R.A.; Hilton-Jones, D. Statin-related myopathies. *Pract. Neurol.* **2018**, *18*, 97–105.
11. Mannino, G.; Di Stefano, V.; Lauria, A.; Pitonzo, R.; Gentile, C. Vaccinium macrocarpon (Cranberry)-Based Dietary Supplements: Variation in Mass Uniformity, Proanthocyanidin Dosage and Anthocyanin Profile Demonstrates Quality Control Standard Needed. *Nutrients* **2020**, *12*, 992. [CrossRef] [PubMed]
12. Gordon, R.Y.; Cooperman, T.; Obermeyer, W.; Becker, D.J. Marked variability of monacolin levels in commercial red yeast rice products: Buyer beware! *Arch. Intern. Med.* **2010**, *170*, 1722–1727. [CrossRef]
13. Cicero, A.F.G.; Fogacci, F.; Banach, M. Red yeast rice for hypercholesterolemia. *Methodist Debakey Cardiovasc. J.* **2019**, *15*, 192.

14. Gerards, M.C.; Terlou, R.J.; Yu, H.; Koks, C.H.W.; Gerdes, V.E.A. Traditional Chinese lipid-lowering agent red yeast rice results in significant LDL reduction but safety is uncertain–a systematic review and meta-analysis. *Atherosclerosis* **2015**, *240*, 415–423. [CrossRef] [PubMed]
15. Endo, A.; Kuroda, M. Citrinin, an inhibitor of cholesterol synthesis. *J. Antibiot.* **1976**, *29*, 841–843. [CrossRef] [PubMed]
16. Ying, J.; Du, L.-D.; Du, G.-H. Lovastatin. In *Natural Small Molecule Drugs from Plants*; Springer: Berlin, Germany, 2018; pp. 93–99.
17. Pasha, M.K.; Muzeeb, S.; Basha, S.J.S.; Shashikumar, D.; Mullangi, R.; Srinivas, N.R. Analysis of five HMG-CoA reductase inhibitors—atorvastatin, lovastatin, pravastatin, rosuvastatin and simvastatin: Pharmacological, pharmacokinetic and analytical overview and development of a new method for use in pharmaceutical formulations analysis and. *Biomed. Chromatogr.* **2006**, *20*, 282–293. [CrossRef]
18. De Oliveira Filho, J.W.G.; Islam, M.T.; Ali, E.S.; Uddin, S.J.; de Oliveira Santos, J.V.; de Alencar, M.V.O.B.; Júnior, A.L.G.; Paz, M.F.C.J.; de Brito, M.D.R.M.; de Castroe e Sousa, J.M.; et al. A comprehensive review on biological properties of citrinin. *Food Chem. Toxicol.* **2017**, *110*, 130–141. [CrossRef] [PubMed]
19. Thompson, P.D.; Panza, G.; Zaleski, A.; Taylor, B. Statin-associated side effects. *J. Am. Coll. Cardiol.* **2016**, *67*, 2395–2410. [CrossRef] [PubMed]
20. Flajs, D.; Peraica, M. Toxicological properties of citrinin. *Arch. Ind. Hyg. Toxicol.* **2009**, *60*, 457–464. [CrossRef]
21. Gentile, C.; Mannino, G.; Palazzolo, E.; Gianguzzi, G.; Perrone, A.; Serio, G.; Farina, V. Pomological, Sensorial, Nutritional and Nutraceutical Profile of Seven Cultivars of Cherimoya (*Annona cherimola* Mill). *Foods* **2021**, *10*, 35. [CrossRef] [PubMed]
22. Mannino, G.; Gentile, C.; Porcu, A.; Agliassa, C.; Caradonna, F.; Bertea, C.M. Chemical Profile and Biological Activity of Cherimoya (Annona cherimola Mill.) and Atemoya (*Annona atemoya*) Leaves. *Molecules* **2020**, *25*, 2612. [CrossRef]
23. Hoang, L.; Beneš, F.; Fenclová, M.; Kronusová, O.; Švarcová, V.; Řehořová, K.; Švecová, E.B.; Vosátka, M.; Hajšlová, J.; Kaštánek, P. Phytochemical Composition and In Vitro Biological Activity of Iris spp. (*Iridaceae*): A New Source of Bioactive Constituents for the Inhibition of Oral Bacterial Biofilms. *Antibiotics* **2020**, *9*, 403. [CrossRef]
24. Maheshwari, V. Phytochemicals effective in lowering Low-Density Lipoproteins. *J. Biol. Eng. Res. Rev.* **2020**, *7*, 16–23.
25. Weeks, A. The Molecular Systematics and Biogeography of the *Burseraceae*. Ph.D. Thesis, University of Texas at Austin, Austin, TX, USA, 1969.
26. Rüdiger, A.L.; Siani, A.C.; Junior, V.F.V. The chemistry and pharmacology of the South America genus Protium Burm. f. (*Burseraceae*). *Pharmacogn. Rev.* **2007**, *1*, 93–104.
27. Da Silva, P.S.D.; Leal, I.R.; Wirth, R.; Tabarelli, M. Spatial distribution and fruiting phenology of Protium heptaphyllum (Burseraceae) determine the design of the underground foraging system of Atta sexdens L. (Hymenoptera. *Formicidae*). *Neotrop. Entomol.* **2012**, *41*, 257–262. [CrossRef]
28. Aragão, G.F.; Carneiro, L.M.V.; Junior, A.P.F.; Vieira, L.C.; Bandeira, P.N.; Lemos, T.L.G.; Viana, G.S.d.B. A possible mechanism for anxiolytic and antidepressant effects of alpha-and beta-amyrin from *Protium heptaphyllum* (Aubl.) March. *Pharmacol. Biochem. Behav.* **2006**, *85*, 827–834. [CrossRef]
29. Aragão, G.F.; Carneiro, L.M.V.; Rota-Junior, A.P.; Bandeira, P.N.; Lemos, T.L.G.; Viana, G.S.d.B. Alterations in brain amino acid metabolism and inhibitory effects on PKC are possibly correlated with anticonvulsant effects of the isomeric mixture of α-and β-amyrin from *Protium heptaphyllum*. *Pharm. Biol.* **2015**, *53*, 407–413. [CrossRef] [PubMed]
30. Lima-Júnior, R.C.P.; Oliveira, F.A.; Gurgel, L.A.; Cavalcante, Í.J.M.; Santos, K.A.; Campos, D.A.; Vale, C.A.L.; Silva, R.M.; Chaves, M.H.; Rao, V.S.N. Attenuation of visceral nociception by α-and β-amyrin, a triterpenoid mixture isolated from the resin of Protium heptaphyllum, in mice. *Planta Med.* **2006**, *72*, 34–39. [CrossRef] [PubMed]
31. Oliveira, F.A.; Costa, C.L.S.; Chaves, M.H.; Almeida, F.R.C.; Cavalcante, Í.J.M.; Lima, A.F.; Lima, R.C.P., Jr.; Silva, R.M.; Campos, A.R.; Santos, F.A. Attenuation of capsaicin-induced acute and visceral nociceptive pain by α-and β-amyrin, a triterpene mixture isolated from *Protium heptaphyllum* resin in mice. *Life Sci.* **2005**, *77*, 2942–2952. [CrossRef]
32. Siani, A.C.; Ramos, M.F.d.S.; Menezes-de-Lima, O., Jr.; Ribeiro-dos-Santos, R.; Fernadez-Ferreira, E.; Soares, R.O.A.; Rosas, E.C.; Susunaga, G.S.; Guimarães, A.C.; Zoghbi, M.d.G.B. Evaluation of anti-inflammatory-related activity of essential oils from the leaves and resin of species of Protium. *J. Ethnopharmacol.* **1999**, *66*, 57–69. [CrossRef]
33. Oliveira, F.A.; Vieira-Júnior, G.M.; Chaves, M.H.; Almeida, F.R.C.; Santos, K.A.; Martins, F.S.; Silva, R.M.; Santos, F.A.; Rao, V.S.N. Gastroprotective effect of the mixture of α-and β-amyrin from Protium heptaphyllum: Role of capsaicin-sensitive primary afferent neurons. *Planta Med.* **2004**, *70*, 780–782. [CrossRef]
34. Aragao, G.F.; Pinheiro, M.C.C.; Bandeira, P.N.; Lemos, T.L.G.; Viana, G.S. Analgesic and anti-inflammatory activities of the isomeric mixture of alpha-and beta-amyrin from Protium heptaphyllum (Aubl.) March. *J. Herb. Pharmacother.* **2008**, *7*, 31–47. [CrossRef]
35. Pinto, L.M.S.; Helena, C.M.; Santos, F.A.; Rao, V. Anti-inflammatory effect of alpha, beta-Amyrin, a pentacyclic triterpene from Protium heptaphyllum in rat model of acute periodontitis. *Inflammopharmacology* **2007**, *15*, 1–5.
36. Faustino, C.G.; de Medeiros, F.A.; Ribeiro Galardo, A.K.; Lobato Rodrigues, A.B.; Lopes Martins, R.; de Medeiros Souza Lima, Y.; Fechine Tavares, J.; Alves de Medeiros, M.A.; dos Santos Cruz, J.; Almeida, S.S.M.d.S. Larvicide Activity on Aedes aegypti of Essential Oil Nanoemulsion from the Protium heptaphyllum Resin. *Molecules* **2020**, *25*, 5333. [CrossRef] [PubMed]
37. Cabral, R.S.C.; Alves, C.C.F.; Batista, H.R.F.; Sousa, W.C.; Abrahão, I.S.; Crotti, A.E.M.; Santiago, M.B.; Martins, C.H.G.; Miranda, M.L.D. Chemical composition of essential oils from different parts of *Protium heptaphyllum* (Aubl.) Marchand and their in vitro antibacterial activity. *Nat. Prod. Res.* **2020**, *34*, 2378–2383. [CrossRef] [PubMed]

38. De Lima, E.M.; Cazelli, D.S.P.; Pinto, F.E.; Mazuco, R.A.; Kalil, I.C.; Lenz, D.; Scherer, R.; de Andrade, T.U.; Endringer, D.C. Essential oil from the resin of Protium heptaphyllum: Chemical composition, cytotoxicity, antimicrobial activity, and antimutagenicity. *Pharmacogn. Mag.* **2016**, *12*, S42. [PubMed]
39. Santos, F.A.; Frota, J.T.; Arruda, B.R.; de Melo, T.S.; de Castro Brito, G.A.; Chaves, M.H.; Rao, V.S. Antihyperglycemic and hypolipidemic effects of α, β-amyrin, a triterpenoid mixture from Protium heptaphyllum in mice. *Lipids Health Dis.* **2012**, *11*, 98. [CrossRef]
40. Carvalho, K.M.M.B.; de Melo, T.S.; de Melo, K.M.; Quinderé, A.L.G.; de Oliveira, F.T.B.; Viana, A.F.S.C.; Nunes, P.I.G.; da Silva Quetz, J.; de Araújo Viana, D.; Havt, A. Amyrins from Protium heptaphyllum reduce high-fat diet-induced obesity in mice via modulation of enzymatic, hormonal and inflammatory responses. *Planta Med.* **2017**, *83*, 285–291. [CrossRef]
41. Carvalho, K.M.M.B.; Marinho Filho, J.D.B.; de Melo, T.S.; Araújo, A.J.; Quetz, J.d.S.; da Cunha, M.d.P.S.S.; de Melo, K.M.; da Silva, A.A.d.C.A.; Tomé, A.R.; Havt, A. The resin from protium *heptaphyllum* prevents high-fat diet-induced obesity in mice: Scientific evidence and potential mechanisms. *Evid. Based Complement. Altern. Med.* **2015**, *2015*. [CrossRef]
42. de Melo, K.M.; de Oliveira, F.T.B.; Silva, R.A.C.; Quinderé, A.L.G.; Marinho Filho, J.D.B.; Araújo, A.J.; Pereira, E.D.B.; Carvalho, A.A.; Chaves, M.H.; Rao, V.S. α, β-Amyrin, a pentacyclic triterpenoid from Protium heptaphyllum suppresses adipocyte differentiation accompanied by down regulation of PPARγ and C/EBPα in 3T3-L1 cells. *Biomed. Pharmacother.* **2019**, *109*, 1860–1866. [CrossRef]
43. Serajuddin, A.T.M.; Ranadive, S.A.; Mahoney, E.M. Relative lipophilicities, solubilities, and structure-pharmacological considerations of 3-hydroxy-3-methylglutaryl-coenzyme A (HMG-CoA) reductase inhibitors pravastatin, lovastatin, mevastatin, and simvastatin. *J. Pharm. Sci.* **1991**, *80*, 830–834. [CrossRef]
44. McGraw, G.W.; Hemingway, R.W.; Ingram, L.L.; Canady, C.S.; McGraw, W.B. Thermal degradation of terpenes: Camphene, Δ3-carene, limonene, and α-terpinene. *Environ. Sci. Technol.* **1999**, *33*, 4029–4033. [CrossRef]
45. Chern, L.Y. Monoterpenes in plants-a mini review. *Asian J. Plant. Biol.* **2013**, *1*, 15–19.
46. Nogueira, A.O.; Oliveira, Y.I.S.; Adjafre, B.L.; de Moraes, M.E.A.; Aragao, G.F. Pharmacological effects of the isomeric mixture of alpha and beta amyrin from Protium heptaphyllum: A literature review. *Fundam. Clin. Pharmacol.* **2019**, *33*, 4–12. [CrossRef]
47. Büchele, B.; Zugmaier, W.; Simmet, T. Analysis of pentacyclic triterpenic acids from frankincense gum resins and related phytopharmaceuticals by high-performance liquid chromatography. Identification of lupeolic acid, a novel pentacyclic triterpene. *J. Chromatogr. B* **2003**, *791*, 21–30. [CrossRef]
48. Singh, B.; Kumar, R.; Bhandari, S.; Pathania, S.; Lal, B. Volatile constituents of natural Boswellia serrata oleo-gum-resin and commercial samples. *Flavour Fragr. J.* **2007**, *22*, 145–147. [CrossRef]
49. Mahajan, B.; Taneja, S.C.; Sethi, V.K.; Dhar, K.L. Two triterpenoids from Boswellia serrata gum resin. *Phytochemistry* **1995**, *39*, 453–455. [CrossRef]
50. López-Hortas, L.; Pérez-Larrán, P.; González-Muñoz, M.J.; Falqué, E.; Domínguez, H. Recent developments on the extraction and application of ursolic acid. A review. *Food Res. Int.* **2018**, *103*, 130–149. [CrossRef]
51. Volkman, J.K. Sterols and other triterpenoids: Source specificity and evolution of biosynthetic pathways. *Org. Geochem.* **2005**, *36*, 139–159. [CrossRef]
52. Khan, M.S.Y.; Bano, S.; Javed, K.; Mueed, M.A. A comprehensive review on the chemistry and pharmacology of Corchorus species—a source of cardiac glycosides, triterpenoids, ionones, flavonoids, coumarins, steroids and some other compounds. *J. Sci. Ind. Res.* **2006**, *65*, 283–298.
53. Kimura, Y.; Taniguchi, M.; Baba, K. Antitumor and antimetastatic effects on liver of triterpenoid fractions of Ganoderma lucidum: Mechanism of action and isolation of an active substance. *Anticancer Res.* **2002**, *22*, 3309–3318.
54. Alberts, A.W. Discovery, biochemistry and biology of lovastatin. *Am. J. Cardiol.* **1988**, *62*, J10–J15. [CrossRef]
55. Seenivasan, A.; Gummadi, S.N.; Panda, T. Comparison of the elution characteristics of individual forms of lovastatin in both isocratic and gradient modes and HPLC-PDA method development for pure and fermentation-derived lovastatin. *Prep. Biochem. Biotechnol.* **2017**, *47*, 901–908. [CrossRef]
56. Klingelhöfer, I.; Morlock, G.E. Lovastatin in lactone and hydroxy acid forms and citrinin in red yeast rice powders analyzed by HPTLC-UV/FLD. *Anal. Bioanal. Chem.* **2019**, *411*, 6655–6665. [CrossRef]
57. Beltrán, D.; Frutos-Lisón, M.D.; Espín, J.C.; García-Villalba, R. Re-examining the role of the gut microbiota in the conversion of the lipid-lowering statin monacolin K (lovastatin) into its active β-hydroxy acid metabolite. *Food Funct.* **2019**, *10*, 1787–1791. [CrossRef]
58. Tang, J.-J.; Li, J.-G.; Qi, W.; Qiu, W.-W.; Li, P.-S.; Li, B.-L.; Song, B.-L. Inhibition of SREBP by a Small Molecule, Betulin, Improves Hyperlipidemia and Insulin Resistance and Reduces Atherosclerotic Plaques. *Cell Metab.* **2011**, *13*, 44–56. [CrossRef]
59. Prabhakar, P.; Reeta, K.; Maulik, S.K.; Dinda, A.K.; Gupta, Y.K. α-Amyrin attenuates high fructose diet-induced metabolic syndrome in rats. *Appl. Physiol. Nutr. Metab.* **2017**, *42*, 23–32. [CrossRef] [PubMed]
60. Somova, L.O.; Nadar, A.; Rammanan, P.; Shode, F.O. Cardiovascular, antihyperlipidemic and antioxidant effects of oleanolic and ursolic acids in experimental hypertension. *Phytomedicine* **2003**, *10*, 115–121. [CrossRef] [PubMed]
61. Lee, H.K.; Nam, G.W.; Kim, S.H.; Lee, S.H. Phytocomponents of triterpenoids, oleanolic acid and ursolic acid, regulated differently the processing of epidermal keratinocytes via PPAR-alpha pathway. *Exp. Dermatol.* **2006**, *15*, 66–73. [CrossRef] [PubMed]
62. Istvan, E.S.; Deisenhofer, J. Structural mechanism for statin inhibition of HMG-CoA reductase. *Science* **2001**, *292*, 1160–1164. [CrossRef]

63. Tobert, J.A. Lovastatin and beyond: The history of the HMG-CoA reductase inhibitors. *Nat. Rev. Drug Discov.* **2003**, *2*, 517–526. [CrossRef]
64. Schachter, M. Chemical, pharmacokinetic and pharmacodynamic properties of statins: An update. *Fundam. Clin. Pharmacol.* **2005**, *19*, 117–125. [CrossRef] [PubMed]
65. Zolkiflee, N.F.; Majeed, A.B.A. Lovastatin: History, physicochemistry, pharmacokinetics and enhanced solubility. *Int. J. Res. Pharm. Sci.* **2017**, *8*, 90–102.
66. Gesto, D.S.; Pereira, C.; Cerqueira, N.M.F.S.; Sousa, S.F. An Atomic-Level Perspective of HMG-CoA-Reductase: The Target Enzyme to Treat Hypercholesterolemia. *Molecules* **2020**, *25*, 3891. [CrossRef] [PubMed]
67. Kashyap, D.; Sharma, A.; Tuli, H.S.; Punia, S.; Sharma, A.K. Ursolic acid and oleanolic acid: Pentacyclic terpenoids with promising anti-inflammatory activities. *Recent Pat. Inflamm. Allergy Drug Discov.* **2016**, *10*, 21–33. [CrossRef]
68. Lateef, T.; Naeem, S.; Qureshi, S.A. In-silico studies of HMG-Co A reductase inhibitors present in fruits of Withania coagulans Dunal (Solanaceae). *Trop. J. Pharm. Res.* **2020**, *19*, 305–312. [CrossRef]
69. Marahatha, R.; Basnet, S.; Bhattarai, B.R.; Budhathoki, P.; Aryal, B.; Adhikari, B.; Lamichhane, G.; Poudel, D.K.; Parajuli, N. Potential natural inhibitors of xanthine oxidase and HMG-CoA reductase in cholesterol regulation: In silico analysis. *BMC Complement. Med. Ther.* **2021**, *21*, 1–11. [CrossRef]
70. Goswami, S.; Vidyarthi, A.S.; Bhunia, B.; Mandal, T. A review on lovastatin and its production. *J. Biochem. Technol.* **2013**, *4*, 581–587.
71. Alberts, A.W. Lovastatin and simvastatin-inhibitors of HMG CoA reductase and cholesterol biosynthesis. *Cardiology* **1990**, *77*, 14–21. [CrossRef]
72. Mannino, G.; Perrone, A.; Campobenedetto, C.; Schittone, A.; Margherita Bertea, C.; Gentile, C. Phytochemical profile and antioxidative properties of Plinia trunciflora fruits: A new source of nutraceuticals. *Food Chem.* **2020**, *307*, 125515. [CrossRef]
73. Rohit Singh, T.; Ezhilarasan, D. Ethanolic extract of Lagerstroemia Speciosa (L.) Pers., induces apoptosis and cell cycle arrest in HepG2 cells. *Nutr. Cancer* **2020**, *72*, 146–156. [CrossRef]
74. Mannino, G.; Caradonna, F.; Cruciata, I.; Lauria, A.; Perrone, A.; Gentile, C. Melatonin reduces inflammatory response in human intestinal epithelial cells stimulated by interleukin-1β. *J. Pineal Res.* **2019**, *67*. [CrossRef]
75. Caradonna, F.; Consiglio, O.; Luparello, C.; Gentile, C. Science and Healthy Meals in the World: Nutritional Epigenomics and Nutrigenetics of the Mediterranean Diet. *Nutrients* **2020**, *12*, 1748. [CrossRef] [PubMed]
76. Cohen, L.H.; Van Vliet, A.; Roodenburg, L.; Jansen, L.M.C.; Griffigen, M. Pravastatin inhibited the cholesterol synthesis in human hepatoma cell line Hep G2 less than simvastatin and lovastatin, which is reflected in the upregulation of 3-hydroxy-3-methylglutaryl coenzyme A reductase and squalene synthase. *Biochem. Pharmacol.* **1993**, *45*, 2203–2208. [CrossRef]
77. Yu, X.-H.; Zheng, X.-L.; Tang, C.-K. Peroxisome proliferator-activated receptor α in lipid metabolism and atherosclerosis. In *Advances in Clinical Chemistry*; Elsevier: Amsterdam, The Netherlands, 2015; Volume 71, pp. 171–203, ISBN 0065-2423.
78. Choi, Y.-J.; Lee, S.J.; Kim, H.I.; Lee, H.J.; Kang, S.J.; Kim, T.Y.; Cheon, C.; Ko, S.-G. Platycodin D enhances LDLR expression and LDL uptake via down-regulation of IDOL mRNA in hepatic cells. *Sci. Rep.* **2020**, *10*, 1–13. [CrossRef]
79. Zhang, C.; Tian, Y.; Zhang, M.; Tuo, Q.; Chen, J.; Liao, D. IDOL, inducible degrader of low-density lipoprotein receptor, serves as a potential therapeutic target for dyslipidemia. *Med. Hypotheses* **2016**, *86*, 138–142. [CrossRef] [PubMed]
80. Yang, H.; Zhang, M.; Long, S.; Tuo, Q.; Tian, Y.; Chen, J.; Zhang, C.; Liao, D. Cholesterol in LDL receptor recycling and degradation. *Clin. Chim. Acta* **2020**, *500*, 81–86. [CrossRef] [PubMed]
81. Hayakawa, E.H.; Kato, H.; Nardone, G.A.; Usukura, J. A prospective mechanism and source of cholesterol uptake by Plasmodium falciparum-infected erythrocytes co-cultured with HepG2 cells. *Parasitol. Int.* **2020**, *80*, 102179. [CrossRef] [PubMed]
82. Liu, J.; Zhang, F.; Li, C.; Lin, M.; Briggs, M.R. Synergistic activation of human LDL receptor expression by SCAP ligand and cytokine oncostatin M. *Arterioscler. Thromb. Vasc. Biol.* **2003**, *23*, 90–96. [CrossRef]
83. Dong, B.; Wu, M.; Cao, A.; Li, H.; Liu, J. Suppression of Idol expression is an additional mechanism underlying statin-induced up-regulation of hepatic LDL receptor expression. *Int. J. Mol. Med.* **2011**, *27*, 103–110.
84. Hageman, J.; Herrema, H.; Groen, A.K.; Kuipers, F. A role of the bile salt receptor FXR in atherosclerosis. *Arterioscler. Thromb. Vasc. Biol.* **2010**, *30*, 1519–1528. [CrossRef]
85. Tacer, K.F. Oxysterols and bile acid act as signaling molecules that regulate cholesterol homeostasis: Nuclear receptors LXR, FXR, and fibroblast growth factor 15/19. In *Mammalian Sterols*; Springer: Berlin, Germany, 2020; pp. 117–143.
86. Lu, J.; Huang, G.; Hu, S.; Wang, Z.; Guan, S. 1, 3-Dichloro-2-propanol induced hyperlipidemia in C57BL/6J mice via AMPK signaling pathway. *Food Chem. Toxicol.* **2014**, *64*, 403–409. [CrossRef]
87. Cao, Y.; Bei, W.; Hu, Y.; Cao, L.; Huang, L.; Wang, L.; Luo, D.; Chen, Y.; Yao, X.; He, W. Hypocholesterolemia of Rhizoma Coptidis alkaloids is related to the bile acid by up-regulated CYP7A1 in hyperlipidemic rats. *Phytomedicine* **2012**, *19*, 686–692. [CrossRef]
88. Iwayanagi, Y.; Takada, T.; Tomura, F.; Yamanashi, Y.; Terada, T.; Inui, K.; Suzuki, H. Human NPC1L1 expression is positively regulated by PPARα. *Pharm. Res.* **2011**, *28*, 405–412. [CrossRef] [PubMed]
89. Hulsmans, M.; Geeraert, B.; Arnould, T.; Tsatsanis, C.; Holvoet, P. PPAR agonist-induced reduction of Mcp1 in atherosclerotic plaques of obese, insulin-resistant mice depends on adiponectin-induced Irak3 expression. *PLoS ONE* **2013**, *8*, e62253. [CrossRef]
90. Ruttkies, C.; Schymanski, E.L.; Wolf, S.; Hollender, J.; Neumann, S. MetFrag relaunched: Incorporating strategies beyond in silico fragmentation. *J. Cheminform.* **2016**, *8*, 3. [CrossRef]

91. Sherman, W.; Beard, H.S.; Farid, R. Use of an induced fit receptor structure in virtual screening. *Chem. Biol. Drug Des.* **2006**, *67*, 83–84. [CrossRef]
92. Kellici, T.F.; Ntountaniotis, D.; Liapakis, G.; Tzakos, A.G.; Mavromoustakos, T. The dynamic properties of angiotensin II type 1 receptor inverse agonists in solution and in the receptor site. *Arab. J. Chem.* **2019**, *12*, 5062–5078. [CrossRef]
93. Banks, J.L.; Beard, H.S.; Cao, Y.; Cho, A.E.; Damm, W.; Farid, R.; Felts, A.K.; Halgren, T.A.; Mainz, D.T.; Maple, J.R. Integrated modeling program, applied chemical theory (IMPACT). *J. Comput. Chem.* **2005**, *26*, 1752–1780. [CrossRef]
94. Rose, P.W.; Bi, C.; Bluhm, W.F.; Christie, C.H.; Dimitropoulos, D.; Dutta, S.; Green, R.K.; Goodsell, D.S.; Prlić, A.; Quesada, M. The RCSB Protein Data Bank: New resources for research and education. *Nucleic Acids Res.* **2012**, *41*, D475–D482. [CrossRef]
95. Sastry, G.M.; Adzhigirey, M.; Day, T.; Annabhimoju, R.; Sherman, W. Protein and ligand preparation: Parameters, protocols, and influence on virtual screening enrichments. *J. Comput. Aided. Mol. Des.* **2013**, *27*, 221–234. [CrossRef]
96. Friesner, R.A.; Murphy, R.B.; Repasky, M.P.; Frye, L.L.; Greenwood, J.R.; Halgren, T.A.; Sanschagrin, P.C.; Mainz, D.T. Extra precision glide: Docking and scoring incorporating a model of hydrophobic enclosure for protein−ligand complexes. *J. Med. Chem.* **2006**, *49*, 6177–6196. [CrossRef]
97. Friesner, R.A.; Banks, J.L.; Murphy, R.B.; Halgren, T.A.; Klicic, J.J.; Mainz, D.T.; Repasky, M.P.; Knoll, E.H.; Shelley, M.; Perry, J.K. Glide: A new approach for rapid, accurate docking and scoring. 1. Method and assessment of docking accuracy. *J. Med. Chem.* **2004**, *47*, 1739–1749. [CrossRef]
98. Halgren, T.A.; Murphy, R.B.; Friesner, R.A.; Beard, H.S.; Frye, L.L.; Pollard, W.T.; Banks, J.L. Glide: A new approach for rapid, accurate docking and scoring. 2. Enrichment factors in database screening. *J. Med. Chem.* **2004**, *47*, 1750–1759. [CrossRef]
99. Zhong, H.; Tran, L.M.; Stang, J.L. Induced-fit docking studies of the active and inactive states of protein tyrosine kinases. *J. Mol. Graph. Model.* **2009**, *28*, 336–346. [CrossRef] [PubMed]
100. Sherman, W.; Day, T.; Jacobson, M.P.; Friesner, R.A.; Farid, R. Novel procedure for modeling ligand/receptor induced fit effects. *J. Med. Chem.* **2006**, *49*, 534–553. [CrossRef]
101. Hasanat, A.; Chowdhury, T.A.; Kabir, M.S.H.; Chowdhury, M.S.; Chy, M.; Uddin, N.; Barua, J.; Chakrabarty, N.; Paul, A. Antinociceptive activity of Macaranga denticulata Muell. Arg.(Family: Euphorbiaceae): In vivo and in silico studies. *Medicines* **2017**, *4*, 88. [CrossRef]
102. Lauria, A.; Mannino, S.; Gentile, C.; Mannino, G.; Martorana, A.; Peri, D. DRUDIT: Web-based DRUgs DIscovery Tools to design small molecules as modulators of biological targets. *Bioinformatics* **2019**, *36*, 1562–1569. [CrossRef]
103. Wang, H.; Aslanian, R.; Madison, V.S. Induced-fit docking of mometasone furoate and further evidence for glucocorticoid receptor 17α pocket flexibility. *J. Mol. Graph. Model.* **2008**, *27*, 512–521. [CrossRef]
104. Luo, H.-J.; Wang, J.-Z.; Deng, W.-Q.; Zou, K. Induced-fit docking and binding free energy calculation on furostanol saponins from Tupistra chinensis as epidermal growth factor receptor inhibitors. *Med. Chem. Res.* **2013**, *22*, 4970–4979. [CrossRef]
105. Jacobson, M.P.; Pincus, D.L.; Rapp, C.S.; Day, T.J.F.; Honig, B.; Shaw, D.E.; Friesner, R.A. A hierarchical approach to all-atom protein loop prediction. *Proteins Struct. Funct. Bioinform.* **2004**, *55*, 351–367. [CrossRef]
106. Jacobson, M.P.; Friesner, R.A.; Xiang, Z.; Honig, B. On the role of the crystal environment in determining protein side-chain conformations. *J. Mol. Biol.* **2002**, *320*, 597–608. [CrossRef]
107. Campobenedetto, C.; Mannino, G.; Agliassa, C.; Acquadro, A.; Contartese, V.; Garabello, C.; Bertea, C.M. Transcriptome Analyses and Antioxidant Activity Profiling Reveal the Role of a Lignin-Derived Biostimulant Seed Treatment in Enhancing Heat Stress Tolerance in Soybean. *Plants* **2020**, *9*, 1308. [CrossRef] [PubMed]
108. Campobenedetto, C.; Grange, E.; Mannino, G.; Van Arkel, J.; Beekwilder, J.; Karlova, R.; Garabello, C.; Contartese, V.; Bertea, C.M. A Biostimulant Seed Treatment Improved Heat Stress Tolerance During Cucumber Seed Germination by Acting on the Antioxidant System and Glyoxylate Cycle. *Front. Plant. Sci.* **2020**, *11*, 836. [CrossRef]
109. Pfaffl, M.W. A new mathematical model for relative quantification in real-time RT–PCR. *Nucleic Acids Res.* **2001**, *29*, e45. [CrossRef]
110. Siani, A.C.; Moraes, R.; Veiga Junior, V.F. Toward Establishing the Productive Chain for Triterpene-Based Amazonian Oleoresins as Valuable Non-Timber Forest Products. *Open J. For.* **2017**, *7*, 188–208. [CrossRef]

International Journal of *Molecular Sciences*

MDPI

Article

Methyl P-Coumarate Ameliorates the Inflammatory Response in Activated-Airway Epithelial Cells and Mice with Allergic Asthma

Ji-Won Park [1,†], Jinseon Choi [1,2,†], Juhyun Lee [1,3], Jin-Mi Park [1,2], Seong-Man Kim [1], Jae-Hong Min [4], Da-Yun Seo [1,2], Soo-Hyeon Goo [1], Ju-Hee Kim [1], Ok-Kyoung Kwon [1,5], Kihoon Lee [6], Kyung-Seop Ahn [1], Sei-Ryang Oh [1,5,*] and Jae-Won Lee [1,*]

1 Natural Medicine Research Center, Korea Research Institute of Bioscience and Biotechnology, Cheongju 28116, Republic of Korea
2 College of Pharmacy, Chungbuk National University, Cheongju 28160, Republic of Korea
3 College of Pharmacy, Chungnam National University, Daejeon 34134, Republic of Korea
4 Laboratory Animal Resources Division, Toxicological Evaluation and Research Department, National Institute of Food and Drug Safety Evaluation, Ministry of Food and Drug Safety, Osong Health Technology Administration Complex, Cheongju 28159, Republic of Korea
5 Natural Product Central Bank, Korea Research Institute of Bioscience and Biotechnology, Cheonju 28116, Republic of Korea
6 Laboratory Animal Resource Center, Korea Research Institute of Bioscience and Biotechnology, Cheongju 28116, Republic of Korea
* Correspondence: seiryang@kribb.re.kr (S.-R.O.); suc369@kribb.re.kr (J.-W.L.)
† These authors contributed equally to this work.

Citation: Park, J.-W.; Choi, J.; Lee, J.; Park, J.-M.; Kim, S.-M.; Min, J.-H.; Seo, D.-Y.; Goo, S.-H.; Kim, J.-H.; Kwon, O.-K.; et al. Methyl P-Coumarate Ameliorates the Inflammatory Response in Activated-Airway Epithelial Cells and Mice with Allergic Asthma. *Int. J. Mol. Sci.* **2022**, *23*, 14909. https://doi.org/10.3390/ijms232314909

Academic Editor: Michael Roth

Received: 8 October 2022
Accepted: 23 November 2022
Published: 28 November 2022

Publisher's Note: MDPI stays neutral with regard to jurisdictional claims in published maps and institutional affiliations.

Abstract: Methyl p-coumarate (methyl p-hydroxycinnamate) (MH) is a natural compound found in a variety of plants. In the present study, we evaluated the ameliorative effects of MH on airway inflammation in an experimental model of allergic asthma (AA). In this in vitro study, MH was found to exert anti-inflammatory activity on PMA-stimulated A549 airway epithelial cells by suppressing the secretion of IL-6, IL-8, MCP-1, and ICAM-1. In addition, MH exerted an inhibitory effect not only on NF-κB (p-NF-κB and p-IκB) and AP-1 (p-c-Fos and p-c-Jun) activation but also on A549 cell and EOL-1 cell (eosinophil cell lines) adhesion. In LPS-stimulated RAW264.7 macrophages, MH had an inhibitory effect on TNF-α, IL-1β, IL-6, and MCP-1. The results from in vivo study revealed that the increases in eosinophils/Th2 cytokines/MCP-1 in the bronchoalveolar lavage fluid (BALF) and IgE in the serum of OVA-induced mice with AA were effectively inhibited by MH administration. MH also exerted a reductive effect on the immune cell influx, mucus secretion, and iNOS/COX-2 expression in the lungs of mice with AA. The effects of MH were accompanied by the inactivation of NF-κB. Collectively, the findings of the present study indicated that MH attenuates airway inflammation in mice with AA, suggesting its potential as an adjuvant in asthma therapy.

Keywords: asthma; methyl p-coumarate; Th2 cytokines; eosinophil; NF-κB

1. Introduction

The prevalence of allergic asthma (AA) is increasing globally [1,2]. The sustained airway inflammation induced by allergic sensitization is a major pathophysiological characteristic of AA [3]. Airway epithelial cell-derived cytokines, chemokines, and adhesion molecules (IL-6, IL-8, MCP-1, and ICAM-1) are associated with the development of AA [4,5]. Macrophage-released cytokines and chemokines (TNF-α, IL-1β, IL-6, and MCP-1) affect airway inflammation in asthma pathogenesis [6,7]. Th2 cytokines (IL-4, -5, and -13) and eosinophils play an important role in the development of airway inflammation and mucus secretion in subjects with AA [8,9]. This increase in mucus secretion is a feature of AA and is related to airflow limitations [10,11]. It has also been reported that the increase in the expression of inducible nitric oxide (NO) synthase [iNOS] may affect mucus secretion by inducing NO production in subjects with AA [12,13]. The inhibition of cyclooxygenase-2 (COX-2) has been shown to improve asthmatic symptoms in a clinical study [14] and in mice with AA [15]. The increased activation of NF-κB causes bronchial inflammation, inducing cytokine/chemokine/mediator expression in an experimental models of AA [16–18]. Phorbol 12-myristate 13-acetate (PMA) has been used in in vitro asthma studies as it is known to promote the expression of cytokines/chemokines/adhesion molecules and the activation of NF-κB/AP-1 in A549 airway cells [15,19]. Ovalbumin (OVA) has been used in in vivo asthma studies as it induces bronchial inflammation by accelerating eosinophil influx, Th2 cytokine/IgE production, mucus secretion, iNOS/COX-2 expression, and NF-κB activation [20–22].

The beneficial effects of plant phenolic compounds on the improvement of inflammatory disorders have been reported in both in vitro and in vivo studies [23–25]. Methyl p-coumarate (methyl p-hydroxycinnamate) (MH), an esterified derivative of p-coumaric acid, was discovered in the flower of the medicinal plant *Trixis michuacana var longifolia* (D. Dow) C. [26], the roots and stems of *Comptonia peregrine* [27], the bark of *Melicope latifolia* [28], the leaves of Hainan *Morinda citrifolia* (Noni) [29], the fruit of Jujube (*Ziziphus jujuba* Mill.) [30], and in aloe vera [31]. Zhang et al. reported that MH has antibacterial properties [29], and Kubo et al. showed that MH has a suppressive effect on the formation of melanin in murine melanoma cell line B16 [32]. In addition, the experimental results from Lee et al. demonstrated that MH exerts an anti-inflammatory effect by suppressing the secretion of NO and the expression of iNOS in lipopolysaccharide (LPS)-stimulated RAW264.7 macrophages, and it exerts protective effects in mice with LPS-induced acute respiratory distress syndrome (ARDS) by inhibiting the secretion of TNF-α/IL-1β and the expression of iNOS [33,34]. However, to date and to the best of our knowledge, it has not been determined whether MH exerts an anti-asthmatic effect. Based on the inhibitory effect of MH on iNOS, which is an important factor in AA pathogenesis, protective effects of MH against AA are expected. Thus, the present study was undertaken to examine the protective effects of MH in an experimental model of AA.

2. Results

2.1. Suppressive Effects of MH against the PMA-Stimulated Inflammatory Response in A549 Cells

CytoX assay was used to assess the cytotoxicity of MH (5, 10, 25, 50, and 100 μM) in A549 cells or 10 nM PMA-stimulated A549 cells. The results showed that significant cell death did not occur until 100 μg/mL MH was administered (Figure 1a,b). Thus, doses were chosen for in vitro anti-inflammatory effects. ELISA results indicated that 10 nM PMA led to significant upregulation of IL-6 secretion (Figure 1c), whereas pretreatment with MH suppressed this upregulation in PMA-stimulated A549 cells. MH pretreatment inhibited the secretion of IL-8, MCP-1, and ICAM-1 in A549 cells following PMA administration (Figure 1d–f). In addition, MH suppressed the PMA-stimulated phosphorylation of IκBα and p65 in A549 cells (Figure 2a,b). It also exerted suppressive effects on NF-κB nuclear translocation in PMA-stimulated A549 cells (Figure 2c,d). In addition, MH exerted an inhibitory effect not only against c-Jun/c-Fos phosphorylation but also c-Jun/c-Fos nuclear translocation in PMA-stimulated A549 cells (Figure 2e–h). Collectively, the results from these in vitro studies confirmed that MH exerts anti-inflammatory effects in PMA-stimulated A549 cells by suppressing the secretion of cytokine/chemokine/adhesion molecules and the activation of NF-κB/AP-1.

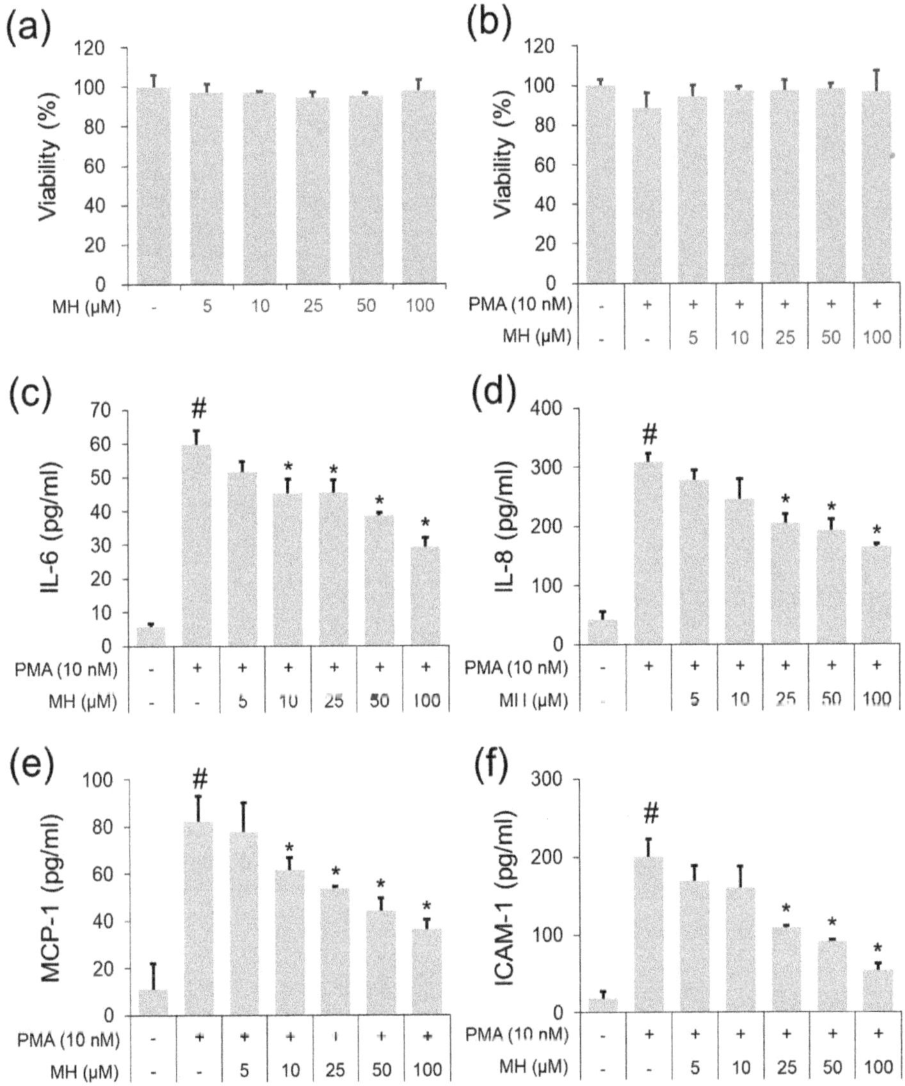

Figure 1. Effects of MH on secretion of cytokines, chemokines, and adhesion molecules in PMA−stimulated A549 cells. Cells were treated with MH (5, 10, 25, 50, and 100 μM) 1 h prior to incubation with 10 nM PMA for 18 h. (**a**,**b**) Cell viability was evaluated using CytoX assay. (**c**–**f**) The levels of IL−6, IL−8, MCP−1, and ICAM−1 were detected using ELISA kits. Data are expressed as the mean ± SD (n = 3) ($^{\#}$ $p < 0.05$ for comparison with control; * $p < 0.05$ for comparison with 10 nM PMA).

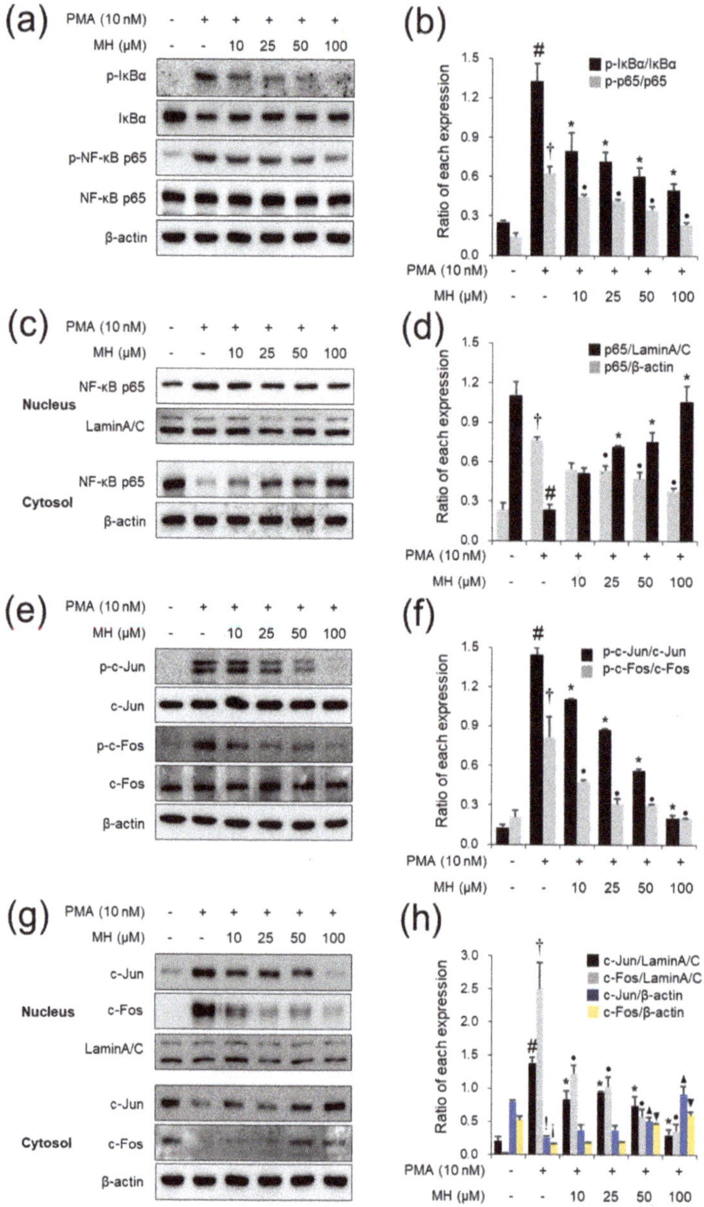

Figure 2. Effects of MH on activation of NF−κB and AP−1 in PMA−stimulated A549 cells. Cells were treated with MH (10, 25, 50, and 100 μM) 1 h prior to incubation with 10 nM PMA for 1 h. (**a**,**b**) The levels of p−IκBα and p−NF-κB p65 in whole cell lysate were detected using Western blotting. (**c**,**d**) The levels of NF−κB p65 in cytoplasmic cell fraction lysate and nuclear cell fraction lysate were detected using Western blotting. (**e**,**f**) The levels of p−c−Jun and p−c−Fos in whole cell lysate were detected using Western blotting. (**g**,**h**) The levels of c−Jun and c−Fos in cytoplasmic cell fraction lysate and nuclear cell fraction lysate were detected using Western blotting. Data are expressed as the mean ± SD (n = 3) ($^{\#,\dagger,!,¡}$ $p < 0.05$ for comparison with control; $^{*,\bullet,\blacktriangle,\blacktriangledown}$ $p < 0.05$ for comparison with 10 nM PMA).

2.2. Suppressive Effects of MH on Adhesion of Eosinophil and Airway Epithelial Cells

The infiltration of eosinophils into the airway epithelium is an important event in asthma development [35]. It has been proven that the promotion of ICAM-1 secretion in airway epithelial cells causes the recruitment of eosinophils [4,15]. Here, the inhibitory ability of MH against ICAM-1 secretion was notable in activated A549 airway epithelial cells (Figure 1f). Thus, we examined whether MH could affect the adhesion of airway epithelial cells and eosinophils. As shown in Figure 3a,b, the increase in fluorescence intensity that indicates the adhesion of eosinophil cell line EOL-1 and A549 cells was remarkably reversed by MH pretreatment.

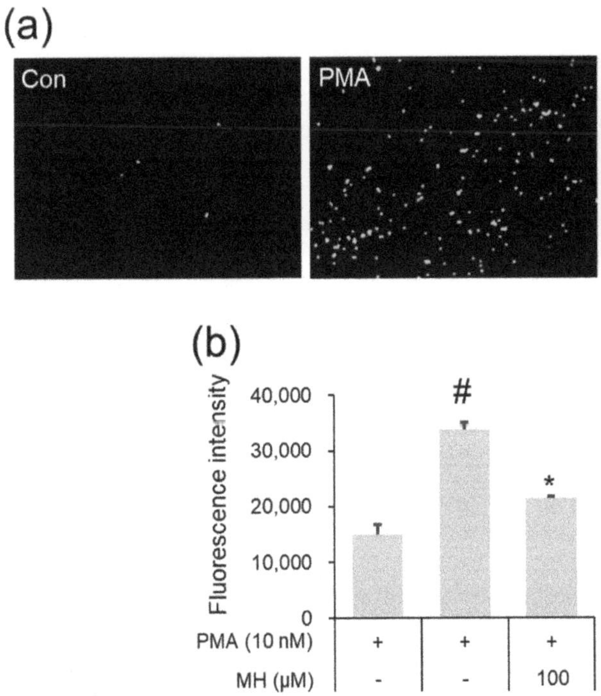

Figure 3. Effect of MH on the adhesion of A549 and EOL−1 cells. A549 cells were treated with 100 μM MH 1 h prior to incubation with 10 nM PMA for 5 h. Subsequently, EOL−1 cells stained with calcein AM were incubated with A549 cells for 1 h. (**a**) Green fluorescence, which indicated A549 and EOL−1 cell adhesion, was observed using fluorescence microscopy (magnification, ×100; scale bar, 100 μM). (**b**) The levels of EOL−1 cells adhering to A549 cells were quantified by detecting the fluorescence intensity. Data are expressed as the mean ± SD (n = 3) ($^{\#}$ $p < 0.05$ for comparison with control; * $p < 0.05$ for comparison with 10 nM PMA).

2.3. Suppressive Effects of MH against the LPS-Induced Inflammatory Response in RAW264.7 Cells

Previous observations indicated the protective properties of MH in the LPS-stimulated inflammatory response in RAW264.7 macrophages by showing its suppressive effect on the secretion of NO/PGE2 and the activation of NF-κB [33]. However, the inhibitory ability of MH on cytokines and chemokines, such as IL-6 and MCP-1, which are important factors in AA development [6,7], was not confirmed in that study. Interestingly, the effects of MH on the suppression of molecules (TNF-α, IL-1β, IL-6, and MCP-1) were notable in LPS-stimulated RAW264.7 cells (Figure 4a–d).

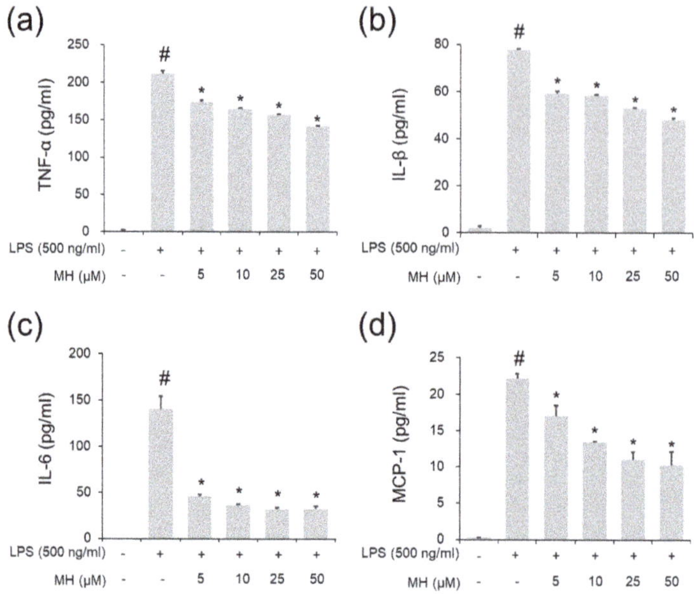

Figure 4. Effects of MH on secretion of cytokines and chemokines in LPS−stimulated RAW264.7 cells. Cells were treated with MH (5, 10, 25, 50, and 100 μM) 1 h prior to incubation with 500 ng/mL LPS for 18 h. (**a–d**) The levels of TNF−α, IL−1β, IL−6, and MCP−1 were detected using ELISA kits. Data are expressed as the mean ± SD (n = 3) ([#] $p < 0.05$ for comparison with control; * $p < 0.05$ for comparison with 500 ng/mL LPS).

2.4. Inhibitory Effects of MH on Recruitment of Immune Cells in Mice with AA

Based on in vitro anti-inflammatory ability, we examined the ameliorative ability of MH on airway inflammation in OVA-induced mice with AA (Figure 5). It is well - known that the recruitment of immune cells (eosinophils and macrophages) accelerates the inflammatory response in the airway [9]; therefore, the suppressive ability of MH on eosinophil/macrophage influx was examined. A significant upregulation in these cells was discovered in the BALF of the AA group compared to the NC group through Diff-Quik staining and cell counting (Figure 6a,b). However, a notable decrease in these cells was confirmed in the AA group with a 5 mg/kg administration of MH. Interestingly, its ability was comparable to that of a 1 mg/kg administration of DEX, which was used as a positive control.

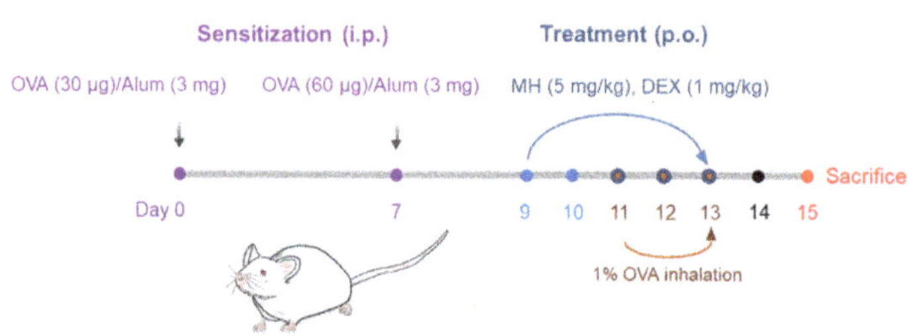

Figure 5. The establishment of the mice with AA and the administration of MH. BALB/c mice were randomly divided into four groups, sensitized with OVA/alum mixture on days 0 and 7, and exposed to OVA inhalation on days 11–13. Oral administration of MH or DEX was performed on days 9–13. On day 15, the mice were sacrificed, and the collection of BALF, serum, and lungs was performed for analysis.

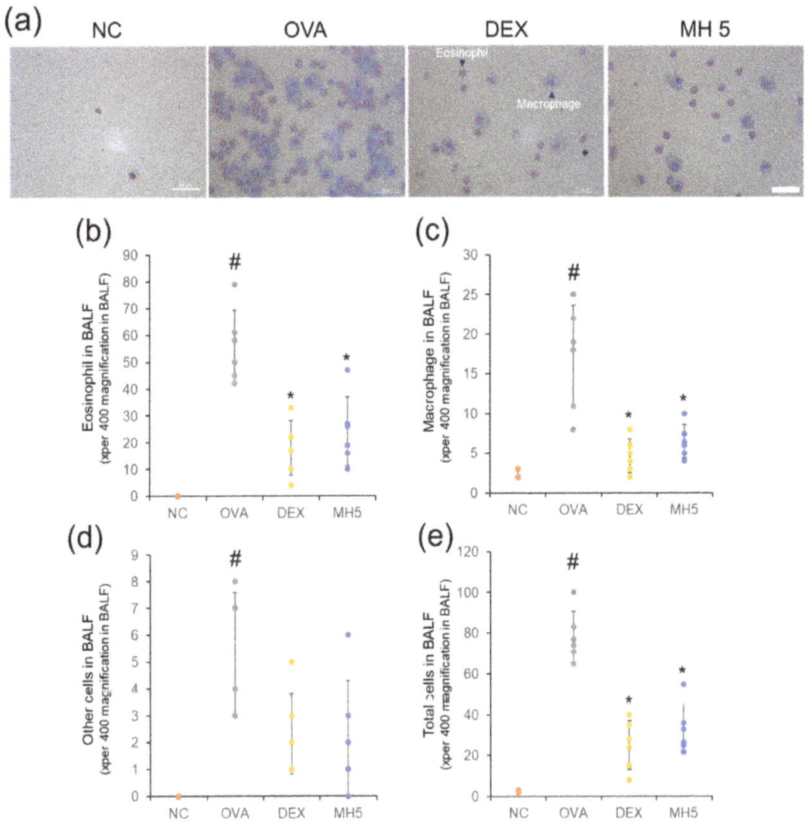

Figure 6. Effects of MH on recruitment of eosinophils and macrophages in mice with AA. (**a**) Microscope images of immune cells (magnification, ×400; scale bar, 25 μm). (**b–e**) The numbers of immune cells in BALF were determined using cell counting. NC: normal control mice; OVA: OVA-sensitized/challenged mice; DEX: 1 mg/kg DEX-treated OVA mice; and MH 5: 5 mg/kg MH-treated OVA mice group. Data are expressed as the mean ± SD (n = 6) ([#] $p < 0.05$ for comparison with normal control; * $p < 0.05$ for comparison with OVA group).

2.5. Suppressive Effects of MH on Secretion of Th2 Cytokines, MCP-1, and IgE in Mice with AA

Next, we focused on MH suppression of Th2 cytokines, MCP-1, and IgE. ELISA results showed the remarkable upregulation of Th2 cytokines (IL−4, −5, and −13) in BALF of mice with AA (Figure 7a–c). This upregulation was significantly decreased in mice with AA and MH administration. Based on the role of the key chemokine, MCP-1, in eosinophil/macrophage infiltration [36], the inhibitory effect of MH on this molecule was examined. MH administration inhibited the increased MCP-1 level in BALF of the AA group (Figure 7d). In addition, MH resulted in the downregulation of OVA-specific IgE in the serum of mice with AA (Figure 7e). Generally, its inhibitory ability against IL-4, IL-5, MCP-1, and IgE secretion was similar to that of DEX.

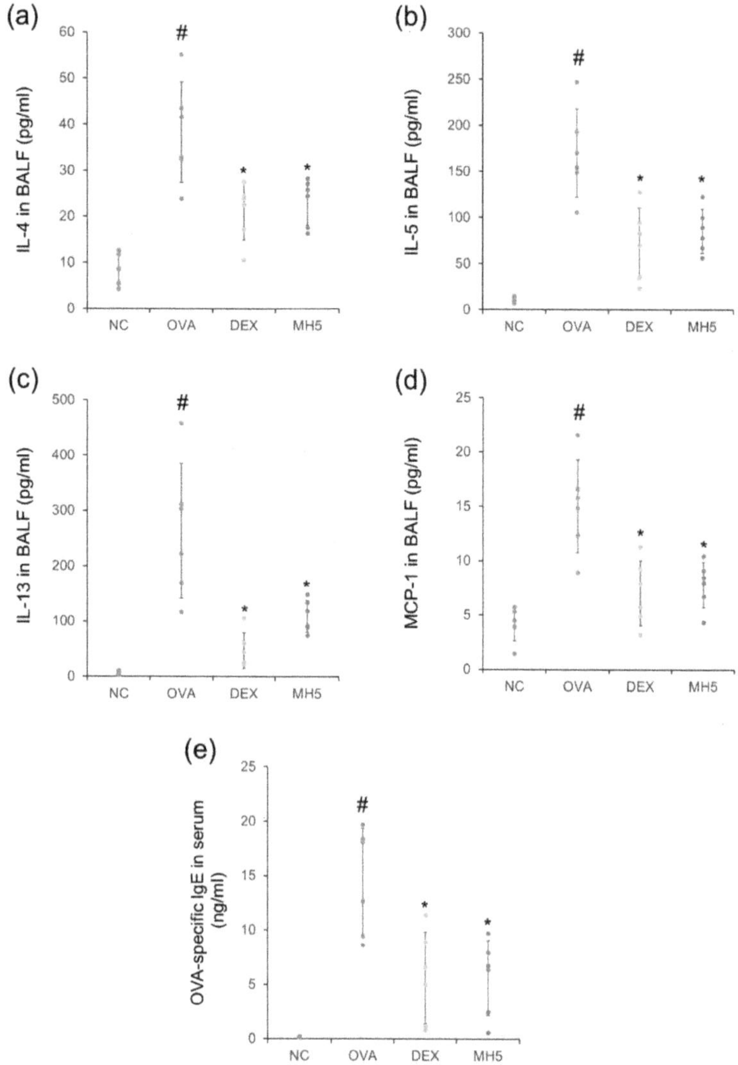

Figure 7. Effects of MH on the production of Th2 cytokines, MCP−1, and IgE in mice with AA. (**a–d**) The levels of Th2 cytokines (IL−4, −5, and −13) and MCP−1 in BALF were determined using ELISA assays. (**e**) The level of OVA−specific IgE in the serum was determined using ELISA assays. Data are expressed as the mean ± SD (n = 6) ($^{\#}$ $p < 0.05$ for comparison with normal control; * $p < 0.05$ for comparison with OVA group).

2.6. Suppressive Effects of MH on Immune Cell Influx and Mucus Secretion in Mice with AA

To detect immune cell influx in lung tissue, we performed H&E staining. The results indicated an increased influx of immune cells around the airway in the AA group (Figure 8a). This trend was effectively ameliorated in both the MH- and DEX-treated groups. To confirm the secretion level of mucus in the lungs, PAS staining was performed. As shown in Figure 8b, the results indicated that mucus secretion was markedly increased in the airway epithelium of mice with AA (Figure 8b). However, this increase was attenuated with the administration of MH or DEX.

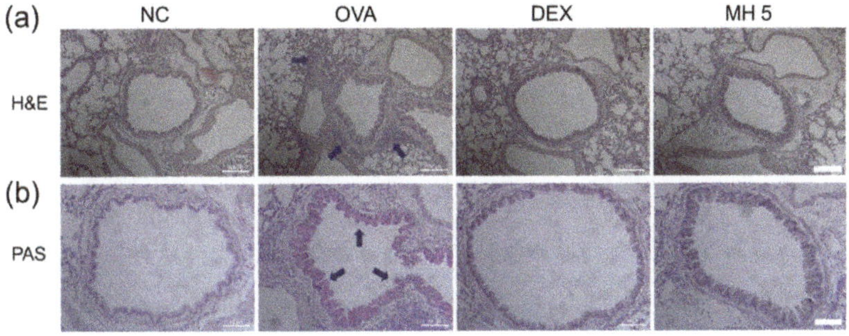

Figure 8. Effects of MH on immune cell influx and mucus generation in mice with AA. (**a**) The existence of immune cells surrounding the airway was confirmed using H&E staining (magnification, ×100; scale bar, 100 μM). (**b**) Airway mucus was detected using PAS staining (magnification, ×200; scale bar, 50 μM).

2.7. Suppressive Effects of MH on iNOS and COX-2 Expression in Mice with AA

The expression changes of iNOS and COX-2 in the lungs of mice were determined using Western blotting. As shown in Figure 9a,b, an increase in iNOS expression was observed in the lung tissue lysates of mice with AA compared to the control group, and this increase was suppressed in the AA group with MH administration. Subsequently, the expression level of COX-2 was examined, and the results revealed that the elevated expression of COX-2 in the lung tissue lysates of the AA group was reduced with MH administration. The effect of MH (5 mg/kg) on the suppression of these molecules was comparable to that of DEX (1 mg/kg).

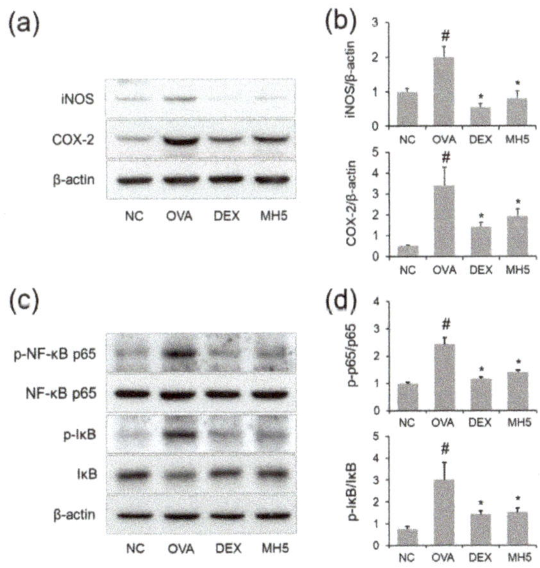

Figure 9. Effects of MH on iNOS/COX−2 expression and NF−κB activation in mice with AA. (**a**,**b**) The levels of iNOS and COX−2 in lung tissues were detected using Western blotting, and quantitative analysis was performed using ImageJ software. (**c**,**d**) The levels of p−NF−κB p65 and p−IκBα in lung tissues were detected using Western blotting, and quantitative analysis was performed using ImageJ software. Data are expressed as the mean ± SD (n = 3) (# $p < 0.05$ for comparison with normal control; * $p < 0.05$ for comparison with OVA group).

2.8. Effects of MH on NF-κB Inactivation in Mice with AA

As shown in Figure 9c,d, increased p65 phosphorylation was confirmed in the lungs of the AA group, whereas this tendency was reversed by MH administration. In addition, MH had a reductive effect on IκBα phosphorylation in the AA group (Figure 9c,d). The ability of MH to suppress NF-κB and IκBα activation was similar to that of DEX.

3. Discussion

The airway epithelium is the main source of inflammatory-associated molecules [37], and airway epithelial cell-derived IL-6 affects T helper cell (Th cell)-derived cytokine production [5]. The upregulation of IL-8 has been reported in the airway epithelium of asthma patients [38,39]. Bronchial epithelial cell-derived MCP-1 is known to induce bronchial remodeling by affecting monocyte/macrophage recruitment against allergen exposure in subjects with AA [40]. In addition, airway epithelial cell-derived ICAM-1 is closely associated with the promotion of airway inflammation in subjects with AA through eosinophil recruitment [4,15]. Macrophages isolated from patients with asthma generate more IL-6 than those in patients without asthma [41]. Macrophages are well-known producers of MCP-1, and the amelioration of airway inflammation was related to MCP-1 suppression in an experimental mouse model of OVA-induced AA [16,42]. Collectively, the inhibition of airway epithelial cell- and macrophage-released molecules may ameliorate airway inflammation in subjects with AA. In the present study, MH notably inhibited not only IL-6/IL-8/MCP-1/ICAM-1 secretion in activated-airway epithelial cells but also airway cell/eosinophil adhesion. Furthermore, MH decreased the secretion of TNF-α, IL-1β, IL-6, and MCP-1 in activated macrophages. Since these results showed that MH has anti-inflammatory effects on both airway epithelial cells and macrophages, we evaluated its protective effect against airway inflammation in an AA animal model.

Cumulative evidence indicates a pivotal role of Th2 cytokines, MCP-1, and IgE in the development of AA. IL-4 causes Th2 cell differentiation and macrophage/B-cell activation [43]. IL-5 promotes eosinophilic inflammation in the pathogenesis of asthma [44]. IL-4 and -5 lead to MCP-1 upregulation in bronchial epithelial cells by affecting macrophage recruitment [40,45]. IL-13 affects not only B-cell activation, IgE production, and mast cell activation but also mucus secretion in subjects with AA [46]. In addition, modulating Th2 cytokines and IgE has been suggested as a therapeutic approach in asthma therapy [47–49]. This information highlights the importance of Th2 cytokines and IgE regulation in subjects with AA. In this study, MH exerted a regulatory effect on Th2 cytokines/IgE secretion, as well as immune cell recruitment, in mice with AA. Similar to the results from an in vitro study, the regulatory ability of MH on MCP-1 was discovered in an in vivo study. Our results indicated that MH may ameliorate asthma symptoms by modulating Th2 cytokines, MCP-1, and IgE.

As mentioned earlier, the abnormal production of mucus can obstruct airflow [10,11], and IL-13 affects goblet cell hyperplasia and mucus production [50]. As the ability of MH to suppress IL-13 was confirmed in an in vivo study, its suppressive ability in mucus secretion was also examined. Interestingly, its ability was similar to that of the positive control of DEX. These results reflect that MH has a regulatory effect on both IL-13 secretion and mucus production, suggesting that MH may improve the airflow obstruction in subjects with AA. Further confirmation of the inhibitory ability of MH on another factor affecting mucus production could highlight the anti-asthmatic effect of MH.

The upregulation of various molecules, including cytokines, chemokines, and adhesion molecules, has been reported in the pathogenesis of asthma [36,51,52] and in experimental models of inflammatory lung diseases, including AA [53–55]. Accumulated evidence has indicated that the activation of NF-κB plays a pivotal role in the expression of inflammatory molecules in lung epithelial cells and macrophages [5,13,15,56,57]. Chauhan et al. suggested that a reduction in NF-κB activation can ameliorate OVA-induced airway inflammation, leading to the suppression of inflammatory molecules [15,18,58]. Thus, targeting NF-κB pathways could be a promising therapeutic approach for AA. The in vitro and in vivo results from this study revealed that MH has an inhibitory ability against various molecules and NF-κB activation. Therefore, these results reflect that the protective effects of MH against airway inflammation in experimental asthma may be associated with the modulating ability of MH against NF-κB activation. However, further studies on the mechanisms of asthma improvement are needed before it can be developed as an adjuvant.

The information obtained from in vitro and in vivo studies has shown the biological properties of natural compounds from plants. Their regulatory effects against inflammatory molecules and NF-κB activation have been proposed for the treatment of inflammatory lung diseases, including AA [15,53,59,60]. Interestingly, in this study, the plant phenolic compound MH inhibited the production of various molecules and the activation of NF-κB in both in vitro and in vivo studies of experimental asthma, indicating the potential of natural compounds for asthma therapy. Compared with the anti-asthma effect of the existing natural compounds (40 or 200 mg/kg) [59,60], MH showed a superior effect at a low dose (5 mg/kg). Therefore, it is thought to be advantageous for development as an anti-asthma adjuvant.

4. Material and Methods

4.1. Reagents and Cell Culture

Methyl p-coumarate (methyl p-hydroxycinnamate) (MH) was obtained from SynQuest Laboratories, Inc. Airway epithelial cell line A549 and macrophage cell line RAW264.7 were obtained from the American Type Cell Culture (ATCC, Rockville, MD, USA), and were respectively grown in RPMI 1640 (contains 100 units/mL of penicillin in 100 μg/mL) and DMEM (contains 1% antibiotic antimycotic solution), supplemented with 10% FBS, at 37 °C in a CO_2 incubator.

To determine the cell viability, A549 cells were placed into 96-well plates at an average density of 2×10^4 cells per well and were maintained with MH (5–100 μg/mL) for 1 h. Then, the cells were incubated with 10 nM PMA or without PMA for 24 h. Subsequently, cell viability was measured in triplicate using a CytoX cell viability kit (LPS Solution, Inc., Daejeon, Korea).

To analyze the secretion levels of inflammatory molecules from A549 and RAW264.7 cells, A549 cells were placed in 96-well plates at an average density of 0.5×10^5 cells per well, maintained with MH for 1 h, and incubated with PMA (10 nM) for 18 h. RAW264.7 cells were placed in 96-well plates at an average density of 0.5×10^5 cells per well, maintained with MH for 1 h, and incubated with 500 ng/mL LPS for 18 h. Subsequently, the culture supernatants were assayed for cytokine, chemokine, and adhesion molecules using ELISA kits (BD Biosciences, Franklin Lakes, NJ, USA and R&D Systems, Inc., Minneapolis, MN, USA).

4.2. Nuclear and Cytoplasmic Extraction

The isolation of the nuclear and cytoplasmic fractions was performed based on previous protocols [61]. In brief, A549 cells were moved to a 60 mm cell culture dish at an average density of 2.5×10^5 cells per dish and maintained with MH (10–100 μM) for 1 h. Subsequently, the cells were incubated with 10 nM PMA for 1 h. Finally, cell harvesting was performed using mechanical scraping, and the isolation of the nuclear and cytoplasmic fractions was performed using a NucBuster™ Protein Extraction Kit (cat. no. 71183, Merck, Darmstadt, Germany).

4.3. Cell Adhesion Assays

Cell adhesion assays were used to determine the inhibitory ability of MH against the adhesion of airway cells and eosinophils based on previous protocols [15]. In brief, A549 cells were placed in 96-well plates at an average density of 1×10^4 cells per well, maintained with MH for 1 h, and subsequently incubated with PMA for 5 h. Eosinophilic leukemia EOL-1 cells stained with calcein AM (cat. no. 4892-010-02, Bio-Techne Corp, Minneapolis, MN, USA), a cell-permeable dye that has been used for cell adhesion research [4], were placed at an average density of 5×10^4 cells per well in 96-well plates containing A549 cells.

After 1 h, the co-cultured cells were washed with PBS, and the degree of adhesion was quantitatively analyzed through the fluorescence intensity, measured using a fluorescence microscope (Eclipse Ti-U, Nikon, Tokyo, Japan, 490/520 nm).

4.4. Establishment of the OVA-Induced AA mouse model

Female BALB/c mice (6 weeks old, Koatech Co., Ltd., Pyeongtaek, Korea) were divided into four groups (n = 6 per group) as follows: (i) NC, normal control group; (ii) OVA, ovalbumin/alum-sensitized group; (iii) DEX, OVA + dexamethasone (DEX, 1 mg/kg)-treated group; (iv) MH, OVA + methyl p-hydroxycinnamate (MH, 5 mg/kg)-treated group.

The experimental AA mouse model was developed based on previous protocols [21]. In brief, the intraperitoneal injection (i.p.) of the OVA/alum mixture was performed on BALB/c mice on days 0 and 7. They were exposed to 1% OVA for 1 h daily on days 11 to 13. The oral administration (p.o.) of 5 mg/kg MH or 1 mg/kg DEX (positive control) was performed on days 9 to 13.

4.5. Immune Cell Count and ELISA

BALF collection to detect the immune cell count and Th2 cytokine production was performed under anesthesia, induced by the mixture of 30 mg/kg Zoletil and 5 mg/kg xylazine (i.p.) on day 15, as previously described [34]. Diff-Quik staining was performed using BALF cells to distinguish the morphology of immune cells, and the immune cells, such as eosinophils and macrophages, were counted using a light microscope (magnification, ×400) [62]. BALF supernatant was assayed for Th2 cytokines/MCP-1, and serum was assayed for IgE using ELISA kits (R&D Systems, Inc., Minneapolis, MN, USA and Biolegend, Inc., San Diego, CA, USA).

4.6. Western Blot Analysis

The lysates of cell culture and mouse lungs were prepared in a lysis buffer in the presence of protease/phosphatase inhibitors, and BCA protein quantification was then performed. Denatured protein samples were loaded on 10–15% SDS-PAGE gels and transferred to PVDF membranes. After blocking the membranes with 5% skimmed milk in TBST, the membranes were incubated with the following primary antibodies: phosphorylated (p)-IκBα (cat. no. 9246), p-NF-κB p65 (3033), p-c-Jun (32740), c-Jun (9165), p-c-Fos (5348), and c-Fos (2250) (all 1:1000; Cell Signaling Technology, Inc., Danvers, MA, USA); IκBα (cat. no. MA5-15132) and β-actin (MA5-15739) (both 1:1000; Invitrogen; Thermo Fisher Scientific, Inc., Rockford, IL, USA); NF-κB p65 (cat. no. sc-8008), iNOS (sc-651), COX-2 (sc-1747), and Lamin A/C (sc-376248) (all 1:1000; Santa Cruz Biotechnology, Inc., CA, USA). Then, the membranes were maintained with the corresponding secondary antibodies, washed with PBS, and exposed to ECL solution to visualize each band.

4.7. Histological Analysis

The lung tissues were washed, fixed with 10% formalin, and finally embedded in paraffin. The paraffin-embedded lung tissues were then sectioned at a thickness of 4 μm using a rotary microtome and stained with H&E and PAS staining solution [63,64].

4.8. Ethics Statement

The procedures for the animal experiments were approved by the IACUC on 23 March 2022 of the Korea Research Institute of Bioscience and Biotechnology (KRIBB, Chungbuk, Korea; KRIBB-AEC-22074). The procedure was also performed in compliance with the NIH Guidelines for the Care and Use of Laboratory Animals, as well as the Korean national laws for animal welfare.

4.9. Statistical Analysis

Values are expressed as the mean ± SD ≥ 3 independent experiments. One-way ANOVA followed by Tukey's multiple comparison test were performed to analyze the differences between groups using SPSS 20.0 software (IBM Corp, Armonk, NY, USA). $p < 0.05$ was considered to indicate a statistically significant difference.

5. Conclusions

The modulatory effects of MH on airway cell/macrophage-derived molecules, airway cell/eosinophil adhesion, Th2 cytokine/MCP-1/IgE secretion, eosinophil/macrophage recruitment, mucus production, and iNOS/COX-2 expression were, in general, significant. These effects were accompanied by the suppression of NF-κB activation. These results demonstrate the anti-inflammatory effects of MH on airway inflammation and suggest that these effects may be based on NF-κB inactivation. Thus, it is suggested that MH may hold promise as an adjuvant in the treatment of AA. Further experiments confirming whether MH acts directly on the proteins may clarify the effect and mechanism of MH.

Author Contributions: J.-W.P. carried out the biological experiments for the in vitro study and analyzed the data. J.C. and J.L. carried out the biological experiments for the in vivo study and analyzed the data. J.-M.P., S.-M.K. and J.-H.M. carried out the biological experiments for the in vivo study. D.-Y.S., S.-H.G., J.-H.K., O.-K.K. and K.L. carried out the biological experiments for the in vitro study. K.-S.A. wrote and reviewed the manuscript. S.-R.O. and J.-W.L. wrote, reviewed, and edited the manuscript. All authors have read and agreed to the published version of the manuscript.

Funding: This research was funded by the KRIBB Research Initiative Program (grant nos. KGM5522211 & KGS1232221) of Korea.

Institutional Review Board Statement: Not applicable.

Informed Consent Statement: Not applicable.

Data Availability Statement: Not applicable.

Conflicts of Interest: The authors declare no conflict of interest.

References

1. Bezzaoucha, A. Epidemiology of asthma in children and young adults in Algiers. *Rev. Mal. Respir.* **1992**, *9*, 417–423. [PubMed]
2. Loftus, P.A.; Wise, S.K. Epidemiology of asthma. *Curr. Opin. Otolaryngol. Head Neck Surg.* **2016**, *24*, 245–249. [CrossRef]
3. Lambrecht, B.N.; Hammad, H. The immunology of asthma. *Nat. Immunol.* **2015**, *16*, 45–56. [CrossRef] [PubMed]
4. Kwon, O.K.; Lee, J.W.; Xuezhen, X.; Harmalkar, D.S.; Song, J.G.; Park, J.W.; Hwang, D.; Min, J.H.; Kim, J.H.; Han, H.K.; et al. DK-1108 exerts anti-inflammatory activity against phorbol 12-myristate 13-acetate-induced inflammation and protective effect against OVA-induced allergic asthma. *Biomed. Pharmacother.* **2020**, *132*, 110950. [CrossRef] [PubMed]
5. Rincon, M.; Irvin, C.G. Role of IL-6 in asthma and other inflammatory pulmonary diseases. *Int. J. Biol. Sci.* **2012**, *8*, 1281–1290. [CrossRef]
6. Sharma, N.; Akkoyunlu, M.; Rabin, R.L. Macrophages-common culprit in obesity and asthma. *Allergy* **2018**, *73*, 1196–1205. [CrossRef]
7. Ishioka, S. Role of cytokines in pathophysiology of asthma. *Nihon Naika Gakkai Zasshi* **1996**, *85*, 184–188. [PubMed]
8. Truyen, E.; Coteur, L.; Dilissen, E.; Overbergh, L.; Dupont, L.J.; Ceuppens, J.L.; Bullens, D.M. Evaluation of airway inflammation by quantitative Th1/Th2 cytokine mRNA measurement in sputum of asthma patients. *Thorax* **2006**, *61*, 202–208. [CrossRef] [PubMed]
9. Lee, J.W.; Chun, W.; Lee, H.J.; Min, J.H.; Kim, S.M.; Seo, J.Y.; Ahn, K.S.; Oh, S.R. The Role of Macrophages in the Development of Acute and Chronic Inflammatory Lung Diseases. *Cells* **2021**, *10*, 897. [CrossRef] [PubMed]
10. Takezawa, K.; Ogawa, T.; Shimizu, S.; Shimizu, T. Epidermal growth factor receptor inhibitor AG1478 inhibits mucus hypersecretion in airway epithelium. *Am. J. Rhinol. Allergy* **2016**, *30*, e1–e6. [CrossRef] [PubMed]
11. Dunican, E.M.; Elicker, B.M.; Gierada, D.S.; Nagle, S.K.; Schiebler, M.L.; Newell, J.D.; Raymond, W.W.; Lachowicz-Scroggins, M.E.; Di Maio, S.; Hoffman, E.A.; et al. Mucus plugs in patients with asthma linked to eosinophilia and airflow obstruction. *J. Clin. Investig.* **2018**, *128*, 997–1009. [CrossRef] [PubMed]
12. Prado, C.M.; Martins, M.A.; Tiberio, I.F. Nitric oxide in asthma physiopathology. *ISRN Allergy* **2011**, *2011*, 832560. [CrossRef] [PubMed]
13. Kim, S.M.; Ryu, H.W.; Kwon, O.K.; Hwang, D.; Kim, M.G.; Min, J.H.; Zhang, Z.; Kim, S.Y.; Paik, J.H.; Oh, S.R.; et al. *Callicarpa japonica* Thunb. ameliorates allergic airway inflammation by suppressing NF-kappaB activation and upregulating HO-1 expression. *J. Ethnopharmacol.* **2021**, *267*, 113523. [CrossRef] [PubMed]

14. Daham, K.; James, A.; Balgoma, D.; Kupczyk, M.; Billing, B.; Lindeberg, A.; Henriksson, E.; FitzGerald, G.A.; Wheelock, C.E.; Dahlen, S.E.; et al. Effects of selective COX-2 inhibition on allergen-induced bronchoconstriction and airway inflammation in asthma. *J. Allergy Clin. Immunol.* **2014**, *134*, 306–313. [CrossRef]
15. Park, J.W.; Kim, S.M.; Min, J.H.; Kim, M.G.; Kwon, O.K.; Hwang, D.; Oh, J.H.; Park, M.W.; Chun, W.; Lee, H.J.; et al. 3,4,5-Trihydroxycinnamic acid exerts anti-asthmatic effects in vitro and in vivo. *Int. Immunopharmacol.* **2020**, *88*, 107002. [CrossRef]
16. Lee, J.W.; Ryu, H.W.; Kim, D.Y.; Kwon, O.K.; Jang, H.J.; Kwon, H.J.; Kim, S.Y.; Lee, S.U.; Kim, S.M.; Oh, E.S.; et al. Biflavonoid-rich fraction from *Daphne pseudomezereum* var. koreana Hamaya exerts anti-inflammatory effect in an experimental animal model of allergic asthma. *J. Ethnopharmacol.* **2021**, *265*, 113386. [CrossRef]
17. Xu, T.; Ge, X.; Lu, C.; Dai, W.; Chen, H.; Xiao, Z.; Wu, L.; Liang, G.; Ying, S.; Zhang, Y.; et al. Baicalein attenuates OVA-induced allergic airway inflammation through the inhibition of the NF-kappaB signaling pathway. *Aging* **2019**, *11*, 9310–9327. [CrossRef]
18. Chauhan, P.S.; Singh, D.K.; Dash, D.; Singh, R. Intranasal curcumin regulates chronic asthma in mice by modulating NF-kB activation and MAPK signaling. *Phytomedicine* **2018**, *51*, 29–38. [CrossRef]
19. Patil, R.H.; Kumar, M.N.; Kumar, K.K.M.; Nagesh, R.; Kavya, K.; Babu, R.L.; Ramesh, G.T.; Sharma, S.C. Dexamethasone inhibits inflammatory response via down regulation of AP-1 transcription factor in human lung epithelial cells. *Gene* **2018**, *645*, 85–94. [CrossRef]
20. Lee, J.W.; Min, J.H.; Kim, M.G.; Kim, S.M.; Kwon, O.K.; Oh, T.K.; Lee, J.K.; Kim, T.Y.; Lee, S.W.; Choi, S.; et al. *Pistacia weinmannifolia* root exerts a protective role in ovalbumininduced lung inflammation in a mouse allergic asthma model. *Int. J. Mol. Med.* **2019**, *44*, 2171–2180. [CrossRef]
21. Kim, M.G.; Kim, S.M.; Min, J.H.; Kwon, O.K.; Park, M.H.; Park, J.W.; Ahn, H.I.; Hwang, J.Y.; Oh, S.R.; Lee, J.W.; et al. Anti-inflammatory effects of linalool on ovalbumin-induced pulmonary inflammation. *Int. Immunopharmacol.* **2019**, *74*, 105706. [CrossRef] [PubMed]
22. Maruthamuthu, V.; Henry, L.J.K.; Ramar, M.K.; Kandasamy, R. *Myxopyrum serratulum* ameliorates airway inflammation in LPS-stimulated RAW 264.7 macrophages and OVA-induced murine model of allergic asthma. *J. Ethnopharmacol.* **2020**, *255*, 112369. [CrossRef] [PubMed]
23. Zeghbib, W.; Boudjouan, F.; Vasconcelos, V.; Lopes, G. Phenolic Compounds' Occurrence in Opuntia Species and Their Role in the Inflammatory Process: A Review. *Molecules* **2022**, *27*, 4763. [CrossRef] [PubMed]
24. Ko, J.W.; Kwon, H.J.; Seo, C.S.; Choi, S.J.; Shin, N.R.; Kim, S.H.; Kim, Y.H.; Kim, J.C.; Kim, M.S.; Shin, I.S. 4-Hydroxycinnamic acid suppresses airway inflammation and mucus hypersecretion in allergic asthma induced by ovalbumin challenge. *Phytother. Res.* **2020**, *34*, 624–633. [CrossRef] [PubMed]
25. Min, J.H.; Kim, M.G.; Kim, S.M.; Park, J.W.; Chun, W.; Lee, H.J.; Oh, S.R.; Ahn, K.S.; Lee, J.W. 3,4,5-Trihydroxycinnamic acid exerts a protective effect on pulmonary inflammation in an experimental animal model of COPD. *Int. Immunopharmacol.* **2020**, *85*, 106656. [CrossRef] [PubMed]
26. Satooka, H.; Nihei, K.-I.; Kubo, I. Intracellular oxidation of methyl p-coumarate is involved in anti-melanogenic and cytotoxic activities against melanoma cells. *Phytochem. Lett.* **2022**, *50*, 89–94. [CrossRef]
27. Hooperb, S.N.J.; Jurgens, T.; Chandler, R.F.; Stevens, M.F. Methyl p-coumarate: A cytotoxic constituent from *Comptonia peregrina*. *Phytochemistry* **1984**, *23*, 2096–2097. [CrossRef]
28. Quek, A.; Kassim, N.K.; Lim, P.C.; Tan, D.C.; Mohammad Latif, M.A.; Ismail, A.; Shaari, K.; Awang, K. alpha-Amylase and dipeptidyl peptidase-4 (DPP-4) inhibitory effects of *Melicope latifolia* bark extracts and identification of bioactive constituents using in vitro and in silico approaches. *Pharm. Biol.* **2021**, *59*, 964–973. [CrossRef]
29. Zhang, W.M.; Wang, W.; Zhang, J.J.; Wang, Z.R.; Wang, Y.; Hao, W.J.; Huang, W.Y. Antibacterial Constituents of Hainan *Morinda citrifolia* (Noni) Leaves. *J. Food Sci.* **2016**, *81*, M1192–M1196. [CrossRef]
30. Yuan, S.; Li, W.; Li, Q.; Wang, L.; Cao, J.; Jiang, W. Defense Responses, Induced by p-Coumaric Acid and Methyl p-Coumarate, of Jujube (*Ziziphus jujuba* Mill.) Fruit against Black Spot Rot Caused by *Alternaria alternata*. *J. Agric. Food Chem.* **2019**, *67*, 2801–2810. [CrossRef]
31. Yang, Y.; Wang, H.; Guo, L.; Chen, Y. Determination of three compounds in *Aloe vera* by capillary electrophoresis. *Biomed. Chromatogr.* **2004**, *18*, 112–116. [CrossRef]
32. Kubo, I.; Nihei, K.; Tsujimoto, K. Methyl p-coumarate, a melanin formation inhibitor in B16 mouse melanoma cells. *Bioorg. Med. Chem.* **2004**, *12*, 5349–5354. [CrossRef] [PubMed]
33. Vo, V.A.; Lee, J.W.; Shin, S.Y.; Kwon, J.H.; Lee, H.J.; Kim, S.S.; Kwon, Y.S.; Chun, W. Methyl p-Hydroxycinnamate Suppresses Lipopolysaccharide-Induced Inflammatory Responses through Akt Phosphorylation in RAW264.7 Cells. *Biomol. Ther.* **2014**, *22*, 10–16. [CrossRef]
34. Kim, S.M.; Min, J.H.; Kim, J.H.; Choi, J.; Park, J.M.; Lee, J.; Goo, S.H.; Oh, J.H.; Kim, S.H.; Chun, W.; et al. Methyl phydroxycinnamate exerts antiinflammatory effects in mouse models of lipopolysaccharideinduced ARDS. *Mol. Med. Rep.* **2022**, *25*, 37. [CrossRef]
35. Burke-Gaffney, A.; Hellewell, P.G. A CD18/ICAM-1-dependent pathway mediates eosinophil adhesion to human bronchial epithelial cells. *Am. J. Respir. Cell Mol. Biol.* **1998**, *19*, 408–418. [CrossRef] [PubMed]
36. Deshmane, S.L.; Kremlev, S.; Amini, S.; Sawaya, B.E. Monocyte chemoattractant protein-1 (MCP-1): An overview. *J. Interf. Cytokine Res.* **2009**, *29*, 313–326. [CrossRef] [PubMed]

37. Poynter, M.E.; Irvin, C.G. Interleukin-6 as a biomarker for asthma: Hype or is there something else? *Eur. Respir. J.* **2016**, *48*, 979–981. [CrossRef] [PubMed]
38. Amin, K.; Ludviksdottir, D.; Janson, C.; Nettelbladt, O.; Bjornsson, E.; Roomans, G.M.; Boman, G.; Seveus, L.; Venge, P. Inflammation and structural changes in the airways of patients with atopic and nonatopic asthma. BHR Group. *Am. J. Respir. Crit. Care Med.* **2000**, *162*, 2295–2301. [CrossRef] [PubMed]
39. Hoshi, H.; Ohno, I.; Honma, M.; Tanno, Y.; Yamauchi, K.; Tamura, G.; Shirato, K. IL-5, IL-8 and GM-CSF immunostaining of sputum cells in bronchial asthma and chronic bronchitis. *Clin. Exp. Allergy* **1995**, *25*, 720–728. [CrossRef]
40. Lee, Y.G.; Jeong, J.J.; Nyenhuis, S.; Berdyshev, E.; Chung, S.; Ranjan, R.; Karpurapu, M.; Deng, J.; Qian, F.; Kelly, E.A.; et al. Recruited alveolar macrophages, in response to airway epithelial-derived monocyte chemoattractant protein 1/CCl2, regulate airway inflammation and remodeling in allergic asthma. *Am. J. Respir. Cell Mol. Biol.* **2015**, *52*, 772–784. [CrossRef] [PubMed]
41. Nie, H.; Wang, A.; He, Q.; Yang, Q.; Liu, L.; Zhang, G.; Huang, Y.; Ding, X.; Yu, H.; Hu, S. Phenotypic switch in lung interstitial macrophage polarization in an ovalbumin-induced mouse model of asthma. *Exp. Ther. Med.* **2017**, *14*, 1284–1292. [CrossRef]
42. Jiang, S.; Wang, Q.; Wang, Y.; Song, X.; Zhang, Y. Blockade of CCL2/CCR2 signaling pathway prevents inflammatory monocyte recruitment and attenuates OVA-Induced allergic asthma in mice. *Immunol. Lett.* **2019**, *214*, 30–36. [CrossRef]
43. Junttila, I.S. Tuning the Cytokine Responses: An Update on Interleukin (IL)-4 and IL-13 Receptor Complexes. *Front. Immunol.* **2018**, *9*, 888. [CrossRef] [PubMed]
44. Nagase, H.; Ueki, S.; Fujieda, S. The roles of IL-5 and anti-IL-5 treatment in eosinophilic diseases: Asthma, eosinophilic granulomatosis with polyangiitis, and eosinophilic chronic rhinosinusitis. *Allergol. Int.* **2020**, *69*, 178–186. [CrossRef]
45. Ip, W.K.; Wong, C.K.; Lam, C.W. Interleukin (IL)-4 and IL-13 up-regulate monocyte chemoattractant protein-1 expression in human bronchial epithelial cells: Involvement of p38 mitogen-activated protein kinase, extracellular signal-regulated kinase 1/2 and Janus kinase-2 but not c-Jun NH2-terminal kinase 1/2 signalling pathways. *Clin. Exp. Immunol.* **2006**, *145*, 162–172. [CrossRef] [PubMed]
46. Lambrecht, B.N.; Hammad, H.; Fahy, J.V. The Cytokines of Asthma. *Immunity* **2019**, *50*, 975–991. [CrossRef] [PubMed]
47. Leon, B.; Ballesteros-Tato, A. Modulating Th2 Cell Immunity for the Treatment of Asthma. *Front. Immunol.* **2021**, *12*, 637948. [CrossRef] [PubMed]
48. Pelaia, G.; Canonica, G.W.; Matucci, A.; Paolini, R.; Triggiani, M.; Paggiaro, P. Targeted therapy in severe asthma today: Focus on immunoglobulin E. *Drug Des. Devel Ther.* **2017**, *11*, 1979–1987. [CrossRef] [PubMed]
49. Chipps, B.E.; Marshik, P.L. Targeted IgE Therapy for Patients With Moderate to Severe Asthma. *Biotechnol. Healthc.* **2004**, *1*, 56–61. [PubMed]
50. Rael, E.L.; Lockey, R.F. Interleukin-13 signaling and its role in asthma. *World Allergy Organ. J.* **2011**, *4*, 54–64. [CrossRef] [PubMed]
51. Bagnasco, D.; Ferrando, M.; Varricchi, G.; Passalacqua, G.; Canonica, G.W. A Critical Evaluation of Anti-IL-13 and Anti-IL-4 Strategies in Severe Asthma. *Int. Arch. Allergy Immunol.* **2016**, *170*, 122–131. [CrossRef] [PubMed]
52. Pelaia, C.; Heffler, E.; Crimi, C.; Maglio, A.; Vatrella, A.; Pelaia, G.; Canonica, G.W. Interleukins 4 and 13 in Asthma: Key Pathophysiologic Cytokines and Druggable Molecular Targets. *Front. Pharmacol.* **2022**, *13*, 851940. [CrossRef]
53. Ryu, H.W.; Lee, J.W.; Kim, M.O.; Lee, R.W.; Kang, M.J.; Kim, S.M.; Min, J.H.; Oh, E.S.; Song, Y.N.; Jung, S.; et al. Daphnodorin C isolated from the stems of *Daphne kiusiana* Miquel attenuates airway inflammation in a mouse model of chronic obstructive pulmonary disease. *Phytomedicine* **2022**, *96*, 153848. [CrossRef]
54. Kim, S.M.; Ryu, H.W.; Kwon, O.K.; Min, J.H.; Park, J.M.; Kim, D.Y.; Oh, S.R.; Lee, S.J.; Ahn, K.S.; Lee, J.W. Protective Effect of *Paulownia tomentosa* Fruits in an Experimental Animal Model of Acute Lung Injury. *Microbiol. Biotechnol. Lett.* **2022**, *50*, 310–318. [CrossRef]
55. Qian, J.; Ma, X.; Xun, Y.; Pan, L. Protective effect of forsythiaside A on OVA-induced asthma in mice. *Eur. J. Pharmacol.* **2017**, *812*, 250–255. [CrossRef]
56. Li, J.; Kartha, S.; Iasvovskaia, S.; Tan, A.; Bhat, R.K.; Manaligod, J.M.; Page, K.; Brasier, A.R.; Hershenson, M.B. Regulation of human airway epithelial cell IL-8 expression by MAP kinases. *Am. J. Physiol. Lung Cell. Mol. Physiol.* **2002**, *283*, L690–L699. [CrossRef] [PubMed]
57. Shin, N.R.; Lee, A.Y.; Song, J.H.; Yang, S.; Park, I.; Lim, J.O.; Jung, T.Y.; Ko, J.W.; Kim, J.C.; Lim, K.S.; et al. *Scrophularia buergeriana* attenuates allergic inflammation by reducing NF-kappaB activation. *Phytomedicine* **2020**, *67*, 153159. [CrossRef]
58. Lu, X.; Xu, C.; Yang, R.; Zhang, G. Ganoderic Acid A Alleviates OVA-Induced Asthma in Mice. *Inflammation* **2021**, *44*, 1908–1915. [CrossRef]
59. Oh, S.W.; Cha, J.Y.; Jung, J.E.; Chang, B.C.; Kwon, H.J.; Lee, B.R.; Kim, D.Y. Curcumin attenuates allergic airway inflammation and hyper-responsiveness in mice through NF-kappaB inhibition. *J. Ethnopharmacol.* **2011**, *136*, 414–421. [CrossRef] [PubMed]
60. Lu, Y.; Cai, S.; Nie, J.; Li, Y.; Shi, G.; Hao, J.; Fu, W.; Tan, H.; Chen, S.; Li, B.; et al. The natural compound nujiangexanthone A suppresses mast cell activation and allergic asthma. *Biochem. Pharmacol.* **2016**, *100*, 61–72. [CrossRef] [PubMed]
61. Park, J.W.; Ryu, H.W.; Ahn, H.I.; Min, J.H.; Kim, S.M.; Kim, M.G.; Kwon, O.K.; Hwang, D.; Kim, S.Y.; Choi, S.; et al. The Anti-Inflammatory Effect of *Trichilia martiana* C. DC. in the Lipopolysaccharide-Stimulated Inflammatory Response in Macrophages and Airway Epithelial Cells and in LPS-Challenged Mice. *J. Microbiol. Biotechnol.* **2020**, *30*, 1614–1625. [CrossRef] [PubMed]
62. Kim, M.O.; Lee, J.W.; Lee, J.K.; Song, Y.N.; Oh, E.S.; Ro, H.; Yoon, D.; Jeong, Y.H.; Park, J.Y.; Hong, S.T.; et al. Black Ginseng Extract Suppresses Airway Inflammation Induced by Cigarette Smoke and Lipopolysaccharides In Vivo. *Antioxidants* **2022**, *11*, 679. [CrossRef] [PubMed]

63. Lee, J.W.; Kim, M.O.; Song, Y.N.; Min, J.H.; Kim, S.M.; Kang, M.J.; Oh, E.S.; Lee, R.W.; Jung, S.; Ro, H.; et al. Compound K ameliorates airway inflammation and mucus secretion through the regulation of PKC signaling in vitro and in vivo. *J. Ginseng Res.* **2022**, *46*, 496–504. [CrossRef] [PubMed]
64. Park, H.A.; Kwon, O.K.; Ryu, H.W.; Min, J.H.; Park, M.W.; Park, M.H.; Paik, J.H.; Choi, S.; Paryanto, I.; Yuniato, P.; et al. *Physalis peruviana* L. inhibits ovalbumininduced airway inflammation by attenuating the activation of NFkappaB and inflammatory molecules. *Int. J. Mol. Med.* **2019**, *43*, 1830–1838. [CrossRef] [PubMed]

International Journal of *Molecular Sciences*

MDPI

Article

A Lupin (*Lupinus angustifolius*) Protein Hydrolysate Exerts Anxiolytic-Like Effects in Western Diet-Fed ApoE$^{-/-}$ Mice

Guillermo Santos-Sánchez [1,2,†], Eduardo Ponce-España [1,2,†], Juan Carlos López [3], Nuria Álvarez-Sánchez [1,2,‡], Ana Isabel Álvarez-López [1,2], Justo Pedroche [4], Francisco Millán [4], María Carmen Millán-Linares [2,4], Patricia Judith Lardone [1,2], Ignacio Bejarano [1,2], Ivan Cruz-Chamorro [1,2,*] and Antonio Carrillo-Vico [1,2,*]

1 Instituto de Biomedicina de Sevilla, IBiS (Universidad de Sevilla, HUVR, Junta de Andalucía, CSIC), 41013 Seville, Spain
2 Departamento de Bioquímica Médica y Biología Molecular e Inmunología, Facultad de Medicina, Universidad de Sevilla, 41009 Seville, Spain
3 Departamento de Psicología Experimental, Universidad de Sevilla, 41009 Seville, Spain
4 Department of Food & Health, Instituto de la Grasa, CSIC, Ctra, Utrera Km 1, 41013 Seville, Spain
* Correspondence: icruz-ibis@us.es (I.C.-C.); vico@us.es (A.C.-V.); Tel.: +34-955923106 (A.C.-V.); Fax: +34-954907048 (A.C.-V.)
† These authors contributed equally to this work.
‡ Present addresses: Keenan Research Centre for Biomedical Science of St. Michael's Hospital, Toronto, ON M5S 1B2, Canada.

Citation: Santos-Sánchez, G.; Ponce-España, E.; López, J.C.; Álvarez-Sánchez, N.; Álvarez-López, A.I.; Pedroche, J.; Millán, F.; Millán-Linares, M.C.; Lardone, P.J.; Bejarano, I.; et al. A Lupin (*Lupinus angustifolius*) Protein Hydrolysate Exerts Anxiolytic-Like Effects in Western Diet-Fed ApoE$^{-/-}$ Mice. *Int. J. Mol. Sci.* **2022**, *23*, 9828. https://doi.org/10.3390/ijms23179828

Academic Editor: Akiyoshi Saitoh

Received: 25 July 2022
Accepted: 25 August 2022
Published: 29 August 2022

Abstract: Anxiety is the most prevalent psychiatric disorder worldwide, causing a substantial economic burden due to the associated healthcare costs. Given that commercial anxiolytic treatments may cause important side effects and have medical restrictions for prescription and high costs, the search for new natural and safer treatments is gaining attention. Since lupin protein hydrolysate (LPH) has been shown to be safe and exert anti-inflammatory and antioxidant effects, key risk factors for the anxiety process and memory impairment, we evaluated in this study the potential effects of LPH on anxiety and spatial memory in a Western diet (WD)-induced anxiety model in ApoE$^{-/-}$ mice. We showed that 20.86% of the 278 identified LPH peptides have biological activity related to anxiolytic/analgesic effects; the principal motifs found were the following: VPL, PGP, YL, and GQ. Moreover, 14 weeks of intragastrical LPH treatment (100 mg/kg) restored the WD-induced anxiety effects, reestablishing the anxiety levels observed in the standard diet (SD)-fed mice since they spent less time in the anxiety zones of the elevated plus maze (EPM). Furthermore, a significant increase in the number of *head dips* was recorded in LPH-treated mice, which indicates a greater exploration capacity and less fear due to lower levels of anxiety. Interestingly, the LPH group showed similar thigmotaxis, a well-established indicator of animal anxiety and fear, to the SD group, counteracting the WD effect. This is the first study to show that LPH treatment has anxiolytic effects, pointing to LPH as a potential component of future nutritional therapies in patients with anxiety.

Keywords: lupin; peptides; protein hydrolysates; anxiety; ApoE$^{-/-}$; functional foods; peptidomics

1. Introduction

Anxiety disorders (AnxDs), characterized by anxiety and fear, are the most common mental disorder worldwide [1]. They affect 33.7% of the global population during their lifetime, generating an important economic burden due to enormous healthcare expenditure [2]. AnxDs have serious consequences on physical and mental health (headache, irritability, breathing problems, depression, fatigue, etc.), affecting the course of normal daily activities of patients and reducing their quality of life [3]. Numerous studies have shown a strong relationship between anxiety and the consumption of diets rich in refined sugars and saturated fats [4,5]. Furthermore, the intake of these types of diet is the main risk factor for the generation of chronic diseases (diabetes, high blood pressure, cardiovascular diseases (CVDs), Alzheimer's disease, and chronic obstructive pulmonary disease), which in turn have been shown to be closely related to the presence of anxiety [6–11]. Thus, anxiety is highly prevalent in patients with chronic diseases and can also increase the risk of worsening functional impairment, comorbidities, and mortality [12–14]. Within chronic diseases, there is a close connection between anxiety and memory loss. Thus, several pieces of evidence have shown how acute stress can hinder the memorization process [15,16]. For these reasons, anxiety is postulated to be a modifiable risk factor for chronic diseases.

Nutrition is considered a major risk factor for chronic diseases. Scientific evidence supports the view that diet changes have positive or negative effects on health [17]. Thus, fine control of the diet can be useful in preventing the onset of some diseases. In this regard, dietary supplementation has been considered a strategy to modulate different metabolic pathways [18,19]. In particular, nutritional psychiatry, based on diet improvement for the prevention and treatment of mental disorders, including anxiety, is gaining attention in the scientific community, which uses animal models to assess the influence of new nutritional strategies and pharmacological interventions [20].

There are several commercial treatments to reduce anxiety (selective serotonin reuptake inhibitors, barbiturates, benzodiazepines, analogues of benzodiazepine, etc.); however, many of them have important side effects that affect quality of life, such as drowsiness, sedation, confusion, and headache [21]. Therefore, the search for new natural and safer treatments has been of great interest over the last few years. The dietary supplementation with proteins and peptides have shown beneficial effects in human health modulating and/or optimizing several physiological processes and diseases such as hypertension, obesity, atherosclerosis, neurological dysfunctions, and other metabolic disorders [22–27]. There are many peptides from different foods that have also shown anxiolytic and anti-amnesic activity [28]. Soymorphin-5 (YPFVV), soymorphin-6 (YPFVVN), and soymorphin-7 (YPFVVNA) [29], derived from soybean β-conglycinin, as well as rubiscolin-6 (YPLDLF) and rubimetide (MRW) [30], obtained from ribulose-1,5-bisphosphate carboxylase-oxygenase (RuBisCO) [31], have been shown to possess anxiolytic-like effects in mouse models. Moreover, ovolin (VYLPR) [32] from ovoalbumine, and peptides from αs1-casein [33,34] and β-lactoglobulin [35], have also been shown to reduce anxiety. Numerous peptides with antiamnesic effects from β-lactoglobulin have also been identified [36].

On the other hand, several studies have reported high levels of anxiety and spatial cognitive deficits (memory loss) in apolipoprotein E (ApoE) knockout mice ($ApoE^{-/-}$) compared to wild-type mice [37–39]. ApoE deficiency results in an age-dependent dysregulation of the hypothalamic-pituitary-adrenal (HPA) axis through a mechanism that affects primarily the adrenal gland. The HPA axis regulates the secretion of glucocorticoids (GCs), which play important roles in several brain functions, including cognition. Dysregulation of the HPA axis has also been associated with behavioral alterations. Thus, $ApoE^{-/-}$ mice show higher anxiety values than wild-type animals by using the elevated plus maze (EPM) test [40]. In addition, anxiety and memory loss can be accelerated and increased in $ApoE^{-/-}$ fed a high-fat diet by oxidant and inflammatory effects [37,38]. Moreover, recent studies have shown a strong link between high cholesterol levels and anxiety [4,41].

Oxidative stress and inflammation play a key role in the anxiety process and memory impairment. In fact, alteration in redox balance, increased reactive oxygen species (ROS) production and high circulating inflammatory cytokines such as interleukin (IL)-1, IL-6, and tumor necrosis factor (TNF) have been detected both in anxiety patients and stressed animal models of anxiety [42–44]. In this line, our group has previously described that a *Lupinus angustifolius* protein hydrolysate (LPH) exerts hypocholesterolemic, anti-inflammatory and antioxidant effects in in vitro [45] and in vivo [46,47] models. In light of these considerations, this study aimed to identify LPH peptides with potential anxiolytic and antiamnesic effects and to evaluate the potential effects of LPH.

2. Results

2.1. Characterization of LPH

2.1.1. Chemical Analysis of LPH

The molecular weights of the detected peptides ranged from 0.76 to 3.11 kDa. Specifically, the percentage of peptides identified with a molecular weight of <1 kDa, 1–1.5 kDa, 1.5–2.0 kDa, 2.0–2.5 kDa, and >2.5 kDa were 5.05%, 41.00%, 39.20%, 11.15%, and 3.60%, respectively (Figure 1A). LPH contained peptides with hydrophobicity <+10 kcal/mol (10.43%), +10–15 kcal/mol (19.42%), +15–20 kcal/mol (36.33%), +20–30 kcal/mol (21.94%), and >+30 kcal/mol (11.87%) (Figure 1A). Furthermore, the peptides consisted of 7–26 amino acid (aa) residues, being the most frequent peptides (83.10%) containing between 10–19 aa (Figure 1B). Regarding the aa composition, glutamic acid, leucine, and isoleucine were the most represented (12.90%, 10.30% and 8.00%, respectively), while tryptophan, and methionine were the least (0.4%, and 0.5%) (Table 1).

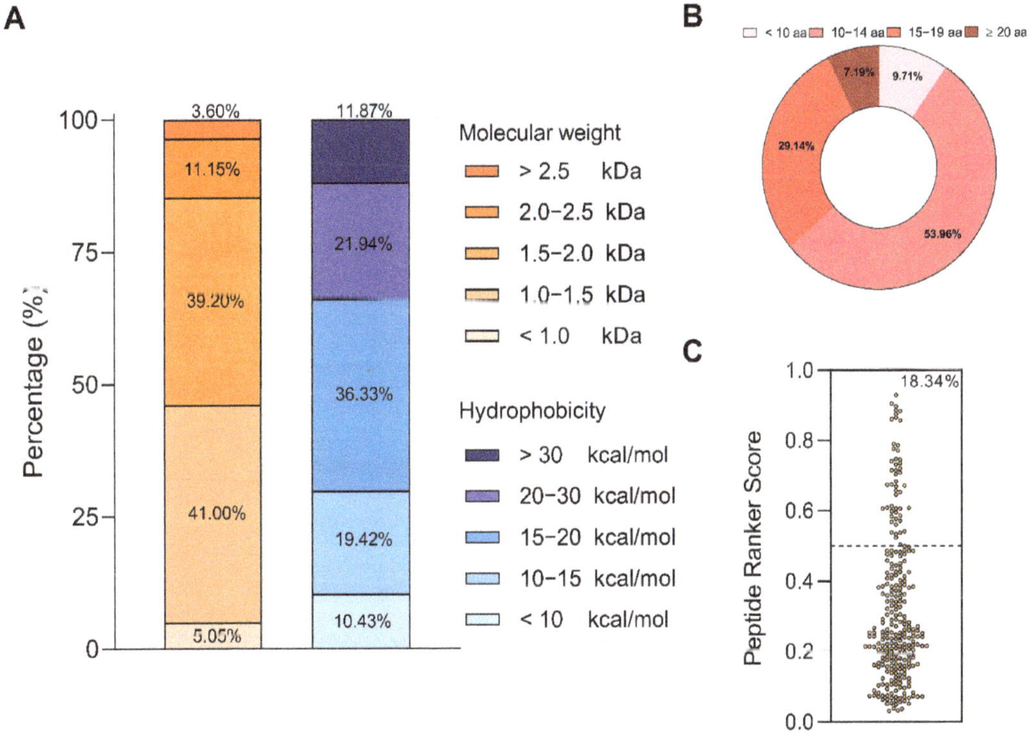

Figure 1. Percentage of molecular weight and hydrophobicity (**A**), length distribution (**B**), and predicted potential bioactivity (**C**) of the LPH peptides.

Table 1. Amino acid composition of the identified peptides in the LPH.

Amino Acid	No.	%
Glu (E)	493	12.9
Leu (L)	396	10.3
Ile (I)	307	8.0
Pro (P)	307	8.0
Arg (R)	287	7.5
Asp (D)	275	7.2
Val (V)	249	6.5
Gly (G)	224	5.8
Ser (S)	202	5.3
Gln (Q)	196	5.1
Asn (N)	185	4.8
Lys (K)	151	3.9
Ala (A)	130	3.4
Thr (T)	130	3.4
Phe (F)	122	3.2
Tyr (Y)	84	2.2
His (H)	60	1.6
Trp (W)	19	0.5
Met (M)	16	0.4
Cys (C)	0	0.0

Ala, alanine; Arg, arginine; Asn, asparagine; Asp, aspartic acid; Cys, cysteine; Gln, glutamine; Glu, glutamic acid; Gly, glycine; His, histidine; Ile, isoleucine; Leu, leucine; Lys, lysine; Met, methionine; Phe, phenylalanine; Pro, proline; Ser, serine; Thr, threonine; Trp, tryptophan; Tyr, tyrosine; Val, valine.

Finally, the bioactivity analysis of the LPH showed that 51 of the 278 sequences (18.34%) possess a score value greater than 0.5 threshold (Figure 1C).

2.1.2. LPH Contains Peptides with Anxiolytic and Antiamnesic Effects

There were 278 peptides with an area greater than 107 identified in LPH (Supplementary Table S3). These peptides belonged mainly to conglutins, the main storage protein in lupin seed. Of the 278 identified sequences, 58 peptides (20.86%) with potential biological activity related to anxiolytic/analgesic effects were identified. In particular, 49 (17.62%) sequences contained a demonstrated antiamnesic motif and 9 (3.24%) sequences contained a demonstrated anxiolytic motif (Table 2).

Table 2. The number of identified LPH peptides with anti-amnesic and anxiolytic activity.

Effect	Bioactive Peptide Motif [a]	BIOPEP-UWM ID [b]	Origin Protein [c]	Accession Number [c]	N. Peptides	Reference
anti-amnesic	VPL	3166	Non-conglutin proteins		1	[48]
	PGP PG GP	3459 3460 3461	α-Conglutin	F5B8V7	3	[49]
			β-Conglutin	F5B8W1 F5B8W2 F5B8W3	14	
			Non-conglutin proteins		31	
anxiolytic	YL	8310	α-Conglutin	F5B8V6	4	[50]
			Non-conglutin proteins		1	
	GQ	2890	α-Conglutin	F5B8V6 F5B8V7	3	[51]
			Non-conglutin proteins		1	
TOTAL					58	

[a] 1-letter amino acid code. [b] ID number present in the BIOPEP-UWM database [52]. [c] Accession number present in "UniProtKB" (http://www.uniprot.org/, accessed on 1 April 2022).

Of these 58 peptides, 41.38% were peptides from conglutin proteins, whereas 58.62% were from non-conglutin proteins. The tripeptides VPL and PGP, and the dipeptides PG and GP, were the sequences related to antiamnesic effects and identified with the BIOPEP-UWM IDs 3166, 3459, 3460, and 3461, respectively. The dipeptides YL and GQ were the sequences associated with anxiolytic effects and identified with the following IDs, 8310 and 2890, respectively. The physicochemical properties and primary structures of the identified motifs are shown in Figure 2.

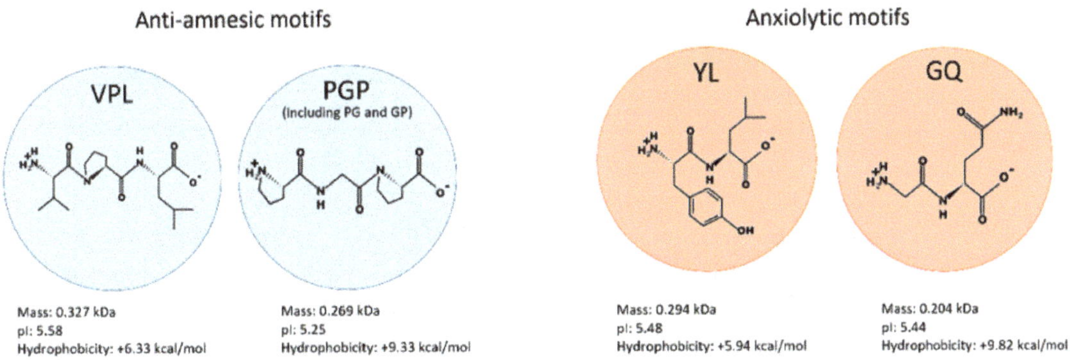

Figure 2. Physicochemical properties and primary structures of the anti-amnesic and anxiolytic motifs. pI, isoelectric point.

2.2. *In Vivo Experiments*

2.2.1. LPH Treatment Does Not Alter the Body Weight of Mice

To find the differences in weight changes between mice fed different diets and treated or not with LPH, the body weights of the mice were measured throughout the experiment. As shown in Table 3, there were no significant differences in the baseline body weight (BBW) at the beginning of the experiment among the experimental groups. Furthermore, after 16 weeks of diet, there were no differences in the final body weight (FBW) and in the body weight gain (BWG) between the groups fed WD and SD. In addition, 14 weeks of LPH treatment did not generate changes in the FBW and BWG of the mice, compared to the groups fed WD or SD.

Table 3. Body weight parameters.

Parameter (g)	Experimental Group		
	SD	WD	WD + LPH
BBW	20.35 ± 0.41	20.98 ± 0.36	20.88 ± 0.49
FBW	26.20 ± 0.87	26.50 ± 0.54	27.15 ± 0.69
BWG	5.85 ± 1.18	5.53 ± 0.65	6.28 ± 1.09

Baseline body weight (BBW), final body weight (FBW) and body weight gain (BWG) in ApoE$^{-/-}$ mice. Values are shown as the mean and standard error of the mean of each group. SD, standard diet fed-mice; WD, Western diet-fed mice; WD + LPH, Western diet-fed mice treated with LPH. No statistical differences were observed between the groups for each weight parameter.

2.2.2. LPH Palliates the Anxious Effects Induced by WD Ingestion

As shown in Figure 3B, WD-fed mice spent significantly less time in the open arms of the elevated plus maze and more in the closed arms compared to the SD group. This effect was overcome by LPH treatment. Furthermore, the time spent in the center was significantly shorter in WD compared to SD and WD + LPH. Representative images of the tracks of the mice in EPM are shown in Figure 3A. Other anxiety-related behaviors, such as *head dips* and rears, were also evaluated. As shown in Figure 3C, the number of *head dips* was significantly lower in the WD group compared to the SD and WD + LPH groups, while no differences in the rears were observed among groups.

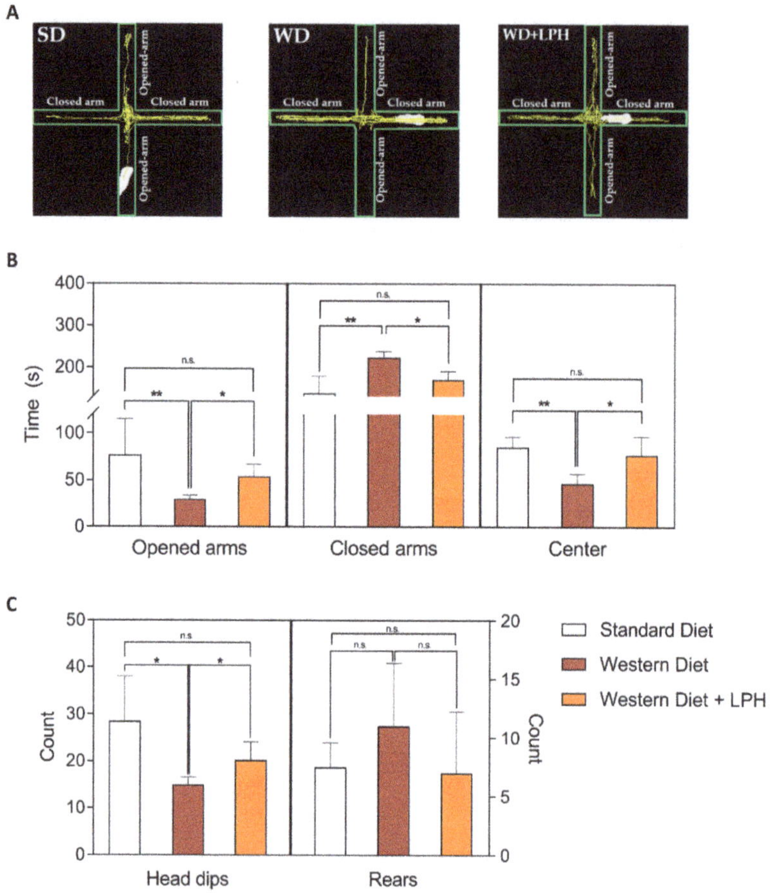

Figure 3. Representative images of the tracks of mice in the elevated plus maze (**A**). Time spent in opened arms, closed arms, and center zone (**B**), *head dips* and rears (**C**). Values are shown as the mean and standard deviation of each group. * $p \leq 0.05$; ** $p \leq 0.01$; n.s., not significant; SD, standard diet fed-mice; WD, Western diet-fed mice; WD + LPH, Western diet-fed mice treated with LPH; LPH, lupin protein hydrolysate.

2.2.3. LPH Treatment Does Not Improve Spatial Memory but Modulates WD-Induced Thigmotaxis, an Anxiety-Related Behavior

To study spatial learning and memory, the platform in the Morris water maze (MWM) was placed according to the Figure 4A. During nonvisible platform sessions (days 1–5), all groups learned to reach the submerged platform, due to the decrease in the mean latency over the consecutive five days of the learning period in all groups (Figure 4B). There were no significant differences between the groups in the latency time. After the removal of the platform (trial phase), there were also no differences in the time spent in the platform zone among the groups (Figure 4D), but curiously, there was a decrease in the total distance traveled for the WD-fed mice compared to the SD-fed mice (Figure 4E). This effect was overcome by the LPH treatment. Moreover, thigmotaxis was significantly higher in the WD diet group compared to the SD group, while LPH was able to reverse this increase, reducing thigmotaxis to values significantly different to the WD group, both considering the number of times animals approached the walls of the pool (Figure 4F) and the time spent in the outer area of the pool (Figure 4G). Representative images of the tracks of mice in the trial phase are shown in Figure 4C.

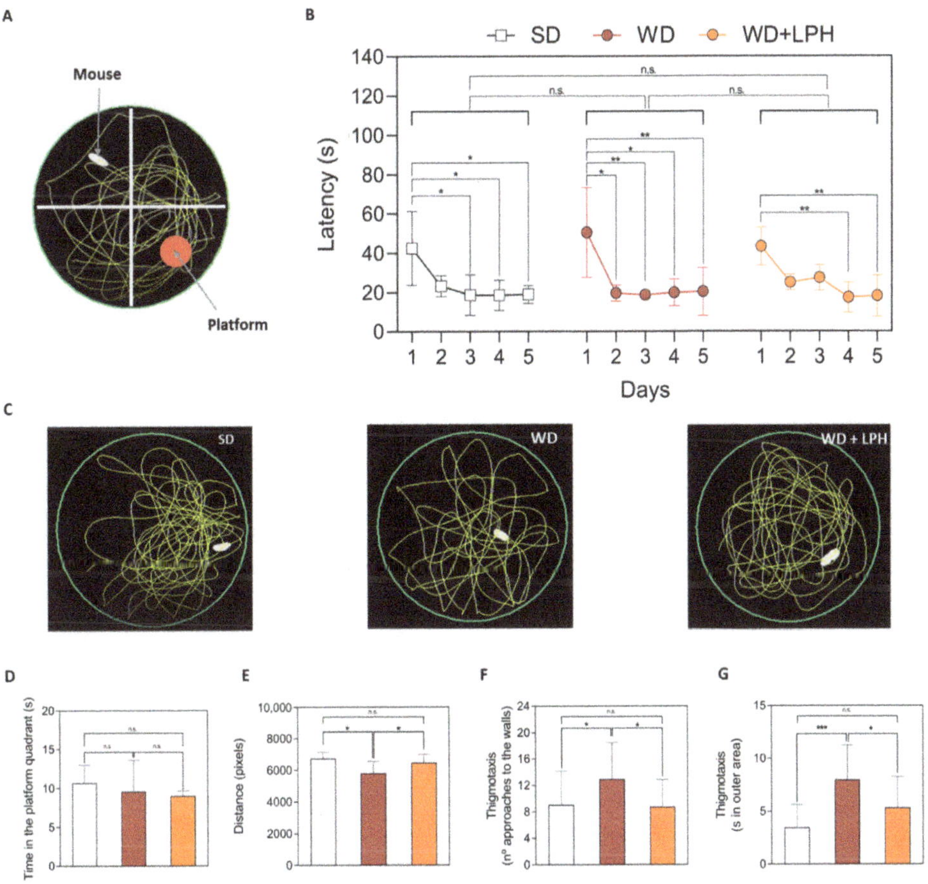

Figure 4. Representative image of the acquisition phase of the Morris water maze (**A**), latency of the mice during the five days (**B**). Representative images of the trial phase (**C**); time in the platform zone (**D**), distance traveled (**E**) and thigmotaxis (**F**,**G**). Values are shown as the mean and standard deviation of each group. * $p \leq 0.05$; ** $p \leq 0.01$; *** $p \leq 0.001$; n.s., not significant; SD, standard diet-fed mice; WD, Western diet-fed mice; WD + LPH, Western diet-fed mice treated with LPH; LPH, lupine protein hydrolysate.

3. Discussion

LPH is a mixture of low molecular weight peptides obtained after hydrolysis of *L. angustifolius* proteins with Alcalase®, which have shown beneficial effects on oxidant and inflammatory status in different models [45–47]. Due to inflammation and oxidative stress are key processes in anxiety and memory impairment, the present work aimed to study the potential anxiolytic and antiamnesic effects of LPH. To achieve this goal, a multidisciplinary study has been conducted using a combination of analytical, molecular, biochemical, and behavioral techniques.

Since the bioactivity of food-derived peptides depends on their physicochemical features, such as length, hydrophobicity, and amino acid sequence, our first objective was to identify the composition of LPH peptides. Mass spectrometry analysis revealed the presence of 278 sequences from the *L. angustifolius* database in LPH, of which 58% derive from conglutins, the main seed storage proteins in lupin [53]. Furthermore, we found that approximately 20% of LPH peptides are potentially bioactive. In addition, the physical-chemical analysis showed that LPH mainly contains small-sized and hydrophobic peptides. Both are important features of peptides that determine their interaction with several physiological targets and their bioactivities, and these factors have also been demonstrated to influence peptide self-assembly, emulsifying capacity, and other properties, including biostability and potentially bioavailability [54].

The peptides analysis allowed us to identify sequences containing some known anxiolytic and antiamnesic motifs. Specifically, we identified 4 peptides (VPL, PGP, PG, and GP) present in 49 different sequences with antiamnesic effects and 2 peptides (YL and GQ) in 9 different sequences with anxiolytic effects. LPH contained 5 different sequences that present the YL dipeptide, which is able to activate the 5-hydroxytryptamine (serotonin) receptor 1A, the dopamine D1 receptor, and the type A receptor of c-amino butyric acid in mice, which play a pivotal role in anxiety. On the other hand, dipeptide YL has shown comparable effects to diazepam in equal doses [50], while PGP, PG, and GP have been demonstrated to enhance memory consolidation processes in the central nervous system [49]. In accordance with these data, the present study reports the beneficial effects of 14 weeks of LPH treatment on WD consumption-induced anxiety in ApoE$^{-/-}$ mice. In fact, WD-fed ApoE$^{-/-}$ mice have previously been demonstrated to successfully reproduce spatial cognitive deficits (memory loss) and anxiety status through a dysregulation of the HPA axis that regulates GCs synthesis, which plays an important role in several brain functions [38,39]. LPH exhibited anxiolytic-like activity, with no differences in learning or spatial memory, and its effects were not related to change in body weight, since mice belonging to different groups did not show a significant difference in BWG.

It is well known that high-fat and high-free-sugar diets are part of the environmental factors that can aggravate or favor the development of anxiety [55,56]; many reports have shown that a high-fat diet accelerates cognitive deficits and anxiety in ApoE$^{-/-}$ mice [37]. To study anxiety, we used the EPM, a well-established test to evaluate anxiolytic/anxiety-like behaviors. In the EPM, mice experience the natural conflict between exploring a new place and the tendency to avoid a dangerous area [57]. We observed that WD significantly increases anxiety behavior since WD-fed mice remained less time on the opened arms and the center of the platform in the EPM compared to SD-fed mice. Opened arms and center areas are considered anxiety zones because rodents have an innate fear of elevated open spaces and tend to spend less time in them [58,59]. Thus, mice treated with anxiolytic drugs (i.e., diazepam) remained longer in the opened arms and in the center of the EPM [60]. Interestingly, WD-fed mice treated with LPH remained longer in the opened arms and in the center zone, and less time in the closed arms compared to the WD group. In addition, mice fed with WD showed fewer *head dips* in comparison to the control group. This behavior, which consists of lowering the head over the sides of the opened arms toward the floor, is considered exploratory and is related to a lower level of anxiety and fear [61]. These results are consistent with previous studies in humans [62] and mice [63], in which the anxiogenic power of a high-fat diet is also demonstrated. Interestingly, a significant increase in the number of *head dips* was recorded in LPH-treated mice, pointing to a higher exploration capacity and less fear, all caused by lower levels of anxiety.

The results obtained in the MWM revealed no impairment in memory or spatial learning after WD consumption. There were no differences in latency time or time spent in targeted section between mice fed with SD and WD. This fact could be associated with the age of the mice and the time of WD consumption. Janssen et al. concluded that ApoE$^{-/-}$ mice perform MWM with better results than wild-type ones and demonstrated that WD does not alter the results in ApoE$^{-/-}$ mice [64]. Furthermore, Champagne et al.

showed that older ApoE$^{-/-}$ mice obtain the worst results in MWM [65]. Apart from that, the present study shows that WD-fed mice covered less distance than mice from SD and LPH groups. Several pieces of evidence have shown that changes in distance may be due to alterations in the motivation to find the platform and greater capacity for exploration [66], but also due to lower activity or worse fitness [67]. Furthermore, WD-treated mice exhibited more *thigmotaxis* than the LPH group. Moreover, LPH-treated mice showed similar *thigmotaxis* to the SD group. This behavior is a well-established indicator of animal anxiety and fear [68,69]. This fact is consistent with the results observed in the EPM, strengthening the protective effect of LPH on WD-induced anxiety.

Although bioactive peptides from white eggs [70], salmon [71], bovine casein [72], or soy [29] have been described to exert anti-anxiety activity, to our knowledge, this is the first study to report the anxiolytic-like properties of a protein hydrolysate from lupin. High levels of oxidative stress and inflammation in the brain have been widely reported to be two of the main contributing factors involved in the development of anxiety [42–44]. Moreover, recent studies have shown a strong link between high cholesterol levels and anxiety [69]. Our group has previously shown that LPH exerts anti-inflammatory, antioxidant, and lipid-lowering effects both in ApoE$^{-/-}$ mice [47,73] and humans [45,46]. Therefore, we suggest that these LPH properties may also be directly or indirectly responsible for the anxiolytic-like effects. In addition, the presence of peptides in the LPH with already demonstrated anxiolytic effects similar to those of diazepam, such as YL and GQ, could also be the cause of the demonstrated anxiolytic effects. However, the presence of other peptides in the LPH that have not yet proved their anxiolytic effects cannot be ruled out.

As in each study, this has certain solvable limitations. The number of mice used was limited (n = 4 per group); however, i) a small number of mice was sufficient to achieve significant differences, ii) two different anxiety analyses were performed to confirm the effect, and iii) the Cohen's test analysis shows a large size effect on each variable studied (Supplementary Table S2). We also consider important to highlight that an SD + LPH group has not been included in the study, since SD mice do not exhibit anxious behaviors. In fact, the only reason we used an SD group was to check that WD consumption generates anxiety.

The main strength of this work is the multidisciplinary strategy used. First, a detailed chemical characterization of the LPH composition was performed by using nano-HPLC-MS/MS and UHPLC-HRMS to identify its peptide composition. Afterward, an in silico study was carried out for the identification of anxiolytic and antiamnesic peptides. Finally, an in vivo study confirmed through two different tests (EPM and thigmotaxis during the MWM) that LPH treatment palliates the anxious effects generated by the ingestion of WD. This study is the first to show the in vivo anxiolytic-like effect of a plant-derived total protein hydrolysate.

4. Materials and Methods

4.1. LPH Preparation

LPH was produced at the Instituto de la Grasa (CSIC, Seville, Spain), as previously described [45]. Briefly, the lupin protein isolate was resuspended in distilled water (10% w/v) and hydrolyzed in a bioreactor at pH 8 and temperature 50 °C using Alcalase® 2.4 L (2.4 AU/g; Novozymes, Bagsvaerd, Denmark) for 15 min. The enzyme was inactivated by heating at 85 °C for 15 min; after centrifugation at 8000 rpm for 15 min, the supernatant containing LPH was collected and lyophilized. Finally, it was dissolved in 0.9% saline solution to obtain the LPH necessary for the duration of the experiment, filtered, autoclaved, aliquoted, and stored at −80 °C. The chemical stability and characterization of LPH were checked out at the several steps of this process through HPLC, no differences were observed (data not shown).

4.2. Purification and Concentration of Peptides

An amount of 1 mg of LPH was acidified with aqueous trifluoroacetic acid (TFA) at pH 2.5, loaded into the Bond Elut C18 EWP cartridge (Aligent, Santa Clara, CA, USA) (previously washed with acetonitrile (ACN) and conditioned with 0.1% TFA), and washed with 3 mL of 0.1% TFA. The elution was carried out with 0.5 mL ACN/H_2O (50:50, *v/v*) containing 0.1% TFA, and the peptides were dried in a Speed Vac SC250 Express (Thermo Savant, Holbrook, NT, USA). The dry residue was reconstituted in 150 µL of 0.1% formic acid in H_2O.

4.3. Peptides' Analysis and Identification by Mass Spectrometry

The peptides were studied by nano-HPLC using an Ultimate 3000 coupled to an Orbitrap Elite mass spectrometer (Thermo Fisher Scientific, Bremen, Germany), as previously described [74]. The preconcentration of the samples (20 µL) was performed on a µ-precolumn (Thermo, 300 µm i.d. 5 mm Acclaim PepMap 100 C18, 5 µm particle size, 100 Å pore size) using H_2O/ACN (99:1 *v*/*v*) with 0.1% TFA (*v*/*v*) at a flow rate of 10 µL/min. The peptides were dispersed on an EASY-Spray column (Thermo, 15 cm × 75 µm i.d. PepMap C18, 3 µm particles, 100 Å pore size).

The peptide spectra were obtained using the same parameters described in our previous work [73]. The protein sequence database of *L. angustifolius* (31,386 sequences) was downloaded from UnitProt and used for the identification of raw data spectra using Proteome Discoverer v1.3 (Thermo) in combination with the Mascot search engine v2.3.02. Precursor ion tolerance and the fragment ion tolerance were 10 ppm and 0.05 Da, respectively; no enzyme was used for digestion and methionine oxidation was considered as dynamic modification. The decoy function, set at 1%, was used for false discovery rate calculations.

4.4. Bioactivities Peptide Analysis

The physicochemical properties (molecular weight, amino acid composition, and hydrophobicity) of the peptides were obtained using the open access ProtParam tool (https://web.expasy.org/protparam/, accessed on 1 April 2022) [75]. The peptide Ranker tool (http://distilldeep.ucd.ie/PeptideRanker/, accessed on 1 April 2022) was used to predict the bioactivity of LPH [76]. It provides scores in the range of 0−1, being 1 the most active. The threshold was fixed at 0.5; therefore, peptides with scores above 0.5 were labeled as 'bioactive'. To identify sequences with demonstrated bioactive motifs, the peptides were analyzed using the BIOPEP-UWM database (http://www.uwm.edu.pl/biochemia/index.php/pl/biopep/, accessed on 1 April 2022) [52]. In addition, the primary structure of the motifs was drawn using the PepDraw tool (https://pepdraw.com/, accessed on 1 June 2022).

4.5. Animals and Experimental Design

The experimental design is shown in Supplementary Figure S1. Twelve male ApoE$^{-/-}$ mice (B6.129P2-ApoEtm1Unc/J) were housed in the animal facility of the Faculty of Psychology (University of Seville, Seville, Spain) under specific pathogen-free conditions in a room with controlled temperature (22 ± 2 °C), humidity (<55%), and a 12-h light–dark cycle with free access to water and food. The mice were housed in a sealsafe® 1285L cage (Tecniplast, Italy) [77] with a floor area of 542 cm^2 and a maximum air speed at the animal level of 0.05 m/s. Four mice were housed per cage. The particular characteristics of these cages allow no air drafts at the animal level, avoiding the risk of stress and heat loss. The animals were initially classified into two groups: mice fed a standard diet (SD, n = 4, Teklad Global 14% Protein Rodent Maintenance Diet, ENVIGO, Indianapolis, IN, USA) [78] and mice fed a Western diet (WD, n = 8, 58V8-45 kcal% fat, TestDiet, St. Louis, MO, USA) [79] from the Special Diets Production Section of the University of Granada (Granada, Spain). The composition of each diet is specified in Supplementary Table S1.

Six-week-old mice from the WD group were randomly divided into two groups and treated intragastrically with LPH (100 mg/kg, n = 4) or vehicle (n = 4) for 14 weeks, respectively. Thus, the experimental groups were set as follows: SD-fed mice group (SD, n = 4), WD-fed group (WD, n = 4), and WD-fed and LPH-treated (100 mg/kg) mice group (WD + LPH, n = 4). SD-fed mice were also intragastrically treated with vehicle. The dose of LPH was selected based on our previous studies [45–47,73]. Individual body weight was measured and recorded weekly. Behavioral tests were performed at the Laboratory of Animal Behavior & Neuroscience (a specific installation inside the Animal Facility of the Faculty of Psychology), where the animals were placed a week earlier for their habituation. The tests were carried out with a 10-day inter-test interval.

The experimental procedures were approved by the Ethics Committee of the Virgen Macarena-Virgen del Rocío University Hospital (reference number 21/06/2016/105) and were carried out under Spanish legislation and the EU Directive 2010/63/EU for animal experiments.

4.6. Behavioral Tests

4.6.1. Elevated Plus Maze

Anxiety-like behavior was evaluated using the EPM test. It was performed as previously described [32]. Briefly, the maze consists of four arms made out of polyvinyl chloride; two non-consecutive opened arms (30 cm long × 5 cm wide) and two closed arms that generate a common center zone (5 × 5 cm). The EPM was placed 60 cm above the floor in the center of a room (286 × 288 × 320 cm; w-l-h respectively) illuminated by four 100-W halogen lamps. The characteristics of the EPM are shown in Figure 5. In order to minimize exploratory behavior and facilitate habituation to the context, mice were placed in the room for 45 min prior to the test. To start the test, each mouse was placed in one of the opened arms facing the opposite direction to the center and was free to move for 5 min. All sessions were recorded using a camera located over the maze. For the trials, the experimenter remained in an adjoining zone to control the video tracking system. Additionally, the observer could see the performance of the animal in real time on a monitor. Other anxiety-related behaviors, such as *head dips* and rears, and the number of times that mice showed them, were also annotated and recorded. The test started at 11:30 a.m. during the light phase of the light-dark cycle, and none of the researchers stayed in the room while the test took place. The floor of the elevated plus maze apparatus was cleaned with 10% ethanol between tests. Subsequently, the recording was processed using the Animal Tracker plugin for ImageJ v. 1.53k software (National Institutes of Health-NIH-, Bethesda, MD, USA) and the time spent in the arms and center of the maze was measured. Opened arms and center areas are considered anxiety zones according to [58,59]. Analyses were carried out under blind conditions by three investigators. Representative videos are available in Videos S1–S3.

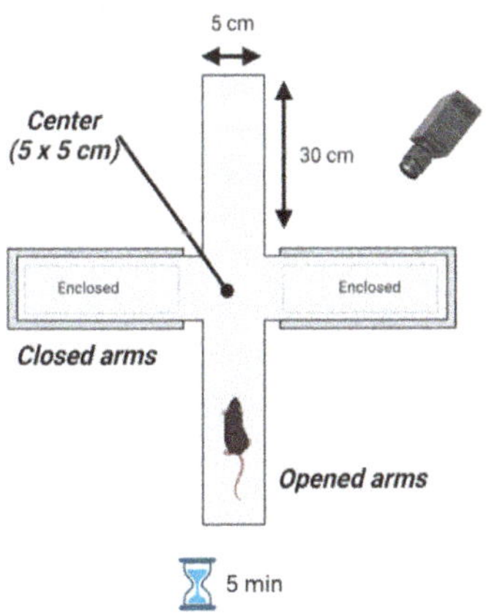

Figure 5. Characteristics of the Elevated Plus Maze. The maze consists of four arms: two non-consecutive open arms (30 cm long × 5 cm wide) and two closed arms that generate a common center zone (5 × 5 cm). The EPM was placed 60 cm above the floor. To start the test, each mouse was placed in one of the opened arms facing the opposite direction of the center and was free to move for 5 min. All sessions were recorded using a camera located over the maze. Figure created by BioRender.com.

4.6.2. Morris Water Maze

The MWM was designed as a method to study spatial memory and learning processes. The experimental procedures were performed as described by Janseen et al. [64]. Briefly, the test consists of a circular pool (100 cm in diameter) filled with water (at 25 °C) and a circular platform (8 cm in diameter, 20 cm in height) located in a specific quadrant of the pool. It was virtually divided into four different sections, and different visual clues were located on the walls of the room (characteristics of the MWM are shown in Figure 6). The test was carried out for 5 consecutive days. To avoid the use of possible intramaze cues to solve the task, the experimental apparatus was randomly rotated between sessions. On day 0, mice received two habituation trainings; animals were located in two different sections and allowed to swim for 90 s until they reached the visible platform (2 cm above the water surface). Once on the platform, the mice stand there for 15 s. On days 1–5, animals were placed in each section and allowed to swim for 90 s or until they reached the non-visible platform. In this phase, the water was opaque by adding a white dye (lime) and the time between tests was 45 min. Finally, on the fifth day, the platform was removed, and the mice were placed in the pool for 90 s (the scheme of the Morris Water Maze protocol is shown in Supplementary Figure S2). All sessions were recorded with a video tracking system that overlooked the pool from above. The test started at 11:30 a.m. during the light phase of the light–dark cycle, and the experimenter stayed in an adjoining zone for the test. The latency time, distance traveled, and time spent in each quadrant were analyzed using the Animal Tracker plugin for ImageJ software (NIH). In addition, *thigmotaxis*, considered as the times the animal approaches the walls of the pool and the time spent in the outer area (15% of the apparatus) of the pool, was calculated. Analyses were carried out under blind conditions by three investigators. Representative videos are available in Videos S4–S6.

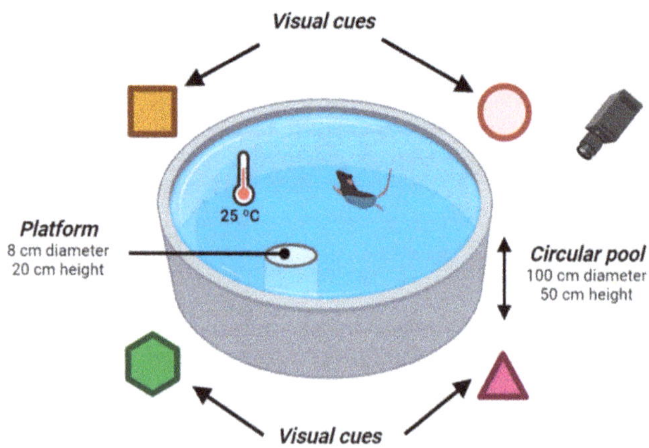

Figure 6. Characteristics of the Morris Water Maze. The test consists of a circular pool (100 cm in diameter) filled with water (at 25 °C) and a circular platform (8 cm in diameter, 20 cm in height) placed in a specific zone of the pool. The pool was virtually divided into four different quadrants and different visual clues were located on the walls of the room. All sessions were recorded with a video monitoring system that overlooks the pool from above. Figure created by BioRender.com.

4.7. Statistical Analysis

All results were presented as mean ± standard deviation, and the statistical analysis was carried out using one-way ANOVA followed by Dunn's post hoc test using Jeffreys's Amazing Statistics Program (JASP v. 0.16.3, Amsterdam, The Netherlands). A difference with a p-value ≤ 0.05 was considered statistically significant. The size effect was analyzed using Cohen's test, and a d-value > 0.80 was considered as 'large effect size'.

5. Conclusions

In conclusion, this is the first study to show the in vivo anxiolytic effects of a lupin protein hydrolysate. Moreover, several sequences containing peptide motifs associated with anxiolytic effects were identified within the LPH mixture. Future studies will be needed to investigate the molecular mechanisms that cause the anxiolytic effect of LPH, as well as to compare this effect with an anxiolytic drug such as diazepam. In addition, several strategies, such as the incorporation of peptides into biocompatible vehicles to enhance their stability and bioavailability during transepithelial transport, are recommended for future investigation. The present study confirms the pleiotropic effects of the peptide mixture, including anxiolytic effects, pointing to LPH as a potential component of future nutritional therapies in patients with anxiety, being a possible strategy to reduce the consumption of drugs with side effects.

Supplementary Materials: The supporting information can be downloaded at: https://www.mdpi.com/article/10.3390/ijms23179828/s1.

Author Contributions: The following are the authors' contributions: Conceptualization: A.C.-V., I.C.-C., J.C.L., P.J.L., N.Á.-S. and I.B.; Methodology: G.S.-S., E.P.-E., I.C.-C., J.C.L., N.Á.-S. and A.I.Á.-L.; Resources: J.C.L., J.P., F.M., M.C.M.-L., P.J.L. and A.C.-V.; Formal analysis: G.S.-S., E.P.-E. and I.C.-C.; Drafting of the manuscript: G.S.-S., E.P.-E., I.C.-C. and A.C.-V.; Supervision: A.C.-V., P.J.L., N.Á.-S. and I.C.-C. Funding acquisition: A.C.-V., J.P., F.M. and P.J.L. All authors have read and agreed to the published version of the manuscript.

Funding: This research was funded by the Spanish Government, Ministerio de Economía y Competitividad (AGL2012-40247-C02-01, and AGL2012-40247-C02-02), the Andalusian Government Ministry of Health (PC-0111-2016-0111, and PEMP-0085-2020) and the PAIDI Program from the Andalusian Government (CTS160). G.S.-S. was supported by a FPU grant from the Spanish Ministerio de Educación, Cultura y Deporte (FPU16/02339). E.P.-E. and I.B. were supported by the VI Program of Inner Initiative for Research and Transfer of University of Seville (VI PPIT-US). I.C.-C. was supported by the VI Program of Inner Initiative for Research and Transfer of the University of Seville (VIPPIT-2020-II.4) and by a postdoctoral fellowship from the Andalusian Government Ministry of Economy, Knowledge, Business, and University (DOC_00587/2020). N.A.-S. was supported by a fellowship from the National Net RETICEF for Aging Studies (RD12/0043/0012 from the Instituto de Salud Carlos III, Spanish Ministerio de Ciencia e Innovación). A.I.Á.-L. was funded by Andalusian Government Ministry of Health (PI-0136-2019).

Institutional Review Board Statement: The experiments were performed under the Spanish legislation and the EU Directive 2010/63/EU for animal experiments and was approved by the virgen Macarena and Virgen del Rocío University Hospitals ethical committee (reference 21/06/2016/105).

Informed Consent Statement: Not applicable.

Data Availability Statement: Data are contained within the article and Supplementary Materials.

Acknowledgments: We thank the staff from the IBiS Animal Facility for their valuable assistance.

Conflicts of Interest: The authors declare no conflict of interest.

References

1. Otte, C. Cognitive behavioral therapy in anxiety disorders: Current state of the evidence. *Dialogues Clin. Neurosci.* **2011**, *13*, 413–421. [PubMed]
2. Perna, G.; Alciati, A.; Sangiorgio, E.; Caldirola, D.; Nemeroff, C.B. Personalized clinical approaches to anxiety disorders. *Anxiety Disord.* **2020**, *1191*, 489–521.
3. Canuto, A.; Weber, K.; Baertschi, M.; Andreas, S.; Volkert, J.; Dehoust, M.C.; Sehner, S.; Suling, A.; Wegscheider, K.; Ausín, B. Anxiety disorders in old age: Psychiatric comorbidities, quality of life, and prevalence according to age, gender, and country. *Am. J. Geriatr. Psychiatry* **2018**, *26*, 174–185. [PubMed]
4. Gainey, S.J.; Kwakwa, K.A.; Bray, J.K.; Pillote, M.M.; Tir, V.L.; Towers, A.E.; Freund, G.G. Short-term high-fat diet (HFD) induced anxiety-like behaviors and cognitive impairment are improved with treatment by glyburide. *Front. Behav. Neurosci.* **2016**, *10*, 156.
5. Clark, T.D.; Crean, A.J.; Senior, A.M. Obesogenic diets induce anxiety in rodents: A systematic review and meta-analysis. *Obes. Rev.* **2022**, *23*, e13399.

6. Clarke, D.M.; Currie, K.C. Depression, anxiety and their relationship with chronic diseases: A review of the epidemiology, risk and treatment evidence. *Med. J. Aust.* **2009**, *190*, S54–S60.
7. Gerontoukou, E.-I.; Michaelidoy, S.; Rekleiti, M.; Saridi, M.; Souliotis, K. Investigation of anxiety and depression in patients with chronic diseases. *Health Psychol. Res.* **2015**, *3*, 2123. [CrossRef]
8. Ng, R.; Sutradhar, R.; Yao, Z.; Wodchis, W.P.; Rosella, L.C. Smoking, drinking, diet and physical activity—modifiable lifestyle risk factors and their associations with age to first chronic disease. *Int. J. Epidemiol.* **2020**, *49*, 113–130.
9. Scoditti, E.; Massaro, M.; Garbarino, S.; Toraldo, D.M. Role of diet in chronic obstructive pulmonary disease prevention and treatment. *Nutrients* **2019**, *11*, 1357.
10. Dong, C.; Bu, X.; Liu, J.; Wei, L.; Ma, A.; Wang, T. Cardiovascular disease burden attributable to dietary risk factors from 1990 to 2019: A systematic analysis of the Global Burden of Disease study. *Nutr. Metab. Cardiovasc. Dis.* **2022**, *32*, 897–907.
11. Ruan, Y.; Tang, J.; Guo, X.; Li, K.; Li, D. Dietary fat intake and risk of Alzheimer's disease and dementia: A meta-analysis of cohort studies. *Curr. Alzheimer Res.* **2018**, *15*, 869–876.
12. Celano, C.M.; Daunis, D.J.; Lokko, H.N.; Campbell, K.A.; Huffman, J.C. Anxiety disorders and cardiovascular disease. *Curr. Psychiatry Rep.* **2016**, *18*, 101.
13. Naicker, K.; Johnson, J.A.; Skogen, J.C.; Manuel, D.; Øverland, S.; Sivertsen, B.; Colman, I. Type 2 diabetes and comorbid symptoms of depression and anxiety: Longitudinal associations with mortality risk. *Diabetes Care* **2017**, *40*, 352–358.
14. Hofmeijer-Sevink, M.K.; Batelaan, N.M.; van Megen, H.J.G.M.; Penninx, B.W.; Cath, D.C.; van den Hout, M.A.; van Balkom, A.J.L.M. Clinical relevance of comorbidity in anxiety disorders: A report from the Netherlands Study of Depression and Anxiety (NESDA). *J. Affect. Disord.* **2012**, *137*, 106–112.
15. Shields, G.S.; Sazma, M.A.; McCullough, A.M.; Yonelinas, A.P. The effects of acute stress on episodic memory: A meta-analysis and integrative review. *Psychol. Bull.* **2017**, *143*, 636.
16. Vogel, S.; Schwabe, L. Learning and memory under stress: Implications for the classroom. *Npj Sci. Learn.* **2016**, *1*, 16011.
17. Who, J.; Consultation, F.E. Diet, nutrition and the prevention of chronic diseases. *World Health Organ. Tech. Rep. Ser.* **2003**, *916*, 1–149.
18. Yamazaki, Y.; Zhao, N.; Caulfield, T.R.; Liu, C.-C.; Bu, G. Apolipoprotein E and Alzheimer disease: Pathobiology and targeting strategies. *Nat. Rev. Neurol.* **2019**, *15*, 501–518.
19. Kara, S.; Anton, F.; Solle, D.; Neumann, M.; Hitzmann, B.; Scheper, T.; Liese, A. Fluorescence spectroscopy as a novel method for on-line analysis of biocatalytic C–C bond formations. *J. Mol. Catal. B Enzym.* **2010**, *66*, 124–129.
20. Aucoin, M.; LaChance, L.; Naidoo, U.; Remy, D.; Shekdar, T.; Sayar, N.; Cardozo, V.; Rawana, T.; Chan, I.; Cooley, K. Diet and Anxiety: A Scoping Review. *Nutrients* **2021**, *13*, 4418.
21. Panes, A.; Verdoux, H.; Fourrier-Réglat, A.; Berdaï, D.; Pariente, A.; Tournier, M. Misuse of benzodiazepines: Prevalence and impact in an inpatient population with psychiatric disorders. *Br. J. Clin. Pharmacol.* **2020**, *86*, 601–610.
22. Chakrabarti, S.; Guha, S.; Majumder, K. Food-derived bioactive peptides in human health: Challenges and opportunities. *Nutrients* **2018**, *10*, 1738.
23. Daliri, E.B.-M.; Oh, D.H.; Lee, B.H. Bioactive peptides. *Foods* **2017**, *6*, 32.
24. Ulug, S.K.; Jahandideh, F.; Wu, J. Novel technologies for the production of bioactive peptides. *Trends Food Sci. Technol.* **2021**, *108*, 27–39.
25. Giannetto, A.; Esposito, E.; Lanza, M.; Oliva, S.; Riolo, K.; Di Pietro, S.; Abbate, J.M.; Briguglio, G.; Cassata, G.; Cicero, L. Protein hydrolysates from anchovy (*Engraulis encrasicolus*) waste: In vitro and in vivo biological activities. *Mar. Drugs* **2020**, *18*, 86.
26. Abbate, J.M.; Macrì, F.; Arfuso, F.; Iaria, C.; Capparucci, F.; Anfuso, C.; Ieni, A.; Cicero, L.; Briguglio, G.; Lanteri, G. Anti-atherogenic effect of 10% supplementation of anchovy (*Engraulis encrasicolus*) waste protein hydrolysates in ApoE-deficient mice. *Nutrients* **2021**, *13*, 2137.
27. Abbate, J.M.; Macrì, F.; Capparucci, F.; Iaria, C.; Briguglio, G.; Cicero, L.; Salvo, A.; Arfuso, F.; Ieni, A.; Piccione, G. Administration of protein hydrolysates from anchovy (*Engraulis encrasicolus*) waste for twelve weeks decreases metabolic dysfunction-associated fatty liver disease severity in $ApoE^{-/-}$mice. *Animals* **2020**, *10*, 2303.
28. Hafeez, Z.; Benoit, S.; Cakir-Kiefer, C.; Dary, A.; Miclo, L. Food protein-derived anxiolytic peptides: Their potential role in anxiety management. *Food Funct.* **2021**, *12*, 1415–1431.
29. Ohinata, K.; Agui, S.; Yoshikawa, M. Soymorphins, novel μ opioid peptides derived from soy β-conglycinin β-subunit, have anxiolytic activities. *Biosci. Biotechnol. Biochem.* **2007**, *71*, 2618–2621. [CrossRef]
30. Zhao, H.; Sonada, S.; Yoshikawa, A.; Ohinata, K.; Yoshikawa, M. Rubimetide, humanin, and MMK1 exert anxiolytic-like activities via the formyl peptide receptor 2 in mice followed by the successive activation of DP1, A2A, and GABAA receptors. *Peptides* **2016**, *83*, 16–20.
31. Hirata, H.; Sonoda, S.; Agui, S.; Yoshida, M.; Ohinata, K.; Yoshikawa, M. Rubiscolin-6, a δ opioid peptide derived from spinach Rubisco, has anxiolytic effect via activating σ1 and dopamine D1 receptors. *Peptides* **2007**, *28*, 1998–2003. [PubMed]
32. Oda, A.; Kaneko, K.; Mizushige, T.; Lazarus, M.; Urade, Y.; Ohinata, K. Characterization of ovolin, an orally active tryptic peptide released from ovalbumin with anxiolytic-like activity. *J. Neurochem.* **2012**, *122*, 356–362. [PubMed]
33. dela Pena, I.J.I.; Kim, H.J.; de la Pena, J.B.; Kim, M.; Botanas, C.J.; You, K.Y.; Woo, T.; Lee, Y.S.; Jung, J.-C.; Kim, K.-M. A tryptic hydrolysate from bovine milk αs1-casein enhances pentobarbital-induced sleep in mice via the GABAA receptor. *Behav. Brain Res.* **2016**, *313*, 184–190.

34. Mizushige, T.; Sawashi, Y.; Yamada, A.; Kanamoto, R.; Ohinata, K. Characterization of Tyr-Leu-Gly, a novel anxiolytic-like peptide released from bovine αS-casein. *FASEB J.* **2013**, *27*, 2911–2917.
35. Yamada, A.; Mizushige, T.; Kanamoto, R.; Ohinata, K. Identification of novel β-lactoglobulin-derived peptides, wheylin-1 and-2, having anxiolytic-like activity in mice. *Mol. Nutr. Food Res.* **2014**, *58*, 353–358. [PubMed]
36. Ohinata, K.; Sonoda, S.; Inoue, N.; Yamauchi, R.; Wada, K.; Yoshikawa, M. β-Lactotensin, a neurotensin agonist peptide derived from bovine β-lactoglobulin, enhances memory consolidation in mice. *Peptides* **2007**, *28*, 1470–1474.
37. Yuan, T.; Chu, C.; Shi, R.; Cui, T.; Zhang, X.; Zhao, Y.; Shi, X.; Hui, Y.; Pan, J.; Qian, R. ApoE-Dependent Protective Effects of Sesamol on High-Fat Diet-Induced Behavioral Disorders: Regulation of the Microbiome-Gut–Brain Axis. *J. Agric. Food Chem.* **2019**, *67*, 6190–6201. [CrossRef]
38. Chen, K.; Yuan, R.; Geng, S.; Zhang, Y.; Ran, T.; Kowalski, E.; Liu, J.; Li, L. Toll-interacting protein deficiency promotes neurodegeneration via impeding autophagy completion in high-fat diet-fed ApoE$^{-/-}$mouse model. *Brain Behav. Immun.* **2017**, *59*, 200–210.
39. Raber, J. Role of apolipoprotein E in anxiety. *Neural Plast.* **2007**, *2007*, 091236.
40. Raber, J.; Akana, S.F.; Bhatnagar, S.; Dallman, M.F.; Wong, D.; Mucke, L. Hypothalamic–Pituitary–Adrenal Dysfunction inApoe$^{-/-}$Mice: Possible Role in Behavioral and Metabolic Alterations. *J. Neurosci.* **2000**, *20*, 2064–2071.
41. Strekalova, T.; Evans, M.; Costa-Nunes, J.; Bachurin, S.; Yeritsyan, N.; Couch, Y.; Steinbusch, H.M.; Köhler, S.E.; Lesch, K.-P.; Anthony, D.C. Tlr4 upregulation in the brain accompanies depression-and anxiety-like behaviors induced by a high-cholesterol diet. *Brain Behav. Immun.* **2015**, *48*, 42–47. [PubMed]
42. Felger, J.C. Imaging the role of inflammation in mood and anxiety-related disorders. *Curr. Neuropharmacol.* **2018**, *16*, 533–558. [PubMed]
43. Fedoce, A.d.G.; Ferreira, F.; Bota, R.G.; Bonet-Costa, V.; Sun, P.Y.; Davies, K.J. The role of oxidative stress in anxiety disorder: Cause or consequence? *Free. Radic. Res.* **2018**, *52*, 737–750. [PubMed]
44. Barbieri, S.S.; Sandrini, L.; Musazzi, L.; Popoli, M.; Ieraci, A. Apocynin prevents anxiety-like behavior and histone deacetylases overexpression induced by sub-chronic stress in mice. *Biomolecules* **2021**, *11*, 885.
45. Cruz-Chamorro, I.; Álvarez-Sánchez, N.; del Carmen Millán-Linares, M.; del Mar Yust, M.; Pedroche, J.; Millán, F.; Lardone, P.J.; Carrera-Sánchez, C.; Guerrero, J.M.; Carrillo-Vico, A. Lupine protein hydrolysates decrease the inflammatory response and improve the oxidative status in human peripheral lymphocytes. *Food Res. Int.* **2019**, *126*, 108585.
46. Cruz-Chamorro, I.; Álvarez-Sánchez, N.; Álvarez-Ríos, A.I.; Santos-Sánchez, G.; Pedroche, J.; Millán, F.; Sánchez, C.C.; Fernández-Pachón, M.S.; Millán-Linares, M.C.; Martinez-Lopez, A. Safety and Efficacy of a Beverage Containing Lupine Protein Hydrolysates on the Immune, Oxidative and Lipid Status in Healthy Subjects: An Intervention Study (the Lupine-1 Trial). *Mol. Nutr. Food Res.* **2021**, *65*, 2100139.
47. Santos-Sánchez, G.; Cruz-Chamorro, I.; Álvarez-Ríos, A.I.; Fernández-Santos, J.M.; Vázquez-Román, M.V.; Rodríguez-Ortiz, B.; Álvarez-Sánchez, N.; Álvarez-López, A.I.; Millán-Linares, M.d.C.; Millán, F. Lupinus angustifolius protein hydrolysates reduce abdominal adiposity and ameliorate metabolic associated fatty liver disease (MAFLD) in Western diet fed-ApoE$^{-/-}$mice. *Antioxidants* **2021**, *10*, 1222.
48. Maruyama, S.; Ohmori, T.; Nakagami, T. Prolylendopeptidase inhibitory activity of a glial fibrillary acidic protein fragment and other proline-rich peptides. *Biosci. Biotechnol. Biochem.* **1996**, *60*, 358–359.
49. Ashmarin, I.; Karazeeva, E.; Lyapina, L.; Samonina, G. The simplest proline-containing peptides PG, GP, PGP, and GPGG: Regulatory activity and possible sources of biosynthesis. *Biochem. Biokhimiia* **1998**, *63*, 119–124.
50. Kanegawa, N.; Suzuki, C.; Ohinata, K. Dipeptide Tyr-Leu (YL) exhibits anxiolytic-like activity after oral administration via activating serotonin 5-HT1A, dopamine D1 and GABAA receptors in mice. *FEBS Lett.* **2010**, *584*, 599–604.
51. Parish, D.; Smyth, D.; Normanton, J.; Wolstencroft, J. Glycyl glutamine, an inhibitory neuropeptide derived from β-endorphin. *Nature* **1983**, *306*, 267–270. [PubMed]
52. Minkiewicz, P.; Iwaniak, A.; Darewicz, M. BIOPEP-UWM Database of Bioactive Peptides: Current Opportunities. *Int. J. Mol. Sci.* **2019**, *20*, 5978.
53. Burgos-Díaz, C.; Opazo-Navarrete, M.; Wandersleben, T.; Soto-Añual, M.; Barahona, T.; Bustamante, M. Chemical and nutritional evaluation of protein-rich ingredients obtained through a technological process from yellow lupin seeds (*Lupinus luteus*). *Plant Foods Hum. Nutr.* **2019**, *74*, 508–517. [PubMed]
54. Zaky, A.A.; Simal-Gandara, J.; Eun, J.-B.; Shim, J.-H.; Abd El-Aty, A. Bioactivities, applications, safety, and health benefits of bioactive peptides from food and by-products: A review. *Front. Nutr.* **2021**, *8*, 815640.
55. Norwitz, N.G.; Naidoo, U. Nutrition as Metabolic Treatment for Anxiety. *Front. Psychiatry* **2021**, *12*, 598119.
56. Ljungberg, T.; Bondza, E.; Lethin, C. Evidence of the importance of dietary habits regarding depressive symptoms and depression. *Int. J. Environ. Res. Public Health* **2020**, *17*, 1616.
57. Qubty, D.; Schreiber, S.; Rubovitch, V.; Boag, A.; Pick, C.G. No Significant Effects of Cellphone Electromagnetic Radiation on Mice Memory or Anxiety: Some Mixed Effects on Traumatic Brain Injured Mice. *Neurotrauma Rep.* **2021**, *2*, 381–390.
58. Rodgers, R.; Dalvi, A. Anxiety, defence and the elevated plus-maze. *Neurosci. Biobehav. Rev.* **1997**, *21*, 801–810.
59. Walf, A.A.; Frye, C.A. The use of the elevated plus maze as an assay of anxiety-related behavior in rodents. *Nat. Protoc.* **2007**, *2*, 322–328.

60. Liu, J.; Scott, B.W.; Burnham, W.M. Effects of cannabidiol and Δ9-tetrahydrocannabinol in the elevated plus maze in mice. *Behav. Pharmacol.* **2022**, *33*, 206–212.
61. Grahn, R.E.; Kalman, B.A.; Vlasaty, J.A.; Perna, J.A.; Nevins-Herbert, C.; Patton, S.M.; Barison, L.K. Effects of plus-maze experience and chlordiazepoxide on anxiety-like behavior and serotonin neural activity in the dorsal raphe nucleus in rats. *Behav. Pharmacol.* **2019**, *30*, 208–219. [PubMed]
62. Baker, K.D.; Loughman, A.; Spencer, S.J.; Reichelt, A.C. The impact of obesity and hypercaloric diet consumption on anxiety and emotional behavior across the lifespan. *Neurosci. Biobehav. Rev.* **2017**, *83*, 173–182. [PubMed]
63. Dutheil, S.; Ota, K.T.; Wohleb, E.S.; Rasmussen, K.; Duman, R.S. High-fat diet induced anxiety and anhedonia: Impact on brain homeostasis and inflammation. *Neuropsychopharmacology* **2016**, *41*, 1874–1887. [CrossRef] [PubMed]
64. Janssen, C.I.; Jansen, D.; Mutsaers, M.P.; Dederen, P.J.; Geenen, B.; Mulder, M.T.; Kiliaan, A.J. The Effect of a High-Fat Diet on Brain Plasticity, Inflammation and Cognition in Female ApoE4-Knockin and ApoE-Knockout Mice. *PLoS ONE* **2016**, *11*, e0155307. [CrossRef]
65. Champagne, D.; Dupuy, J.-B.; Rochford, J.; Poirier, J. Apolipoprotein E knockout mice display procedural deficits in the Morris water maze: Analysis of learning strategies in three versions of the task. *Neuroscience* **2002**, *114*, 641–654.
66. Belviranli, M.; Atalik, K.; Okudan, N.; Gökbel, H. Age and sex affect spatial and emotional behaviors in rats: The role of repeated elevated plus maze test. *Neuroscience* **2012**, *227*, 1–9.
67. Pietrelli, A.; Lopez-Costa, J.; Goñi, R.; Brusco, A.; Basso, N. Aerobic exercise prevents age-dependent cognitive decline and reduces anxiety-related behaviors in middle-aged and old rats. *Neuroscience* **2012**, *202*, 252–266.
68. Alveal-Mellado, D.; Giménez-Llort, L. Thigmotaxis helps to differentiate normal and pathological aging processes in a mice model for Alzheimer's Disease. *Med. Sci. Forum* **2021**, *8*, 2.
69. Mayagoitia, K.; Shin, S.D.; Rubini, M.; Siebold, L.; Wilson, C.G.; Bellinger, D.L.; Figueroa, J.D.; Soriano, S. Short-term exposure to dietary cholesterol is associated with downregulation of interleukin-15, reduced thigmotaxis and memory impairment in mice. *Behav. Brain Res.* **2020**, *393*, 112779.
70. Yu, Z.; Zhao, W.; Ding, L.; Yu, Y.; Liu, J. Anxiolytic effects of ACE inhibitory peptides on the behavior of rats in an elevated plus-maze. *Food Funct.* **2016**, *7*, 491–497.
71. Belhaj, N.; Desor, F.; Gleizes, C.; Denis, F.M.; Arab-Tehrany, E.; Soulimani, R.; Linder, M. Anxiolytic-like effect of a salmon phospholipopeptidic complex composed of polyunsaturated fatty acids and bioactive peptides. *Mar. Drugs* **2013**, *11*, 4294–4317. [CrossRef]
72. Miclo, L.; Perrin, E.; Driou, A.; Papadopoulos, V.; Boujrad, N.; Vanderesse, R.; Boudier, J.-F.; Desor, D.; Linden, G.; Gaillard, J.-L. Characterization of α-casozepine, a tryptic peptide from bovine αs1-casein with benzodiazepine-like activity. *FASEB J.* **2001**, *15*, 1780–1782. [CrossRef]
73. Santos-Sánchez, G.; Cruz-Chamorro, I.; Bollati, C.; Bartolomei, M.; Pedroche, J.; Millán, F.; Millán-Linares, M.d.C.; Capriotti, A.; Cerrato, A.; Laganà, A.; et al. A Lupinus angustifolius protein hydrolysate exerts hypocholesterolemic effect in western diet-fed-ApoE$^{-/-}$mice through the modulation of LDLR and PCSK9 pathways. *Food Funct.* **2022**, *13*, 4158–4170. [CrossRef]
74. Fercha, A.; Capriotti, A.L.; Caruso, G.; Cavaliere, C.; Samperi, R.; Stampachiacchiere, S.; Laganà, A. Comparative analysis of metabolic proteome variation in ascorbate-primed and unprimed wheat seeds during germination under salt stress. *J. Proteom.* **2014**, *108*, 238–257. [CrossRef]
75. Gasteiger, E.; Hoogland, C.; Gattiker, A.; Wilkins, M.R.; Appel, R.D.; Bairoch, A. Protein identification and analysis tools on the ExPASy server. In *The Proteomics Protocols Handbook*; Springer Protocols Handbooks; Humana Press: Totowa, NJ, USA, 2005; pp. 571–607.
76. Mooney, C.; Haslam, N.J.; Pollastri, G.; Shields, D.C. Towards the improved discovery and design of functional peptides: Common features of diverse classes permit generalized prediction of bioactivity. *PLoS ONE* **2012**, *7*, e45012. [CrossRef]
77. Tecniplast. Sealsafe®1285L Cage. Available online: https://www.tecniplast.it/uk/product/sealsafe.html (accessed on 1 July 2022).
78. ENVIGO. Teklad Global 14% Protein Rodent Maintenance Diet. Available online: https://insights.envigo.com/hubfs/resources/data-sheets/2014s-datasheet-0915.pdf (accessed on 1 July 2022).
79. TestDiet. 58V8-45 kcal % Fat. Available online: https://www.testdiet.com/Diets/High-Fat-DIO/index.html (accessed on 1 July 2022).

International Journal of *Molecular Sciences*

MDPI

Article

Rosemary Extract-Induced Autophagy and Decrease in Accumulation of Collagen Type I in Osteogenesis Imperfecta Skin Fibroblasts

Joanna Sutkowska-Skolimowska [1], Justyna Brańska-Januszewska [2], Jakub W. Strawa [3], Halina Ostrowska [2], Malwina Botor [4], Katarzyna Gawron [4] and Anna Galicka [1,*]

1 Department of Medical Chemistry, Medical University of Bialystok, Mickiewicza 2A, 15-222 Bialystok, Poland
2 Department of Biology, Medical University of Bialystok, Mickiewicza 2A, 15-222 Bialystok, Poland
3 Department of Pharmacognosy, Medical University of Bialystok, Mickiewicza 2A, 15-230 Bialystok, Poland
4 Department of Molecular Biology and Genetics, Faculty of Medical Sciences in Katowice, Medical University of Silesia, Medykow 18, 40-475 Katowice, Poland
* Correspondence: angajko@umb.edu.pl

Abstract: Osteogenesis imperfecta (OI) is a heterogeneous connective tissue disease mainly caused by structural mutations in type I collagen. Mutant collagen accumulates intracellularly, causing cellular stress that has recently been shown to be phenotype-related. Therefore, the aim of the study was to search for potential drugs reducing collagen accumulation and improving OI fibroblast homeostasis. We found that rosemary extract (RE), which is of great interest to researchers due to its high therapeutic potential, at concentrations of 50 and 100 μg/mL significantly reduced the level of accumulated collagen in the fibroblasts of four patients with severe and lethal OI. The decrease in collagen accumulation was associated with RE-induced autophagy as was evidenced by an increase in the LC3-II/LC3-I ratio, a decrease in p62, and co localization of type I collagen with LC3-II and LAMP2A by confocal microscopy. The unfolded protein response, activated in three of the four tested cells, and the level of pro-apoptotic markers (Bax, CHOP and cleaved caspase 3) were attenuated by RE. In addition, the role of RE-modulated proteasome in the degradation of unfolded procollagen chains was investigated. This study provides new insight into the beneficial effects of RE that may have some implications in OI therapy targeting cellular stress.

Keywords: rosemary extract; collagen type I; unfolded protein response; autophagy; proteasome; skin fibroblasts; apoptosis; osteogenesis imperfecta

Citation: Sutkowska-Skolimowska, J.; Brańska-Januszewska, J.; Strawa, J.W.; Ostrowska, H.; Botor, M.; Gawron, K.; Galicka, A. Rosemary Extract-Induced Autophagy and Decrease in Accumulation of Collagen Type I in Osteogenesis Imperfecta Skin Fibroblasts. *Int. J. Mol. Sci.* **2022**, 23, 10341. https://doi.org/10.3390/ijms231810341

Academic Editors: Antonio Carrillo-Vico and Ivan Cruz-Chamorro

Received: 14 August 2022
Accepted: 6 September 2022
Published: 7 September 2022

1. Introduction

Osteogenesis imperfecta (OI) is a rare hereditary bone disease, with a frequency of 1 in 15,000 to 20,000 live births, characterized by phenotypic and genotypic heterogeneity [1,2]. The most common symptoms include bone fragility, skeletal deformities, reduced bone mineral density, short stature, blue sclera, dentinogenesis imperfecta, hearing impairment, joint hypermobility, skin fragility, muscle weakness, and cardio-respiratory defects [1–3].

Despite recent discoveries of mutations in many genes, the most common are mutations in genes encoding type I collagen, which cause the majority (85%) of cases of OI [1–5]. Mutations in the *COL1A1* or *COL1A2* genes encoding the α1(I) and α2(I) chains of type I collagen, respectively, are dominant and cause quantitative or structural disturbances in collagen. The classification of OI due to collagen mutations includes four types, the phenotype of which ranges from mild (type I) and moderate (type IV) to severe (type III) and perinatal lethal in OI type II [6]. The molecular defect in OI type I is a null *COL1A1* allele due to premature stop codons, either directly or through frame shifts resulting in reduced synthesis of functional collagen type I [4,5,7]. OI types II-IV are mainly caused by substitutions of glycine residues by another amino acid (80%), but also by splicing site mutations as well as small triplet deletion or duplication mutations, which shift the register of α-chains in the helix [2,4,5,8]. Most of them result in the synthesis of mutant misfolded collagen molecules. In recent years, many other causative genes associated with recessive and X-linked forms of the disease have been detected. Most of these genes code for type I collagen-related proteins that play an important role in folding and post-translational modifications, secretion as well as quality control of collagen synthesis. Mutations of proteins unrelated to collagen type I play an important role in osteoblast maturation and bone mineralization [2,3,5,8–10].

Synthesis of type I collagen is a complex process including the intracellular and extracellular steps preceding the formation of mature collagen fibrils. Two proα1 and one proα2 chain are synthesized in the endoplasmic reticulum (ER) and undergo important post-translational modifications prior to triple helix folding, including hydroxylation of proline (at C-4 and C-3) and lysine residues [11,12], which have fundamental importance for the stability of the helix. Proper folding and post-translational modifications in the ER determine the effectiveness of collagen secretion to the extracellular matrix (ECM) [11,12]. Glycine substitutions are responsible for the delay in the formation of the triple helix. Prolonged exposure of procollagen chains to post-translational modifying enzymes leads to increased hydroxylation of proline and lysine residues and glycosylation of hydroxylysine residues, causing the synthesis of collagen molecules with the abnormal structure [2,4,5,8,13]. The mutated collagen is secreted into ECM, but it may be partially retained in the ER causing cellular stress, which may be related to the clinical outcome [14–18]. It has been reported that some OI cells activate an unfolded protein response (UPR) to restore cell homeostasis [14,17–21]. The best-studied three ER membrane receptors of UPR include inositol requiring enzyme 1 (IRE1), PKR-like endoplasmic reticulum kinase (PERK), and activating transcription factor 6 (ATF6) [19,20,22]. Under normal conditions, the chaperone—binding immunoglobulin protein (BiP) binds all three sensor proteins in their ER luminal domain and keeps them inactive, while under stressful conditions it binds preferentially to misfolded proteins leading to activation of UPR pathways. ATF6 moves to the Golgi apparatus, where it is cleaved by different proteases and then, as an active transcription factor, enters the nucleus and activates the promoter of its related target genes. IRE1 and PERK are activated by autophosphorylation and oligomerization. Activated IRE1 forms an alternative spliced variant of the X-Box binding protein 1 (XBP1s), which, as a transcription factor, increases the expression of various chaperones and proteins involved in the proteasomal of ER-associated degradation (ERAD). The activation of PERK inhibits global protein synthesis through phosphorylation of eukaryotic translation initiation factor (eIF2α) but favors the translation of some mRNAs, such as the activating transcription factor 4 (ATF4), which is involved in both cell survival and ER stress-dependent apoptosis. During chronic stress, ER promotes apoptosis by upregulating genes such as the homologous protein of the CCAAT enhancer binding protein (CHOP) [20,22,23]. If conformation of mutated collagen is not improved by chaperones, it is destined for degradation most often via autophagy [24].

Autophagy is a complex system regulated by more than 30 autophagy-related gene (ATG) proteins. The most studied markers associated with autophagy are beclin 1, a microtubule-associated protein 1 light chain 3 (LC3), and sequestosome 1 (SQSTM1/p62), later referred to as p62 [25]. In some cases, especially the mutations occurring in C-propeptide, that most affect the trimer assembly, the retrotranslocation of misfolded procollagen chains into the cytosol may occur and result in their degradation by the proteasome [26,27].

So far, anti-catabolic bisphosphonates, denosumab as a synthetic parathyroid hormone and growth hormone have been used in the therapy of OI [1,3,10]. Experimental OI therapy strategies such as genetically engineered stem cell transplantation methods, reprogramming of somatic cells into pluripotent stem cells as well as anti-transforming growth factor (TGF-β) therapy have been summarized by us recently [28]. The major disadvantages of these therapies are their poor efficacy, or cytotoxic side effects. Research is still underway to find new and more effective drugs. According to the latest research, the chemical 4-phenylbutyrate chaperone (4-PBA) is of great interest, the molecular target of which is ER stress caused by intracellular retention of mutant collagen in osteoblasts and fibroblasts [16–18,29,30].

Medicinal plants containing significant amounts of polyphenols are of great interest to researchers due to their high therapeutic potential. Polyphenols are widely studied as regulators of fundamental biological processes, including cell proliferation, apoptosis, and autophagy [31,32]. Rosemary (*Rosmarinus officinalis* L., Lamiaceae) is a rich source of many bioactive polyphenol compounds with a wide range of biological activities, such as antioxidant, antimicrobial, anti-inflammatory, antidiabetic and anticancer [33,34]. In our previous study, we showed a beneficial effect of rosemary extract (RE) on the biosynthesis of type I collagen in OI type I fibroblasts with quantitative type I collagen defect [35]. This time we examined whether RE could reduce the accumulation of mutant collagen in the fibroblasts of two patients with severe type III and two patients with lethal type II OI carrying mutations in α(I) chain. In addition, possible mechanisms involved in the action of RE were also investigated.

2. Results

2.1. Steady-State Collagen Analysis in OI Fibroblasts

To identify mutated collagen, the mobility of the α1(I) and α2(I) bands was analyzed by SDS-urea polyacrylamide gel electrophoresis (SDS-urea PAGE) (Figure 1). All cells with glycine substitution (G910S and G1448V in patients 1 and 2 with severe type III, and G691C and G352S in patients 1 and 2 with lethal type II OI) in the α1(I) showed delayed migration of α1(I) and α2(I) chains and intracellular retention. In OI type II (patient 1) cells, an additional band $[\alpha1(I)]_2$ dimer was identified between the α1(I) monomers and the α1(III) trimer, confirming the substitution of glycine with cysteine (G691C).

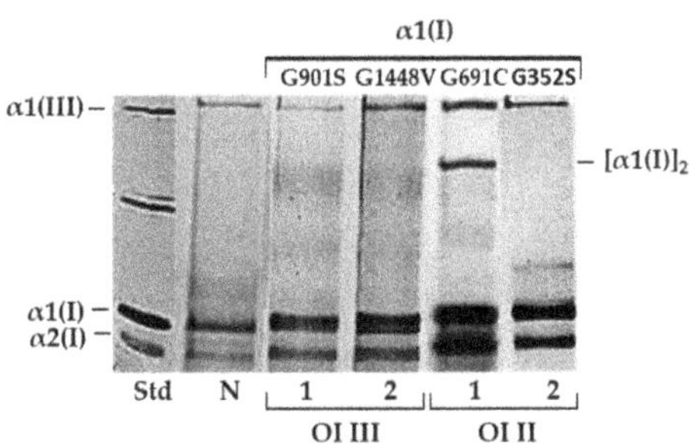

Figure 1. SDS-urea polyacrylamide gel electrophoresis (SDS-urea PAGE), in non-reducing conditions, of collagen type I in fibroblasts of patients 1 and 2 OI types III and II with glycine substitutions in α1(I) chain. To detect collagen after electrophoretic separation, silver staining was used; N—normal cells, Std—standard (bovine collagen type I) (Biocolor Life Science, UK).

2.2. Reduced Accumulation of Collagen Type I in OI Fibroblasts Treated with Rosemary Extract

In untreated OI fibroblasts, the accumulation of collagen type I in all four cell lines was confirmed by Western blot (Figure 2). When OI cells were treated with RE at the concentrations of 1–100 μg/mL, the level of accumulated collagen in cell layers was significantly reduced at 25, 50 and 100 μg/mL RE in OI type III patient 1, and at the concentrations of 50 and 100 μg/mL RE in the rest of the cells. Moreover, in cells of OI III 1 (at 50 μg/mL RE) and OI III 2 (at 50 and 100 μg/mL RE) the level of collagen type I was normalized (Figure 2a). The level of collagen secreted by OI fibroblasts, apart from the increase in the medium of OI III 1 fibroblasts treated with 100 μg/mL RE, remained unchanged. It should be added that RE at these concentrations did not significantly affect the viability of OI cells (Supplementary Figure S1).

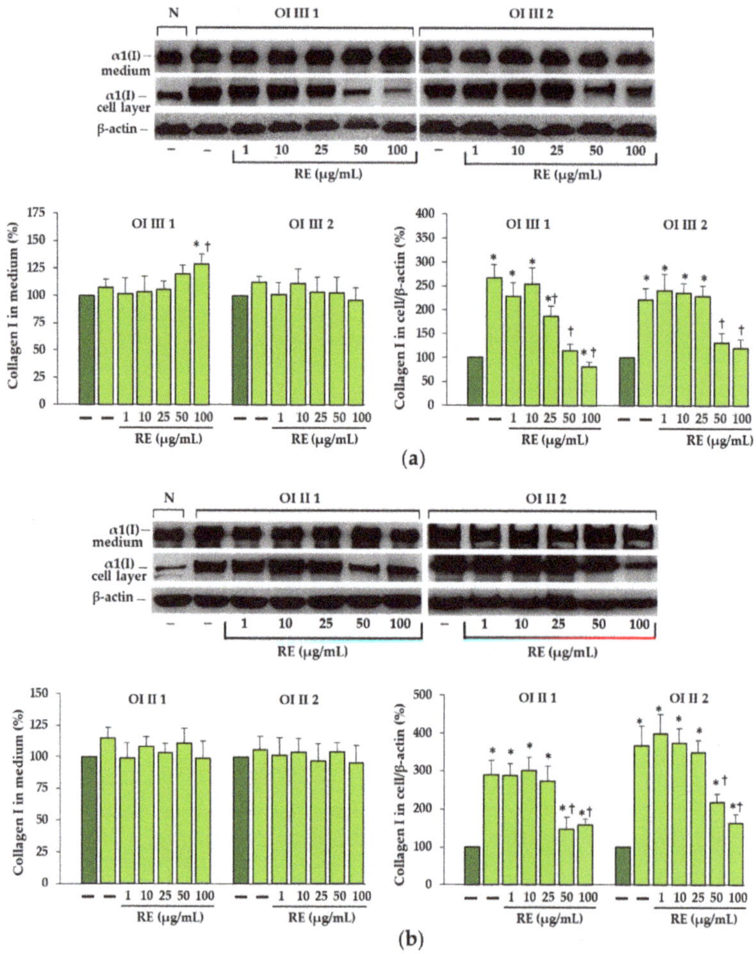

Figure 2. The effect of rosemary extract (RE) on the levels of secreted (in medium) and intracellularly accumulated type I collagen in OI type III (**a**), and OI type II (**b**) cells; β-actin was used as cell protein loading control. The bars represent the results of the gel densitometry as the mean values from three independent experiments; * $p < 0.05$, OI cells vs. normal (N) cells; † $p < 0.05$, OI treated cells vs. OI untreated cells. The data are expressed as a percentage of the normal sample taken as 100%; dark green and light green bars represent normal and OI, respectively.

Since rosmarinic acid (RA) is an essential component of RE, we also examined its effect on the mutant collagen retention. RA in a wide range of concentrations (1–100 μM), apart from lowering the accumulation of collagen in OI II 1 cells at its highest concentration (100 μM), had no effect in the remaining cells (Supplementary Figure S2). In the subsequent studies, we focused on the mechanisms of action of RE at the concentrations of 50 and 100 μg/mL.

2.3. Activation of Unfolded Protein Response in OI Cells

To determine whether retained mutant procollagen activates UPR, the expression of BiP, protein disulfide isomerase (PDI), ATF4, AFT6 and XBP-1s was evaluated (Figure 3).

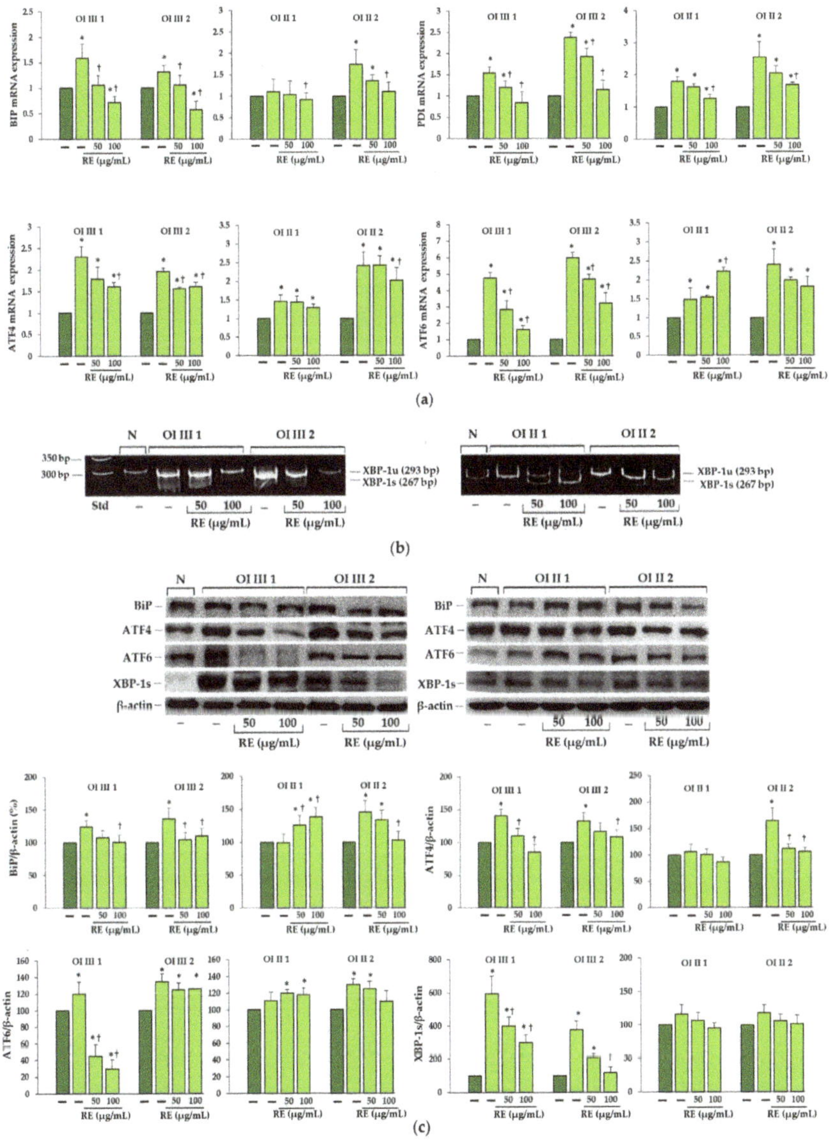

Figure 3. (**a**) The influence of rosemary extract (RE) on mRNA relative expression of binding immunoglobulin protein (BiP), protein disulfide isomerase (PDI), and activating transcription factors 4 and 6 (ATF4, ATF6) in OI types III and II cells. The bars represent the results of the mean values from three independent experiments; * $p < 0.05$, OI cells vs. normal cells; † $p < 0.05$, OI treated cells vs. OI untreated cells; dark green and light green bars represent normal and OI, respectively. (**b**) Unspliced and spliced X-box binding protein 1 (Xbp-1u and Xbp-1s) RT-PCR products were analyzed on 7% polyacrylamide gel. (**c**) Western blot analysis of BiP, ATF4, ATF6 and XBP-1s in OI type III and II cells; β-actin was used as cell protein loading control. The bars represent the results as the mean values from three independent experiments; * $p < 0.05$, OI cells vs. normal (N) cells; † $p < 0.05$, OI treated cells vs. OI untreated cells. The data are expressed as a percentage of the normal sample taken as 100%; dark green and light green bars represent normal and OI, respectively.

Expression of the chaperone BiP, which is an activator of UPR sensors, was significantly increased in untreated OI cells at the mRNA level beyond OI II 1 cells where it remained unchanged compared to the normal cells (Figure 3a). PDI, which catalyzes the formation and isomerization of disulfide bonds and acts as a collagen chaperone, was significantly increased in all untreated OI cells at mRNA level (Figure 3a). In the presence of RE (50 and 100 μg/mL), the expression of both chaperone transcripts was mostly reduced to a different extent compared to the untreated OI cells (Figure 3a).

ATF4, which is the effector in the PERK branch, was according to the analysis of its transcript expression, upregulated in untreated OI cells. RE either did not change or it decreased the expression of this factor mRNA, dependently on RE concentration, in comparison to the untreated OI cells (Figure 3a).

Expression of ATF6 was also investigated, and the level of its mRNA in untreated OI cells increased compared to the control. In cells exposed to RE (50 and 100 μg/mL) the decrease in ATF6 gene expression was noted in OI III, no effect in OI II 2 or stimulating effect of 100 μg/mL RE in OI II 1 (Figure 3a).

Activation of the IRE1α branch was studied by determining the level of XBP-1 expression in which splicing is mediated by this protein. In normal cells, the splicing form of XBP-1 (XBP-1s) was absent, while in OI it was manifested mainly in type III OI and was barely detectable in OI II cells (Figure 3b). Under the influence of RE at a concentration of 100 μg/mL, the spliced form disappeared in OI type III and only the unspliced (XBP-1u) form was observed, while in RE-treated OI II cells a slower migrating band appeared, with greater intensity in OI II 1 (Figure 3b). To find out whether the changes in the expression of proteins included in the UPR in OI cells reflect the expression of their genes, Western blots were performed (Figure 3c).

The results of the Western blot analysis revealed a similar increase in BIP at the protein level as was at the mRNA level, and no increase in untreated OI II 1 cells where no upregulation of the BIP gene was shown. Interestingly, RE at both concentrations (50 and 100 μg/mL) increased BIP expression in OI II 1 cells, and in others with upregulation of this chaperone protein, the RE induced decrease in relation to untreated OI cells was noted (Figure 3c). The expression of ATF4 and ATF6 transcription factors at the protein level was increased as at the mRNA level, except for OI II 1 cells, where their genes were upregulated. In RE-treated cells, the normalization of ATF4 level was achieved, while the ATF-6 level was either unaffected or significantly decreased by RE in OI III 1 cells (Figure 3c). XBP-1s at the protein level was almost undetectable in normal age-matched cells for OI type III, while its expression increased almost 6-fold and 4-fold in OI III 1 and 2 cells, respectively, and significantly decreased under the influence of RE. In contrast, in untreated OI II cells, only an upward trend and no RE effect on the expression of this spliced form of the factor was observed (Figure 3c).

2.4. Rosemary Extract-Induced Autophagy in OI Cells

ATG5 and beclin 1, which are important in initiation of autophagosome formation, increased in all untreated OI cells at the mRNA level and their expression intensified in the presence of RE (Figure 4a). In addition, the expression levels of autophagy related markers were determined by Western blot as well. Beclin 1 level increased in untreated OI III and OI II 2 and even more in RE treated cells consistent with gene expression, while there was no change in untreated and RE treated OI II 1 cells compared to normal cells (Figure 4b). LC3-II as a marker of the final fusion of the autophagosome with the lysosome was increased in untreated OI cells apart for OI III 1. In the presence of RE (50 and 100 μg/mL), no effect in OI III 2 and significant stimulation of LC3-II expression in the remaining cells was demonstrated as compared to untreated cells (Figure 4b).

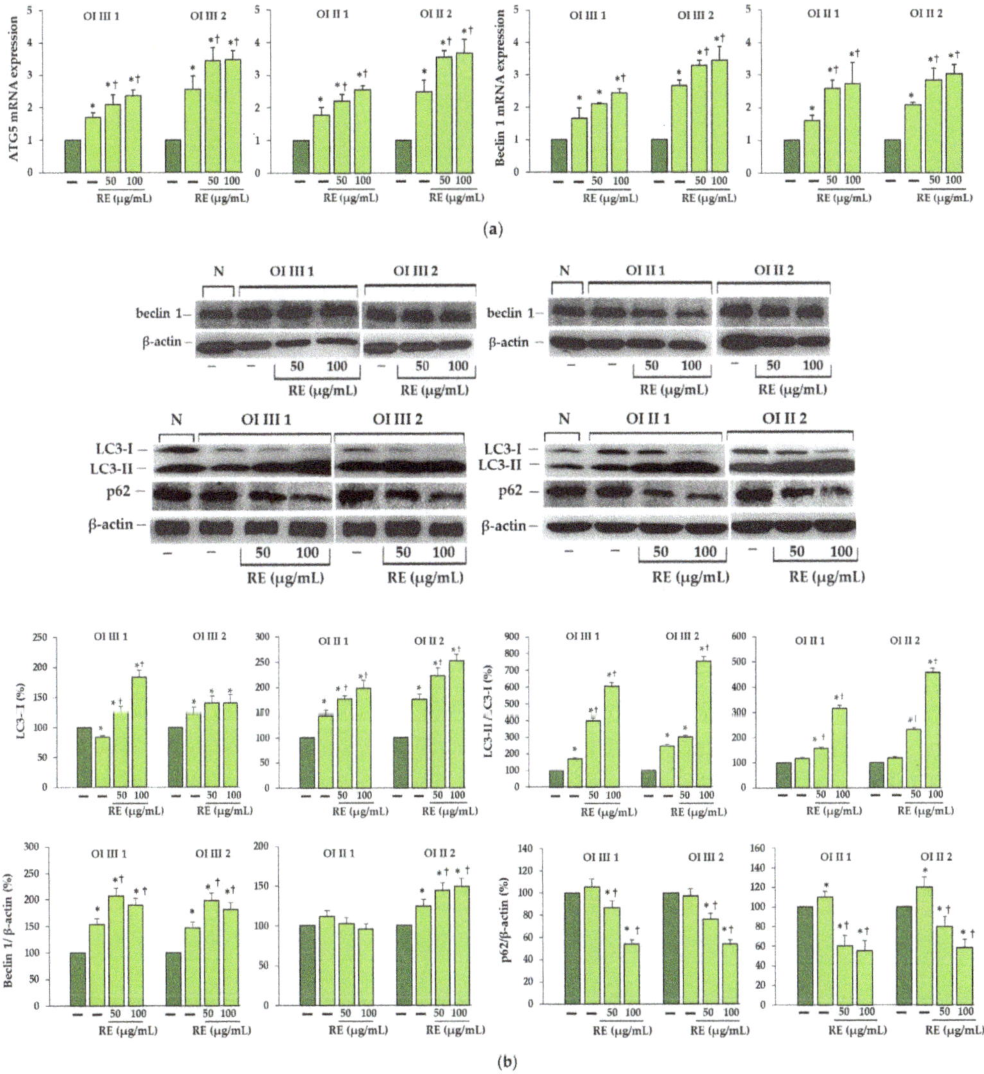

Figure 4. (**a**) The influence of rosemary extract (RE) on mRNA relative expression of autophagy-related gene (ATG5) and beclin 1 in OI types III and II cells. The bars represent the results of the mean values from three independent experiments; * $p < 0.05$, OI cells vs. normal cells; † $p < 0.05$, OI treated cells vs. OI untreated cells; dark green and light green bars represent normal and OI, respectively. (**b**) Western blot analysis of beclin 1, microtubule-associated protein 1 light chain 3 (LC3-I/LC3-II), and sequestosome 1 (SQSTM1/p62) in OI type III and II cells; β-actin was used as cell protein loading control. The bars represent the results as the mean values from three independent experiments; * $p < 0.05$, OI cells vs. normal (N) cells; † $p < 0.05$, OI treated cells vs. OI untreated cells. The data are expressed as a percentage of the normal sample taken as 100%; dark green and light green bars represent normal and OI, respectively.

In order to assess the dynamic process of autophagy, the ratio of LC3-II to LC3-I was determined, which in untreated cells either increased (in OI III 1 and 2) or remained unchanged (in OI II 1 and 2) compared to normal, while RE significantly induced autophagic flow in all OI cells (Figure 4b). The expression of p62, which is an indicator of autophagic degradation, was unchanged in OI III and increased in OI II cells. We examined whether RE facilitates degradation of p62. Treatment of cells with RE significantly decreased the p62 level, which suggests RE-induced autophagic degradation (Figure 4b). Since p62 and LC3-II are also degraded along with the digestion of cell components by autophagy; therefore, in order to further confirm that the increase in LC3-II induced by RE was not indicative of accumulation of this protein, ammonium chloride (NH_4Cl) as a lysosome proteolysis inhibitor was used. In all OI cells with inhibited lysosomal protein degradation, not only the LC3-II level but also p62 were significantly higher than in cells treated with RE alone (Supplementary Figure S3).

To confirm that collagen type I is degraded in the lysosomal pathway, a lysosome-enriched cell fraction was prepared and tested for its level in the presence and absence of lysosomal inhibitor in RE treated cells. We found that collagen type I was localized to a very small extent in the lysosomal fraction of normal cells and to a much greater extent in untreated OI cells (Figure 5). In OI cells treated with RE alone the level of type I collagen was lower than in the presence of inhibitor of lysosomes NH_4Cl, which indicates its lysosomal degradation.

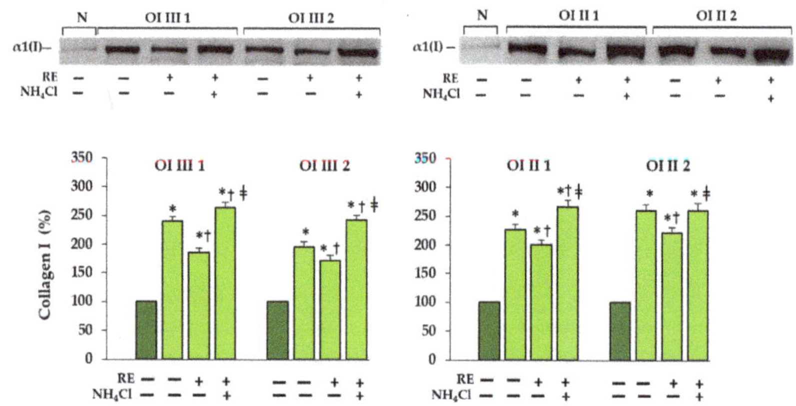

Figure 5. Western blot analysis of type I collagen in the lysosome-enriched cellular fraction, prepared as described in Methods, in OI types III and II cells treated with 50 μg/mL RE and 50 mM NH_4Cl. Equal amounts of proteins were loaded on 10% polyacrylamide gel. The bars represent the results of the gel densitometry as the mean values from three independent experiments; * $p < 0.05$, OI cells vs. normal (N) cells; † $p < 0.05$, OI cells treated with RE vs. OI untreated cells; ‡ $p < 0.05$, OI cells treated with RE + NH_4Cl vs. OI cells treated with RE alone. The data are expressed as a percentage of the normal sample taken as 100%; dark green and light green bars represent normal and OI, respectively.

Using confocal fluorescence microscopy with immunofluorescence staining, representative images of which are presented in Figure 6, we assessed the localization of endogenous collagen I, LC3-II and lysosomal-associated membrane protein 2A (LAMP2A) in normal and OI cells. It was observed that compared to normal cells, collagen type I staining intensity (red) in untreated OI cells increased, and after treatment with RE it decreased, whereas LC3-II (green) inversely decreased in untreated OI cells and increased in the presence of RE. Similar to the Western blot results, collagen type I staining decreased in the presence RE; moreover, it was co-localized with LC3-II as evidenced by the merged images (Figure 6a). We additionally assessed co-localization of collagen type I and LAMP2A, which is a lysosomal marker. As shown in Figure 6b, co-immunofluorescence staining for collagen type I (red) and LAMP2A (green) showed the increase in their co-localization after treatment of OI cells with RE. The above results suggest a stimulating effect of RE on the intracellular collagen degradation mediated by the autophagolysosomes.

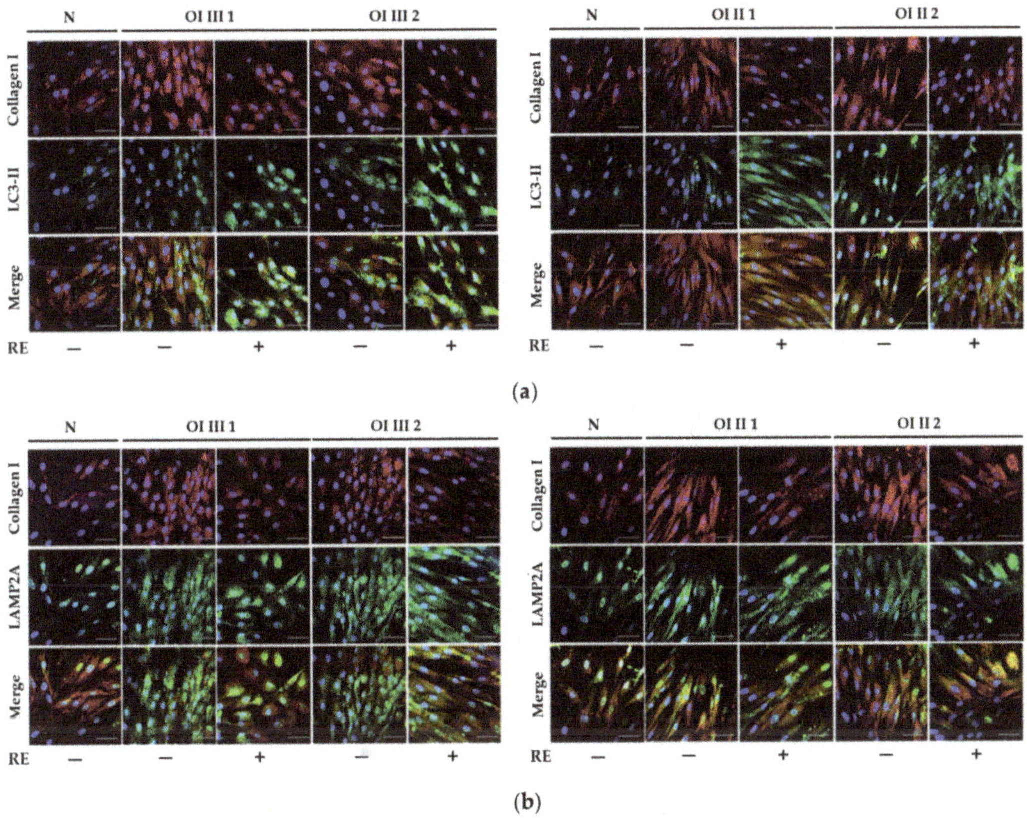

Figure 6. Colocalization of collagen type I (red)/LC3-II (green) (**a**), and of collagen type I (red)/LAMP2 (green) (**b**), and changes in the intensity of staining of these proteins in normal (N) and OI types III and II cells, treated with 50 µg/mL rosemary extract (RE), revealed by confocal microscopy analysis. A merge signals of colocalizing collagen type I/LC3-II and collagen type I (/LAMP2 were detectable, with greater intensity in RE-treated cells. Nuclei were stained with DAPI (blue). Scale bar, 50 µm. Representative immunofluorescent images are shown.

2.5. Involvement of Proteasome in Degradation of Unfolded Procollagen Chains in OI III 2 Cells

Since it has been reported that unfolded procollagen chains can be degraded by the ERAD pathway [27], we checked the polyubiquinination of type I procollagen in OI cells following SDS-PAGE under non-reducing conditions. As it turned out, under these conditions, polyubiquitinated proteins corresponding to the molecular weight of procollagen chains were detected only in OI III 2 cells, and the increase in this modification was noted in the presence of RE (Supplementary Figure S4). In order to determine the way of their degradation, a Western blot of polyubiquitinated proteins (Figure 7a) and SDS-PAGE of silver-stained proteins (Figure 7b) in the presence of proteasome and autophagy inhibitors were performed. Three polyubiquitinated protein bands corresponded to monomers, dimers and trimers of type I procollagen. Dimers and trimers are stabilized by the formation of inter-chain disulfide bonds within the C-propeptide. There was the increased level of unfolded proα1(I) chains and a decreased level of trimers in OI cells compared to the control cells (Figure 7b), while the level of monomer polyubiquitination was slightly reduced in OI cells (Figure 7a). In the presence of bortezomib (BR), a proteasome inhibitor, increased levels of polyubiquitinated monomers correlated with increased level of unfolded procollagen chains. Treatment with autophagy inhibitors chloroquine (CQ) and 3-methyladenine (3-MA) remained without effect on the level of polyubiquitination, but interestingly, the level of unfolded procollagen chains in the presence of CQ slightly increased and under the influence of 3-MA decreased as compared to untreated OI cells. Rosemary extract caused the increase in the polyubiquitination of monomers and dimers (Figure 7a), which was accompanied by a marked decrease in the level of unfolded collagen (Figure 7b).

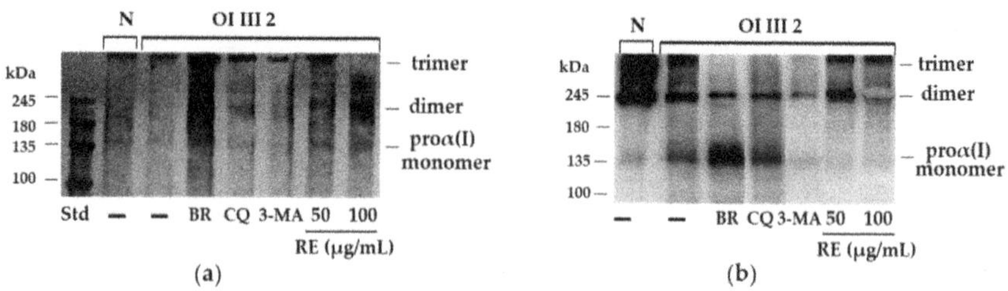

Figure 7. Polyubiquitynation (**a**) and SDS-PAGE of silver-stained (**b**) procollagen type I monomers, dimers and trimers, performed under non-reducing conditions in normal (N) and OI III 2 cells treated with 50 and 100 μg/mL rosemary extract (RE), 50 nM bortezomib (BR), 50 μM chloroquine (CQ), and 5 mM 3-methyladenine (3-MA).

Additionally, the effect of RE on proteasomal activities: chymotrypsin-like (Ch-L), trypsin-like (T-L) and caspase-like (C-L) in untreated and RE-treated OI fibroblasts was examined. As shown in Figure 8a, the decrease in T-L and C-L activities and no change in ChT-L activity were found in untreated OI cells, while in the presence of RE all activities were lowered as compared to untreated cells. The effectiveness of BR was confirmed by inhibitory effect on all proteasome activities (Figure 8b). The use of autophagy inhibitors, in turn, resulted in a slight stimulation of ChT-L activity by 3-MA and an inhibitory effect of CQ on all proteasome activities (Figure 8b).

To check whether the inhibition of proteasome activity by RE does not lead to the accumulation of non-collagen proteins, the level of silver-stained lysate proteins separated on 10% gels under reducing conditions was assessed and additionally the level of polyubiquitination of these proteins (Supplementary Figure S5). The increase in the level of polyubiquitinated total proteins (Supplementary Figure S5a) coincided with the increase in the level of total protein expression in OI untreated cells (Supplementary Figure S5b) compared to the normal, which can be explained by the disruption of the autophagy and proteasome degradation processes. Treatment of OI cells with RE did not lead to the total protein accumulation, as was the case with the use of proteasome (BR and MG132) or autophagy (CQ, 3-MA and NH_4Cl) inhibitors, their levels were comparable to the normal (Supplementary Figure S5b).

2.6. *Expression and Activity of Collagen Type I Degrading MMPs*

Additionally, we assessed the expression of matrix metalloproteinases MMP-1 and MMP-2, degrading extracellular collagen I, in untreated and RE exposed OI cells. MMP-1 mRNA in OI cells was significantly upregulated compared to normal cells, while RE significantly decreased it. MMP-2 was upregulated at the mRNA level in OI type II and was normalized in the presence of both RE concentrations (50 and 100 μg/mL) in OI II 1 and at 100 μg/mL in OI II 2 (Supplementary Figure S6a).

The zymography allowed to identify mainly pro-MMP-2 in media of normal cells and additionally its active form MMP-2 in OI, while RE, beyond OI II 1, at a concentration of 100 μg/mL, showed a lowering effect, the highest in OI III 1 (Supplementary Figure S6b).

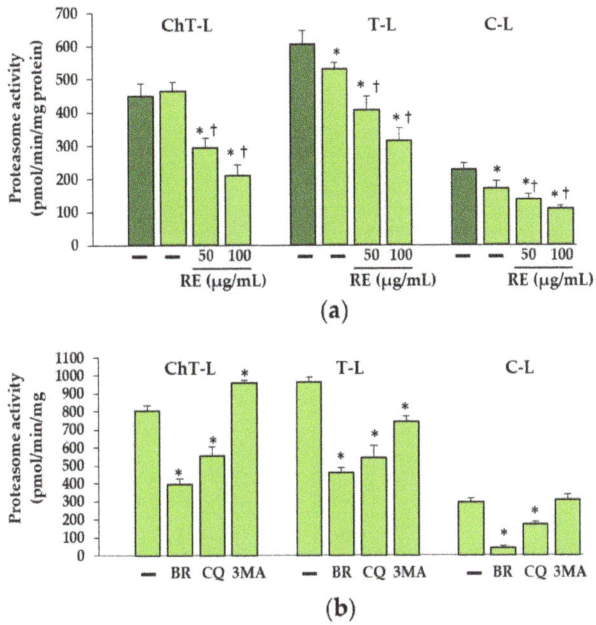

Figure 8. Proteasomal activities: chymotrypsin-like (ChT-L), trypsin-like (T-L) and caspase-like (C-L) in OI III 2 cells treated with 50 and 100 μg/mL rosemary extract (RE) (**a**), 5 nM bortezomib (BR), 50 μM chloroquine (CQ) and 5 mM 3-methyladenine (3-MA) (**b**). The bars represent the mean values from three independent experiments. (**a**) * $p < 0.05$, OI cells vs. normal cells; † $p < 0.05$, OI cells treated with RE vs. OI untreated cells; dark green and light green bars represent normal and OI, respectively; (**b**) * $p < 0.05$, OI cells treated with inhibitors vs. OI untreated cells.

2.7. Rosemary Extract Reduces the Expression of Apoptosis Markers in OI Cells

The results presented in Figure 9 showed the significant increase in the expression of proapoptotic proteins such as Bax and CHOP at both mRNA and the protein levels as well as cleaved caspase-3 protein. Under the influence of RE at both concentrations (50 and 100 μg/mL), downregulation of these markers was mostly observed; even if the levels of Bax and CHOP mRNA did not change in OI II 1 cells, a significant decrease was observed at the protein level.

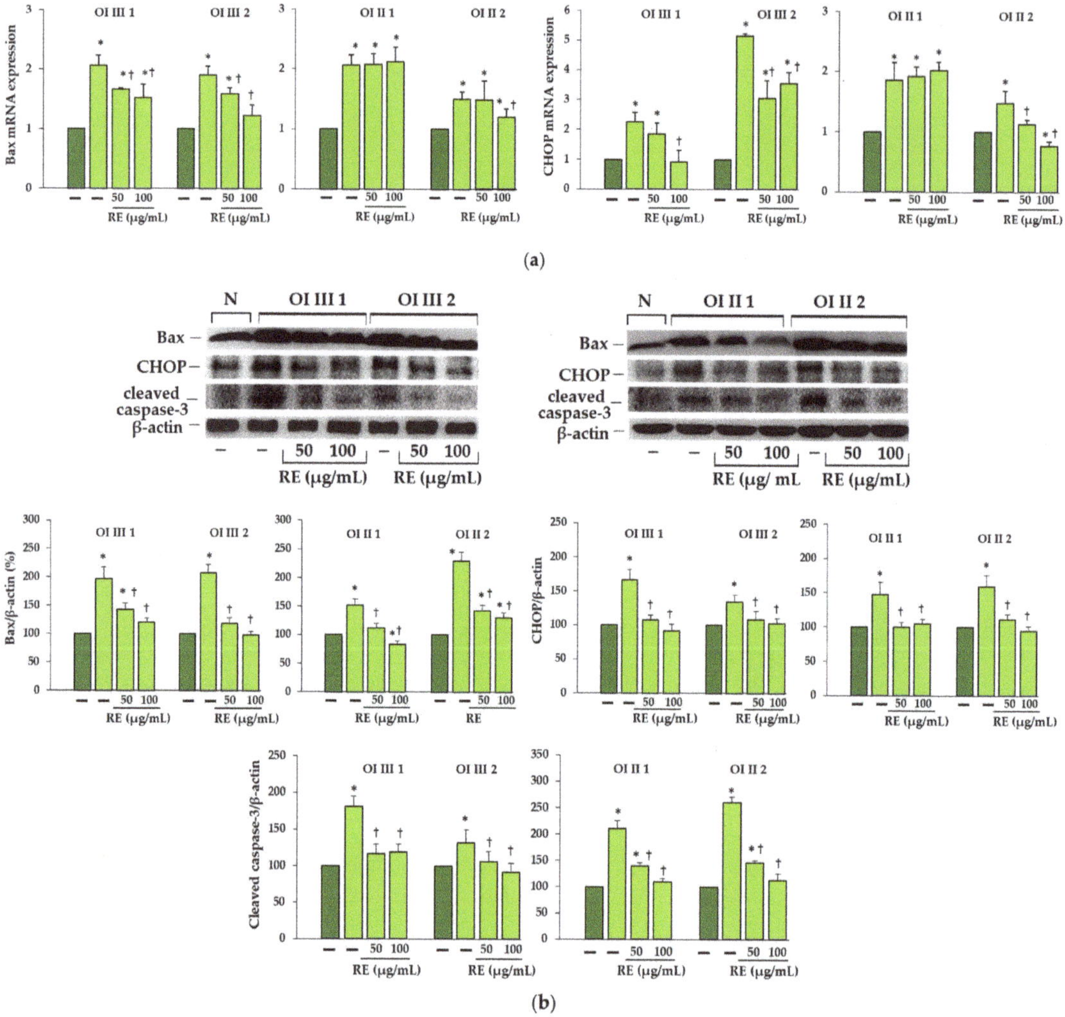

Figure 9. (**a**) The effect of rosemary extract (RE) on mRNA relative expression of Bax and CCAAT enhancer binding protein (CHOP) in OI types III and II cells. The bars represent the results of the mean values from three independent experiments; * $p < 0.05$, OI cells vs. normal cells; † $p < 0.05$, OI treated cells vs. OI untreated cells; dark green and light green bars represent normal and OI, respectively. (**b**) Western blot analysis of Bax, CHOP and cleaved caspase-3 in OI type III and II cells; β-actin was used as cell protein loading control. The bars represent the results as the mean values from three independent experiments; * $p < 0.05$, OI cells vs. normal (N) cells; † $p < 0.05$, OI treated cells vs. OI untreated cells. The data are expressed as a percentage of the normal sample taken as 100%; dark green and light green bars represent normal and OI, respectively.

3. Discussion

OI is a genetically and phenotypically heterogeneous group of connective tissue disorders caused mainly by an autosomal dominant mutations in collagen type I, which is the most abundant protein in bone and skin ECM [1–10]. The discovery in the last two decades of new causative genes (around twenty) that are involved in the regulation and function of type I collagen, but also in other aspects of bone biology [2,3,5,8–10] confirms the complexity of the underlying mechanisms and complicates understanding the relationship between mutations and phenotype as well as finding an effective and a universal method of treating the disease. Recently, ER stress caused by intracellular retention of mutant collagen in osteoblasts and fibroblasts was found as an attractive target of OI therapy [14–21]. The use of the chemical chaperone 4-PBA, approved by *Food and Drug Administration*, reduced ER stress and restored cell homeostasis in human fibroblasts of OI patients carrying dominant mutations in α1 and α2 collagen type I chains [17] as well as recessive mutations in cartilage-associated protein (*CRTAP*), prolyl-3-hydroxylase 1 (*P3H1*) and cyclophilin B (*PPIB*) impairing prolyl-3 hydroxylation of collagen type I [18]. Moreover, it was found that 4-PBA normalizes the overproduction of type I collagen and improves the misfolding of the type I collagen helix in OI fibroblasts due to glycine substitution, and also improves the impaired mineralization of osteoblasts differentiated from OI induced pluripotent stem cells [30]. Administration of this drug to the OI dominant zebrafish model, carrying typical glycine substitution G574D in *COL1A1*, alleviated cellular stress and improved bone mineralization in larvae and skeletal deformity in adults. This was accompanied by the reduction of the ER cisternae size and promoting the secretion of collagen [29]. Similarly, in osteoblasts of two murine OI models carrying G349C mutation in *COL1A1* (Brtl mouse) and G610C in *COL1A2* (Amish mouse), 4-PBA prevented collagen type I accumulation through increased its secretion and reduction of aggregates in mutant cells [16]. In addition, increased collagen incorporation into the matrix and improved mineral deposition in osteoblasts was observed in both murine models, which convinces about the influence of ER stress on the phenotype. Therefore, as well as the discovery of the therapeutic potential of this chemical chaperone, finding of other safe compounds to reduce cellular stress and restore cell homeostasis may be a new strategy for treating this disease, or at least some OI cases.

This study provides, for the first time, evidence of the beneficial effects of rosemary extract on fibroblasts with mutations in the collagen triple helix resulting in intracellular collagen accumulation. Moreover, the presented results explain the likely mechanisms of reducing cellular stress, mainly by enhancing the degradation of mutant collagen through autophagy. For our study, we chose glycine substitutions in the α1(I) chains of two patients with severe OI type III (G901S and G1448V), and two with lethal type II (G352S and G691C). As we demonstrated by SDS-urea-PAGE, all cells showed delayed migration of α1 and α2 collagen chains and intracellular accumulation of mutant collagen.

Collagen type I is characterized by a unique right-handed triple helical structure, and each of three left-handed polyproline-like helices contains a repeated sequence (Gly-X-Y) in which X and Y are often proline and hydroxyproline [36]. The triple helical domain is flanked on both sides by N- and C-terminal propeptides. Procollagen folding takes place in the ER, after which the protein is transported to the Golgi apparatus and secreted, where the cleavage of propeptides takes place and mature collagen is formed. Folding is a very complex process involving many chaperones, such as BiP, PDI, prolyl 4-hydroxylase, various peptidyl-prolyl *cis-trans* isomerases, and heat shock protein 47, whose role is to provide stabilization of the structure and to protect aggregation of unfolded chains [11,12,37].

We found increased BiP expression in two OI type III (G901S and G1448V) and one OI type II (G352S) cells, and a consequent activation of UPR pathways, whereas in OI II 1 cells with G691C the expression of this chaperone remained unchanged compared to the normal cells. According to Besio et al. [17] increased BiP was detected in three out of five tested fibroblasts with glycine substitution in α1 and α2 chains. Expression of PDI, which catalyzes the formation and isomerization of disulfide bonds and acts as a collagen chaperone by interacting with collagen α single chains, was upregulated in all α1(I) mutant cells, while in the study of Besio et al. [17] in four out five cells with mutations in α1(I) and it remained unchanged in cells with mutations in α2(I). The same authors reported the activation of mainly the PERK pathway and increased ATF4 expression in cells with mutations in α1 and α2 as well as IRE1α pathway with a predominance in cells with mutations in α2, but there was no difference in the level of activated ATF6 [17]. In contrast, in our study, along with the increase in BiP expression, activation of both transcription factors ATF4 (effector of PERK pathway) and ATF6 in cells with BiP upregulation was found, while expression of spliced forms of XBP1 (effector of IRE1α pathway) was predominant in OI type III as confirmed by real-time PCR and polyacrylamide gel electrophoresis of (reverse transcriptase) RT-PCR product. It should be added that in our studies, the lack of increase in BiP expression in cells with G691C mutation was consistently associated with the lack of activation of UPR proteins, while in the case of some collagen and non-collagen mutations, activation of some UPR pathways was also observed in the absence of BiP [17,18]. It is possible that other factors or regulatory mechanisms are involved in the activation of the UPR during ER stress. In the osteoblasts of the mouse OI model with the substitution of glycine by cysteine (G610C), closely located to the one we studied (G691C) but in α2(I), no conventional UPR was detected also, only enhanced autophagy [15]. Moreover, as it turned out in our studies, the upregulation of genes of transcription factors ATF4 and ATF6 did not coincide with the increased amount of the protein, which means that their expression is regulated by post-transcriptional mechanisms and the expression of the genes themselves cannot be compared without determining the expression at the protein level.

Based on the obtained results, we can say that RE, if not completely eliminated, largely reduced ER stress caused by the accumulation of mutant collagen. This was evidenced by the decreased expression of chaperone proteins BiP and PDI as well as effectors of UPR branches along with the decreased level of intracellular collagen. Interestingly, in OI II 1 cells where we did not detect increased BiP expression and activation of UPR pathways, upregulation of pro-survival factors (BiP, ATF6 and XBP-1s), but not ATF4, was found in RE-treated cells. Activation of the PERK pathway leads to inhibition of global protein translation by inhibition of eIF2α except for ATF4, which in the active form can upregulate both the survival (autophagy) and the apoptotic (CHOP) pathway genes. In turn, the active transcription factors ATF6 a and XBP1s enhance the expression of chaperones, and also ATF6 of genes for proteins involved in ERAD [19,20,22]. While it is still unknown how activation of individual UPR pathways and their effectors directly affects collagen, one study found that forced XBP1s expression in cells with glycine substitution (G425S) in α1 (I) chain enhanced the folding/assembly and secretion of mutant type I collagen [38].

Since the cell response to ER stress caused by intracellular accumulation of mutant collagen is most often autophagy or, less frequently, ERAD, we investigated the activation of these two degradation systems in untreated and RE exposed OI cells.

Autophagy, is a dynamic tightly regulated lysosomal pathway of degradation of intracellular components, including soluble proteins, aggregated proteins and damaged cell organelles. It is an evolutionarily conserved process, capable of responding to stress to limit cell damage. The autophagy process is regulated by several ATG core proteins, of which LC3 plays essential role in the formation and maturation of the autophagosome, [25,39]. Cytosolic form LC3-I is converted into an active membrane-bound form LC3-II during the formation of the autophagosomes, while the final degradation of the cargo takes place after fusion of autophagosome with lysosomes. It is strictly dependent on the p62, which apart from the ratio LC3-II/LC3I, is an important marker of effective autophagic flux [40]. Even though the LC3-II level was increasing in lethal untreated cells as compared to the aged-matched control, the ratio LC3-II/LC3-I remained unchanged, which along with the p62 increase indicated a lack of activation of autophagy. OI III cells showed an increase in the LC3-II/LC3-I ratio, but the p62 level remained unchanged, which also did not suggest an increase in autophagic activity. p62 with LC3 recognition sequence binds to LC3-II and after the formation of the autophagosome and its fusion with the lysosome, is degraded inside the autophagolysosome, that is why the decrease in p62 expression may indicate an active process of autophagy.

A markedly increased autophagic activity was observed in all OI cells after RE exposure, as shown by a dose-dependent increase in LC3-I to LC3-II conversion, along with accelerating p62 degradation. In addition, the stimulation of ATG5 mRNA and beclin 1 mRNA and protein, which initiate autophagosome formation, was demonstrated in RE-treated cells. Another evidence of the stimulating effect of RE on the degradation of mutant collagen with involvement of lysosomal pathway was the confirmation of the presence of collagen type I in the lysosomal fraction and the increase in its level in cells treated additionally with ammonium chloride, which raises the pH and thus inactivates lysosomes. The same results were obtained in the presence of chloroquine—another autophagy inhibitor (not shown). Moreover, the immunofluorescence microscopy studies have showed that collagen type I collocation increase in RE-treated OI cells with both the marker of autophagosome (LC3-II) and marker of lysosomes (LAMP2A). These results clearly indicate the RE-mediated degradation of mutated collagen type I in the autophagolysosomal process, although digestion of collagen by lysosomes regardless of autophagy cannot be ruled out. Omari et al. [41] showed that in addition to autophagy, accumulated in OI collagen type I can be digested in a noncanonical autophagy process.

As reported earlier the proteasome may be involved in the removal of unfolded procollagen chains [27]. Misfolded proteins or unfolded procollagen chains are retranslocated from the ER to the cytosol for degradation by the 26S proteasome after modification with polyubiquitin chains. As expected, polyubiquitination of unfolded proα1(I) chains was demonstrated in OI III 2 with a C-propeptide mutation (G1448V). The increase in the amount of unfolded procollagen chains in untreated OI cells can be explained by decreased C-L and T-L proteasome activities. The greater accumulation of these chains in the presence of BR, which inhibited to a much greater extent all three activities (ChT-L, T-L and C-L) of the proteasome confirms the proteasome's contribution to the removal of these chains. However, while it turned out that RE also decreased ChT-L, T-L and C-L activities, this decrease did not coincide with the accumulation of unfolded chains. On the contrary, there were lower levels of them than in untreated cells. At this stage of the study, it is difficult to explain the mechanism of action of RE in these cells, but it is very likely that in the case of inhibition of the proteasome activity, unfolded chains may be partially degraded in the process of RE-activated autophagy or by the proteasome, as the activity of the proteasome was only partially inhibited. It was also noted that, despite the reduction in proteasome activity by CQ, which inhibits autophagy, there was no additional accumulation of unfolded chains. On the other hand, another inhibitor of autophagy 3-MA that block autophagy at the initiation and maturation stages by acting on phosphoinositol 3 phosphate kinase (PI3K), caused a decrease in the level of unfolded chains compared to untreated OI cells, perhaps due to its stimulating effect on ChT-L activity. Since the two protein degradation systems (proteasome and autophagy) appear to be mechanically linked, it is suspected that when the proteasome is inhibited, autophagy may be activated to remove polyubiquitinated/unfolded protein aggregates and promote cell survival [42]. It is possible that a decrease in proteasome activity, also noted in other RE-treated OI cells used in this study (results not shown), triggers autophagy, protecting cells from the toxic long-term stress leading to cell apoptosis. The increase in the expression of pro-apoptotic proteins (Bax, CHOP and active caspase 3) in OI cells and their significant reduction or even normalization in the presence of RE, may unequivocally indicate that the accumulation of mutant type I collagen caused such stress and despite the activation of UPR (with the exception of OI II), the degradation processes mediated by autophagy and the proteasome were disturbed. The cell's response to inhibited intracellular degradation processes could be a significant upregulation of extracellular MMP-1 and MMP-2 genes, lowered in the presence of RE, while the importance of these upregulation requires further study. Moreover, the observed increase in the level of extracellular type I collagen only in the presence of 100 μg/mL RE in OI III 1 can be explained by its strong inhibitory effect on the activity of MMP-2 and not by an increase in collagen secretion. Negative correlations between the activity of lysosomal enzymes and MMPs have been reported [43] and are worth studying further, but our research focuses on the intracellular degradation processes.

It is also possible that, under the influence of RE, collagen folding is improved due to reduced over-modification of free procollagen chains as a RE-induced decrease in mRNA expression of one of the enzymes (β(1-O) galactosyltransferase (GLT25D1)) involved in these modifications was observed (data not shown).

Phytochemicals, due to their natural origin, low toxicity, as well as many valuable biological and pharmacological properties, are of wide interest among researchers as potential drugs with a high therapeutic and preventive potential for many diseases. *R. officinalis* is a polyphenol-rich source constituents such as luteolin and apigenin derivatives, caffeic acid derivative (rosmarinic acid) and other such as diterpenes (rosmanol isomers), detailed qualitative analysis of which was presented earlier [35].

In our previous studies, we have shown a stimulatory effect of rosemary and lemon balm extracts, RA as well as some flavonoids on collagen biosynthesis in OI type I [35,44] as well as normal human skin fibroblasts [45–47]. In these studies, the use of RA alone in a wide range of concentrations did not bring the expected effects, which may suggest the participation of other components of the rosemary extract in stimulating autophagy or its other effects. It has been reported by other authors that luteolin 7-*O*-glucoside (one of the identified RE components) protects against damage to the heart muscle induced by starvation by enhancing autophagy through inhibition of mechanistic target of rapamycin (mTOR) and extracellular signal-regulated kinase (ERK) signaling pathway [48]. It has also been shown that apigenin increases the expression of LC3-II, the formation of autophagolysosomal vacuoles and triggers autophagic flow in hepatocellular carcinoma cells [49]. It is widely accepted that a variety of plant extracts and dietary phytochemicals including resveratrol, curcumin, epigallocatechin-3-gallate, punicalagin, oleuropein, myricetin and rosmarinic, norhydroguaiaretic, and ferulic acids may stimulate autophagy [31,50]. These compounds remove protein aggregates, stimulate the antioxidant defense and ameliorate the ER stress, resulting in increased cell survival [31–34,50]. Pierzynowska et al. [51] reported on the removal of mutant huntingtin aggregates in the transfected HEK293 cells via genistein-induced autophagy, which may be the basis for the development of an effective therapy for this inherited neurodegenerative disease. Interestingly, like RE in OI cells, genistein showed significant inhibition of all protesome activities in the fibroblasts of patients with all types of mucopolysaccharidosis, which according to the authors, may lead to the stabilization of lysosomal enzymes and constitute a new approach in the treatment of this genetic disease [52].

A synergistic effect of several different compounds present in the rosemary extract may also be likely. While the biological properties (e.g., antioxidant) of polyphenols were previously related mainly to the structure of these compounds, now a more convincing explanation is modulating the activity and/or expression of key proteins for signaling cascades by interacting with them or modulating epigenetic regulation of gene expression. Therefore, it is believed that pleiotropic mechanisms and specific polyphenol-protein interactions are involved in their beneficial effects [53,54].

Finally, the limitations of this study should also be mentioned. Firstly, experiments were conducted on fibroblasts, which is related to the availability of biological material, and the disease mainly affects the skeletal system. On the other hand, collagen type I is a major component of skin and bone and, with a few exceptions, is similarly expressed in fibroblasts and osteoblasts. It is also worth noting that the activity of the proteasome may vary with age and even tissues or may be different in OI patients; therefore, more detailed studies are needed to understand the molecular mechanisms of RE-induced changes in proteasome activity. Although at this stage of our research we did not focus on explaining the consequences of proteasome inhibition by RE, the lack of accumulation of non-collagen proteins was shown.

4. Materials and Methods

4.1. Chemicals

Dulbecco's minimal essential medium (DMEM), fetal bovine serum (FBS) and phosphate-buffered saline (PBS) were obtained from Gibco (Thermo Fisher Scientific, Waltham, MA, USA); penicillin, streptomycin, and glutamine were purchased from Quality Biologicals Inc. (Gaithersburg, MD, USA). Radioimmunoprecipitation assay (RIPA) buffer, protease inhibitor cocktail (P8340), magnesium L-ascorbate, sodium dodecyl sulfate (SDS), dimethyl sulfoxide (DMSO), [3-(4,5-dimethylthiazol-2-yl)-2,5-diphenyltetrazolium bromide] (MTT), bovine serum albumin (BSA), pepsin, gelatin, NH_4Cl, CQ, 3-MA, and MG132 were provided by Sigma-Aldrich Corp. (St. Louis, MO, USA). BR was a product of Selleck Chemicals (Houston, TX, USA). Proteasome substrates: N-Suc-LLVY-AMC (7-amido-4-methylcoumarin) was purchased from Sigma-Aldrich Corp. (St. Louis, MO, USA), Bz-VGR-AMC and Z-LLE-AMC were obtained from Enzo Life Sciences, Inc. (Farmingdale, NY, USA). Rosemary extract (RE) was prepared and characterized using LC-MS technique according to procedure described in our previous study [35]. RA was a product of BIOKOM (Warsaw, Poland).

4.2. Fibroblast Culture and Treatment

The study was performed on skin fibroblasts derived from two patients with severe OI type III and mutations in *COL1A1*: Gly901Ser (patient 1) and Gly1448Val (patient 2), and two patients with lethal OI type II and mutations in *COL1A1*: Gly691Cys (patient 1) and Gly352Ser (patient 2) as well as two age matched normal cells. The normal skin fibroblast line used as a control for OI type III was CRL-1474 obtained from American Type Culture Collection (Manassas, VA, USA), and as a control for OI type II the normal line was derived from the foreskin on the 7th day of life of the donor. Fibroblasts from skin biopsy of OI patients and healthy control were obtained after informed consent in accordance with the Declaration of Helsinki and was approved by Bioethical Committee of the Jagiellonian University in Kraków, Poland (KBET/108/B/2007).

Fibroblasts were cultured in DMEM supplemented with 10% FBS, 2 mM glutamine, penicillin (50 U/mL) and streptomycin (50 μg/mL) at 37 °C in a humidified incubator in atmosphere containing 5% CO_2. For experiments, fibroblasts were grown to 90% confluence and the cultured medium was replaced with fresh DMEM without serum, supplemented with 25 μg/mL of magnesium ascorbate, before addition of compounds. Compounds were stored at 4 °C as the concentrated stock solutions in DMSO and were diluted in medium prior to addition to cell cultures. Fibroblasts were treated with RE at the concentration of 1–100 μg/mL and RA at the concentrations of 1–100 μM for 24 h. In addition, cells were treated with autophagy inhibitors: 50 μM CQ, 50 mM NH_4Cl and 5 mM 3-MA or proteasome inhibitors: 50 nM BR and 2.5 μM MG132, all of which were dissolved in DMSO and appropriately diluted before adding to cell cultures. In all experiments the concentration of DMSO did not exceed 0.05% (v/v).

4.3. MTT Test to Determine Viability of Treated Cells

Fibroblasts (1×10^4 cells per well) were treated with RE (1–200 μg/mL) for 24 h. Then cells were washed three times with PBS and MTT solution (0.5 mg/mL) was added for 4 h. After removing MTT solution 1 mL of 0.1 M HCl in absolute isopropanol was added to dissolve formazan crystals by thoroughly shaking on a plate shaker (BioSan, Riga, Latvia), and the absorbance at 570 nm was measured using a microplate reader (TECAN, Männedorf, Switzerland).

4.4. Quantitative Real-Time PCR

Total RNA was isolated from cultured cells using a Total RNA Mini Plus concentrator (A&A Biotechnology, Gdynia, Poland) and the concentration of RNA was determined using NanoDrop 2000 spectrophotometer (Thermo Fisher Scientific, Waltham, MA, USA). The equal amounts (1 μg) of total RNA were used to the synthesis of complementary DNA (cDNA) with the use of cDNA Synthesis Kit (Bioline, London, UK). Quantitative Real-time PCR (qRT-PCR) analysis was performed in the CFX96 Real-Time System thermal cycler (Bio-Rad, Hercules, CA, USA) using the SensiFAST™ SYBR kit (Bioline, London, UK). The expression of desired gene was normalized to the level of glyceraldehyde-3-phosphate dehydrogenase (GAPDH) and changes were calculated by the ΔΔCt method.

The sequences of primers (Genomed, Warsaw, Poland) are shown in Supplementary Table S1. The qRT-PCR parameters were as follows: 30 s at 95 °C followed by 40 cycles: 10 s at 95 °C, 10 s at 60–62 °C and 20 s at 72 °C. The reaction products were verified by analysis of their melting curves.

4.5. XBP1 Splicing Analysis

PCR mixture contained 1 μg of isolated RNA and primers (0.3 μM each): sense (5′-TCAG CTT TTA CGA GAG AAA ACT CAT GGC CT-3′) and antisense (5′-AGA ACA TGT GTG TCG TCC AAG TGT GTC GTC CAA GTG TG-3′) purchased in Genomed (Warsaw, Poland). Samples were incubated 30 min at 50 °C followed by 30 cycles at 94 °C, 60 °C, and 72 °C for 30 s each in the CFX96 Real-Time System thermal cycler (Bio-Rad, Hercules, CA, USA). Reaction products were analyzed by electrophoresis on 7% polyacrylamide gel and visualized with ethidium bromide.

4.6. Western Blot

Cell layers were harvested using RIPA buffer (Sigma-Aldrich Corp., St. Louis, MO, USA) and protease inhibitor cocktail (P8340) (Sigma-Aldrich Corp., St. Louis, MO, USA). The conditioned media were collected and concentrated 10 times with Centrifugal Filter Units (10K) (Merck Millipore Ltd., Carrigtwohill, County Cork, Ireland). The concentration of total protein in cell lysates and media was measured using BCA Protein Assay Kit (Pierce, Rockford, IL, USA) and Coomassie Plus—The Better Bradford Assay Reagent (ThermoFisher Scientific, Rockford, IL, USA), respectively. For Western blot an equal amount of protein (20 μg) was loaded on polyacrylamide gel (7.5%, 10% or 12% depending on the molecular mass of protein). Proteins were transferred from gels onto Immobilon-P Transfer membranes (Merck Millipore Ltd., Tullagreen, Carrigtwohill, County Cork, Ireland), which were blocked with 5% (w/v) non-fat dried milk diluted in 50 mM Tris-HCl, pH 7.5, 500 mM NaCl, 0.05% (v/v) Tween 20 (TBS-T) for 1 h at room temperature. Then, membranes were washed with TBS-T and incubated overnight at 4 °C with solutions of the following primary monoclonal antibodies: mouse anti-collagen type I (1:1000; Santa Cruz Biotechnology Inc., Santa Cruz, CA, USA), rabbit anti-ATF4 (1:1000; Abcam, Cambridge, UK), rabbit anti-ATF6 (1.1000; Abcam, Cambridge, UK), rabbit Bax (1:1000; Cell Signaling Technology, Danvers, MA, USA), mouse Beclin-1 (1:1000; Cell Signaling Technology, Danvers, MA, USA), rabbit BiP (1:1000; Cell Signaling Technology, Danvers, MA, USA), mouse cleaved caspase-3 (1:1000; Santa Cruz Biotechnology Inc., Santa Cruz, CA, USA), mouse CHOP (1:1000, Cell Signaling Technology, Danvers, MA, USA), rabbit anti-procollagen I (1:1000; Abcam, Cambridge, UK), rabbit LC3 (1:1000, Cell Signaling Technology, Danvers, MA, USA), mouse anti-p62 (1:1000; Abcam, Cambridge, UK), mouse XBP-1s (1:1000; Cell Signaling Technology, Danvers, MA, USA), mouse poly ubiquitinylated proteins, multi ubiquitin chains (1:500; Biomol Int., Plymouth Meeting, PA, USA), and rabbit anti- β-actin (1:1000; Sigma-Aldrich Corp., St. Louis, MO, USA) as a loading control. In the next step the appropriate horseradish peroxidase conjugated secondary antibody: anti-mouse immunoglobulin G (IgG) (whole molecule) (1:2000; Sigma-Aldrich Corp., St. Louis, MO, USA), anti-rabbit antibodies (1:2000; Cell Signaling Technology, Danvers, MA, USA), anti-rabbit immunoglobulin G (IgG), Fc, HRP conjugate antibodies (1:2000; EMD Millipore Corp., Temecula, CA, USA) or anti-mouse IgG (whole molecule)—alkaline phosphatase antibody (1:2000; Sigma-Aldrich Corp., St. Louis, MO, USA) was added for 1 h with gentle shaking. After washing with TBS-T membranes were subjected to Westar Supernova Chemiluminescent Substrate for Western Blotting (Cyanagen, Bologna, Italy) and analyzed by densitometry (G:BOX, Syngene, Cambridge, UK). The intensity of analyzed proteins were normalized to β-actin which was a loading control. The data were expressed as a percentage of the normal sample taken as 100%. Determination of polyubiquitinated proteins was performed using Sigma Fast BCIP/NBT Alkaline Phosphatase Substrate (Sigma-Aldrich, St. Louis, MO, USA) by colorimetric detection (Gel Doc XR and Gel Documentation System; Molecular Imager Gel Doc XR, Bio-Rad Laboratories Inc., Hercules, CA, USA).

4.7. Immunofluorescence

Fibroblasts grown on cover-slips were fixed in 4% paraformaldehyde in PBS for 10 min at room temperature. After fixation, the cells were permeabilized in PBS containing 0.2% Triton-X100 for 5 min and blocked in 5 % normal donkey serum (Sigma-Aldrich Corp., St. Louis, MO, USA) at room temperature for 60 min to block non-specific reactions. Then cells were incubated with mouse monoclonal anti-collagen type I antibody (1:250, Santa Cruz Biotechnology Inc., Santa Cruz, CA, USA) and rabbit monoclonal anti-LC3B antibody (1:2000, Cell Signaling Technology, Danvers, MA, USA) or rabbit polyclonal anti-LAMP2A antibody (1:100, Abcam) for 60 min at room temperature. After incubation, the cells were washed three times with PBS and incubated in donkey anti-mouse IgG conjugated with Alexa Fluor 543 (1:200, Molecular Probes) or donkey anti-rabbit IgG conjugated with Alexa Fluor 488 (1:200, Molecular Probes) at room temperature for 1 h. Then, the cells were washed three times in PBS and stained with 4′,6′-diamidino-2-phenylindole (DAPI, Sigma-Aldrich Corp., St. Louis, MO, USA) for 10 min to indicate cell nuclei. The samples were washed twice with PBS and embedded in fluorescent medium (Medium Coverquick, Hygeco, OH, USA), dried overnight and stored in the dark until assessment. The immune labeled cells were analyzed using Nikon Digital Sight DS-Fi1 camera and a fluorescence microscope Nikon ECLIPSE Ti/C1 Plus, equipped with three filters DAPI (blue), FITC (green), and TRITC (red) (excitation wavelength/emission filter: 405/450 nm, 488/515 nm, 543/605 nm, respectively). No fluorescence signal was detected when cells were incubated with secondary antibodies alone (data not shown). At least five pictures of different areas of each treatment group were taken, independently analyzed and one representative image for each study group was presented.

4.8. Steady-State Analysis of Type I Collagen

Procollagens was extracted from cell lysates by precipitation overnight at 4 °C with ammonium sulfate (176 mg/mL). To obtain collagen, procollagen was subjected to digestion with pepsin (50 μg/mL) for 4 h at 4 °C. For electrophoretic analysis of migration of collagen chains, SDS-urea-PAGE (5% polyacrylamide gel) and silver staining were used.

4.9. Subcellular Fractionation

Cells were suspended in buffer containing 40 mM KCl, 5 mM $MgCl_2$, 2 mM EGTA, 10 mM HEPES, pH 7.5 for 30 min on ice. They were then homogenized by shearing 30 times through a 28.5-gauge needle and centrifuged at 1000× g for 10 min. The pellet was collected as the nuclear fraction, while the supernatant was subjected to centrifugation at 12,000× g for 10 min. The lysosome enriched pellet was washed using isotonic buffer (150 mM NaCl, 5 mM $MgCl_2$, 2 mM EGTA, 10 mM HEPES pH 7.5) and dissolved in lysis buffer (1% Triton X-100, 150 mM NaCl, 50 mM Tris-HCl pH 7.5). The presence of collagen type I in this fraction was analyzed by Western blot.

4.10. Determination of Proteasome Activities

Cells were sonicated in lysis buffer containing 50 mM Tris-HCl (pH 7.5), 150 mM NaCl, 1 mM EDTA,1 mM EGTA, 0.5% Triton X, and centrifuged at 12,000× g for 15 min at 40 °C. The total protein concentration in supernatants was determined by the Bradford method in BioPhotometer (Eppendorf, Hamburg, Germany), using the Bio-Rad Protein Assay Dye Reagent Concentrate (Bio-Rad Laboratories Inc., Hercules, CA, USA) with BSA as a standard. The supernatants were diluted to concentration of 1.5 mg protein/mL in the lysis buffer. The reaction mixture (total volume of 50 μL) contained 30 μL of assay buffer (100 mM Tris/HCL, pH 7.5, 1 mM EDTA, 1 mM EGTA pH 7.5), 10 μL of cell lysate supernatant and 10 μL of fluorogenic peptide-AMC substrates: Suc-LLVY-AMC (Sigma Aldrich Corp., St. Louis, MO, USA) for chymotrypsin-like activity, Bz-VGR-AMC (Enzo Life Sciences, Inc., Farmingdale, NY, USA) for trypsin-like activity or Z-LLE-AMC (Enzo Life Sciences, Inc., Farmingdale, NY, USA) for caspase-like activity in a final concentration of 100 μM each [55]. The 96-well black plates (Corning Inc., Corning, NY, USA) were used and the assays were performed at 37 °C in FLUOStar OPTIMA (BMG Labtech Gmbh, Offenburg, Germany) over 30 min with one reading every 2 min, at 355 nm for excitation and 460 nm for emission. One unit of the proteasome activity was expressed as the amount of AMC released from the substrate per minute (pmol/min). The activity was calculated for the amount of total protein (U/mg). All assays were performed in triplicates.

4.11. Zymography for the Determination of MMP Activity

The conditioned media were collected and subjected non-reducing SDS–PAGE. The substrate for MMP was gelatin (1 mg/mL) (Sigma-Aldrich Corp., St. Louis, MO, USA). In order to remove SDS, gels were incubated in 2.5% Triton X-100 solution for 30 min at room temperature. In the next step, gels were incubated overnight in 50 mM Tris-HCl, pH 8.0, 5 mM $CaCl_2$, 5 μM $ZnCl_2$ and 0.02% NaN_3) at 37 °C with gentle shaking. After that, they were stained with Commassie blue R-250 solution and destained until the appearance of white stripes on a dark blue background. Images of the zymograms were analyzed by densitometry (G:BOX, Syngene, Cambridge, UK).

4.12. Statistical Analysis

The results were statistically analyzed using the Statistica 12 software (StatSoft, Tulsa, OK, USA) and presented as the mean ± standard deviation (SD). Statistical differences were estimated by the use of one-way ANOVA followed by Tukey's test and values of $p < 0.05$ were considered as significant.

5. Conclusions

This study provides new insight into the effects of rosemary extract on OI fibroblasts with mutant collagen retention. The data presented here shows, for the first time, the pro-autophagy effect and the protective action of RE against the apoptosis of OI fibroblasts. Our findings could have important implications in OI treatment trials as RE removed accumulated mutant collagen and unfolded procollagen chains, improving cell homeostasis as indicated by decreased expression of UPR proteins. Reduction of proteasome activity by RE did not result in additional accumulation of non-collagen proteins. Although the exact mechanisms of the autophagy stimulatory and the proteasome and MMP (-1 and -2) inhibitory effects of RE require elucidation, the obtained results of our research are promising and worth continuing in order to understand the molecular pathways involved in the pathology of OI and beneficial effects of RE.

Supplementary Materials: The following supporting information can be downloaded at: https://www.mdpi.com/article/10.3390/ijms231810341/s1.

Author Contributions: Conceptualization, J.S.-S. and A.G.; methodology, J.S.-S., J.W.S., J.B.-J., H.O. and A.G.; formal analysis, K.G., H.O. and A.G.; investigation, J.S.-S., J.B.-J., J.W.S., M.B.; writing—original draft preparation, J.S.-S. and A.G.; writing—review and editing, A.G.; K.G.; visualization, J.S.-S.; supervision, A.G.; project administration, J.S.-S. and A.G. All authors have read and agreed to the published version of the manuscript.

Funding: The research was funded by the project № POWR.03.02.00-00-I051/16 from European Union funds, POWER 2014-2020, grant № 04/IMSD/G/2021.

Institutional Review Board Statement: The study was conducted according to the guidelines of the Declaration of Helsinki, and approved by the Bioethical Committee of the Jagiellonian University in Kraków, Poland (KBET/108/B/2007, 31 May 2011).

Informed Consent Statement: Informed consent was obtained from all subjects involved in the study.

Acknowledgments: The authors would like to thank Aleksandra Auguściak-Duma and Joanna Witecka from Department of Molecular Biology and Genetics, Faculty of Medical Sciences in Katowice, Medical University of Silesia, Katowice, Poland for genetic analysis of the OI skin fibroblasts provided in the study.

Conflicts of Interest: The authors declare no conflict of interest.

Abbreviations

3-MA	3-Methyladenine
4-PBA	4-Phenylbutyrate
ATF4	Activating transcription factor 4
ATF6	Activating transcription factor 6
BiP	Binding-Immunoglobulin Protein
BR	Bortezomib
C-L	caspase-like activity
Ch-L	chymotrypsin-like activity
CHOP	CCAAT enhancer binding protein
CQ	Chloroquine
DMEM	Dulbecco's minimal essential medium
DMSO	Dimethyl sulfoxide
ECM	Extracellular matrix
eIF2	Eucaryotic translation initiation factor
ER	Endoplasmic reticulum
ERAD	Endoplasmic reticulum (ER)-associated degradation
FBS	Fetal bovine serum
IRE1	Inositol requiring enzyme 1
LAPM2A	Lysosomal-associated membrane protein 2A
LC3	Microtubule associated protein 1 Light Chain 3
MMPs	Matrix Metalloproteinases
NH_4Cl	Ammonium chloride
OI	Osteogenesis imperfecta
P3H1	Prolyl 3-Hydroxylase 1
PBS	Phosphate-buffered saline
PERK	PKR-like endoplasmic reticulum kinase
PDI	Protein disulfide isomerase
RA	Rosmarinic acid
RE	Rosemary extract
RIPA	Radioimmunoprecipitation assay buffer
SDS	Sodium dodecyl sulfate
SDS-urea PAGE	SDS-urea polyacrylamide gel electrophoresis
SQSTM1/p62	Sequestosome-1
TGF-β	Transforming Growth Factor Beta
T-L	trypsin-like activity
UPR	Unfolded protein response
XBP-1s	X-Box binding protein 1

References

1. Forlino, A.; Cabral, W.A.; Barnes, A.M.; Marini, J.C. New perspectives on osteogenesis imperfecta. *Nat. Rev. Endocrinol.* **2011**, *7*, 540–557. [CrossRef] [PubMed]
2. Marini, J.C.; Forlino, A.; Bächinger, H.P.; Bishop, N.J.; Byers, P.H.; Paepe, A.; Fassier, F.; Fratzl-Zelman, N.; Kozloff, K.M.; Krakow, D.; et al. Osteogenesis imperfecta. *Nat. Rev. Dis. Primers* **2017**, *3*, 17052. [CrossRef] [PubMed]
3. Marom, R.; Rabenhorst, B.M.; Morello, R. Osteogenesis imperfecta: An update on clinical features and therapies. *Eur. J. Endocrinol.* **2020**, *183*, R95–R106. [CrossRef] [PubMed]
4. Gajko-Galicka, A. Mutations in type I collagen genes resulting in osteogenesis imperfecta in humans. *Acta Biochim. Pol.* **2002**, *49*, 433–441. [CrossRef]
5. Lim, J.; Grafe, I.; Alexander, S.; Lee, B. Genetic causes and mechanisms of Osteogenesis Imperfecta. *Bone* **2017**, *102*, 40–49. [CrossRef] [PubMed]
6. Van Dijk, F.S.; Sillence, D.O. Osteogenesis imperfecta: Clinical diagnosis, nomenclature and severity assessment. *Am. J. Med. Genet. A* **2014**, *164*, 1470–1481. [CrossRef]
7. Willing, M.C.; Deschenes, S.P.; Scott, D.A.; Byers, P.H.; Slayton, R.L.; Pitts, S.H.; Arikat, H.; Roberts, E.J. Osteogenesis imperfecta type I: Molecular heterogeneity for COL1A1 null alleles of type I collagen. *Am. J. Hum. Genet.* **1994**, *55*, 638–647.

8. Jovanovic, M.; Guterman-Ram, G.; Marini, J.C. Osteogenesis Imperfecta: Mechanisms and signaling pathways connecting classical and rare OI types. *Endocr. Rev.* **2022**, *43*, 61–90. [CrossRef] [PubMed]
9. Kang, H.; Aryal, A.C.S.; Marini, J.C. Osteogenesis imperfecta: New genes reveal novel mechanisms in bone dysplasia. *Transl. Res.* **2017**, *181*, 27–48. [CrossRef] [PubMed]
10. Etich, J.; Leßmeier, L.; Rehberg, M.; Sill, H.; Zaucke, F.; Netzer, C.; Semler, O. Osteogenesis imperfecta-pathophysiology and therapeutic options. *Mol. Cell Pediatr.* **2020**, *7*, 9. [CrossRef]
11. Ishikawa, Y.; Bächinger, H.P. A molecular ensemble in the rER for procollagen maturation. *Biochim. Biophys. Acta* **2013**, *1833*, 2479–2491. [CrossRef] [PubMed]
12. Claeys, L.; Storoni, S.; Eekhoff, M.; Elting, M.; Wisse, L.; Pals, G.; Bravenboer, N.; Maugeri, A.; Micha, D. Collagen transport and related pathways in Osteogenesis Imperfecta. *Hum. Genet.* **2021**, *140*, 1121–1141. [CrossRef] [PubMed]
13. Taga, Y.; Kusubata, M.; Ogawa-Goto, K.; Hattori, S. Site-specific quantitative analysis of overglycosylation of collagen in osteogenesis imperfecta using hydrazide chemistry and SILAC. *J. Proteome Res.* **2013**, *12*, 2225–2232. [CrossRef]
14. Lisse, T.S.; Thiele, F.; Fuchs, H.; Hans, W.; Przemeck, G.K.; Abe, K.; Rathkolb, B.; Quintanilla-Martinez, L.; Hoelzlwimmer, G.; Helfrich, M.; et al. ER stress-mediated apoptosis in a new mouse model of osteogenesis imperfecta. *PLoS Genet.* **2008**, *4*, e7. [CrossRef]
15. Mirigian, L.S.; Makareeva, E.; Mertz, E.L.; Omari, S.; Roberts-Pilgrim, A.M.; Oestreich, A.K.; Phillips, C.L.; Leikin, S. Osteoblast malfunction caused by cell stress response to procollagen misfolding in α2(I)-G610C mouse model of Osteogenesis Imperfecta. *J. Bone Miner. Res.* **2016**, *31*, 1608–1616. [CrossRef] [PubMed]
16. Garibaldi, N.; Contento, B.M.; Babini, G.; Morini, J.; Siciliani, S.; Biggiogera, M.; Raspanti, M.; Marini, J.C.; Rossi, A.; Forlino, A.; et al. Targeting cellular stress in vitro improves osteoblast homeostasis, matrix collagen content and mineralization in two murine models of osteogenesis imperfecta. *Matrix Biol.* **2021**, *98*, 1–20. [CrossRef]
17. Besio, R.; Iula, G.; Garibaldi, N.; Cipolla, L.; Sabbioneda, S.; Biggiogera, M.; Marini, J.C.; Rossi, A.; Forlino, A. 4-PBA ameliorates cellular homeostasis in fibroblasts from osteogenesis imperfecta patients by enhancing autophagy and stimulating protein secretion. *Biochim. Biophys. Acta Mol. Basis Dis.* **2018**, *1864*, 1642–1652. [CrossRef]
18. Besio, R.; Garibaldi, N.; Leoni, L.; Cipolla, L.; Sabbioneda, S.; Biggiogera, M.; Mottes, M.; Aglan, M.; Otaify, G.A.; Temtamy, S.A.; et al. Cellular stress due to impairment of collagen prolyl hydroxylation complex is rescued by the chaperone 4-phenylbutyrate. *Dis. Model Mech.* **2019**, *12*, dmm038521. [CrossRef]
19. Bateman, J.F.; Shoulders, M.D.; Lamandé, S.R. Collagen misfolding mutations: The contribution of the unfolded protein response to the molecular pathology. *Connect. Tissue Res.* **2022**, *63*, 210–227. [CrossRef] [PubMed]
20. Boot-Handford, R.P.; Briggs, M.D. The unfolded protein response and its relevance to connective tissue diseases. *Cell Tissue Res.* **2010**, *339*, 197–211. [CrossRef] [PubMed]
21. Sutkowska-Skolimowska, J.; Galicka, A. ER stress in osteogenesis imperfecta (OI)-causative mutations and potential treatment. In *Advances in Biomedical Research—Cancer and Miscellaneous*; Młynarczuk-Biały, I., Biały, Ł., Eds.; TYGIEL Scientific Publisher: Lublin, Poland; Medical University of Warsaw: Warsaw, Poland, 2021; pp. 133–143.
22. Hetz, C.; Chevet, E.; Oakes, S.A. Proteostasis control by the unfolded protein response. *Nat. Cell Biol.* **2015**, *17*, 829–838. [CrossRef]
23. Tsang, K.Y.; Chan, D.; Bateman, J.F.; Cheah, K.S. In vivo cellular adaptation of ER stress: Survival strategies with double-edged consequences. *J. Cell Sci.* **2010**, *123*, 2145–2154. [CrossRef]
24. Ishida, Y.; Yamamoto, A.; Kitamura, A.; Lamandé, S.R.; Yoshimori, T.; Bateman, J.F.; Kubota, H.; Nagata, K. Autophagic elimination of misfolded procollagen aggregates in the endoplasmic reticulum as a means of cell protection. *Mol. Biol. Cell* **2009**, *20*, 2744–2754. [CrossRef] [PubMed]
25. Yang, Z.; Klionsky, D.J. An overview of the molecular mechanism of autophagy. *Curr. Top Microbiol. Immunol.* **2009**, *335*, 1–32. [CrossRef]
26. Lamandé, S.R.; Chessler, S.D.; Golub, S.B.; Byers, P.H.; Chan, D.; Cole, W.G.; Sillence, D.O.; Bateman, J.F. Endoplasmic reticulum-mediated quality control of type I collagen production by cells from osteogenesis imperfecta patients with mutations in the pro alpha 1 (I) chain carboxyl-terminal propeptide which impair subunit assembly. *J. Biol. Chem.* **1995**, *270*, 8642–8649. [CrossRef]
27. Fitzgerald, J.; Lamandé, S.R.; Bateman, J.F. Proteasomal degradation of unassembled mutant type I collagen pro-alpha1(I) chains. *J. Biol. Chem.* **1999**, *274*, 27392–27398. [CrossRef] [PubMed]
28. Botor, M.; Fus-Kujawa, A.; Uroczynska, M.; Stepien, K.L.; Galicka, A.; Gawron, K.; Sieron, A.L. Osteogenesis imperfecta: Current and prospective therapies. *Biomolecules* **2021**, *11*, 1493. [CrossRef] [PubMed]
29. Gioia, R.; Tonelli, F.; Ceppi, I.; Biggiogera, M.; Leikin, S.; Fisher, S.; Tenedini, E.; Yorgan, T.A.; Schinke, T.; Tian, K.; et al. The chaperone activity of 4PBA ameliorates the skeletal phenotype of Chihuahua, a zebrafish model for dominant osteogenesis imperfecta. *Hum. Mol. Genet.* **2017**, *26*, 2897–2911. [CrossRef] [PubMed]
30. Takeyari, S.; Kubota, T.; Ohata, Y.; Fujiwara, M.; Kitaoka, T.; Taga, Y.; Mizuno, K.; Ozono, K. 4-Phenylbutyric acid enhances the mineralization of osteogenesis imperfecta iPSC-derived osteoblasts. *J. Biol. Chem.* **2021**, *296*, 100027. [CrossRef]
31. García-Aguilar, A.; Palomino, O.; Benito, M.; Guillén, C. Dietary polyphenols in metabolic and neurodegenerative diseases: Molecular targets in autophagy and biological effects. *Antioxidants* **2021**, *10*, 142. [CrossRef] [PubMed]
32. Fraga, C.G.; Croft, K.D.; Kennedy, D.O.; Tomás-Barberán, F.A. The effects of polyphenols and other bioactives on human health. *Food Funct.* **2019**, *10*, 514–528. [CrossRef]

33. de Oliveira, J.R.; Camargo, S.E.A.; de Oliveira, L.D. *Rosmarinus officinalis* L. (rosemary) as therapeutic and prophylactic agent. *J. Biomed. Sci.* **2019**, *26*, 5. [CrossRef] [PubMed]
34. González-Minero, F.J.; Bravo-Díaz, L.; Ayala-Gómez, A. *Rosmarinus officinalis* L. (Rosemary): An ancient plant with uses in personal healthcare and cosmetics. *Cosmetics* **2020**, *7*, 77. [CrossRef]
35. Sutkowska, J.; Hupert, N.; Gawron, K.; Strawa, J.W.; Tomczyk, M.; Forlino, A.; Galicka, A. The stimulating effect of rosmarinic acid and extracts from rosemary and lemon balm on collagen type I biosynthesis in Osteogenesis Imperfecta type I skin fibroblasts. *Pharmaceutics* **2021**, *13*, 938. [CrossRef]
36. Kirkness, M.W.; Lehmann, K.; Forde, N.R. Mechanics and structural stability of the collagen triple helix. *Curr. Opin. Chem. Biol.* **2019**, *53*, 98–105. [CrossRef]
37. Makareeva, E.; Leikin, S. Collagen structure, folding and function. In *Osteogenesis Imperfecta*, 2nd ed.; Shapiro, J.R., Byers, P.H., Glorieux, F.H., Sponseller, P.D., Eds.; Academic Press: London, UK, 2014; pp. 71–84. [CrossRef]
38. DiChiara, A.S.; Doan, N.D.; Bikovtseva, A.A.; Rowley, L.; Butty, V.L.; Weis, M.E.; Eyre, D.R.; Lamandé, S.R.; Bateman, J.F.; Shoulders, M.D. XBP1s-mediated endoplasmic reticulum proteostasis network enhancement can selectively improve folding and secretion of an osteogenesis imperfecta—causing collagen—I variant. *bioRxiv* **2021**. [CrossRef]
39. Mizushima, N. A brief history of autophagy from cell biology to physiology and disease. *Nat. Cell Biol.* **2018**, *20*, 521–527. [CrossRef] [PubMed]
40. Sánchez-Martín, P.; Komatsu, M. p62/SQSTM1-steering the cell through health and disease. *J. Cell Sci.* **2018**, *131*, jcs222836. [CrossRef]
41. Omari, S.; Makareeva, E.; Roberts-Pilgrim, A.; Mirigian, L.; Jarnik, M.; Ott, C.; Lippincott-Schwartz, J.; Leikin, S. Noncanonical autophagy at ER exit sites regulates procollagen turnover. *Proc. Natl. Acad. Sci. USA* **2018**, *115*, E10099–E10108. [CrossRef]
42. Wojcik, S. Crosstalk between autophagy and proteasome protein degradation systems: Possible implications for cancer therapy. *Folia Histochem. Cytobiol.* **2013**, *51*, 249–264. [CrossRef] [PubMed]
43. Qin, D.; Ren, R.; Jia, C.; Lu, Y.; Yang, Q.; Chen, L.; Wu, X.; Zhu, J.; Guo, Y.; Yang, P.; et al. Rapamycin protects skin fibroblasts from ultraviolet B-induced photoaging by suppressing the production of reactive oxygen species. *Cell Physiol. Biochem.* **2018**, *46*, 1849–1860. [CrossRef]
44. Galicka, A.; Nazaruk, J. Stimulation of collagen biosynthesis by flavonoid glycosides in skin fibroblasts of osteogenesis imperfecta type I and the potential mechanism of their action. *Int. J. Mol. Med.* **2007**, *20*, 889–895. [CrossRef]
45. Nazaruk, J.; Galicka, A. The influence of selected flavonoids from the leaves of Cirsium palustre (L.) Scop. on collagen expression in human skin fibroblasts. *Phytother. Res.* **2014**, *28*, 1399–1405. [CrossRef]
46. Matwiejczuk, N.; Galicka, A.; Zareba, I.; Brzóska, M.M. The protective effect of rosmarinic acid against unfavorable influence of methylparaben and propylparaben on collagen in human skin fibroblasts. *Nutrients* **2020**, *12*, 1282. [CrossRef] [PubMed]
47. Galicka, A.; Sutkowska-Skolimowska, J. The beneficial effect of rosmarinic acid on benzophenone-3-induced alterations in human skin fibroblasts. *Int. J. Mol. Sci.* **2021**, *22*, 11451. [CrossRef] [PubMed]
48. Yao, H.; Zhou, L.; Tang, L.; Guan, Y.; Chen, S.; Zhang, Y.; Han, X. Protective effects of luteolin-7-O-glucoside against starvation-induced injury through upregulation of autophagy in H9c2 cells. *Biosci. Trends* **2017**, *11*, 557–564. [CrossRef]
49. Yang, J.; Pi, C.; Wang, G. Inhibition of PI3K/Akt/mTOR pathway by apigenin induces apoptosis and autophagy in hepatocellular carcinoma cells. *Biomed. Pharmacother.* **2018**, *103*, 699–707. [CrossRef] [PubMed]
50. Yessenkyzy, A.; Saliev, T.; Zhanaliyeva, M.; Masoud, A.R.; Umbayev, B.; Sergazy, S.; Krivykh, E.; Gulyayev, A.; Nurgozhin, T. Polyphenols as caloric-restriction mimetics and autophagy inducers in aging research. *Nutrients* **2020**, *12*, 1344. [CrossRef]
51. Pierzynowska, K.; Gaffke, L.; Hać, A.; Mantej, J.; Niedziałek, N.; Brokowska, J.; Węgrzyn, G. Correction of Huntington's disease phenotype by genistein-induced autophagy in the cellular model. *Neuromolecular Med.* **2018**, *20*, 112–123. [CrossRef] [PubMed]
52. Pierzynowska, K.; Gaffke, L.; Jankowska, E.; Rintz, E.; Witkowska, J.; Wieczerzak, E.; Podlacha, M.; Węgrzyn, G. Proteasome composition and activity changes in cultured fibroblasts derived from mucopolysaccharidoses patients and their modulation by genistein. *Front. Cell Dev. Biol.* **2020**, *8*, 540726. [CrossRef] [PubMed]
53. Hrelia, S.; Angeloni, C. New mechanisms of action of natural antioxidants in health and disease. *Antioxidants* **2020**, *9*, 344. [CrossRef] [PubMed]
54. Lacroix, S.; Klicic Badoux, J.; Scott-Boyer, M.P.; Parolo, S.; Matone, A.; Priami, C.; Morine, M.J.; Kaput, J.; Moco, S. A computationally driven analysis of the polyphenol-protein interactome. *Sci. Rep.* **2018**, *8*, 2232. [CrossRef]
55. Rodgers, K.J.; Dean, R.T. Assessment of proteasome activity in cell lysates and tissue homogenates using peptide substrates. *Int. J. Biochem. Cell Biol.* **2003**, *35*, 716–727. [CrossRef]

International Journal of
Molecular Sciences

Review

Lycium barbarum Berries (Solanaceae) as Source of Bioactive Compounds for Healthy Purposes: A Review

Filipa Teixeira, Ana Margarida Silva, Cristina Delerue-Matos and Francisca Rodrigues *

REQUIMTE/LAQV, ISEP, Polytechnic of Porto, Rua Dr. António Bernardino de Almeida, 4249-015 Porto, Portugal
* Correspondence: francisca.rodrigues@graq.isep.ipp.pt or franciscapintolisboa@gmail.com;
Tel.: +351-22-83-40-500; Fax: +351-22-83-21-159

Abstract: *Lycium barbarum* L. is a species widely used in dietary supplements and natural healthcare products. The berries, also known as goji or wolfberries, mostly grow in China, but recent reports on their outstanding bioactive properties have increased their popularity and cultivation around the world. Goji berries are a remarkable source of phenolic compounds (such as phenolic acids and flavonoids), carotenoids, organic acids, carbohydrates (fructose and glucose), and vitamins (ascorbic acid). Several biological activities, such as antioxidant, antimicrobial, anti-inflammatory, prebiotic, and anticancer activities, have been associated with its consumption. Hence, goji berries were highlighted as an excellent source of functional ingredients with promising applications in food and nutraceutical fields. This review aims to summarize the phytochemical composition and biological activities, along with various industrial applications, of *L. barbarum* berries. Simultaneously, the valorization of goji berries by-products, with its associated economic advantages, will be emphasized and explored.

Keywords: goji berries; pro-healthy effects; phenolics; biological properties; functional ingredients

Citation: Teixeira, F.; Silva, A.M.; Delerue-Matos, C.; Rodrigues, F. *Lycium barbarum* Berries (Solanaceae) as Source of Bioactive Compounds for Healthy Purposes: A Review. *Int. J. Mol. Sci.* **2023**, *24*, 4777. https://doi.org/10.3390/ijms24054777

Academic Editors: Antonio Carrillo Vico and Ivan Cruz-Chamorro

Received: 26 January 2023
Revised: 26 February 2023
Accepted: 28 February 2023
Published: 1 March 2023

1. Introduction

Plants have been used for thousands of years as a source of compounds for traditional medicine, aiming to prevent and treat health problems [1]. Due to a broad number of studies that describe the influence of natural products on the human endogenous defense system [2,3] as well as "curative" effects against a wide spectrum of disorders, including cardiovascular and neurodegenerative diseases, obesity, and certain types of cancer [4], multiple natural products have been used for healthcare proposes. Nowadays, the interest in exotic berry-type fruits has expanded worldwide [5–8]. Society has become more concerned with eating habits, mostly due to the reinforcement of a positive relationship between good eating habits and the prevention of disease development, particularly diabetes and cardiovascular and neurological pathologies [9]. Simultaneously, the emergent awareness of the planet and the impact that various types of industries have on it has boosted the population's concerns and demands for greener formulations with bioactive ingredients recovered from natural sources [5,10]. Therefore, the consumption of natural matrices has increased in recent years, not only as supplements for imbalanced diets but also as an integral part of a normal healthy diet [7,11].

Recently, several reports highlighted the impressive bioactive capacities of goji berries [2,4,10,12–16], not only in cell assays [17–19] but also in animal studies [3,14,20] and human trials [21]. Despite the growing numbers of published papers regarding the bioactive composition of goji berries, few of them give particular emphasis to in vivo studies as well as to the by-products generated during berry production. Therefore, the aim of this work is to provide a comprehensive review of the bioactive compounds of goji berries, along with their biological activity, giving a particular focus to in vivo studies and clinical trials. The various industrial applications of *L. barbarum* berries will also be highlighted, as well as the valorization of goji berries by-products.

2. *Lycium barbarum* L.

Lycium barbarum L. is one of the most common species of the Solanaceae family [13,22]. The berries traditionally grow in China, Tibet, and other parts of Asia [6,13,23]. China is the primary worldwide supplier [24,25], with about 25,000–30,000 tons of dried fruit annually produced in Ningxia, Xinjiang, Gansu, Qinghai, and Mongolia [26]. Asia is the region with the highest production (71.2%), followed by Africa (15.8%), America (8.3%), Oceania (3.4%) and Europe (1.3%) [27]. Due to the rising reports of the positive correlation between the consumption of natural matrices and health improvement [9], the production of fruit has been increasing in the last 20 years all over the world. This is the case of goji berries, whose production has been increasing in the last decades [28], particularly in Europe (Italy, Romania, Bulgaria, Portugal, Greece, Serbia), Northern America, and Australia. Currently, Romania has the largest cultivated area of *L. barbarum* in the European Union [4,10]

The goji berries can be divided into different classes according to their ripening stage, dimension, weight, color, firmness, solid soluble content, pH, and titratable acidity [22]. The mature fruit (Figure 1) is between 1 and 2 cm long, presenting an ellipsoid shape and a bright orange-red color, similar to a mature mini-tomato, and contains between 20 and 40 tiny seeds per fruit [6,13,22].

Figure 1. *L. barbarum* L. berries [29].

The berries have a sweet taste [16] and are widely used as a dietary supplement and natural health product [4,23,25]. Although mostly consumed fresh in the regions of cultivation [9], around the world, goji berries are essentially consumed dried [9,12] or transformed into alimentary products, such as juices, herbal teas, yogurt products, granola, powders, and tablets, among others [9,12,30]. The most commonly sold goji berry-based products are beverages, wine, juice, tea, and concentrates [4,14,23]. A popular goji berry juice brand, "GoChi", has shown increasing effectiveness as an antioxidant, giving rise to subjective feelings of general well-being as well as improvements in neurologic/psychologic performance and human gastrointestinal functions [23], which has led to huge popularity among consumers.

For thousands of years, goji berries have been used as herbal medicine in Asian countries [24]. Based on their rich nutritional value and medical properties, such as their antioxidant, antimicrobial, immunomodulatory, and anti-inflammatory effects [15,23,31], the fruit has been employed as an anti-aging treatment, tranquilizer, and thirst-quenching treatment [2,13]. As a folk medicine, *L. barbarum* fruits have been employed by the local population for blood nourishing, in the treatment of early onset diabetes, tuberculosis, dizziness, and chronic cough, and for the protection of eye health [15].

All these pro-healthy properties have attracted the attention of consumers to goji juices and fruits, transforming goji berries into one of the most popular functional food ingredients/supplements worldwide [13]. Nevertheless, the consumption of natural supplements needs to be balanced to avoid negative effects related to overuse or interaction with other medical treatments [7]. Therefore, risk/benefit evaluations are urgently needed when used in foods or health-promoting formulations in order to avoid negative impacts [4].

3. Bioactive Compounds and Chemical Composition of *L. barbarum* L.

The bioactive composition of plants is influenced by many factors, such as variety, ripeness, geographic location, and climatic conditions. Since during fruit growth, physiological, biochemical, and molecular changes occur, ripeness might be the most influential factor in the fruit's bioactive composition [31,32]. Therefore, deep knowledge about the ripening stage of goji berries is needed to determine the best stage and obtain the most adequate bioactive compounds. Yet, in general, various researchers have reported that goji berries have a remarkable concentration of antioxidants, fat, dietary fibers, essential amino acids, valuable trace minerals, and vitamins [3,6,13,23,30,33,34]. In the next subsections, the different classes of bioactive compounds present in goji berries will be deeply analyzed.

3.1. Phenolic Compounds

Phenolic compounds act as a defense mechanism to provide adaptation and survival capacity in adverse environmental conditions to plants, such as protection against ultraviolet radiation (UV), pathogen aggression, parasites, and predators [5,35]. These compounds usually add nutritional and functional value, contributing to the fruit's organoleptic characteristics, such as astringency, bitterness, and aroma. Simultaneously, they guarantee outstanding biological activities and pro-healthy properties against oxidative stress [5,13,31], being capable of delaying, preventing, and inhibiting oxidation by scavenging free radicals and, therefore, reducing oxidative stress [35].

Reactive oxygen species (ROS) naturally occur in living organisms, being involved in processes such as proliferation and apoptosis [25]. However, when the quantities of ROS, such as superoxide radicals ($O_2^{\cdot -}$), hydrogen peroxide (H_2O_2), and hydroxyl radicals ($OH^{\cdot -}$), overcome the activity of endogenous antioxidant mechanisms, namely antioxidant enzymes (e.g., superoxide dismutase (SOD), catalase (CAT), glutathione peroxidase (GPx)) and non-enzymatic molecules (e.g., glutathione (GSH), ascorbic acid, α-tocopherol), a state of oxidative stress is initiated [10,25,35].

Oxidative stress is implicated in the aging process as well as in several pathologies (e.g., cardiovascular dysfunction, various typologies of cancer, inflammation, rheumatism, diabetes, rheumatoid arthritis, pulmonary emphysema, dermatitis, cataract, neurodegenerative diseases, endothelial cell dysfunction, and several autoimmune diseases linked to degenerative processes of aging) that frequently lead to invalidity or death [3,9,13]. When the endogenous antioxidant mechanisms are not enough to stop and prevent oxidative stress, supplementation is needed to strengthen the antioxidant state and the cell defense mechanisms of organisms [25]. Fruit phenolic compounds are excellent candidates to perform this scavenging effect against ROS and other radical species [35].

Phenolics, flavonoids, and carotenoids are intrinsically related to nutritional and health promotion capacity; therefore, their quantification may help to clarify some activities of natural products [36]. Table 1 summarizes the total phenolic compounds (TPC), total flavonoid content and total carotenoid content (TCC) of goji berries, according to different authors.

Table 1. Total phenolic compounds (TPC), total flavonoid content and total carotenoid content (TCC) of goji berries according to different studies.

TPC	TFC	TCC	Reference
31.6 mg GAE/100 g dw	28.3 mg CAE/100 g dw	23.30 mg CAE/100 g dw	[2]
145.2 mg GAE/100 g dw	74.5 mg QE/100 g dw	-	[3]
1160–1570 mg GAE/100 g dw	-	-	[4]
1413 mg GAE/100 g dw	-	-	[37]
3000 mg GAE/100 g dw	2480 mg QE/100 g dw	-	[12]
268.5 mg GAE/100 g fw	-	-	[13]
449–778 mg GAE/100 g dw	-	400–950 mg/100 g dw	[28]
162.4 mg GAE/100 g fw	214.2 mg HE/100 g fw	41.71 mg/100 g fw	[33]
97.23 mg/100 g dw	-	212.94 mg/100 g dw	[36]

GAE—gallic acid equivalents; CAE—catechin equivalents; HE—hyperoside equivalent; QE—quercetin equivalents; TPC—total phenolic content; TFC—total flavonoid content; TCC—total carotenoid content; fw—fresh weight; dw—dry weight.

As can be observed in Table 1, the values achieved for the different assays (TPC, TFC, and TCC) are slightly different between studies. For example, the TPC varied between 31.6 mg GAE/100 g dw [2] and 3000 mg GAE/100 g dw [12]. These variations are mostly due to the different postharvest techniques used that mainly affect the bioactive components present, as well as the extraction solvents employed and the samples' geographical origin. Islam et al. [2] studied the phenolic profile, antioxidant capacity, and carotenoid content of dried goji berries harvested in China and extracted with a mixture of acetone/water/acetic acid (70:29.5:0.5). According to the authors, the TPC achieved a value of 31.6 mg GAE/100 g dw, while the TFC was 28.3 mg CAE/100 g dw. These values were considerably lower than the ones reported by Magalhães et al. [12]. Despite the same geographical origin, the authors lyophilized the samples and extracted them for 5 days using methanol (80%), which could justify the huge differences observed.

A study conducted by Donno et al. [13] using goji berries supplied by a farm located in Alzate di Momo, Northern Italy, reported the presence of organic acids (4461.02 mg/100 g fw) and polyphenolic compounds (12,697.90 mg/100 g fw). Meanwhile, in a study [31] using goji berries provided by a planting base in Gansu Province, China, the researchers identified nine phenolic compounds by UPLC-MS/MS, including quercetin, isoquercitrin, chlorogenic acid, ferulic acid, *p*-coumaric acid, caffeic acid, isorhamnetin, cinnamic acid, and rutin, being rutin, isoquercitrin, and chlorogenic acid. Similarly, Pires et al. [30] extracted goji berries supplied by a Portuguese company and reported the presence of nineteen phenolic compounds (71 mg/g dw): eight flavonols (27.6 mg/g dw), seven phenolic acid derivatives (32.7 mg/g dw), one flavan-3-ol (10.4 mg/g dw), and three chlorogenic acids (25.07 mg/g dw). According to the authors, the principal phenolic compounds quantified were quercetin-3-*O*-rutinoside (16.6 mg/g dw) and *p*-coumaric acid (12.3 mg/g dw). Nardi et al. [3] conducted a phytochemical analysis on a methanolic extract of goji berries, revealing the presence of phenolic compounds (142.2 mg/100 g of extract) and flavonoids (74.5 mg/100 g of extract), while the qualitative analysis identified rutin and quercetin as the principal compounds. Wojdyło et al. [36] highlighted the carotenoid content of goji berries from new cultivars in Poland (212.94 mg/100 g dw). The authors identified different types of carotenoids, namely zeaxanthin (84.54 mg/100 g dw), β-carotene (19.35 mg/100 g dw), neoxanthin (16.04 mg/100 g dw) and cryptoxanthin (72.29 mg/100 g dw).

3.2. Nutritional Composition

A study carried out by Bora et al. [23] attested to the presence of carbohydrates (46 g/100 g of fw), dietary fibers (16 g/100 g fw), proteins (13 g/100 g fw) and fat (1.5 g/100 g fw) in goji berries. In another study, Ilić et al. [33] reported the presence of moisture (75.32 g/100 g of fw), carbohydrates (16.93 g/100 g fw,), dietary fiber (3.63 g/100 g fw), protein (1.98 g/100 g fw), fat (1.15 g/100 g fw), and ash (0.84 g/100 g fw) in *L. barbarum* berries. Similarly, Pires et al. [30] evaluated the presence of carbohydrates (87 g/100 g dw), proteins (5.3 g/100 g dw), fat (4.1 g/100 g dw), and ash (3.21 g/100 g dw) in dried goji berry fruits and stems. The same authors also reported the presence of soluble sugars (27.9 g/100 g dw), such as fructose (12.7 g/100 g dw), glucose (14.4 g/100 g dw), and sucrose (0.8 g/100 g dw). However, different authors [10,16,20,21,38–41] attested that polysaccharides are the major carbohydrates present in goji berries, being the principal active ingredients isolated from the fruit. Among them, water-soluble polysaccharides, homogeneous polysaccharides, pectin polysaccharides, acidic heteropolysaccharides, and arabinogalactans (composed of arabinose, glucosamine, galactose, glucose, xylose, mannose, fructose, ribose, galacturonic acid, and glucuronic acid) are the most prevalent.

Regarding organic acids, Pires et al. [30] stated that citric, succinic, and oxalic acids (respectively, 1.29 g/100 g dw, 0.77 g/100 g dw, and 0.010 g/100 g dw) were detected, as well as tocopherols, namely α-tocopherol and δ-tocopherol (0.23 mg/100 g dw and 0.09 mg/100 g dw, respectively). The authors also determined the fatty acids content (4.1 g/100 g dw), more specifically detecting sixteen fatty acids, with polyunsaturated fatty acids being the predominant group, namely linoleic acid (53.4%), oleic acid (16.5%) and palmitic acid (12.77%). In parallel, Ilić et al. [33] reported that the most abundant fatty acids were linoleic (52.1%), oleic (23.6%) and palmitic (17.6%) acids, accounting for 95% of the total fatty acids, which is in concordance with Skenderidis et al. [32], who reported concentrations of 37.89–43.96%, 16.71–20.07%, and 15.08–21.79%, respectively.

Concerning the mineral content, numerous studies showed that the principal minerals present in goji berries are potassium, sodium, and calcium. Bora et al. [23] reported that 100 g of goji berries contain 434 mg of potassium, 60 mg of calcium, 5.4 mg of iron, and 1.5 mg of zinc. Llorent-Martínez et al. [6] reported a higher amount of potassium (1460 mg/100 g), sodium (550 mg/100 g), and calcium (50 mg/100 g), while Ilić et al. [33] quantified potassium (445.12 mg/100 g dw), phosphor (231.52 mg/100 g dw), sodium (74.57 mg/100 g dw), and calcium (29.02 mg/100 g dw).

In what concerns vitamins, ascorbic acid (48.94 mg/100 g fw) and tocopherols (0.33 mg/100 g dw) have been described in goji berries [13,14,30,33,41]. Vitamin E, also known as α-tocopherol, is the major liposoluble antioxidant present in the cells' antioxidant defense system, being able to inhibit membrane lipidic peroxidation [5,42], while vitamin C, also known as ascorbic acid, is an important antioxidant compound of goji berries [10].

Table 2 summarizes the nutritional composition and phenolic composition of goji berries according to the data reported by different authors.

Table 2. Phytochemical and nutritional composition of *L. barbarum* berries, according to different studies.

Compounds	Amount	Reference
Phenolic compounds	12,697.90 mg/100 g fw	[3,9,13,30]
• Phenolic acids	• 32.70 mg/g	
- Chlorogenic acid	- 25.07–1.07 mg/g	
- *p*-coumaric acid	- 12.30 mg/g	
- Ferulic acid	- 0.98–0.93 mg/g	
- Caffeic acid	- 0.93–8.99 mg/g	
• Flavanols	• 27.60 mg/g	
- Rutin	- 16.60–12.84 mg/g	
- Quercetin	- 10.23–9.41 mg/g	
• Flavan-3-ol	• 10.40 mg/g	
- Catechin	- 1.34–1.13 mg/g	
- Epicatechin	- 2.18–1.98 mg/g	
Organic acids	4461.02 mg/100 g fw	[13]
• Citric	• 254.09 mg/100 g fw	
• Malic	• 601.43 mg/100 g fw	
• Oxalic	• 13.41 mg/100 g fw	
• Quinic	• 2011.73 mg/100 g fw	
• Tartaric	• 1580.35 mg/100 g fw	
Carotenoids	212.94 mg/100 g dw	[36]
• Zeaxanthin	• 84.54 mg/100 g dw	
• β-Carotene	• 19.35 mg/100 g dw	
• Cryptoxanthin	• 72,29 mg/100 g dw	
Vitamins		
• Ascorbic acid	• 2.39–48.94 mg/100 g fw	[13,34,36]
• Tocopherol	• 0.33 mg/100 g dw	[30]

Table 2. *Cont.*

Compounds	Amount	Reference
Carbohydrates	77.1–87 g/100 g dw	
• Total sugars	• 45.60–67.83 g/100 g dw	[30,34]
• Soluble sugar	• 27.09 g/100 g dw	[26,34]
- Glucose	- 14.40–17.32 g/100 g dw	[28,30]
- Fructose	- 12.70–21.71 g/100 g dw	
- Sucrose	- 0.80–1.48 g/100 g dw	
Dietary fibers	3.63–16 g/100 g fw	
• Soluble	• 0.90–5.5 g/100 g dw	[23,33]
• Insoluble	• 2.73–11.7 g/100 g dw	[30]
Proteins	5.3–14.3 g/100 g dw	
• Essential amino acids	• 2.139 g/100 g dw	[30,34]
• Non-essential amino acids	• 6.728 g/100 g dw	[34]
Fatty acids	0.39–4.1 g/100 g dw	
• Linoleic acid	• 37.89–53.4%	
• Oleic acid	• 16.5–23.6%	[30,34]
• Palmitic acid	• 12.77–21.79%	
Ash	0.78–3.21 g/100 g dw	[30,34]
Minerals		
• Potassium	• 434–1460 mg/100 g fw	
• Calcium	• 29–60 mg/100 g fw	
• Sodium	• 75–550 mg/100 g fw	[6,23,33]
• Iron	• 5.4 mg/100 g fw	
• Phosphor	• 232 mg/100 g fw	

fw—fresh weight; dw—dry weight.

The variations observed between the different studies may be explained by several reasons. As mentioned earlier, a broad number of biological variables interfere with the bioactive composition of fruits, such as ripeness, geographic origin, or climatic conditions [26,31,32,36]. On the other hand, the extraction techniques employed, the extractor solvents used, and the quantification techniques employed, may lead to different results [9,43]. Generally, mature fruits are associated with huge amounts of bioactive compounds with pro-healthy effects.

4. Biological Activities of *L. barbarum* L.

Goji berries have been used for thousands of years as herbal medicines in Asian countries due to their rich nutritional value, medical properties, and biological activities [2,13,23,31]. Several studies highlighted the pro-healthy effects of goji berries, particularly in regarding their antioxidant [2,4,12,13], anti-tumor [4,10,15,40], antimicrobial [10,24], hypoglycemic [10], hypolipidemic [10,14], anti-mutagenic [40], immunomodulatory [16], prebiotic [10,23,24], anti-aging [10], anti-fatigue [10] and neuroprotective activities [12]. These biological activities have been closely related to the fruits' phenolic composition, particularly phenolic acids, flavonoids, carotenoids, and tannins, which are associated with different biological effects [2,5,23,31,42]. This correlation led to reports of health benefits associated with liver, kidney, eyesight, immune system, circulation, and longevity disorders [4,19,24,30,36]. The following sections will discuss each biological activity in detail.

4.1. Antioxidant Activity

Oxidative stress is a phenomenon that occurs due to an imbalance between pro-oxidants and antioxidants, being a consequence of the excessive production of reactive species [3]. In some situations, external antioxidant supplementation is required to reestablish the balance, and fruits' phenolic compounds are well-known for this capacity [35]. The main contributors to the antioxidant capacity of food, especially fruits and vegetables, are phenolic compounds, particularly phenolic acids and flavonoids, carotenoids, tocopherol, polysaccharides, ascorbic acid, and condensed tannins [2,5,10,36,42]. These molecules stimulate the antioxidant defenses by delaying, inhibiting, or preventing the free radicals from damaging proteins, DNA, and lipids, as well as by scavenging free radicals by hydrogen atom transfer or electron donation or enhancing endogenous antioxidant defenses, such as antioxidant enzymes (SOD, CAT, GPx, ...) [3,10,13,35].

Different assays are usually employed to evaluate the radical and antioxidant scavenging capacity. The most frequently applied include the 2,2-diphenyl-1-picrylhydrazyl radical ($DPPH^{\bullet}$), the 2,20-azino-bis(3-ethylbenzothiazoline-6-sulphonic acid) radical cation ($ABTS^{\bullet+}$), scavenging activity, and ferric reducing antioxidant power (FRAP) [10,43]. Table 3 summarizes the reported data on the antioxidant and antiradical activities of goji berries.

Table 3. Antioxidant and antiradical activities of goji berries according to different authors.

DPPH Assay	ABTS Assay	FRAP Assay	References
4.526 µmol TE/g	129 µmol TE/g	5.324 µmol TE/g	[33]
18.5–13.9 µmol VCE/g	61–54 µmol/g	-	[9]
16.65 µmol TE/g	59.14 µmol TE/g	35.1675 mmol Fe^{2+}E/g	[2]
-	16.0–68.3 µmol TE/g	14.4–63.0 µmol TE/g	[36]
8.79–9.35 mg TE/g	24.86–25.12 mg TE/g	16.91–19.52 mg TE/g	[4]
3.12 mg TE/g	-	-	[22]
-	-	42.10 µmol TE/g dw	[28]
-	-	19.36 µmol Fe^{2+}E/g	[13]

TE—Trolox equivalent; VCE—vitamin C equivalent; IC_{50}—half-maximal inhibitory concentration; Fe^{2+}E—Fe^{2+} equivalent; AAE—ascorbic acid equivalent; fw—fresh weight; dw—dry weight.

A study conducted by Islam et al. [2] assessed the TPC (3.16 mg GAE/g dw), TFC (2.83 mg CAE/g dw), condensed tannin content (CTC; 1.08 mg CAE/g dw), and monomeric anthocyanin content (MAC; 0.24 mg MAC/g dw) of goji berries, along with DPPH (16.65 µmol TE/g dw), ABTS (59.14 µmol TE/g dw), and FRAP (3516.75 mmol Fe^{2+}E/g dw) capacity. The results showed a positive linear correlation between DPPH, ABTS, FRAP, and phenolic compounds (0.786, 0.643 and 0.856, respectively), flavonoids (0.857, 0.714 and 0.786, respectively), condensed tannin (0.429, 0.714 and 0.643, respectively) and anthocyanin content (0.643, 0.786 and 0.857, respectively), supporting the idea that phenolic compounds are the main contributors to the antioxidant/antiradical activities of goji berries, which is in line with previous studies [35]. Another study [13] evaluated the TPC (268.5 mg GAE/100 g fw), FRAP (19.36 µmol Fe^{2+}E/g fw), and total bioactive compound content (TBCC; 5806.80 mg/100 g fw) of a methanolic extract of goji berries and identified and quantified the principal bioactive compounds present, namely polyphenols (12,697.90 mg/100 g fw) and organic acids (4461.01 mg/100 g fw). Afterwards, the authors verified the correlation between the antioxidant activity and the variables evaluated, reporting a strong positive correlation between the antioxidant capacity and phenols (0.8290), organic acids (0.8606), TPC (0.9996) and TBCC (0.8363), which attests to the antioxidant capacity of goji berries.

Similarly, after identifying and quantifying the polyphenols (97.23 mg/100 g dw), carotenoids (212.94 mg/100 g dw), organic acids (24.7%), and flavonols (75.3%), Wojdyło et al. [36] determined the antioxidant/antiradical activities of goji berries by FRAP (1.44–6.30 mmol TE/100 g fw) and ABTS assays (1.60–6.83 mmol TE/100 g fw). The Pearson correlation allowed the verification of a linear correlation between ABTS or FRAP and the total polyphenolic compounds (0.523 and 0.038, respectively), phenolic acids (0.277 and 0.328, respectively), and flavonols (0.531 and 0.409, respectively). However, goji berries have a high content of carotenoids; therefore, the Pearson correlation between the ABTS or FRAP and the total carotenoids (0.462 and 0.409, respectively), zeaxanthin (0.381 and 0.315, respectively), β-carotene (0.316 and 0.277, respectively), neoxanthin (0.546 and 0.527, respectively) and cryptoxanthin (0.411 and 0.379, respectively) proved the importance of these compounds for the antioxidant power of this fruit.

Meanwhile, Pehlİvan Karakaçs et al. [20] evaluated the levels of antioxidant enzymes (namely SOD, CAT, and GPx) and malondialdehyde (MDA), a characteristic molecule of the oxidative state, when a *L. barbarum* polysaccharide (LBP) extract was orally administrated to rats for 4 weeks. The LBP extract was prepared by crushing 200 g of dried fruits and performing 2 extractions with 600 mL of a chloroform:methanol (2:1) solution at 80 °C. The results showed that the SOD, CAT, and GPx levels increased, while the MDA levels decreased in the blood serum of rats treated with LBP extract, supporting the in vivo antioxidant power of goji berries.

Skenderidis et al. [25] also demonstrated the antioxidant activity of an aqueous extract of *L. barbarum* berries cultivated in Greece and extracted by ultrasound-assisted extraction (UAE). The results of the XTT assay performed in C2C12 muscle cells showed cytotoxicity in concentrations higher than 125 μg/mL. The levels of GSH were also evaluated by flow cytometry after exposing C2C12 muscle cells to the extract (25 μg/mL and 100 μg/mL), and the results revealed an increase of 127.5% and 189.5%, respectively, when compared to the control. As for the thiobarbituric acid reactive substance (TBARS) levels, a marker of lipid peroxidation, the decreases of 21.8% and 9.4% after exposure to 25 μg/mL and 100 μg/mL, respectively, attested to the antioxidant capacity of goji berries extracts.

4.2. Anticancer Activity

Despite the extensive research and the advances in cancer treatments made in recent years, cancer is still a worldwide problem [17,40]. The knowledge that several polyphenol-rich extracts from natural matrices have been suggested as promising anticancer agents with few side effects, being associated with lower risks of cancer and cancer mortality, has increased consumers' interest in fruits and natural matrices [5,6,35].

Kwaśnik et al. [17] performed a 3-(4,5-dimethylthiazol-2-yl)-2,5-diphenyltetrazolium bromide (MTT) assay with human natural killer cells (NK-92) against human colon cancer cell line LS180 after 48 h of incubation in the presence or absence of an ethanolic extract of goji berries in different concentrations (1–500 μg/mL). According to the authors, in the absence of the extract, the strongest anticancer effect obtained (with the elimination of 91% of the cancer cells) occurs when LS180 and NK-92 were used in a ratio of 1:1. In the presence of extract, using the same ratio, 94.8% of the LS180 cells were eliminated. Therefore, goji berries extract reduced by 5%, 12.8% and 20.6%, respectively, the proliferation of LS180 using concentrations of 1, 2.5 and 5 μg/mL. When LS180 and NK-92 cells, in a ratio of 2:3, are exposed to goji berries extract in concentrations of 2.5 and 5 μg/mL, the viability of LS180 cells is reduced to 96.5% and 98.1%, respectively. These results support the chemopreventive properties of goji berry extract, depending on the dose and number of lymphocytes used. Probably, this is due to the immunomodulatory properties of goji berries, which increase the viability and proliferation of NK cells and promote their recognition and elimination capacity.

Another study assessed the proliferation and apoptotic and necrotic effects of different concentrations of an ethanolic extract of goji berries in the T47D human breast cancer cell line [44]. The MTT assay results attested to a strong decrease in cell proliferation after exposure to the highest extract concentration (1 mg/mL) (70%, 55.7%, and 51.4% after 24, 48, and 96 h, respectively). The bromodeoxyuridine (BrdU) cell proliferation assay supported these results, showing a sharp decrease in the proliferation of T47D cells, similar to the Neutral Red (NR) cell viability assay, which demonstrated a slight decrease in the T47D cell viability. The Western blotting analyses employed to evaluate the expression of p21 and p53 proteins, cyclin-dependent kinase 6 (CDK6), and cyclin D1 highlighted the significant increase in p21 and p53 protein expression as well as a slight decrease in CDK6 and cyclin D1 expression. Regarding T47D cells, an increase in the apoptotic percentage was detected (37%, 61.6%, and 88% when treated with 0.1, 0.5, and 1 mg/mL of extract, respectively), as well as a significant necrotic change at 0.5 mg/mL of extract for propidium iodide and Hoechst solution staining. These results attested that the anticancer activity of goji berries extract is due to apoptotic effects through the mitochondrial pathway. To confirm these results using Western blotting, the authors demonstrated a dose-dependent significant increase in pro-apoptotic Bax protein expression and a decrease in anti-apoptotic BclxL protein expression after treatment of T47D cells with the extract for 48 h when compared to the control.

A study using hepatoma cells (SMMC-7721 and HepG2), cervical cancer cells (HeLa), gastric carcinoma cells (SGC-7901), and human breast cancer cells (MCF-7) determined the influence of *L. barbarum* fruits' crude polysaccharides on the inhibition of cancer cell growth via cell arrest and apoptosis [40]. According to the authors, the fruits' crude polysaccharides were obtained by water extraction, alcohol precipitation and deproteinization, and fractional precipitation with gradient concentrations of ethanol (30%, 50%, and 70%), originating different fractions (LBGP-I-1, LBGP-I-2, and LBGP-I-3). Through an MTT assay, all *L. barbarum* fruit polysaccharide fractions showed a remarkable inhibition of SMMC-7721, HeLa, and MCF-7 cell growth in a dose-dependent manner, with MCF-7 cells being more sensitive to the LBGP-I-3 fraction (with a cell viability reduction to 48.96%). To study the inhibitory mechanism behind this result, the cell cycle of MCF-7 cells treated with LBGP-I-3 at 1000 μg/mL was analyzed by flow cytometry. The results showed a percentage of 72.76%, 24.21%, and 3.04% of cells in the G0/G1, S, and G2/M phase, respectively, suggesting that LBGP-I-3 arrested the MCF-7 cell cycle at the G0/G1 phase. AO-EB and DAPI staining were used to detect basic morphological changes in apoptotic cells, with the results attesting that the LBGP-I-3 fraction induces the MCF-7 apoptosis. The results also supported that the apoptotic effect was caused by the increased expression of pro-apoptotic Caspase-3, 8, and 9 proteins as well as the decrease in the Bcl-2/Bax ratio and mitochondrial membrane potential. A significant decrease in T-SOD and CAT activities, as well as GSH-Px activity and GSH content, coupled with the enhancement of the MAPK signaling pathway, down-regulating p-ErK1/2 levels and up-regulating p-JKN and p-p38 levels, was also observed.

4.3. Antimicrobial Activity

The phytochemicals present in plants, particularly phenolic acids, flavonoids, and tannins, exert their antimicrobial effects by complexing with extracellular and soluble proteins, leading to the disruption of the microbial membrane, metal ion deprivation, and interactions with enzymes [30,45]. Another way of explaining the antimicrobial activity of polyphenols is their structural features, as well as the pH and sodium chloride concentration, which results in physiological changes in the microorganisms and, eventually, cell death [45].

A study carried out by Kabir et al. [45] confirmed that chlorogenic acid, one of the principal phenolic acids detected in goji berries, exhibited a bacteriostatic and bactericidal effect against *Escherichia coli*. The bacteriostatic effect was assessed by measuring the optical density by determining the growth inhibition of chlorogenic acid when added to a culture of *E. coli*. Regarding the bactericidal effect, it was assessed by using mid-logarithmic phase cell cultures (10^6 cells/mL) at different doses (2.5, 5.0, and 10 mM), treatment times (0, 1, 3, and 6 h), temperatures (20, 37, 45, and 50 ^{0}C), and pH conditions (8.0, 7.0, 6.0, 5.0, and 4.0). The overall results demonstrated a synergetic antimicrobial effect expressed by chlorogenic acid and related compounds and a dose, temperature, and time-dependent bactericidal effect. The authors stated that the bactericidal effect shown by chlorogenic acid may be due to the capacity to promote physiological changes on the microbial cell membrane that result in cell death. This study corroborated the results achieved by Ilić et al. [33], which evaluated the antimicrobial activity of goji berries from Serbia. According to the authors, the methanolic extract of goji berries presented mild antimicrobial activity against Gram-positive and Gram-negative bacteria as well as yeast, with remarkable activity against *Klebsiella pneumoniae, Salmonella abony,* and *Pseudomonas aeruginosa.* In another study, Mocan et al. [46] also explored the antimicrobial activity of *L. barbarum* flowers. According to the authors, the antimicrobial activity was mild against Gram-positive (namely *Staphylococcus aureus, Bacillus subtilis,* and *Listeria monocytogenes*) and Gram-negative (*Salmonella typhimurium*) bacteria and lacking against *E. coli*, oppositely to what was observed for goji berries.

4.4. Anti-Inflammatory Activity

The inflammatory response is initiated by the stimulation of innate immunity, the release of immune effectors, the inhibition of enzymes, and the production of pro-inflammatory mediators, such as tumor necrosis factor-α (TNF-α), interleukins (IL), and transcription factor nuclear factor kappa B (NF-kB) [3]. If over-expressed, inflammation can cause numerous complications in patients, being an important hallmark of diseases or pathological conditions, such as cancer, neurodegenerative diseases, respiratory pathologies, and several autoimmune diseases [3,5]. Natural extracts, typically rich in alkaloids, flavonoids, terpenoids and tannins, have been associated with anti-inflammatory effects [5]. A study conducted by Nardi et al. [3] addressed the anti-inflammatory effect of *L. barbarum* berries using paw edema carrageenan, an acute model of inflammation widely employed to evaluate the anti-inflammatory effect of natural products. Briefly, the right paw of mice was injected with 450 μg of carrageenan, and the left paw was injected with the same volume of sterile phosphate-buffered saline (PBS) (control). After established conditions, the animals were orally administered with a methanolic extract of goji berries in doses of 50 mg/kg and 200 mg/kg, 12 h after 12 h, for 10 days. The authors stated that after the carrageenan injection, neutrophils migrate to the inflamed paw and release enzymes, such as myeloperoxidase, which increase ROS production and induce an inflammatory state. The results attested to the anti-inflammatory capacity of the compounds present in goji berries, with the doses of 50 mg/kg and 200 mg/kg leading to a reduction of paw edema in 38% and 63.8% of mice, respectively.

4.5. Immunomodulatory Activity

Immune response suppression is usually associated with the development of immunological diseases. The actual research is focused on finding immunomodulating agents capable of preventing and treating these disorders [47]. Goji berries have been reported as a support to the immune system [6,47], either by regulating the expression of immune factors, such as cytokines and cell adhesion molecules [16,39], or by promoting the proliferation and activity of immune cells, such as natural killer cells [17] and lymphocytes [47].

Different authors [16,24,39,47] stated that the immunomodulatory effects of *L. barbarum* fruits are greatly related to LBP, one of the major active ingredients isolated from berries. LBP is a mixture of proteoglycans and polysaccharides that mainly consists of arabinose, galactose, glucose, xylose, and a small amount of rhamnose, mannose, and galacturonic acid as its glycosidic part. It was reported that LBP is an adjuvant that improves the immune responses against vaccines and increases humoral immunity [39], probably due to an increased expression of IL-2 and TNF-α, molecules involved in immunomodulation [16]. To assess the immune function alteration during the administration of LBP, Zhu et al. [39] divided fourteen mice into two groups and administered a dose of 0.1 mL/10 g body weight of LBP powder for 14 days in the first group, the experimental group, and the same volume of physiological saline via intragastric administration in the second group, the control group. After the experimental period, blood samples were collected, and the animals were sacrificed to excise the spleen and thymus. The viscera indices were measured (thymus or spleen index = weight of thymus or spleen (mg)/body weight (1000 mg)), and the concentrations of immune factors (namely, transforming growth factor-β (TGF-β), interferon-γ (IFN-γ), and IL-6) in the serum were detected using a commercial enzyme-linked immunosorbent assay (ELISA) kit. The results supported the theory that LBP can enhance the innate immune response since the spleen and thymus index of mice from the experimental group (5.12 mg/1000 mg and 4.12 mg/1000 mg, respectively) were significantly higher ($p < 0.05$) when compared to the control group (namely 4.19 mg/1000 mg and 2.86 mg/1000 mg). Moreover, the concentration of the immune factors TGF-β and IL-6 in the experimental group (namely, > 500 pg/mL and ≈ 100 pg/mL) were also significantly higher ($p < 0.05$) when compared to the control group.

It has also been reported that phenolic amides from the nonpolysaccharide fraction of *L. barbarum* fruits can not only in vitro modulate the proliferation of B and T cells but also enhance the immune cell factors (IFN-γ, IL-2, and IL-10) [47]. Prednisone-induced immunodeficient mice were intragastrically administered with a phenolic amid extract, showing a significant spleen cell proliferation ($p < 0.01$) when compared to the prednisone-induced immunodeficient mice intragastrically administered with an ethanolic extract of goji berries or an LBP extract. The immunological cytokines (IFN-γ, IL-2, and IL-10) were also significantly enhanced ($p < 0.01$) in mice treated with the ethanolic goji berries extract, LBP extract, and phenolic amide extract when compared to the control. Overall, the total phenolic amides had the strongest immunity-activating effects.

Since natural killer cells are one of the first elements from the innate immune system to act, Kwaśnik et al. [17] studied the cell viability and proliferation of human natural killer cells (NK-92) in the presence of an ethanolic goji berry extract. The authors highlighted that goji berry extract has the capacity to enhance NK cell proliferation by 61.0% in concentrations ranging between 1 and 250 μg/mL.

4.6. Prebiotic Activity

Prebiotics are non-digestible ingredients that can selectively promote the growth and activity of specific bacteria, modulating the gut microbiota [39]. Goji berries have demonstrated a gut microbiota positive modulating effect [18,39], probably due to phenolic compounds or polysaccharides present, such as LBP [43]. The prebiotic activity of goji berries was attested to by Skenderidis et al. [18], who cultivated strains of *Bifidobacterium* and *Lactobacillus* in the presence and absence of different encapsulated goji berry extracts. According to the authors, the extract with higher amounts of polyphenols and polysaccharides stimulated the growth and proliferation of probiotic strains on a larger scale, mostly *Bifidobacterium* strains, when the cultures were submitted to a gastrointestinal environment (simulated gastric and intestinal juices). Another study [39] assessed the prebiotic activity of LBP in vitro through the growth of *L. acidophilus* and *B. longum*. Both strains were properly cultivated, and different concentrations of LBP powder (experimental group) and glucose (control group) were administered. These experiments highlighted the positive effect of LBP on the growth of both strains when compared to the control group. The authors also evaluated the effects of LBP intake on the composition of cecal gut microbiota [39]. After mice were administered a dose of 0.1 mL/10 g body weight of LBP powder for 14 days, the cecal content samples were collected, and the microbial DNA was extracted and amplified. The results showed a clear modification of the gut microbiota after treatment, with the dominant bacterial communities being *Firmicutes* and *Proteobacteria*. Furthermore, the percentage of beneficial bacteria, such as *Akkermansia*, *Lactobacillus* and *Prevotellaceae*, significantly increased when compared to the control.

4.7. Neuroprotective Activity

Since neurological damage may occur as a consequence of oxidative stress and inflammation, goji berries have been associated with neuroprotective effects [43]. A study conducted by Fernando et al. [19] assessed the neuroprotective activity of a goji berry powder (GBP) against the development of Alzheimer's disease (AD). The authors observed that diets rich in antioxidants can directly affect amyloid beta levels and influence the development of AD. To test this theory, human neuroblastoma BE(2)-M17 cells were treated with and without 20 μM of amyloid beta 42 and exposed to different concentrations of GBP (0.6, 0.9, 1.2, 1.5 and 1.8 μg/mL). The GBP significantly increased the cell viability up to 105% (GBP at 1.2 μg/mL), while the Western blot analysis showed a significant reduction in the amyloid beta up to 20% (GBP at 1.5 μg/mL). Furthermore, the ELISA attested to a 17% reduction of the amyloid beta in amyloid beta-induced neuronal cells when compared to the control (GBP at 1.5 μg/mL). Other authors [16,24] also stated that the major active ingredient isolated from goji berries, LBP, has ocular and neuroprotective effects [20]. In another study, Pehlİvan Karakaçs et al. [20] reported that LBP has a neuroprotection effect in female rats by performing behavioral tests and immunohistochemistry analyses. The animals were submitted to ovariectomy and daily treated with oral low and high doses of polysaccharides obtained from *L. barbarum* fruits (20 and 200 mg/kg) for 4 weeks. According to the behavioral test performed, the LBP present in goji berries can decrease stress-induced anxious behavior in rats. The behavioral disruption could be caused by the accumulation of ROS in the brain, which increases oxidative stress, leading to mitochondrial dysfunction in neuronal cells and apoptosis. However, behavioral disruption can also be caused by variations in neurotransmitter levels in some brain regions, such as the hippocampus. The neuroprotective effect of LBP was demonstrated by the increased levels of brain-derived neurotrophic factor, which promotes neuronal proliferation, survival, and maintenance of neuron structure and function in the hippocampus region.

4.8. Antihyperglycemic Activity

Diabetes is a complex chronic disease caused by metabolic disorders of carbohydrates, lipids, lipoproteins, and increased oxidative stress [43]. It is characterized by the inability to maintain blood glucose levels within healthy ranges, leading to hyperglycemia [36]. Drug treatment is inevitable; however, it is often expensive or unavailable [36]. In these cases, a change in patient lifestyle and eating habits is extremely important [21,38,43]. Therefore, industries are encouraged to find alternative compounds with antihyperglycemic properties, mostly from natural sources, such as goji berries [5,10]. Enzyme inhibitors are good candidates for the treatment of non-insulin-dependent diabetic patients [36]. Since pancreatic α-amylase and intestinal α-glucosidase are enzymes responsible for the hydrolysis of carbohydrates into monosaccharides, such as glucose, the capacity of goji berries to inhibit these enzymes proves the antihyperglycemic effect of this fruit [36]. According to Wojdyło et al. [36], the inhibition of α-amylase and α-glucosidase was, on average, 63% and 10%, respectively.

Most of the experiments that report the antihyperglycemic activity of goji berries have attributed this property to LBP [21,25,38]. Cai et al. [21] performed a clinical trial to study the antidiabetic efficacy of LBP. Sixty-seven patients diagnosed with type II diabetes were divided into two groups and treated twice daily, for three consecutive months, with capsules of 300 mg of microcrystalline cellulose in the control group and capsules with 150 mg of LBP power and 150 mg microcrystalline cellulose in the experimental group. The results of the oral metabolic tolerance test reported a significant decrease in serum glucose levels when compared to the control (−7.86% vs. 1.61%), as well as serum insulin (−56.71% vs. −8.73%). Regarding the insulinogenic index (insulin/glucose), the results attested to a noticeable increase (−0.98% vs. 0.04%, respectively, for the experimental and control groups). During the experimental period, the patients were asked not to change their previous therapies in order to compare the antidiabetic effect of LBP with or without the hypoglycemic medicines. After 3 months of treatment, the patients without hypoglycemic drugs showed a significant decrease in serum glucose, serum insulin, and the homeostasis model assessment index of insulin resistance, while the patients with hypoglycemic drugs did not present significant alterations, demonstrating that LBP has profound hypoglycemic efficacy when disassociated from hypoglycemic drugs.

The study developed by Zhao et al. [38] focused on the effect of LBP extract on diabetic rabbits (induced by Alloxan), evaluating the extract's effect on diabetic nephropathy (DN).

The experimental period lasted for 12 weeks, in which twenty-five rabbits were divided into five groups: group (I)—control group; group (II)—DN control group; group (III)—LBP prevention group from developing DN; group (IV)—positive DN control group using Telmisartan drug (10 mg/kg/day in 3 mL); group (V)—LBP treatment group using the LBP (10 mg/kg/day in 3 mL) as DN treatment. The results demonstrated that LBP not only helps after the development of DN but also benefits the prevention of complications since group (V) showed improvements in fasting glucose tolerance when compared to group (II), while group (III) had better results than group (V). This study also highlighted that LBP can be more efficient in the treatment of DN than standard drugs (in group (V), the kidney weight index (KWI) was significantly higher ($p < 0.01$) than in group (IV)). As for kidney function, even though LBP may be able to protect this organ from injuries, it cannot entirely treat it since the serum urea nitrogen and creatinine levels from groups (III), (IV) and (V) were significantly lower ($p < 0.01$) that in group (II). No significant changes were observed regarding group (I).

4.9. Antihyperlipidemic Activity

In addition to the biological activities described above, the antihyperlipidemic effect of goji berries has also been reported. Pai et al. [14] fed murine models with a high-fat diet (5 g of deoxycholic acid and 300 g of warm coconut oil mixed with 700 g of powdered rat chow per day) for 45 days to induce a state of hyperlipidemia. In the last 30 days of the study, a group of animals was orally treated daily with 10 mg/kg and 20 mg/kg of a 50% hydroalcoholic extract of *L. barbarum* fruits. The results showed a positive antihyperlipidemic activity, with the extract promoting a significant decrease in the triglyceride levels for both doses as well as a significant decrease in very low-density lipoprotein cholesterol at the highest dose, a significant increase in the high-density lipoprotein cholesterol at the lowest dose, and a dose-dependent decrease in the total cholesterol and low-density lipoprotein cholesterol levels. Additionally, the results obtained with the animals treated with the *L. barbarum* fruits and the animals treated with the standard antihyperlipidemic drug atorvastatin were similar, reinforcing the positive antihyperlipidemic activity of goji berries. The authors explained that this effect may be caused by more than one compound present in goji berries, namely polysaccharides and vitamins, such as vitamin C. Additionally, riboflavin, ascorbic acid, and coumarin and their synergistic effects may contribute to the hypolipidemic effect observed.

5. Potential Applications of *L. barbarum* L.

Goji berries can be used in different industries, such as the food, nutraceutical, or cosmetic industries. Regarding the food industry, it has been reported that the aqueous extracts of goji berries can be employed as potentially prebiotic food additives [10,18].

Goji extract has also been successfully used in meat products to improve their sensory properties and oxidative stability during storage [48]. The addition of 1.0% of goji berry extract and 1.0% of buckwheat flour improved the oxidative stability and quality of modified horse-meat products, resulting in a better smell, taste, and surface color after 21 days of storage [49]. Another study determined the effect of the addition of goji berries or goji berry extract on sausages [48]. The addition of 1% of goji berry extracts effectively suppressed lipolysis and protein/lipid oxidation, reducing the microbial count during storage and preserving the sausages' color, aroma, and taste. Antonini et al. [50] also evaluated the addition of goji berries to meet formulas and determined that the addition of chia seeds and *L. barbarum* puree (2.5% and 5%, respectively) to beef burgers promotes a higher TPC, an increase of up to 70% of the total antiradical capacity (through the oxygen radical absorbance capacity (ORAC), ABTS, and DPPH) and a decrease in the lipid peroxidation of up to 50%.

The confectionery and bakery industries are also using goji berries and their compounds to improve the functional, sensory, and texture properties of various products [10]. Muffins and cookies enriched with different amounts of goji berry powder or by-products showed an increase in the TPC, insoluble, and soluble fiber contents as well as good sensory properties [23].

Nutraceuticals are foods, beverages, or supplements with high concentrations of bioactive compounds with outstanding health-promoting effects on the human body, being incorporated into consumers' daily diets [9]. Due to the bioactive compounds present in their fruits, goji berries are being investigated as a potential nutraceutical ingredient [2,3,5]. As previously mentioned, goji berries have been used in herbal medicine for thousands of years [24], along with hundreds of plants cultivated worldwide for their substances to be used in medicine and pharmaceutical formulations [51]. Based on the long-term traditional use of goji berries, this fruit is now generally recognized as non-toxic [10]. However, adverse effects can occur and, depending on the goji berries' use, a public health risk can be considered [51]. The presence of tropane alkaloids, chemical contaminants, such as pesticides and toxic elements, or some proteins that can cause allergic reactions in sensitive consumers are clearly risks [10]. Therefore, caution and professional guidance should be exercised regarding the safe consumption of natural matrices, such as goji berries, to avoid adverse effects, toxicity, and allergies [52].

Furthermore, the application of this fruit and its components at high levels in different industries, such as the food, pharmaceutical, nutraceutical, or cosmetic industry, can substantially deteriorate the sensory, textural, and overall quality of the final products [23]. More studies need to be conducted to determine to what point the use of this fruit is beneficial for consumers [4]. However, overall, considering the excellent nutritional profile and the positive health effects of goji berries, this fruit can be classified as a "superfruit" [12, 13].

6. Valorization Prospects of *L. barbarum* L.

During the production and processing of food products, a huge amount of waste is generated, in which a significant amount of bioactive compounds may be present [11]. In the European Union (EU), approximately 88 million tons of waste are generated annually, with associated costs estimated at EUR 143 billion [42]. The last data from the Food and Agriculture Organization of the United Nations (FAO) state that fruit and vegetable processing wastes, such as peels, pomace, flowers, stems, leaves, seeds, and pulps, are the 5th highest contributor (8% of total food waste) [42,53].

In the processing of goji berries, significant amounts of wastes are generated (approximately 10 kg of waste per 90 kg of goji beverage) [23]. In addition to the economic costs associated with these wastes' disposal, a serious negative impact on the environment is estimated [23]. As a result, industries face the urgent and necessary challenge of implementing greener, more efficient, sustainable, and eco-friendly processing protocols [5]. Regarding processing protocols, since extraction is a critical step in the isolation and purification of bioactive compounds, industries have focused on overcoming the limitations of conventional extraction methods, such as reduced extraction efficiency, high energy employment, and the use of high amounts of solvent wastes [32]. For this reason, new green extraction technologies have emerged [41], such as hot water extraction (HWE), subcritical water extraction (SWE), supercritical fluids extraction (SFE), ultrasound-assisted extraction (UAE) and microwave-assisted extraction (MAE).

Therefore, the valorization of agro-residues and industrial wastes has proved to be an advantage [23,42]. As previously referred, goji berries are ingested in many forms and bring a panoply of beneficial pro-healthy effects to consumers, from nutritional to medicinal and curative approaches. However, their by-products, including leaves, stems, young shoots, and root bark, can also be consumed as part of a traditional diet, and used for medicinal purposes [33]. For instance, a study developed in Portugal [30] compared the composition of goji stems and berries and reported similar values of phenolic compounds (71.9 vs. 71 mg/g dw), organic acids (2.08 vs. 2.07 g/100 g dw), energy (383 vs. 408 kcal/100 g dw), total carbohydrates (78.1 vs. 87 g/100 g dw) and fat (4.6 vs. 4.1 g/100 g dw). In what concerns tocopherols (3.59 vs. 0.33 mg/100 g dw), proteins (7.4 vs. 5.3 g/100 g dw) and saturated fatty acids (68 vs. 26.1%), the stems revealed a high content. However, stems also had higher antioxidant and antiradical capacities (DPPH scavenging activity (EC_{50} = 0.28 mg/mL), reducing power (EC_{50} = 0.23 mg/mL), β-carotene bleaching inhibition (EC_{50} = 0.26 mg/mL), and TBARS inhibition (EC_{50} = 0.07 mg/mL), as well as antibacterial activity against Gram-negative (*Escherichia coli, Morganella morganii, Pseudomonas aeruginosa, Acinetobacter baumannii*) and Gram-positive bacteria (*Staphylococcus aureus, Listeria monocytogenes, Enterococcus faecalis*) when compared to goji berries [30].

Another study determined the chemical composition (TPC, TFC, and HPLC/MS analysis) and evaluated the antioxidant activity (Trolox equivalent antioxidant capacity (TEAC), hemoglobin/ascorbate peroxidase activity inhibition (HAPX) assay, inhibition of lipid peroxidation catalyzed by cytochrome c and electron paramagnetic resonance (EPR) spectroscopy) of an ethanolic extract of *L. barbarum* flowers [46]. The results show its richness in chlorogenic, *p*-coumaric, and ferulic acids, isoquercitrin, rutin, and quercitrin (similar to what was stated in studies concerning goji berries [3,30,31]) and a TPC and TFC very similar to *L. barbarum* fruits (3.75 and 0.61 mg/g dw vs. 1.45 and 0.75 mg/g dw [3], respectively). Regarding the antioxidant capacity, the results attested to their activity, which was positively correlated with their chemical composition. After an ultrasonic extraction with methanol/water (70:30, *v/v*), an extract of goji berry leaves was evaluated regarding phenolic profile and antioxidant capacity (TEAC and EPR spectroscopy) [54]. The phenolic profile revealed the presence of chlorogenic acid and rutin, as previously reported for goji berries [3,30,31]. Regarding the antioxidant activity, the results showed a high antioxidant power (140 mg TE/g dw and 212 mg FSE/g dw, respectively). Along with the enzyme inhibitory activity (cholinesterase, α-amylase, α-glucosidase and tyrosinase), antimicrobial activity (*S. aureus, Bacillus cereus, Listeria monocytogenes, Enterococcus faecalis, Pseudomonas aeruginosa, Salmonella typhimurium, E. coli, Fusobacterium nucleatum,* and *Peptostreptococcus anaerobius*) and antifungal activity (*Aspergillus flavus, Aspergillus niger, Candida albicans, Candida parapsilosis,* and *Penicillium funiculosum*), this study confirms the potential of goji leaves to be used as a valuable source of bioactive compounds. Apart from that, polyphenols from goji berries leaves were also encapsulated in liposomes to improve their delivery and successfully serve as polyphenol carriers, demonstrating a cytoprotective effect on L-929 mouse fibroblast cells [10].

Based on the results detailed above, it can be stated that goji berry by-products, such as fruits skins, pulps, stems, and seeds, have high amounts of dietary fiber, vitamins, minerals, phytochemicals, and antioxidants, being a valuable ingredient for the food industry. Therefore, the incorporation of goji berry by-products, for example, in the bakery industry, may provide benefits to consumers as well as manufacturers, not only due to their nutritional value but also due to the economic advantage that comes from the elimination of the costs required for its disposal [23]. Nevertheless, safety and security studies should be previously performed considering these potential applications.

7. Conclusions

Goji berries can be classified as a "superfruit" due to their outstanding phytochemical composition, which includes phenolic acids (up to 32.70 mg/g), flavonoids (up to 38.00 mg/g), organic acids (up to 44.61 mg/g fw), carotenoids (up to 2.13 mg/g dw), carbohydrates (up to 87.00 g/100 g dw), and vitamins (up to 48.94 mg/100 g fw). The main phenolic compounds reported in this fruit are phenolic acids (chlorogenic acid and *p*-coumaric acid) and flavonoids (rutin and quercetin). Aside from these compounds, vitamins (ascorbic acid), carbohydrates (polysaccharides), and carotenoids (zeaxanthin, β-carotene, and cryptoxanthin) are the main compounds responsible for the incredible biological activity observed, such as antioxidant, anti-tumor, antimicrobial, hypoglycemic, hypolipidemic, anti-mutagenic, immunomodulatory, and prebiotic activities. In this regard, goji berries can be used in the food, nutraceutical, and pharmaceutical industries as a source of functional ingredients. Despite this, *L. barbarum* leaves, flowers, and steams are valuable sources of bioactive compounds that can be explored for use in different industries. Even though the valorization of goji by-products helps to reduce the food waste produced during food processing, the high energy required, as well as the use of high amounts of solvents in the extraction processes, are still challenges. In the future, the implementation of green extraction techniques is expected at industrial levels, minimizing the wastes produced and maximizing the extraction of bioactive compounds from goji berries and by-products.

Author Contributions: Conceptualization, F.R.; methodology, F.T.; validation, F.R.; investigation, F.T.; resources, F.R.; writing—original draft preparation, F.T.; writing—review and editing, A.M.S., C.D.-M. and F.R.; supervision, F.R.; project administration, F.R.; funding acquisition, F.R. All authors have read and agreed to the published version of the manuscript.

Funding: This research was funded by project EXPL/BAA-GR/0663/2021—Kiwi4Health—Exploring the Eco-Innovative Re-Use of Kiwiberry, supported by national funds by FCT/MCTES and by the projects UIDB/50006/2020 and UIDP/50006/2020 through national funds. This work was also financed by national funds from FCT—Fundação para a Ciência e a Tecnologia, I.P., in the scope of the project UIDP/04378/2020 and UIDB/04378/2020 of the Research Unit on Applied Molecular Biosciences—UCIBIO and the project LA/P/0140/2020 of the Associate Laboratory Institute for Health and Bioeconomy—i4HB.

Institutional Review Board Statement: Not applicable.

Informed Consent Statement: Not applicable.

Data Availability Statement: Not applicable.

Acknowledgments: Filipa Teixeira is thankful for the scholarship from project EXPL/BAA-GR/0663/2021—Kiwi4Health—Exploring the Eco-Innovative Re-Use of Kiwiberry. Ana Margarida Silva (SFRH/BD/144994/2019) is thankful for the Ph.D. grant financed by POPH-QREN and subsidized by the European Science Foundation and Ministério da Ciência, Tecnologia e Ensino Superior. Francisca Rodrigues (CEECIND/01886/2020) is thankful for her contract financed by FCT/MCTES—CEEC Individual Program Contract.

Conflicts of Interest: The authors declare no conflict of interest.

References

1. Mocan, A.; Schafberg, M.; Crișan, G.; Rohn, S. Determination of lignans and phenolic components of *Schisandra chinensis* (Turcz.) Baill. using HPLC-ESI-ToF-MS and HPLC-online TEAC: Contribution of individual components to overall antioxidant activity and comparison with traditional antioxidant assays. *J. Funct. Foods* **2016**, *24*, 579–594. [CrossRef]
2. Islam, T.; Yu, X.; Badwal, T.S.; Xu, B. Comparative studies on phenolic profiles, antioxidant capacities and carotenoid contents of red goji berry (*Lycium barbarum*) and black goji berry (*Lycium ruthenicum*). *Chem. Cent. J.* **2017**, *11*, 59. [CrossRef] [PubMed]
3. Nardi, G.M.; Farias Januario, A.G.; Freire, C.G.; Megiolaro, F.; Schneider, K.; Perazzoli, M.R.; Do Nascimento, S.R.; Gon, A.C.; Mariano, L.N.; Wagner, G.; et al. Anti-inflammatory activity of berry fruits in mice model of inflammation is based on oxidative stress modulation. *Pharmacogn. Res.* **2016**, *8*, S42–S49. [PubMed]
4. Mocan, A.; Moldovan, C.; Zengin, G.; Bender, O.; Locatelli, M.; Simirgiotis, M.; Atalay, A.; Vodnar, D.C.; Rohn, S.; Crisan, G. UHPLC-QTOF-MS analysis of bioactive constituents from two Romanian Goji (*Lycium barbarum* L.) berries cultivars and their antioxidant, enzyme inhibitory, and real-time cytotoxicological evaluation. *Food Chem. Toxicol.* **2018**, *115*, 414–424. [CrossRef]
5. Pinto, D.; Cadiz-Gurrea, M.L.; Vallverdu-Queralt, A.; Delerue-Matos, C.; Rodrigues, F. *Castanea sativa* shells: A review on phytochemical composition, bioactivity and waste management approaches for industrial valorization. *Food Res. Int.* **2021**, *144*, 110364. [CrossRef]
6. Llorent-Martínez, E.J.; Fernández-de Córdova, M.L.; Ortega-Barrales, P.; Ruiz-Medina, A. Characterization and comparison of the chemical composition of exotic superfoods. *Microchem. J.* **2013**, *110*, 444–451. [CrossRef]
7. Attard, E. History, Definition, and Legislation. In *Nonvitamin and Nonmineral Nutritional Supplements*; Nabavi, S.M., Silva, A.S., Eds.; Academic Press: Cambridge, MA, USA, 2019; pp. 3–8.
8. Garcia-Alvarez, A.; Egan, B.; de Klein, S.; Dima, L.; Maggi, F.M.; Isoniemi, M.; Ribas-Barba, L.; Raats, M.M.; Meissner, E.M.; Badea, M.; et al. Usage of plant food supplements across six European countries: Findings from the PlantLIBRA consumer survey. *PLoS ONE* **2014**, *9*, e92265. [CrossRef]
9. Protti, M.; Gualandi, I.; Mandrioli, R.; Zappoli, S.; Tonelli, D.; Mercolini, L. Analytical profiling of selected antioxidants and total antioxidant capacity of goji (Lycium spp.) berries. *J. Pharm. Biomed. Anal.* **2017**, *143*, 252–260. [CrossRef]
10. Vidovic, B.B.; Milincic, D.D.; Marcetic, M.D.; Djuris, J.D.; Ilic, T.D.; Kostic, A.Z.; Pesic, M.B. Health Benefits and Applications of Goji Berries in Functional Food Products Development: A Review. *Antioxidants* **2022**, *11*, 248. [CrossRef]

11. Kumar, K.; Srivastav, S.; Sharanagat, V.S. Ultrasound assisted extraction (UAE) of bioactive compounds from fruit and vegetable processing by-products: A review. *Ultrason. Sonochem.* **2021**, *70*, 105325. [CrossRef]
12. Magalhães, V.; Silva, A.R.; Silva, B.; Zhang, X.; Dias, A.C.P. Comparative studies on the anti-neuroinflammatory and antioxidant activities of black and red goji berries. *J. Funct. Foods* **2022**, *92*, 105038. [CrossRef]
13. Donno, D.; Beccaro, G.L.; Mellano, M.G.; Cerutti, A.K.; Bounous, G. Goji berry fruit (*Lycium* spp.): Antioxidant compound fingerprint and bioactivity evaluation. *J. Funct. Foods* **2015**, *18*, 1070–1085. [CrossRef]
14. Pai, P.G.; Umma Habeeba, P.; Ullal, S.; Ahsan Shoeb, P.; Pradeepti, M.S.; Ramya, K. Evaluation of hypolipidemic effects of *Lycium barbarum* (Goji berry) in a murine model. *J. Nat. Remedies* **2013**, *13*, 4–8.
15. Hsu, H.J.; Huang, R.F.; Kao, T.H.; Inbaraj, B.S.; Chen, B.H. Preparation of carotenoid extracts and nanoemulsions from *Lycium barbarum* L. and their effects on growth of HT-29 colon cancer cells. *Nanotechnology* **2017**, *28*, 135103. [CrossRef]
16. Tang, W.M.; Chan, E.; Kwok, C.Y.; Lee, Y.K.; Wu, J.H.; Wan, C.W.; Chan, R.Y.; Yu, P.H.; Chan, S.W. A review of the anticancer and immunomodulatory effects of *Lycium barbarum* fruit. *Inflammopharmacology* **2012**, *20*, 307–314. [CrossRef]
17. Kwasnik, P.; Lemieszek, M.K.; Rzeski, W. Impact of phytochemicals and plant extracts on viability and proliferation of NK cell line NK-92—A closer look at immunomodulatory properties of goji berries extract in human colon cancer cells. *Ann. Agric. Environ. Med.* **2021**, *28*, 291–299. [CrossRef]
18. Skenderidis, P.; Mitsagga, C.; Lampakis, D.; Petrotos, K.; Giavasis, I. The effect of encapsulated powder of goji berry (*Lycium barbarum*) on growth and survival of probiotic bacteria. *Microorganisms* **2019**, *8*, 57. [CrossRef] [PubMed]
19. Fernando, W.M.A.D.B.; Dong, K.; Durham, R.; Stockmann, R.; Jayasena, V. Effect of goji berry on the formation of extracellular senile plaques of Alzheimer's disease. *Nutr. Healthy Aging* **2021**, *6*, 105–116. [CrossRef]
20. PehlIvan Karaka, S.F.; CoSkun, H.; SoytUrk, H.; Bozat, B.G. Anxiolytic, antioxidant, and neuroprotective effects of goji berry polysaccharides in ovariectomized rats: Experimental evidence from behavioral, biochemical, and immunohistochemical analyses. *Turk. J. Biol.* **2020**, *44*, 238–251. [CrossRef]
21. Cai, H.; Liu, F.; Zuo, P.; Huang, G.; Song, Z.; Wang, T.; Lu, H.; Guo, F.; Han, C.; Sun, G. Practical application of antidiabetic efficacy *of Lycium barbarum* polysaccharide in patients with type 2 diabetes. *Med. Chem.* **2015**, *11*, 383–390. [CrossRef]
22. Fatchurrahman, D.; Amodio, M.L.; Valeria De Chiara, M.L.; Mastrandrea, L.; Colelli, G. Characterization and postharvest behavior of goji berry (*Lycium barbarum* L.) during ripening. *Postharvest Biol. Technol.* **2022**, *191*, 111975. [CrossRef]
23. Bora, P.; Ragaee, S.; Abdel-Aal, E.-S.M. Effect of incorporation of goji berry by-product on biochemical, physical and sensory properties of selected bakery products. *LWT* **2019**, *112*, 108225. [CrossRef]
24. Shah, T.; Bule, M.; Niaz, K. Goji berry (*Lycium barbarum*)—A Superfood. In *Nonvitamin and Nonmineral Nutritional Supplements*; Nabavi, S.M., Silva, A.S., Eds.; Academic Press: Cambridge, MA, USA, 2019; pp. 257–264.
25. Skenderidis, P.; Kerasioti, E.; Karkanta, E.; Stagos, D.; Kouretas, D.; Petrotos, K.; Hadjichristodoulou, C.; Tsakalof, A. Assessment of the antioxidant and antimutagenic activity of extracts from goji berry of Greek cultivation. *Toxicol. Rep.* **2018**, *5*, 251–257. [CrossRef]
26. Lu, Y.; Guo, S.; Zhang, F.; Yan, H.; Qian, D.-W.; Shang, E.-X.; Wang, H.-Q.; Duan, J.-A. Nutritional components characterization of Goji berries from different regions in China. *J. Pharm. Biomed. Anal.* **2021**, *195*, 113859. [CrossRef]
27. FAO. FAOSTAT: Crops and Livestock Products. Available online: https://www.fao.org/faostat/en/ (accessed on 3 January 2023).
28. Kafkaletou, M.; Christopoulos, M.V.; Tsaniklidis, G.; Papadakis, I.; Ioannou, D.; Tzoutzoukou, C.; Tsantili, E. Nutritional value and consumer-perceived quality of fresh goji berries (*Lycium barbarum* L. and *L. chinense* L.) from plants cultivated in Southern Europe. *Fruits* **2018**, *73*, 5–12. [CrossRef]
29. "A Senhora do Monte". Available online: https://asenhoradomonte.com/2021/06/20/como-plantar-bagas-goji/ (accessed on 27 December 2022).
30. Pires, T.C.S.P.; Dias, M.I.; Barros, L.; Calhelha, R.C.; Alves, M.J.; Santos-Buelga, C.; Ferreira, I.C.F.R. Phenolic compounds profile, nutritional compounds and bioactive properties of *Lycium barbarum* L.: A comparative study with stems and fruits. *Ind. Crops Prod.* **2018**, *122*, 574–581. [CrossRef]
31. Zhao, W.-H.; Shi, Y.-P. Comprehensive analysis of phenolic compounds in four varieties of goji berries at different ripening stages by UPLC–MS/MS. *J. Food Compost. Anal.* **2022**, *106*, 104279. [CrossRef]
32. Skenderidis, P.; Lampakis, D.; Giavasis, I.; Leontopoulos, S.; Petrotos, K.; Hadjichristodoulou, C.; Tsakalof, A. Chemical properties, fatty-acid composition, and antioxidant activity of goji berry (*Lycium barbarum* L. and *Lycium chinense* Mill.) Fruits. *Antioxidants* **2019**, *8*, 60. [CrossRef]
33. Ilić, T.; Dodevska, M.; Marčetić, M.; Božić, D.; Kodranov, I.; Vidović, B. Chemical characterization, antioxidant and antimicrobial properties of goji berries cultivated in Serbia. *Foods* **2020**, *9*, 1614. [CrossRef]
34. FoodData Central Search Results. Available online: https://fdc.nal.usda.gov/ (accessed on 23 December 2022).
35. Vasantha Rupasinghe, H.P.; Nair, S.V.G.; Robinson, R.A. Chemopreventive properties of fruit phenolic compounds and their possible mode of actions. In *Studies in Natural Products Chemistry*; Atta ur, R., Ed.; Elsevier: Amsterdam, The Netherlands, 2014; Volume 42, pp. 229–266.
36. Wojdyło, A.; Nowicka, P.; Bąbelewski, P. Phenolic and carotenoid profile of new goji cultivars and their anti-hyperglycemic, anti-aging and antioxidant properties. *J. Funct. Foods* **2018**, *48*, 632–642. [CrossRef]

37. Benchennouf, A.; Grigorakis, S.; Loupassaki, S.; Kokkalou, E. Phytochemical analysis and antioxidant activity of *Lycium barbarum* (Goji) cultivated in Greece. *Pharm. Biol.* **2017**, *55*, 596–602. [CrossRef] [PubMed]
38. Zhao, Q.; Li, J.; Yan, J.; Liu, S.; Guo, Y.; Chen, D.; Luo, Q. *Lycium barbarum* polysaccharides ameliorates renal injury and inflammatory reaction in alloxan-induced diabetic nephropathy rabbits. *Life Sci.* **2016**, *157*, 82–90. [CrossRef] [PubMed]
39. Zhu, W.; Zhou, S.; Liu, J.; McLean, R.J.C.; Chu, W. Prebiotic, immuno-stimulating and gut microbiota-modulating effects of *Lycium barbarum* polysaccharide. *Biomed. Pharmacother.* **2020**, *121*, 109591. [CrossRef] [PubMed]
40. Gong, G.; Liu, Q.; Deng, Y.; Dang, T.; Dai, W.; Liu, T.; Liu, Y.; Sun, J.; Wang, L.; Liu, Y.; et al. Arabinogalactan derived from *Lycium barbarum* fruit inhibits cancer cell growth via cell cycle arrest and apoptosis. *Int. J. Biol. Macromol.* **2020**, *149*, 639–650. [CrossRef] [PubMed]
41. Skenderidis, P.; Petrotos, K.; Giavasis, I.; Hadjichristodoulou, C.; Tsakalof, A. Optimization of ultrasound assisted extraction of of goji berry (*Lycium barbarum*) fruits and evaluation of extracts' bioactivity. *J. Food Process Eng.* **2017**, *40*, e12522. [CrossRef]
42. Silva, A.M.; Costa, P.C.; Delerue-Matos, C.; Latocha, P.; Rodrigues, F. Extraordinary composition of *Actinidia arguta* by-products as skin ingredients: A new challenge for cosmetic and medical skincare industries. *Trends Food Sci. Technol.* **2021**, *116*, 842–853. [CrossRef]
43. Jiang, Y.; Fang, Z.; Leonard, W.; Zhang, P. Phenolic compounds in Lycium berry: Composition, health benefits and industrial applications. *J. Funct. Foods* **2021**, *77*, 104340. [CrossRef]
44. Wawruszak, A.; Czerwonka, A.; Okla, K.; Rzeski, W. Anticancer effect of ethanol *Lycium barbarum* (Goji berry) extract on human breast cancer T47D cell line. *Nat. Prod. Res.* **2016**, *30*, 1993–1996. [CrossRef]
45. Kabir, F.; Katayama, S.; Tanji, N.; Nakamura, S. Antimicrobial effects of chlorogenic acid and related compounds. *J. Korean Soc. Appl. Biol. Chem.* **2014**, *57*, 359–365. [CrossRef]
46. Mocan, A.; Vlase, L.; Vodnar, D.C.; Gheldiu, A.M.; Oprean, R.; Crisan, G. Antioxidant, antimicrobial effects and phenolic profile of *Lycium barbarum* L. flowers. *Molecules* **2015**, *20*, 15060–15071. [CrossRef]
47. Zhu, P.F.; Zhao, Y.L.; Dai, Z.; Qin, X.J.; Yuan, H.L.; Jin, Q.; Wang, Y.F.; Liu, Y.P.; Luo, X.D. Phenolic amides with immunomodulatory activity from the nonpolysaccharide fraction of *Lycium barbarum* fruits. *J. Agric. Food Chem.* **2020**, *68*, 3079–3087. [CrossRef]
48. Fadiloglu, E.E.; Çoban, M.Z. The effects of goji berry (*Lycium barbarum* L.) extract on some chemical, microbiological and sensory characteristics of liquid smoked common carp (*Cyprinus carpio* L., 1758) sausages. *Yuzuncu Yil Univ. J. Agric. Sci.* **2019**, *29*, 702–710.
49. Uzakov, Y.; Kaldarbekova, M.; Kuznetsova, O. Improved technology for new-generation Kazakh national meat products. *Foods Raw Mater.* **2020**, *8*, 76–83. [CrossRef]
50. Antonini, E.; Torri, L.; Piochi, M.; Cabrino, G.; Meli, M.A.; De Bellis, R. Nutritional, antioxidant and sensory properties of functional beef burgers formulated with chia seeds and goji puree, before and after in vitro digestion. *Meat Sci.* **2020**, *161*, 108021. [CrossRef]
51. Salmerón-Manzano, E.; Garrido-Cardenas, J.A.; Manzano-Agugliaro, F. Worldwide research trends on medicinal plants. *Int. J. Environ. Res. Public Health* **2020**, *17*, 3376. [CrossRef]
52. Ferraz, J.d.R.S.; Macedo, J.L.; Silva, D.J.S.; Sampaio, L.V.A. Goji Berry: Nutritional properties and benefits for human health. *Res. Soc. Dev.* **2019**, *8*, e284934. [CrossRef]
53. Rodriguez Garcia, S.L.; Raghavan, V. Green extraction techniques from fruit and vegetable waste to obtain bioactive compounds—A review. *Crit. Rev. Food Sci. Nutr.* **2022**, *62*, 6446–6466. [CrossRef]
54. Mocan, A.; Zengin, G.; Simirgiotis, M.; Schafberg, M.; Mollica, A.; Vodnar, D.C.; Crişan, G.; Rohn, S. Functional constituents of wild and cultivated Goji (*L. barbarum* L.) leaves: Phytochemical characterization, biological profile, and computational studies. *J. Enzyme Inhib. Med. Chem.* **2017**, *32*, 153–168. [CrossRef]

International Journal of
Molecular Sciences

MDPI

Review

Essential Oils from Annonaceae Species from Brazil: A Systematic Review of Their Phytochemistry, and Biological Activities

Márcia Moraes Cascaes [1,*], Odirleny dos Santos Carneiro [2], Lidiane Diniz do Nascimento [3], Ângelo Antônio Barbosa de Moraes [2], Mozaniel Santana de Oliveira [3,*], Jorddy Neves Cruz [3], Giselle Maria Skelding Pinheiro Guilhon [1] and Eloisa Helena de Aguiar Andrade [1,3]

1 Programa de Pós-Graduação em Química, Universidade Federal do Pará, Rua Augusto Corrêa S/N, Guamá, Belém 66075-900, PA, Brazil; giselle@ufpa.br (G.M.S.P.G.); eloisa@museu-goeldi.br (E.H.d.A.A.)

2 Faculdade de Química, Universidade Federal do Pará, Rua Augusto Corrêa S/N, Guamá, Belém 66075-900, PA, Brazil; lenny_carneiro@hotmail.com (O.d.S.C.); angeloquimica17@gmail.com (Â.A.B.d.M.)

3 Laboratório Adolpho Ducke–Coordenação de Botânica, Museu Paraense Emílio Goeldi, Av. Perimetral, 1901, Terra Firme, Belém 66077-830, PA, Brazil; lidianenascimento@museu-goeldi.br (L.D.d.N.); jorddynevescruz@gmail.com (J.N.C.)

* Correspondence: cascaesmm@gmail.com (M.M.C.); mozaniel.oliveira@yahoo.com.br (M.S.d.O.); Tel.: +55-91-982024161 (M.M.C.); +55-91-988647823 (M.S.d.O.)

Citation: Cascaes, M.M.; Carneiro, O.d.S.; Nascimento, L.D.d.; de Moraes, Â.A.B.; de Oliveira, M.S.; Cruz, J.N.; Guilhon, G.M.S.P.; Andrade, E.H.d.A. Essential Oils from Annonaceae Species from Brazil: A Systematic Review of Their Phytochemistry, and Biological Activities. *Int. J. Mol. Sci.* **2021**, *22*, 12140. https://doi.org/10.3390/ijms222212140

Academic Editors: Antonio Carrillo Vico and Ivan Cruz-Chamorro

Received: 13 September 2021
Accepted: 12 October 2021
Published: 9 November 2021

Publisher's Note: MDPI stays neutral with regard to jurisdictional claims in published maps and institutional affiliations.

Abstract: The present work involves a systematic review of the chemical composition and biological effects of essential oils from the Annonaceae species collected in Brazil from 2011 to 2021. Annonaceae is one of the most important botanical families in Brazil, as some species have economic value in the market as local and international fruit. In addition, the species have useful applications in several areas—for instance, as raw materials for use in cosmetics and perfumery and as medicinal plants. In folk medicine, species such as *Annona glabra* L. and *Xylopia sericea* A. St.-Hil. are used to treat diseases such as rheumatism and malaria. The species of Annonaceae are an important source of essential oils and are rich in compounds belonging to the classes of mono and sesquiterpenes; of these compounds, α-pinene, β-pinene, limonene, (*E*)-caryophyllene, bicyclogermacrene, caryophyllene oxide, germacrene D, spathulenol, and β-elemene are the most abundant. The antimicrobial, anti-inflammatory, antileishmania, antioxidant, antiproliferative, cytotoxic, larvicidal, trypanocidal, and antimalarial activities of essential oils from the Annonaceae species in Brazil have been described in previous research, with the most studies on this topic being related to their antiproliferative or cytotoxic activities. In some studies, it was observed that the biological activity reported for these essential oils was superior to that of drugs available on the market, as is the case of the essential oil of the species *Guatteria punctata* (Aubl.) R. A. Howard., which showed a trypanocidal effect that was 34 times stronger than that of the reference drug benznidazol.

Keywords: natural products; Brazilian species; essential oil; applications

1. Introduction

The species of Annonaceae are flowering plants consisting of trees, shrubs, and lianas. These species present a combination of striking characteristics and form one of the most uniform botanical families from both anatomical and structural points of view; they are one of the most primitive families of Angiosperms and belong to the class Magnoliopsida, subclass Magnoliidae, and order Magnoliales [1].

Annonaceae consists of 2106 species and more than 130 genera. The family is concentrated in the tropics, and about 900 species are neotropical, 450 are Afrotropical, and the others are Indomalayan [2]. Annonaceae plays an important ecological role in terms of species diversity, especially in tropical forest ecosystems [3]. In Brazil, the family has confirmed occurrence in all states, with about 380 species being described here, distributed across 32 genera. The Amazon biome contains three quarters of all Annonaceae species, with 268 occurring here. Meanwhile, the Atlantic Forest houses 98 species and the Cerrado has approximately 52 [4].

Some species of Annonaceae are of economic value in the international fresh fruit market, such as *Annona cherimola* Mill. ("Cherimólia") and *Annona squamosa* L. ("pinha") [5]. In Brazil, some *Annona* fruits are highly sought after, such as *Annona crassiflora* Mart. ("Araticum"), *A. squamosa*, and *Annona muricata* L. ("graviola") [3]. In addition, some species are often used as raw materials in the cosmetics and perfumery industries and as medicinal plants [6].

Numerous species of Annonaceae are odoriferous and these fragrances are due to the presence of essential oils (EOs) [7]. In nature, EOs have many important functions, such as attracting insects or allowing allelopathic communication between plants [8]. In addition, they have antibacterial, antiviral, anti-inflammatory, and antifungal properties, among others [9].

According to the review published by Fournier et al. [10], the main volatile constituents of EOs from the Annonaceae species are monoterpene hydrocarbons in fruits and seeds, sesquiterpene hydrocarbons in leaves, and oxygenated sesquiterpenes in bark and roots. After this review (1999), several articles were made available in the literature showing the chemical and biological properties of EOs obtained from Annonaceae species [11–14]. In this context, the present work aims to carry out a systematic review of the essential oils of the Annonaceae species collected in Brazil in the last ten years, evaluating their chemical compositions and their potential biological activities.

2. Essential Oils

EOs are present in various aromatic plants, usually found in tropical and subtropical countries. They are obtained from various parts of aromatic plants, including leaves, flowers, fruits, seeds, buds, rhizomes, roots, and bark [9]. Chemically, EOs are mixtures of 20–60 components in varying concentrations, with some compounds found in high concentrations (20–70%) and others in only small amounts. Most of the components of EOs are designated as lipophilic terpenoids, phenylpropanoids, or derivatives of short-chain aliphatic hydrocarbons of low molecular weight, with the first being the most frequent and characteristic constituents. Among these, acyclic mono- and sesquiterpenoids and mono-, bi-, or tricyclics of different chemical classes constitute the majority of EOs, such as hydrocarbons, ketones, alcohols, oxides, aldehydes, phenols, and esters [15].

Several techniques have been used to obtain EOs. These techniques depend on the part of the plants from which the oil will be extracted, the stability of the oil when faced with heat, and the susceptibility of the oil's constituents to undergoing chemical reactions. Some of the techniques commonly used for EO extraction are hydrodistillation, hydrodiffusion, enfleurage, cold pressing, steam distillation, solvent extraction, microwave-assisted process, and carbon dioxide extraction [16–21].

Essential oils play an important role in plants and act as antibacterials, antivirals, antifungals, and insecticides and protect plants from herbivores. It is possible to list about 3000 EOs, but only 300 are used in perfumes, and makeup products, sanitary products, dentistry, and agriculture; as preservatives and flavor additives for food; as fragrances for household cleaning products; as industrial solvents; and as natural remedies [22].

In recent years, EOs have gained great popularity in the food, cosmetic, and pharmaceutical industries. Consumers have developed increasing interest in the use of natural products as alternatives to artificial additives or pharmacologically relevant agents.

Medical professionals are more interested in the medicinal properties of EOs, as research has shown the antibacterial, fungicidal, relaxant, stimulant, and antidepressant effects of these volatile substances. Furthermore, EOs are known for their therapeutic properties and are therefore used in the treatment of various infections caused by pathogenic and non-pathogenic diseases [16].

Consumer concerns about chemical preservatives have driven the growing interest in some natural antimicrobials, such as EOs [23]. In the food industry, the current trend to reduce the use of food preservatives in favor of natural alternatives makes EOs and their components viable alternatives for this application [24].

In the food industry, limonene monoterpene, a component of many EOs, is used as a flavoring in the production of desserts, ice cream, and non-alcoholic beverages. Thymol, a crystalline substance with an intense odor that is part of the chemical composition of the EO of thyme (*Thymus vulgaris* L. and *T. zygis* L.) and whose content varies between 22% and 50%, is used as a flavoring agent in food products, such as sweets, syrups, and seasoning mixtures [25]. The monoterpene eucalyptol (or cineole) is a colorless liquid with a camphor odor; one of the most abundant sources of 1,8-cineole is *Eucalyptus globulus* Labill. leaves. The EO of this species is used as a flavoring additive in various food products (such as in meat) as well as in beverages. Due to the fresh odor of cineole, this substance is applied in large quantities in oral care products [25].

In the cosmetics industry, EOs are vital assets, as in addition to providing pleasant aromas for various products they are also able to act as preservatives and active agents and, simultaneously, offer several benefits to the skin. EOs with a high added value include are citrus, lavender, eucalyptus, tea tree, and other floral oils, which are used as fragrances, while linalool, geraniol, limonene, citronellol, and citral are very popular fragrance components used in different cosmetics [26].

3. Annonaceae Ethnobotanics

Natural products, especially those derived from plants, have been used to help humanity treat various ailments for many millennia [27]. Plants have played an important role in the survival of many human communities. They have been used in many different ways—e.g., as food, medicines, and ornaments; for mystical and religious purposes; as lumber; and for making handicrafts. Knowledge about the use of plant resources has been transmitted from father to son and from ancient civilizations to the present day.

Brazil is formed from the Amazon Forest biome, the Pantanal, savannah and woodland Cerrado, the semi-arid forest Caatinga, pampas fields, and the Atlantic Forest rainforest. These varied biomes reflect the enormous wealth of flora and the greatest biodiversity on the planet. In addition, there is great cultural diversity in Brazil and the use of medicinal plants results from different knowledge built over time [28].

The Amazon region has approximately 55,000 species of plants, most of which are still little known and many of which are used for medicinal and religious purposes [29]. In this region, indians, caboclos, riverside dwellers, rubber tappers, quilombolas, fishermen, small rural producers, and extractivists hold rich knowledge about plants that is passed from generation to generation through oral tradition. It is also known that the cultural diversity of this country positively influences the ethnobotanical use of medicinal plants and in a way increases its biodiversity, given the inclusion of exotic species in the national flora brought by the different peoples who have come to this country [30].

In this context, ethnobotany is a branch of science that analyzes and studies the knowledge of various peoples about the use of plants. It is through this that the profile of a community and its uses of plants are learned, as each community has its own customs and peculiarities, aiming to extract information that may be beneficial regarding the uses of medicinal plants [30].

Several plants of the Annonaceae family are used in folk medicine due to their pharmacological properties, which are attributed to the presence of secondary metabolites of different classes, such as alkaloids, acetogenins, and flavonoids [31].

The *Annona muricata* L. species, similar to other *Annona* species, including *A. squamosa* L. and *A. reticulata* L., are widely used in traditional medicine against a variety of diseases, especially cancer and parasitic infections. The fruits of *A. muricata* are used as a natural remedy for diseases such as neuralgia, arthritis, diarrhea, dysentery, fever, malaria, parasites, rheumatism, skin rashes, and worms. In addition, many women eat the fruit to increase their production of breast milk after giving birth. The leaves of this species are used to treat cystitis, diabetes, headaches, and insomnia. Extract made from the leaves has anti-rheumatic and neurologic effects, while the cooked leaves are used to treat abscesses and rheumatism. Crushed seeds are believed to be anthelmintic [32].

Among the medicinal species of *Xylopia*, *X. frutescens* Aubl. is found in Central and South America, Africa, and Asia. In Brazil, the plant is popularly known as "embira", "embira-vermelha", and "pau carne", and its seeds are used in folk medicine as a bladder stimulant; to trigger menstruation; and to combat rheumatism, halitosis, caries, and intestinal diseases [33]. The leaves and flowers of the *X. laevigata* (Mart.) R.E.Fr. species are used to treat painful diseases, heart disease, and inflammatory conditions [34]. *Xylopia sericea* A. St.-Hil. is an aromatic plant popularly known as pindaíba, pindaíba vermelha, and/or pimento-de-macaco; it is traditionally used as food and as an antimalarial, similar to other representatives of the *Xylopia* genus [35].

Some species of *Guatteria* are used in traditional medicine; in Northern Brazil, the seeds of *G. ouregou* (Aubl.) Dunal are used to treat dyspepsia, stomach pain, and uterine pain [36].

The species *Duguetia furfuracea* (A.St.-Hil.) Saff. is known as "araticum-seco". In folk medicine, the powder from its seeds is mixed with water for use in the treatment of pediculosis, while an infusion of its leaves and branches is used to treat rheumatism. [37]. *Duguetia lanceolata A.St.-Hil.*, popularly known as pindaíba, beribá, or pinhão, is a perennial species distributed across several states of Brazil; in popular medicine, this plant has been used as an anti-inflammatory, for healing, and as an antimicrobial agent [38]. Table 1 shows the 19 species of Annonaceae used in traditional Brazilian medicine.

Table 1. Ethnobotanical use of Annonaceae species occurring in Brazil.

Scientific Name	Popular Name	Brazil Region	Part of the Plant Used	Medicinal Use	Reference
Annona coriacea (Mart.)	Marolo, araticum and araticum-liso	Southeast	Not specified	Parasites, ulcers, inflammatory processes, rheumatism and anthelmintic	[39]
A. crassiflora Mart.	-	Not specified	Fruits and seeds (infusion)	Diarrhea	[31]
A. crassiflora	Marolo	Cerrado	Seeds	Chronic diarrhea	[40]
A. dioica St. Hil.	Araticum	Cerrado	Seeds, fruit and leaves	Chronic diarrhea, emollient and rheumatism	[40]
A. glabra L.	-	Not specified	Leaves	Rheumatism	[31]
A. glabra	Araticum and araticum do brejo	Northeast	Leaves	Rheumatism and vermifuge	[41]
A. leptopetala (R.E.Fr) H. Rainer	Pinha-brava	Northeast	Not specified	Anti-tumor and anti-inflammatory	[42]
A. montana Macfad	Graviola, araticum-grande and jaca-do-Pará	Northeast	Leaves	Snake bites and obesity	[41]
A. salzmannii A. DC.	-	Northeast	Leaves and bark	Diabetes, tumors, and inflammation	[43]
A. spinescens Mart	-	Not specified	Fruits and seeds	Ulcers	[31]
A. squamosa L.	-	Northeast	Leaves	Stimulants, antispasmodics, sweats, anthelmintics, and insecticides	[44]
A. squamosa	-	Not specified	Leaves	Boils and ulcers	[31]
A. squamosa	Pinha, ata and fruta-de-conde	Northeast	Seeds	Bath to remove lice	[41]

Table 1. *Cont.*

Scientific Name	Popular Name	Brazil Region	Part of the Plant Used	Medicinal Use	Reference
A. sylvatica A. St.-Hil	Araticum, araticum-do-mato, cortiça and cortiça-amarela	Southeast	Leaves	Fever, cough, ulcers caused by syphilis, muscle spasms, and diarrhea	[45]
A. vepretorum Mart.	Pinha da Caatinga	Northeast	Roots and Leaves	Bee and snake stings, inflammation, heart pain, bath for allergies, skin diseases, fungal and bacterial infections	[46]
Duguetia furfuracea (A.St.-Hil.) Saff	Araticum-seco	Cerrado	Seeds	Parasiticidal	[47]
D. furfuracea	Araticum-seco	Southeast	Stem bark	Bath to remove lice	[41]
D. furfuracea	Araticum seco	Cerrado	Branches with leaves	Rheumatism	[40]
D. lanceolata St. Hil.	-	Not specified	Leaves	Anti-inflammatory	[31]
D. lanceolata	Pindaíba, beribá and pinhão	Not specified	Not specified	Anti-inflammatory, healing, and antimicrobial	[38]
Guatteria ouregou (Aubl.) Dunal.	-	North	Seeds	Dyspepsia, stomach and uterine pain	[36]
Rollinia leptopetala R.E.Fr	Pinha-brava	Northeast	Stem bark	Stomachic	[41]
Xylopia aromatica (Lam.) Mart.	Pimenta-de-macaco	Cerrado	Fruits, leaves and stem bark	Digestive and anti-inflammatory	[40]
X. aromatica	Pimenta-de-macaco, pimenta-de-negro	North	Not specified	Carminative and stimulating	[48]
X. frutescens Aubl.	-	Not specified	Barks	Flu	[49]
X. frutescens	Embira and semente-de-embira	Northeast	Seeds and fruits	Digestive	[41]
X. frutescens	Embira, embira-vermelha and pau carne	Northeast	Seeds	Bladder stimulant, triggering menstruation, fighting rheumatism, for halitosis, for tooth decay, and for intestinal diseases	[33]
X. laevigata (Mart.) R. E. Fries	Meiú and pindaíba	Northeast	Leaves and flowers	Painful disorders, heart disease, and inflammatory conditions	[34]
X. sericea A. St.-Hil.	Embiriba and pindaíba	Southeast	Seeds and fruits	Analgesic and anti-inflammatory	[50]
X. sericea	Pindaíba, pindaíba-vermelha and/or pimenta-de-macaco	Southeast	Not specified	Antimalarial	[35]

4. Phytochemistry of Annonaceae Essential Oils

The chemical composition of Annonaceae EOs is varied; in general, mono and sesquiterpenes are the most abundant compounds. In this work, were gathered studies on the chemical composition of EOs from Annonaceae, referring to 38 different species with a geographic distribution in the Brazilian territory. Information regarding the major chemical constituents identified (>5%), yield, collection data, and extraction method used for these EOs is shown in Table 2.

Table 2. Chemical composition of essential oils from Annonaceae species occurring in Brazil.

Species	Collection Place	Collection Date	Part of Plant (Yield)	Extraction Technique	Majority Constituents (% >5); Substance Classes; Total	Reference
Anaxagorea Brevipes Benth	Amazonas	September 2009	Leaves (0.52%)	HD	Guaiol, γ-eudesmol, β-eudesmol and α-eudesmol; M: 3.35%, SH: 13.56%; T: 75.69%	[51]
Annona coriacea Mart.	São Paulo	October 2009	Leaves (0.05%)	HD	Bicyclogermacrene, γ-muurolene and δ-cadinene; M: 20.0%, S: 76.7%; O: 3.3%; T: 96.5%	[39]
A. exsucca DC.	Pará	March 2019	Leaves (NI)	HD	Linalool, β-elemene, (*E*)-caryophyllene, α-humulene, germacrene D and bicyclogermacrene; HM: 2.3%, OM: 11.61%, SH: 80.52%, OS: 4.07%; T: 99.34%	[52]
A. exsucca	Pará	September 2019	Leaves (NI)	HD	*p*-Cymene, sylvestrene, terpinolene, linalool, germacrene D and bicyclogermacrene; HM: 43.36%, OM: 19.39%, SH: 31.29%, OS: 5.10%; T: 99.14%	[52]
A. leptopetala (R.E.Fr.) H. Rainer	Paraíba	August 2016	Leaves (0.04%)	HD	α-Limonene, linalool, α-terpineol, (*E*)-caryophyllene, bicyclogermacrene, spathulenol and guaiol; T: 98.1%	[42]
A. pickelii (Diels) H. Rainer	Sergipe	March 2010	Leaves (0.2%)	HD	Bicyclogermacrene, (*E*)-caryophyllene, α-copaene and germacrene D; M: 0.6%, S: 97.7%; T: 98.3%	[43]
A. pickelii	Sergipe	September 2010	Leaves (0.3%)	HD	Bicyclogermacrene, (*E*)-caryophyllene and α-copaene; T: 99.5%	[53]
A. salzmannii A. DC.	Sergipe	March 2010	Leaves (0.1%)	HD	Bicyclogermacrene, (*E*)-caryophyllene, δ-cadinene, α-copaene, and *allo*-aromadendrene; M: 2.5%, S: 93.7%; T: 96.2%	[43]
A. salzmannii	Sergipe	September 2010	Leaves (0.04%)	HD	δ-cadinene, (*E*)-caryophyllene, α-copaene, bicyclogermacrene and germacrene D; T: 98.7%	[53]
A. squamosa L.	Sergipe	September 2012	Leaves	HD	(*E*)-Caryophyllene, germacrene D and bicyclogermacrene; M: 2.0%; S: 65.1%; T: 99.1%	[44]
A. sylvatica A. St.-Hil Anelise	Mato Grosso do Sul	September 2010	Leaves (0.17%)	HD	Hinesol, (*Z*)-caryophyllene, β-malien, γ-gurjunene; T: 98.97%	[45]
A. vepretorum Mart.	Sergipe	April 2012	Leaves (0.59%)	HD	α-Phellandrene, *o*-cymene, (*E*)-β-ocimene, bicyclogermacrene and spathulenol; M: 30.18%, S: 67.41%, T: 97.59%	[54]
A. vepretorum	Pernambuco	January 2012	Leaves (0.09%)	HD	α-Pinene, limonene, spathulenol and caryophyllene oxide; T: 93.9%	[3]
A. vepretorum	Pernambuco	April 2015	Leaves (NI)	HD	Limonene, (*E*)-β-ocimene, germacrene D and bicyclogermacrene	[55]
A. vepretorum	Sergipe	April 2010	Leaves (NI)	HD	Bicyclogermacrene, spathulenol, α-phellandrene, α-pinene, (*E*)-β-ocimene, germacrene D and *p*-cymene; M: 29.2%, S: 68.9%; T: 98.1%	[56]
A. vepretorum	Sergipe	March 2012	Leaves (0.76%)	HD	Bicyclogermacrene, spathulenol and α-phellandrene; M: 34.0%; S: 65.1%; T: 99.1%	[44]
Bocageopsis multiflora (Mart.) R.E. Fr.	Amazonas	June 2013	Leaves (0.34%)	HD	α-*trans*-Bergamotene, β-bisabolene, spathulenol and β-copaen-4-α-ol; HM: 0.3%, OM: 1.0%, SH: 34.3%, OS: 49.5%, T: 95.0%	[49]
B. multiflora	Rondônia	July 2018	Aerial parts (0.12%)	HD	*cis*-Linalool oxide (furanoid) and 1-*epi*-cubenol	[57]
B. pleiosperma Maas	Amazonas	NI	Leaves (0.28%)	HD	(*E*)-α-Bergamotene, (*E*)-β-farnesene and β-bisabolene; T: 87.64%	[58]
B. pleiosperma	Amazonas	NI	Barks (0.27%)	HD	β-Selinene, α-selinene, β-bisabolene and δ-cadinene; T: 97.11%	[58]
B. pleiosperma	Amazonas	NI	Twigs (0.25%)	HD	β-Bisabolene, (2*Z*,6*Z*)-farnesol and cryptomerone; T: 72.64%	[58]
Cardiopetalum calophyllum (Schltdl.)	Goiás	September 2014	Flowers (NI)	HD	(*E*)-Caryophyllene, germacrene D and germacrene B; M: 0.51%, S: 70.11%	[59]
C. calophyllum	Goiás	December 2014	Fruits (NI)	HD	Germacrene D, germacrene B and spathulenol; M: 0.55%, S: 73.29%	[59]
C. calophyllum	Goiás	March 2014	Leaves (NI)	HD	Spathulenol, viridiflorol, (–)-isolongifolol acetate, and (*Z*,*E*)-farnesol; M: 0.43%, S: 66.04%	[59]

Table 2. *Cont.*

Species	Collection Place	Collection Date	Part of Plant (Yield)	Extraction Technique	Majority Constituents (% >5); Substance Classes; Total	Reference
Duguetia furfuracea (A. St. -Hil.) Saff.	Minas Gerais	August 2016	Stem bark (0.5%)	SD	Cyperene, α-gurjunene, bicyclogermacrene, 2,4,5-trimethoxystyrene and (*E*)-asarone; T: 99.5%	[13]
D. furfuracea	Minas Gerais	August 2016	Leaves (0.8%)	HD	Spathulenol and bicyclogermacrene	[60]
D. furfuracea	Minas Gerais	August 2016	Underground parts (wood) (0.7%)	HD	(*E*)-Asarone, cyperene, 2,4,5-trimethoxystyrene, bicyclogermacrene and α-gurjunene	[60]
D. furfuracea	Minas Gerais	August 2016	Underground parts (trunk) (0.9%)	HD	(*E*)-Asarone and 2,4,5-trimethoxystyrene	[60]
D. lanceolata St. Hil.	Minas Gerais	April 2012	Twigs (0.4%)	HD	β-Elemene, β-caryophyllene, β-selinene, δ-cadinene, caryophyllene oxide, humulene II epoxide, β-eudesmol and cadina-1,4-dien-3-ol; HM: 4.0%, OM: 3.8%, SH: 40.0%, OS: 44.9%; T: 92.9%	[61]
D. lanceolata	Minas Gerais	NI	Leaves (0.4%)	HD	α-Selinene, aristolochene, (*E*)-caryophyllene and (*E*)-calamenene	[60]
D. lanceolata	São Paulo	March 2012	Leaves (0.3%)	HD	*trans*-Muurola-4(14),5-diene, β-bisabolene, 3,4,5-trimethoxy-styrene and 2,4,5-trimethoxy-styrene	[62]
D. lanceolata	Minas Gerais	NI	Barks (0.5%)	HD	β-elemene, caryophyllene oxide and β-selinene; HM: 1.6%, OM: 5.9%, SH: 31.9%, OS: 59.8%, H: 0.4%; T: 99.6%	[38]
D. quitarensis Benth.	Rondônia	June 2018	Aerial parts (0.11%)	HD	4-Heptanol, α-thujene, α-copaene, (*E*)-caryophyllene and germacrene D; M: 21.2%, OM: 2.5%, S: 37.8%, OS: 1.4%; T: 97.3%	[57]
D. gardneriana Mart.	Sergipe	November 2013	Leaves (0.13%)	HD	β-Bisabolene and elemicin; S: 96.0%; T: 96.0%	[11]
Ephedranthus amazonicus R.E. Fr	Amazonas	September 2012	Leaves (0.2%)	HD	Cyclosativene, α-muurolene, spathulenol, caryophyllene oxide and humulene epoxide II; OM: 0.6%, SH: 20.8%, OS: 74.2%; T: 98.0%	[49]
Fusaea longifolia Saff	Rondônia	July 2018	Aerial parts (0.18%)	HD	(*E*)-Caryophyllene, β-selinene, *cis*-β-guayene and (*Z*)-α-bisabolene; M: 0.1%, S: 85.6%, OS: 2.0%; T: 88.5%	[57]
Guatteria australis A. ST.-HIL.	Rio de Janeiro	February 2011	Aerial parts (0.1%)	HD	β-Pinene, *trans*-pinocarveol, *trans*-verbenol, myrtenol, spathulenol and caryophyllene oxide; M: 14.45%, OM: 27.47%, S: 0.76%, OS: 51.89%; T: 94.26%	[63]
G. australis	São Paulo	NI	Leaves (0.16%)	HD	(*E*)-Caryophyllene, germacrene D and germacrene B; M: 17.24%, S: 79.40%; T: 96.64%	[64]
G. blepharophylla Mart.	Amazonas	September 2012	Leaves (0.16%)	HD	Palustrol, spathulenol and caryophyllene oxide; SH: 6.4%, OS: 88.0%; O: 4.6%; T: 99.0%	[49]
G. blepharophylla	Amazonas	January 2008	Leaves (0.3%)	HD	Caryophyllene oxide; M: 0.1%, S: 91.2%; T: 91.3%	[65]
G. elliptica R. E. Fries	São Paulo	NI	Leaves (0.11%)	HD	Spathulenol and caryophyllene oxide; SH: 0.5%, OS: 99.5%; T: 100.0%	[9]
G. elliptica	São Paulo	NI	Leaves (0.21%)	HD	Spathulenol, caryophyllene oxide and β-copaen-α-ol; SH: 9.5%, OS: 91.5%, O: 0.5%; T: 100.0%	[9]
G. friesiana (W.A.Rodrigues) Erkens & Maas	Amazonas	NI	Leaves (1.17%)	HD	γ-Eudesmol, β-eudesmol and α-eudesmol; S: 93.0%; T: 93.0%	[66]
G. friesiana	Amazonas	January 2008	Leaves (0.6%)	HD	β-Eudesmol, γ-eudesmol e α-eudesmol; S: 98.2%; T: 98.2%	[65]
G. hispida (R.E. Fries)	Amazonas	July 2008	Leaves (0.5%)	HD	(*E*)-Caryophyllene; M: 68.4%, S: 31.0%; T: 99.4%	[65]
G. latifolia (Mart.) R.E.Fr.	Rio de Janeiro	February 2011	Aerial parts (0.1%)	HD	Spathulenol and caryophyllene oxide; OM: 6.94%, S: 3.35%, OS: 64.46%; T: 73.24%	[63]
G. megalophylla Diels	Amazonas	September 2018	Leaves (0.12%)	HD	δ-elemene, β-elemene, γ-muurolene, bicyclogermacrene and spathulenol; M: 1.41%, S: 87.30%; T: 88.71	[67]
G. pogonopus Mart.	Sergipe	NI	Leaves (0.22%)	HD	Germacrene D, γ-amorphene and spathulenol; S: 88.4%; T: 88.4%	[66]
G. pogonopus	Sergipe	February 2012	Leaves (0.28%)	HD	α-Pinene, β-pinene, (*E*)-caryophyllene, germacrene D, bicyclogermacrene and γ-patchoulene; M: 23.13%, S: 60.44%; T: 86.19%	[68]
G. punctata (Aubl.) R. A. Howard.	Rondônia	September 2018	Aerial parts (0.39%)	HD	(*E*)-Caryophyllene, germacrene D, *cis*-β-guayene and (*E*)-nerolidol; HO: 2.8%; M: 0.6%; S: 56.8%; OS: 19.1%; T: 79.3%	[57]
G. sellowiana Schltdl	Rio de Janeiro	February 2011	Aerial parts (0.1%)	HD	(*Z*)-β-Farnesene, β-bisabolene, *cis*-α-bisabolene, spathulenol and caryophyllene oxide; OM: 5.16%; S: 6.55%; OS: 78.28%; T: 89.99%	[63]

Table 2. *Cont.*

Species	Collection Place	Collection Date	Part of Plant (Yield)	Extraction Technique	Majority Constituents (% >5); Substance Classes; Total	Reference
G. ferruginea A. St.-Hil.	Rio de Janeiro	February 2011	Aerial parts (0.1%)	HD	*trans*-Pinocarveol, myrtenol, (*E,E*)-α-farnesene, spathulenol and caryophyllene oxide; M: 1.47%; OM: 24.54%; S: 1.91%; OS: 60.41%; T: 88.33%	[63]
Onychopetalum amazonicum R.E.Fr.	Amazonas	March 2015	Leaves (0.18%)	HD	α-Copaene, (*E*)-caryophyllene, bicyclogermacrene, δ-cadinene, spathulenol and caryophyllene oxide; SH: 60.7%, OS: 27.1%; T: 87.8%	[69]
O. amazonicum	Amazonas	March 2015	Trunk bark (0.37%)	HD	α-Gurjunene, *allo*-aromadendrene and α-*epi*-cadinol; SH: 56.9%; OS: 35.3%; T: 92.2%	[69]
O. amazonicum	Amazonas	March 2015	Twigs (0.34%)	HD	α-Gurjunene, α-*epi*-cadinol and cyperotundone; SH: 27.5%; OS: 47.5%; T: 75.0%	[69]
O. periquino (Rusby) D.M. Johnson & N.A. Murray	Acre	March 2017	Leaves (0.24%)	HD	β-elemene, β-selinene and spathulenol; SH: 78.86%; OS: 12.45%; T: 91.31%	[70]
Porcelia macrocarpa R.E. Fries	São Paulo	NI	Leaves	HD	Germacrene D, bicyclogermacrene and phytol; M: 0.39%; S: 76.0%; D: 7.3%; T: 84.0%	[71]
P. macrocarpa	São Paulo	November 2011	Fruits	HD	Neril, geranil formate, γ-muurolene, δ-cadinene, dendrolasin, hexacosane; M: 44.8%; S: 37.1%; D: 0.51%; HC: 10.49%; O: 6.7%; T: 99.6%	[71]
Unonopsis guatterioides (A.DC.) R.E.Fr	Mato Grosso do Sul	March 2005	Leaves (0.15%)	HD	α-Copaene, β-elemene, (*E*)-caryophyllene, α-humulene, *allo*-aromadendrene, germacrene D, bicyclogermacrene and spathulenol	[72]
Xylopia aromatica (Lam.) Mart.	Amazonas	September 2012	Leaves (0.25%)	HD	*trans*-Pinocarveol, α-campholenal, camphor, dihydrocarveol, verbenone and spathulenol; HM: 2.2%; OM: 52.3%; SH: 14.6%; OS: 29.5%; T: 98.6%	[49]
X. aromatica	Goiás	February 2015	Leaves (0.1%)	HD	Bicyclogermacrene, spathulenol, globulol, *cis*-guaia-3,9-dien-11-ol and khusinol; OM: 2.74%, SH: 9.62%, OS: 71.25%, D: 1.2%, O: 13.15%; T: 97.96%	[73]
X. aromatica	Goiás	October 2014	Flowers (0.2%)	HD	Bicyclogermacrene, 7-*epi*-α-eudesmol, khusinol, pentadecan-2-one and *n*-tricosane; OM: 3.44%; SH: 17.24%; OS: 51.7%; D: 6.88%, O: 20.67%; T: 99.93%	[73]
X. frutescens Aubl.	Paraíba	April 2010	Leaves (NI)	HD	(*E*)-Caryophyllene, γ-cadinene, β-ocimene and cadin-4-en-10-ol; T: 90.20%	[74]
X. frutescens	Sergipe	April 2013	Leaves (NI)	HD	(*E*)-β-Ocimene, β-elemene, (*E*)-caryophyllene, germacrene D and bicyclogermacrene	[75]
X. frutescens	Sergipe	July 2011	Leaves (1.0%)	HD	(*E*)-Caryophyllene, bicyclogermacrene, germacrene D, δ-cadinene, viridiflorene and α-copaene; M: 0.41%; S: 96.10%; T: 96.51%	[33]
X. laevigata (Mart.) R. E. Fries	Sergipe	NI	Leaves (1.4%)	HD	Germacrene D, bicyclogermacrene, (*E*)-caryophyllene and germacrene B; T: 98.68%	[76]
X. laevigata	Sergipe	November 2012	Fresh fruits (0.4%)	HD	α-Pinene, β-pinene and limonene; M: 95.0%; S: 4.6%; T: 99.6%	[77]
X. laevigata	Sergipe	April 2013	Leaves	HD	(*E*)-Caryophyllene, γ-muurolene, germacrene D, bicyclogermacrene, δ-cadinene and germacrene B	[75]
X. laevigata	Sergipe	April 2010	Leaves (>1.0%)	HD	γ-Muurolene, δ-cadinene, germacrene D, bicyclogermacrene, α-copaene and (*E*)-caryophyllene; M: 2.14%, S: 95.35%; T: 97.49%	[34]
X. laevigata	Sergipe	March 2010	Leaves (1.58%)	HD	Germacrene D, bicyclogermacrene and (*E*)-caryophyllene; M: 1.15%, S: 98.60%; T: 99.75%	[34]
X. laevigata	Sergipe	July 2010	Leaves (>1.0%)	HD	Germacrene D, bicyclogermacrene, (*E*)-caryophyllene and germacrene B; M: 7.28%, S: 91.18%; D: 0.22 T: 98.68%	[34]
X. langsdorffiana St.-Hil. & Tul.	Paraíba	July 2012	Fresh fruits (0.03%)	HD	α-Pinene, camphene, D-limonene, caryophyllene oxide and esclarene; T: 100.0%	[12]
X. sericea A. St.-Hil.	Minas Gerais	September 2011	Fruits (0.93%)	HD	Germacrene D, spathulenol and guaiol; M: 9.65%; S: 81.5%; D: 7.79%; O: 0.1%; T: 99.04%	[50]
X. sericea	Minas Gerais	July 2012	Leaves (0.5%)	HD	α-Pinene, β-pinene, *o*-cymene and D-limonene	[78]

SD: steam distillation; HD: hydrodistillation; HC: hydrocarbons; D: diterpenes; M: monoterpenes (hydrocarbons and oxygenates); S: sesquiterpenes (hydrocarbons and oxygenates); HM: hydrocarbon monoterpenes; OM: oxygenated monoterpenes; SH: sesquiterpene hydrocarbons; OS: oxygenated sesquiterpenes; O: other class; NI: not informed; T: total identified compounds.

Most studies carried out with EOs of Annonaceae occurring in Brazil, published between the years 2011 and 2021, were conducted with species belonging to the genera *Annona*, *Guatteria*, and *Xylopia* (Figure 1). Collections were mostly carried out in the states of Amazonas and Sergipe (Figure 2). To date, about 100 volatile chemical constituents (>5%) have been obtained from the EOs of Annonaceae species collected in Brazil. Among these compounds, α-pinene, β-pinene, limonene, (*E*)-caryophyllene, bicyclogermacrene, caryophyllene oxide, germacrene D, spathulenol, and β-elemene are the most abundant (Figure 3).

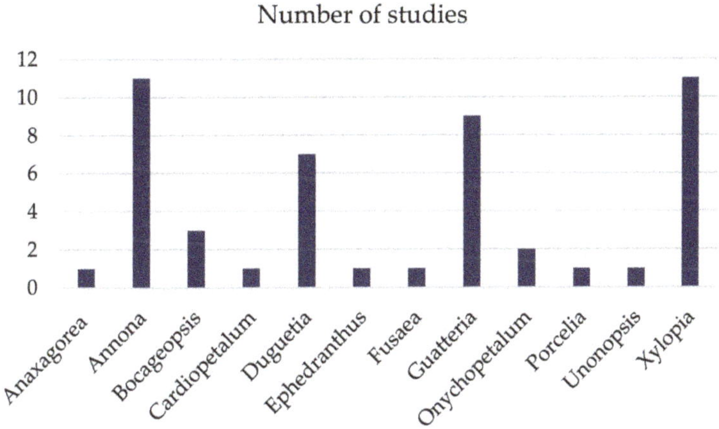

Figure 1. Distribution of studies with essential oils according to the genus of Annonaceae occurring in Brazil from 2011 to 2021.

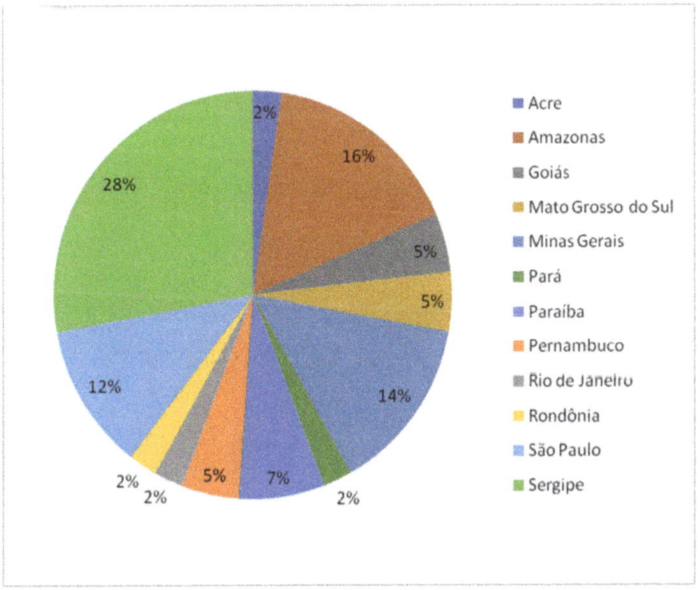

Figure 2. Percentage of studies conducted with essential oils from Annonaceae species collected in Brazil between the years 2011 and 2021.

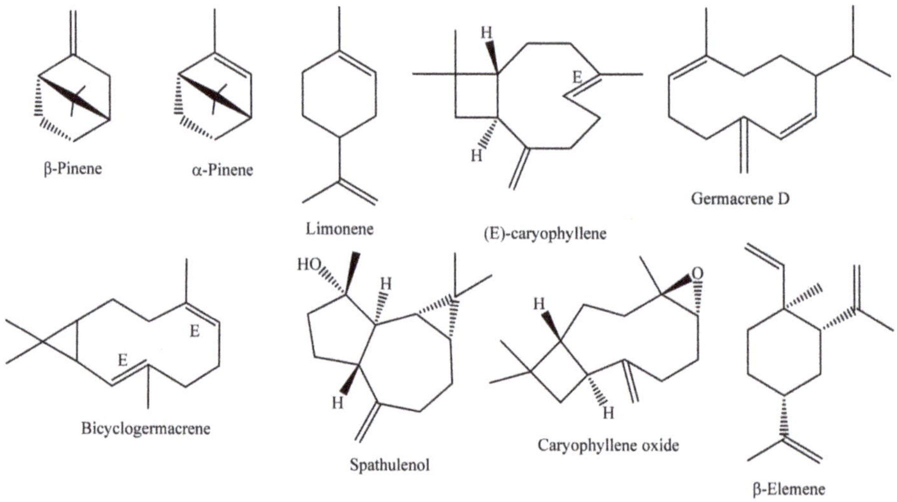

Figure 3. Structure of the major chemical constituents identified in the essential oils of Annonaceae species occurring in Brazil.

5. Biological Activities

It is generally accepted that chemical composition determines the bioactivities of EOs. Annonaceae species have been widely used in folk medicine. Their EOs have been evaluated for several effects, including anti-inflammatory, antitumor, antibacterial, and antioxidant effects [49,61].

A total of 60 studies involving the biological activities of EOs from Annonaceae species collected in the Brazilian territory between the years 2011 and 2021 are described in this work. The bioactivities reported for Annonaceae EOs are represented in Figure 4. Several EOs presented more than one reported biological activity, with the most frequent studies being related to antiproliferative or cytotoxic activities, representing 28% of the results listed here.

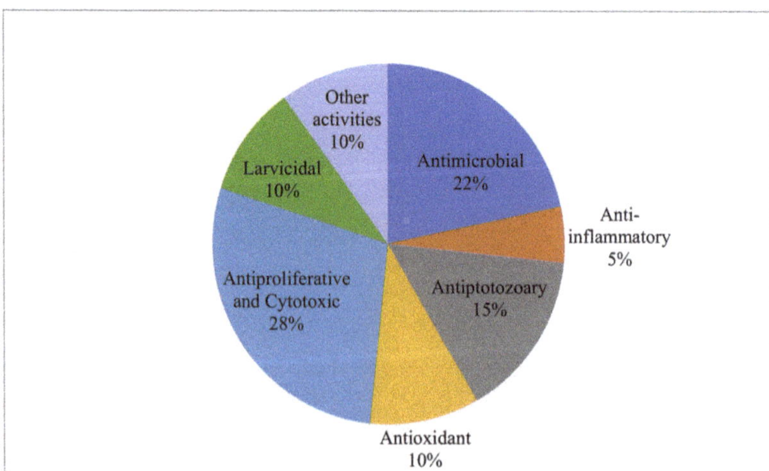

Figure 4. Distribution of studies on the biological activities of essential oils from Annonaceae species occurring in Brazil.

5.1. Antimicrobian Activity

Annonaceae species are an important source of new antimicrobial agents for combating resistant microorganisms; several EOs of this family had their antimicrobial properties evaluated and showed potentially relevant results.

The EOs of *Bocageopsis multiflora*, *Duguetia quitarensis*, *Fusaea longifolia*, and *Guatteria punctata* were evaluated to determine their antibacterial activity [57]. The EO of *B. multiflora*, which is rich in *cis*-linalool oxide (furanoid) (33.1%) and 1-*epi*-cubenol (16.6%), showed antibacterial activity against Gram-negative and Gram-positive strains, with MIC values of 4.68 $\mu g \cdot mL^{-1}$. The EO of *D. quitarensis*, which is mainly composed of 4-heptanol (33.8%), α-thujene (18.4%), (*E*)-caryophyllene (14.4%), germacerne D (6.3%), and α-copaene (5.3%), was found to be active against the Gram-positive microorganisms *Streptococcus mutans* and *Streptococcus pyogenes*, with an MIC value of 37.5 $\mu g \cdot mL^{-1}$. The EO of *F. longifolia*, which is rich in β-selinene (19.3%), *cis*-β-guaiene (18.3%), (*Z*)-α-bisabolene (12.0%), and (*E*)-caryophyllene (7.1%), was found to be active against *Pseudomonas aeruginosa*, *Streptococcus mutans*, and *Staphylococcus aureus* and resistant to methicillin, with an MIC value of 37.5 $\mu g \cdot mL^{-1}$. The EO of *Guatteria punctata*, with a high content of germacrene D (19.8%), (*E*)-nerolidol (9.9%), (*E*)-caryophyllene (8.4%), and *cis*-β-guaiene (5.5%), was found to be active against *S. mutans* and *S. pyogenes*, with an MIC value of 4.68 $\mu g \cdot mL^{-1}$ [57].

The EO from the leaves of *Anaxagorea brevipes*, composed mainly of β-eudesmol (13.16%), α-eudesmol (13.05%), γ-eudesmol (7.54%), and guaiol (5.12%), showed an antibacterial and antifungal inhibitory effect against *Kocuria rhizophila*, penicillinase-negative *Staphylococcus aureus*, *Candida albicans*, and *Candida parapsilosis*, with MIC values ranging from 25.0 to 100.0 $\mu g \cdot mL^{-1}$ [51].

The EOs of *Xylopia aromatica*, which is rich in spathulenol (21.5%), dihydrocarveol (11.6%), and *trans*-pinocarveol (10.2%), as well as *Guatteria blepharophylla*, which is rich in caryophyllene oxide (55.7%), spathulenol (8.9%), and palustrol (6.5%), showed strong activity against the Gram-positive bacteria *Streptococcus sanguinis* (MIC = 0.02 $mg.mL^{-1}$) [49].

The EOs from the leaves, branches, and bark of the trunk of *Onychopetalum amazonicum* were evaluated to determine their antimicrobial activity against four bacterial strains and five pathogenic fungi. The EO from the trunk bark exhibited activity against *Staphylococcus epidermidis*, *E. coli*, and *Kocuria rhizophila*, with an MIC value of 62.5 $\mu g \cdot mL^{-1}$. The observed activity may be associated with the presence of the sesquiterpene *allo*-aromadendreno (21.2%) [69].

The antibacterial activity of the EO of *Xylopia sericea* fruits was investigated and the results showed that this EO, which has a high content of the sesquiterpenes spathulenol (16.42%), guaiol (13.93%), and germacrene D (8.11%), has bacteriostatic effects against *S. aureus* (MIC = 7.8 $\mu g \cdot mL^{-1}$), *Enterobacter cloacae* (MIC = 7.8 $\mu g \cdot mL^{-1}$), *Bacillus cereus* (MIC = 15.6 $\mu g \cdot mL^{-1}$), and *Klebsiella pneumoniae* (MIC = 62.5 $\mu g \cdot mL^{-1}$) [50].

The antimicrobial activity of EOs from *Xylopia aromatica* flowers and leaves was tested against Gram-positive and Gram-negative bacterial strains and fungi. The EO of the flower, which is rich in pentadecan-2-one (16.38%), bicyclogermacrene (9.74%), 7-*epi*-α-eudesmol (7.76%), khusinol (7.23%), *n*-tricosane (6.17%), and heptadecan-2-one (5.83%), and the EO of the leaf, which is rich in spathulenol (27.11%), khusinol (13.04%), bicyclogermacrene (8.52%), globulol (6.47%), and *cis*-guaia-3,9-dien-11-ol (5.98%), exhibited a lower MIC against *S. pyogenes* (200 and 100 $\mu g \cdot mL^{-1}$, respectively) [73].

The EOs from two specimens of *Guatteria elliptica* collected in Paranapiacaba and Caraguatatuba (São Paulo), which have high levels of spathulenol (53.9%) and caryophyllene oxide (40.9%), respectively, showed an inhibition of growth of less than 100% at the highest concentration tested (3 $mg.mL^{-1}$), and MIC values > 3 $mg.mL^{-1}$ against all the microorganisms tested [9].

The EOs from four *Guatteria* species (*G. australis*, *G. ferruginea*, *G. latifolia*, and *G. sellowiana*), which are rich in spathulenol (11.04–40.29%) and caryophyllene oxide (7.74–40.13%), showed a strong antibacterial activity (MIC = 0.062−0.25 $mg.mL^{-1}$) against *Rhodococcus equi* strains [63].

The EO from the leaves of *G. australis*, which is rich in germacrene B (50.6%), germacrene D (22.2%), and (*E*)-caryophyllene (8.9%), had little effect against *S. aureus* and *E. coli* (MIC = 250 μg·mL^{-1}) [64].

The antimicrobial activities of EOs from the leaves, branches, and bark of *Bocageopsis pleiosperma* were evaluated. The EOs obtained from the bark had a moderate effect against *Staphylococcus epidermidis* (MIC = 250 μg·mL^{-1}), while the other EOs did not show antimicrobial activity [58].

The EO from the leaves of *Annona vepretorum*, which is rich in bicyclogermacrene (43.7%), spathulenol (11.4%), α-phelandrene (10.0%), α-pinene (7.1%), (*E*)-β-ocimene (6.8%), germacrene D (5.8%), and *p*-cymene (4.2%), exhibited significant antimicrobial activity against *S. aureus*, *S. epidermidis*, and *Candida tropicalis*, with MIC values below 1000 μg·mL^{-1} [56].

The antimicrobial activities of EO oils from the leaves of *Annona pickelli*, which are rich in bicyclogermacrene (45.4%), (*E*)-caryophyllene (14.6%), and α-copaene (10.6%), and *Annona salzmannii*, with high contents of bicyclogermacrene (20.3%), (*E*)-caryophyllene (19.9%), δ-cadinene (15.3%), α-copaene (10.0%), and *allo*-aromadendreno (5.7%), were evaluated and the results obtained showed that the EO of *A. salzmannii* was more effective, exhibiting significant antimicrobial activity against most of the microorganisms tested [43].

The EO of *D. lanceolata*, which is rich in β-elemene (12.7%), caryophyllene oxide (12.4%), and β-selinene (8.4%), inhibited the growth of *Staphylococcus aureus*, *Streptococcus pyogenes*, *Escherichia coli*, and *Candida albicans*, with MIC values of 60, 20, and 60 μg·mL^{-1}, respectively [38].

5.2. Anti-Inflammatory Activity

Many species of Annonaceae have been used to treat inflammatory diseases in folk medicine. Pharmacological studies have shown that some terpenoids and EOs from this family have significant anti-inflammatory effects, such as caryophyllene oxide and the EO of *Duguetia lanceolata*. The EO from the branches of *D. lanceolata*, which is rich in β-elemene (8.3%), β-caryophyllene (6.2%), caryophyllene oxide (7.7%), β-eudesmol (7.2%), β-selinene (7.1%), and δ-cadinene (5.5%), played a crucial role as a protective factor against carrageenan-induced acute inflammation [61].

The EO from the bark of the underground stem of *Duguetia furfuracea*, which is rich in (*E*)-asarone (21.9%), bicyclogermacrene (16.7%), 2,4,5-trimethoxystyrene (16.1%), α-gurjunene (15.0%), and cyperene (7.8%), was shown to have anti-inflammatory effects [13].

The EO from the leaves of *Annona sylvatica*, which are composed mainly of hinesol (8.16%), (Z)-caryophyllene (7.31%), β-maliene (6.61%), and γ-gurjunene (5.46%), showed anti-inflammatory activity against the persistent inflammation induced by CFA (Complete Freund's Adjuvant) [45].

5.3. Antileishmanial Activity

The EO from the leaves of *Guatteria australis*, which has a high concentration of germacrene B (50.6%), germacrene D (22.2%), and (*E*)-caryophyllene (8.9%), presented anti-leishmania activity against *Leishmania infantum* (IC_{50} = 30.7 μg·mL^{-1}) [64].

The EO of *Annona coriacea*, which has a high percentage of bicyclogermacrene (39.8%), presented antileishmania activity against the promastigote forms of four species of *Leishmania*, being more active against *L. chagasi* (IC_{50} = 39.93 μg·mL^{-1}) [39].

5.4. Antioxidant Activity

Antioxidants are widely used in the food industry for a variety of reasons, including preventing oxidation; neutralizing free radicals; preserving food; and enhancing flavor, aroma, or color. As some synthetic antioxidants exhibit carcinogenic effects and can be toxic to nature, researchers have intensified the search for natural antioxidants [79]. In several studies with EOs, the antioxidant activity is related to compounds such as thymol, carvacrol, α-terpinene, β-terpinene, β-terpinolene, 1,8-cineol, eugenol, and linalool, which have an antioxidant activity similar to that of α-tocopherol [80].

The EOs from the leaves of two specimens of *Guatteria elliptica*, which were collected in Paranapiacaba and Caraguatatuba, showed a low antioxidant potential (EC_{50} = 7.24 and 28.68 $mg.mL^{-1}$ using DPPH assays) for the EOs from Paranapiacaba and Caraguatatuba, respectively [9]. The difference in EC_{50} values can be attributed, at least in part, to the different contents of the main compounds present in EOs. Natural products such as EOs are formed by a complex mixture of organic compounds that act synergistically, increasing biological or even antagonistic activity and thus reducing the verified activity [80,81].

The EO of *Xylopia sericea* was investigated for its antioxidant potential using different methods. The EO of the fruit is rich in spathulenol (16.42%), guaiol (13.93%), and germacrene D (8.11%), and presented significant antioxidant activity through the DPPH (2,2-diphenyl-1-picryl-hydrazyl) methods (IC_{50} 49.1 $\mu g \cdot mL^{-1}$), β-carotene/linoleic acid bleaching (IC_{50} 6.9 $\mu g \cdot mL^{-1}$), TAC (Total Antioxidant Capacity) (IC_{50} 78.2 $\mu g \cdot mL^{-1}$), and TBARS (Thiobarbituric Acid Reactive Substances) (80.0 $\mu g \cdot mL^{-1}$) [50].

The EO from *Duguetia lanceolata* branches showed a high content of β-elemene (8.3%), β-caryophyllene (6.2%), caryophyllene oxide (7.7%), β-eudesmol (7.2%), β-selinene (7.1%), and δ-cadinene (5.5%). Antioxidant effects gained using the DPPH assay (EC_{50} 159.4 $\mu g \cdot mL^{-1}$), Fe^{+3} reduction (EC_{50} 187.8 $\mu g \cdot mL^{-1}$), and the inhibition of lipid peroxidation (41.5%) were considered significant [61].

The antioxidant potential of the EO of *Guatteria australis* leaves, which are rich in germacrene B (50.6%), germacrene D (22.2%), and (*E*)-caryophyllene (8.9%), was evaluated using two methods. Antioxidant capacity was considered either medium (TLC/DPPH, light yellow spot) or small (ORAC assay, 457 $\mu molTE.g^{-1}$) [64].

The EO from the leaves of *Annona vepretorum*, which are rich in bicyclogermacrene (43.7%), spathulenol (11.4%), α-phelandrene (10.0%), α-pinene (7.1%), (*E*)-β-ocimene (6.8%), germacrene D (5.8%), and *p*-cymene (4.2%), was able to capture radicals, but the antioxidant activity was considered weak. In the kinetic method of the ORAC assay, the result obtained was 204.24 $\mu molTE.g^{-1}$, while the TLC produced a yellow spot where the EO was applied due to the DPPH reduction [56].

The Eos from the leaves of *Annona pickelli*, which are rich in bicyclogermacrene (45.4%), (*E*)-caryophyllene (14.6%), and α-copaene (10.6%), as well as *Annona salzmannii*, which have high contents of bicyclogermacrene (20.3%), (*E*)-caryophyllene (19.9%), δ-cadinene (15.3%), α-copaene (10.0%), and *allo*-aromadendrene (5.7%), showed significant antioxidant capacity in the ORAC and DPPH assays [43].

5.5. Antiproliferative and Cytotoxic Activities

The search for new drugs that show activity against different types of cancer has become one of the most interesting subjects to research in the area of natural products. As a result, several EOs from Annonaceae species and their bioactive constituents were evaluated to determine their antiproliferative and cytotoxic properties.

The cytotoxic, mutagenic, and genotoxic profiles of the EO from *Xylopia laevigata* leaves were investigated. The results showed that the EO, which is rich in germacrene D (43.6%), bicyclogermacrene (14.6%), (*E*)-caryophyllene (7.9%), and germacrene B (7.3%), has mutagenic and antiproliferative activities, which can be related to the cytotoxic effect of the main components of the EO [76].

The in vitro cytotoxicity of *Annona vepretorum* EO (pure, microencapsulated with β-cyclodextrin and some of its main constituents) on tumor cell lines of different histotypes was evaluated. Furthermore, the in vivo efficacy of this EO in mice has been described. The results showed that the sesquiterpene spathulenol and EO, which have a high concentration of bicyclogermacrene (35.71%), spathulenol (18.89%), (*E*)-β-ocimene (12.46%), α-phellandrene (8.08%), and *o*-cymene (6.24%), exhibited promising cytotoxicity. The tumor growth in vivo was inhibited by EO treatment (34.46% inhibition) and EO microencapsulation was found to increase tumor growth inhibition (62.66% inhibition) [54].

The antitumor activity and toxicity of the EO of *Annona leptopetala* leaves, which are rich in spathulenol (12.5%) and α-limonene (9.0%), were evaluated. The EO showed antitumor activity in vitro and in vivo, mainly in the leukemia cell line, without major changes seen in the toxicity parameters evaluated [42].

The in vitro cytotoxic activity of the EO from fresh fruits of *Xylopia laevigata* and its main constituents (limonene, α-pinene, and β-pinene) was evaluated against four tumor cell lines (mouse melanoma, human hepatocellular carcinoma, human promyelocytic leukemia, and chronic myelocytic leukemia) and non-tumor cells (human peripheral blood mononuclear cells). Neither the EO nor its major constituents showed cytotoxic activity (IC_{50} > 25.0 μg·mL^{-1}) [77].

The in vitro and in vivo antileukemic potential of the EO from the leaves of *Guatteria megalophylla* was investigated. The in vitro cytotoxic potential of the EO was evaluated in human cancer cell lines (HL-60, MCF-7 CAL27, HSC-3, HepG2, and HCT116) and in non-cancerous human cell lines (MRC-5). The in vivo efficacy was evaluated in C.B17 SCID mice with HL-60 cell xenografts. The results showed that this EO has anti-leukemic potential (with an IC_{50} value of 12.51 μg·mL^{-1} for HL-60 cells), and the main constituents spathulenol (27.7%), γ-muurolene (14.3%), bicyclogermacrene (10.4%), β-elemene (7.4%), and δ-elemene (5.1%) can play a central role in the registered activities [67].

The antiproliferative activity of the EO from the leaves of *Anaxagorea brevipes* was investigated in a number of cancer cell lines and the bioactivity was described against MCF-7 (breast, TGI = 12.8 μg·mL^{-1}), NCI-H460 (lung, TGI) = 13.0 μg·mL^{-1}), and PC-3 (prostate, TGI = 9.6 μg·mL^{-1}). The antiproliferative activity found was attributed to the major constituents of the EO: β-eudesmol (13.16%), α-eudesmol (13.05%), γ-eudesmol (7.54%), and guaiol (5.12%) [51].

The antitumor activity and toxicity of *Xylopia langsdorffiana* EO, which is rich in α-pinene (34.5%) and limonene (31.7%), were evaluated. The EO was found to cause in vitro and in vivo growth inhibition in tumor cells, without major changes seen in the toxicity parameters evaluated [12].

The EOs from two specimens of *Guatteria elliptica* showed important antitumor activity against breast and prostate cancer cells (IC_{50} = 7.0 and 5.5 μg·mL^{-1}, respectively) and a low cytotoxicity against normal fibroblasts (IC_{50} > 22.2 μg·mL^{-1} and IC_{10} = 18.5 μg·mL^{-1}, respectively) [9].

The EO of *Duguetia gardneriana*, which has a high content of β-bisabolene (80.9%), exhibited a cytotoxic effect. The IC_{50} values were obtained for mouse melanoma, human hepatocellular carcinoma, human promyelocytic leukemia, and human chronic myelocytic leukemia cell lines (16.8, 19.1, 13.0 and 19.3 μg·mL^{-1}, respectively). The in vivo antitumor activity was evaluated using C57BL/6 mice inoculated subcutaneously with B16-F10 melanoma cells, revealing tumor growth inhibition rates of 5.37 and 37.52% at doses of 40 and 80 mg/kg/day, respectively [11].

The antiproliferative activity of the EOs of four *Guatteria* species (*G. australis*, *G. ferruginea*, *G. latifolia*, and *G. sellowiana*) was investigated. These EOs contained the oxygenated sesquiterpenes spathulenol (11.04–40.29%) and caryophyllene oxide (7.74–40.13%) as the main constituents. The evaluation of the antiproliferative activity showed a strong selectivity (1.1–4.1 μg·mL^{-1}) against the ovarian cancer tumor lineage, which was even more active than the positive control doxorubicin (11.7 μg·mL^{-1}) [63].

The EO from the leaves of *Guatteria australis*, which had a high concentration of germacrene B (50.6%), germacrene D (22.2%), and (*E*)-caryophyllene (8.9%), had a strong antiproliferative effect against NCI-ADR/RES (ovarian- resistant) and HT-29 (colon). The TGI (Total Growth Inhibition) values were equal to 31.0 and 32.8 μg.mL^{-1}, respectively [64].

The antiproliferative activity of the EO of *Annona sylvatica* leaves, which is rich in hinesol (8.16%), (*Z*)-caryophyllene (7.31%), β-maliene (6.61%), and γ-gurjunene (5.46%), was evaluated in vitro against nine human tumor cell lines: melanoma (UACC-62), breast (MCF-7), lung (NCI-H460), ovary (OVCAR03), prostate (PC-3), colon (HT-29), renal (786-0), resistant ovary (NCI/ADR-Res), and glioma (U251). The results demonstrate that the EO has anticancer activity, with GI_{50} values (concentrations that elicit an inhibition of 50% of the cell growth) in the range of 36.04–5.37 $\mu g \cdot mL^{-1}$, but at the highest concentration cytostatic activity and cytotoxic effects were observed for all cell lines [45].

The EO of *Annona pickelii*¸ which is mainly composed of bicyclogermacrene (38.0%), (*E*)-caryophyllene (27.8%), α-copaene (6.9%), and α-humulene (4.0%), as well as the EO of *Annona salzmannii*, which is rich in δ- cadinene (22.6%), (*E*)-caryophyllene (21.4%), α-copaene (13.3%), bicyclogermacrene (11.3%), and germacrene D (6.9%), exhibited potent antitumor activity. The most significant activity was observed against U251 (glioma, CNS), UACC-62 (melanoma), MCF-7 (breast), NCI-460 (lung), and HT-29 (colon) for the EO of *A. pickelii* e U251, 786-0 (kidney) and NCI-460 for the EO of *A. salzmannii*, all with TGI values below 50 $\mu g \cdot mL^{-1}$ [53].

Xylopia laevigata EO has significant anticancer potential in vitro and in vivo. The cytotoxic effects of EOs from the leaves of three specimens of *X. laevigata* were evaluated against different tumor lines: OVCAR-8 (ovarian carcinoma), SF-295 (GLIOBLASTOMA), HCT-116 (colon carcinoma), HL-60 (promyelocytic leukemia), and PBMC (peripheral lymphoblast). In the in vitro cytotoxic study, different EO samples with similar chemical profiles (γ-muurolene, δ-cadinene, germacrene B, α-copaene, germacrene D, bicyclogermacrene, and (*E*)-caryophyllene) showed cytotoxicity to all the tumor lines tested. In the in vivo antitumor study, the tumor growth inhibition rates were 37.3–42.5% [29].

The EO from *Xylopia sericea* leaves, characterized by α-pinene, β-pinene, *o*-cymene, and D-limonene, showed a low cytotoxicity to HepG2 cells (human hepatocellular carcinoma) (CC50 275.9 $\mu g \cdot mL^{-1}$) [78].

The EO from *Xylopia frutescens* leaves, which are rich in €-caryophyllene (31.48%), bicyclogermacrene (15.13%), germacrene D (9.66%), δ-cadinene (5.44%), viridiflorene (5.09%), and α- copaene (4.35%) showed cytotoxicity against the tumor cell lines NCI-H358M (bronchoalveolar carcinoma of the lung) and PC-3M (metastatic prostate carcinoma), with IC_{50} values ranging from 24.6 to 40.0 $\mu g \cdot mL^{-1}$, respectively. In the in vivo antitumor study, the tumor growth inhibition rates were 31.0–37.5% [33].

The antiproliferative activity of the EO from *Cardiopetalum calophyllum* leaves, which is mainly made up of spathulenol (28.78%), viridiflorol (9.99%), and (*Z,E*)-farnesol (6.51%), was evaluated in different human tumor cell lines: adenocarcinoma of the breast (MCF-7), cervical adenocarcinoma (HeLa), and glioblastoma (M059J), in addition to a normal human cell line (GM07492A, pulmonary fibroblasts). The IC_{50} values ranged from 216.8 to 353.51 $\mu g \cdot mL^{-1}$ and selectivity was not observed [82].

5.6. Larvicidal Activity

The larvicidal effect of EOs from Annonaceae species was tested against several disease vectors. The EOs of two species of *Duguetia* were evaluated against the larvae of *Artemia salina* and *Culex quinquefasciatus*. Essential oils from the leaf, wood, and bark of the underground stem of *D. furfuracea* showed potent activity against *A. salina* larvae (LC_{50} 6.01, 7.79 and 9.98 $\mu g \cdot mL^{-1}$, respectively). The main constituents were spathulenol (47.2%), bicyclogermacrene (26.4%), and caryophyllene oxide (5.2%) in the EO of the leaf, (*E*)-asarone (21.9%), bicyclogermacrene (16.7%), 2,4,5-trimethoxystyrene (16.1%), α-gurjunene (15.0%), and cyperene (7.8%) in the underground stem bark EO, as well as (*E*)-asarone (16.6%), cyperene (15.7%), spathulenol (14.2%), 2,4,5-trimethoxystyrene (13.2%), bicyclogermacrene (8.6%), and α-gurjunene (8.1%) in the wood EO. The EO of *D. lanceolata* leaves, which is rich in α-selinene (11.0%), aristolochene (5.8%), (*E*)-caryophyllene (5.3%), and (*E*)-calamenene (5.2%), also showed potent activity against *A. salina* larvae (LC_{50} 0.89 $\mu g \cdot mL^{-1}$). The EOs of both species were moderately active against *C. quinquefasciatus*, as they exhibited LC_{50} values ranging from 57.8 to 121.7 $\mu g \cdot mL^{-1}$ [60].

The EO of *Onychopetalum periquino*, which has a high concentration of β-elemene (53.16%), spathulenol (11.94%), and β-selinene (9.25%), showed a high larvicidal activity against *Aedes aegypti* larvae, with an LC_{50} of 63.75 $\mu g \cdot mL^{-1}$ reaching 100% mortality at 200 $\mu g \cdot mL^{-1}$ [70].

The EOs of *Xylopia laevigata*, which are rich in germacrene D (27.0%), bicyclogermacrene (12.8%), (*E*)-caryophyllene (8.6%), γ-muurolene (8.6%), and δ-cadinene (6.8%), and of *Xylopia frutescens*, which has high levels of bicyclogermacrene (23.2%), germacrene D (21.2%), (*E*)-caryophyllene (17.4%), β-elemene (6.3%), and (*E*)-β-ocimene (5.2%), did not show larvicidal activity [75].

The larvicidal activity of EOs from *Annona pickelli* leaves, which are rich in bicyclogermacrene (45.4%), (*E*)-caryophyllene (14.6%), and α-copaene (10.6%), and *Annona salzmannii*, which has high contents of bicyclogermacrene (20.3%), (*E*)-caryophyllene (19.9%), δ-cadinene (15.3%), α-copaene (10.0%), and *allo*-aromadendrene (5.7%), was determined against *Aedes aegypti* larvae. However, no larval mortality was detected at concentrations of up to 1000 $\mu g \cdot mL^{-1}$ [43].

The EO of *Duguetia lanceolata*, which is rich in β-elemene (12.7%), caryophyllene oxide (12.4%), and β-selinene (8.4%), was active against *A. salina* larvae with LC_{50} values equal to 49.0 $\mu g \cdot mL^{-1}$ and was about nine times more poisonous than the standard used thymol (LC_{50} = 457.9 $\mu g \cdot mL^{-1}$) [38].

The larvicidal activity of the EOs of *Guatteria blepharophylla*, *Guatteria friesiana*, and *Guatteria hispida* was tested against *A. aegypti* larvae; the lethal concentrations of LC_{50}, LC_{95}, and LC_{99} were, respectively, 85.74, 199.35, and 282.76 ppm for *G. hispida*; 58.72, 107.6, and 138.37 ppm for *G. blepharophylla*; and 52.6, 94.37, and 120.22 ppm for *G. friesiana*. The EO of *G. friesiana*, rich in α-, β- and γ-eudesmol, showed better insecticidal effect [65].

5.7. *Trypanocidal and Antimalarial Activities*

Chagas disease, also known as American trypanosomiasis, is caused by the protozoan parasite *Trypanosoma cruzi*. With a complex pathophysiology and dynamic epidemiological profile, this disease remains an important public health concern and is an emerging disease in non-endemic countries. For its etiological treatment in both the acute and chronic phase, there are two main drugs for the treatment of the disease: benznidazole and nifurtimox [83].

The EOs of *Bocageopsis multiflora*, *Duguetia quitarensis*, *Fusaea longifolia*, and *Guatteria punctata* were evaluated to determine their trypanocidal activity. The results showed that these EOs were active at the concentrations tested. The EO of *G. punctata* was the most active, with an IC_{50} = 0.029 $\mu g \cdot mL^{-1}$, being 34 times more active than the reference drug benznidazole. The authors reported that the strong activity observed for this species can be attributed to the presence of germacrene D (19.8%) and (*E*)-caryophyllene (8.4%) in the composition of the EO of *G. punctata* [57].

Essential oils extracted from the leaves of *Guatteria friesiana*, which have a high content of β-eudesmol (51.9%), γ-eudesmol (18.9%), and α-eudesmol (12.6%), and from the leaves of *Guatteria pogonopus*, which are rich in spathulenol (24.8%), γ-amorphene (14.7%), and germacrene D (11.8%), demonstrated potent trypanocidal and antimalarial activity with IC_{50} values below 41.3 $\mu g \cdot mL^{-1}$ [66].

The EO from the leaves of *Annona vepretorum*, which has high levels of bicyclogermacrene (43.7%), spathulenol (11.4%), α-phellandrene (10.0%), α-pinene (7.1%), (*E*)-β-ocimene (6.8%), germacrene D (5.8%), and *p*-cymene (4.2%), showed potent trypanocidal activity with an IC_{50} value equal to 31.9 $\mu g \cdot mL^{-1}$ [56].

The trypanocidal activity of EOs from *Annona pickelii*, which are rich in bicyclogermacrene (38.0%), (*E*)-caryophyllene (27.8%), α-copaene (6.9%), α-humulene (4.0%), and EOs from *Annona salzmannii*, which are rich in δ-cadinene (22.6%), (*E*)-caryophyllene (21.4%), α-copaene (13.3%), bicyclogermacrene (11.3%), and germacrene D (6.9%), were evaluated. The results showed that the *A. pickelii* EO was the most active, with an IC_{50} value of 27.2 $\mu g \cdot mL^{-1}$, while the IC_{50} value observed for *A. salzmannii* EO was 89.7 $\mu g \cdot mL^{-1}$ [53].

The EOs of *Annona squamosa*, which are rich in (*E*)-caryophyllene (27.4%), germacrene D (17.1%), and bicyclogermacrene (10.8%), and the EOs *Annona vepretorum*, which are rich in bicyclogermacrene (39.0%), spathulenol (14.0%), and α-phellandrene (11.5%), showed potent trypanocidal and antimalarial activity, with IC_{50} values below 20 $\mu g \cdot mL^{-1}$ and a strong inhibition of the proliferation of amastigote forms [44].

The EO of *Annona coriacea*, which has a high percentage of bicyclogermacrene (39.8%), showed trypanocidal activity against trypomastigote forms of *T. cruzi* (IC_{50} 168.50 μg·mL^{-1}) [39].

The antiplasmodic activity of *Xylopia sericea* EO, which is characterized by α-pinene, β-pinene, *o*-cymene, and D-limonene, was evaluated and showed a low growth inhibition (24.0 to 50.0 μg·mL^{-1}) against *Plasmodium falciparum*, a malaria-associated protozoan, in humans [78].

5.8. Other Activities

The EOs *of Xylopia laevigata* and *Xylopia frutescens* showed a low degree of protection against *Aedes aegypti* landings and, therefore, low repellent activity. The EO of *X. laevigata* had a high concentration of germacrene D (27.0%), bicyclogermacrene (12.8%), (*E*)-caryophyllene (8.6%), γ-muurolene (8.6%), and δ-cadinene (6.8%), while high levels of bicyclogermacrene (23.2%), germacrene D (21.2%), (*E*)-caryophyllene (17.4%), β-elemene (6.3%), and (*E*)-β-ocimene (5.2%) were identified in the EO of *X. frutescens* [75].

The anticonvulsant, sedative, anxiolytic, and antidepressant activities of the EO from the leaves of *Annona vepretorum*, which is rich in (*E*)-β-ocimene (42.59%), bicyclogermacrene (18.81%), germacrene D (12.19%), and limonene (10.02%), were investigated in mice. The results showed that acute treatment with the EO of this species has anxiolytic, sedative, antiepileptic, and antidepressant effects [55].

The sesquiterpene caryophyllene oxide and the EO from *Duguetia lanceolata* branches, which is rich in β-elemene (8.3%), caryophyllene oxide (7.7%), β-eudesmol (7.2%), β-selinene (7.1%), β-caryophyllene (6.2%), and δ-cadinene (5.5%), have an antinociceptive effect, as they were shown to reduce abdominal contortions in mice [61].

The insecticidal, antifungal, and antiaflatoxigenic activities of *Duguetia lanceolata* EO were evaluated in stored grain spoilage agents. The main constituents of this EO were β-bisabolene (56.2%) and 2,4,5-trimethoxystyrene (19.1%). The results suggested that the EO has promising grain protection properties against *Sitophilus zeamais* and *Zabrotes subfasciatus*, showing a comparable activity to that of a deltamethrin-based insecticide (positive control) [62].

The antinociceptive effect of a *Duguetia furfuracea* underground stem bark EO, composed mainly of (*E*)-asarone (21.9%), bicyclogermacrene (16.7%), 2,4,5-trimethoxystyrene (16.1%), α-gurjunene (15.0%), and cyperene (7.8%), was investigated. The results showed that the antinociceptive activity of this EO is possibly mediated by adenosinergic and opioidergic pathways and that its properties do not induce effects on motor coordination [13].

The EO from fresh leaves of *Unonopsis guatterioides*, which is rich in α-copaene (15.7%), bicyclogermacrene (15.7%), *trans*-caryophyllene (15.7%), α-humulene (9.0%), *allo*-aromadendreno (8.4%), and spathulenol (7.3%), showed a phytotoxic effect on the germination, growth, and development of monocotyledonous (*Allium cepa*) and dicotyledonous (*Lactuca sativa*) plants [72].

6. Methodology

In this work, a systematic review was carried out to show studies published between the years 2011 and 2021 on the chemical composition and biological properties of EOs of Annonaceae species collected in Brazil, which can serve as a reference for the future research and use of these species. In addition, a section on the ethnobotanical use of these species was also inserted in order to express their importance in traditional Brazilian medicine.

Pubmed, WOS, Scopus, and Scielo were used as virtual databases to search for the peer-reviewed articles that were used to compose the present work. The keywords used for the research were: "Annonaceae", "óleos essenciais", "essential oils", "atividades biológicas", "biological activities", "ethnobotany", and "medicinal use".

The selection of manuscripts to compose this review was based on studies published in peer-reviewed journals; in addition, a careful review was carried out to confirm whether the species studied in the published articles were of Brazilian origin, as reported at www.floradobrasil.jbrj.gov.br (accessed on 29 September 2021). The quality of the reviewed studies is well known—only peer-reviewed articles were included, and we considered only papers in the English language for gathering data regarding the chemical composition and biological properties of EOs of Annonaceae species. However, for the section on the ethnobotanical use of these species, data published in the Portuguese language were also considered. Theses, Ph.D. dissertations, and unpublished articles were not included in this review. Therefore, we focused on phytochemical and/or in vitro, in vivo, and in animal studies, with the aim of providing up-to-date information on the biological properties of EOs from Annonaceae species collected in Brazil.

According to the website CrossRef, from 2013 to 2021, journal articles (1244), components (237), chapters (55), dissertations (32), posted content (10) peer reviews (3), datasets (3), conference papers (3), monographs (1), and books (1) were used, with the year 2018 (201) having the highest number of publications. The main journals that published articles on Annonaceae were ChemInform (53); Phytotaxa (43); Natural Product Research (27); Journal of Essential Oil Research (27); Botanical Journal of the Linnean Society (27); Blumea—Biodiversity, Evolution and Biogeography of Plants (27); Biochemical Systematics and Ecology (27); Nordic Journal of Botany (26); Kew Bulletin (25); and Taxon (24).

In the science direct database, a total of 1888 papers were published, including review and research articles, chapters, and books. The main periodicals were Journal of Ethnopharmacology (270); Biochemical Systematics and Ecology (49); Forest Ecology and Management (49); South African Journal of Botany (45); Phytochemistry (38); Phytochemistry Letters (34); Review of Palaeobotany and Palynology (33); Herbal Medicine (32); Industrial Crops and Products (28); Journal of Herbal Medicine (27); Flora (26); Molecular Phylogenetics and Evolution (23); Studies in Natural Products Chemistry (23); Asian Pacific Journal of Tropical Biomedicine (23); Food Research International (22); European Journal of Medicinal Chemistry (22); Tetrahedron Letters (21); Palaeogeography, Palaeoclimatology, Palaeoecology (21); Biomedicine & Pharmacotherapy (19); Dictionary of Trees, Volume 2: South America, 2014 (18); Brazilian Journal of Pharmacognosy (18); The Alkaloids: Chemistry and Biology (17); Phytom Medicine (16); Bioorganic & Medicinal Chemistry Letters (15); and Food Chemistry (14). By analyzing the numbers of papers published in the two databases, we were able to identify the importance of the topic for the scientific community. Furthermore, this is the first report on a literature review of the Annonaceae species found in Brazil.

7. Conclusions

Studies relating to natural products are important, as they can be sources of new chemically active molecules with potential applications in diverse human activities. In the present review, we note that Brazilian Annonaceae species can be sources of bioactive compounds such as α-pinene, β-pinene, limonene, (*E*)-caryophyllene, bicyclogermacrene, caryophyllene oxide, germacrene D, spathulenol, and β-elemene, which are present in the essential oils of the plants. Furthermore, the potential use of these EOs in terms of their antimicrobial, antiproliferative, cytotoxic, larvicidal, antioxidant, anti-inflammatory activities, etc., was also described. In some cases, it was possible to observe that the biological activity reported for the essential oil (EO) was superior to that of drugs available on the market, such as the EO of the species *Guatteria punctata*, which showed a trypanocidal effect that was 34 times more active than that of the reference drug benznidazole. This and other studies demonstrate that it is necessary to expand research to the EOs of Annonaceae, especially species occurring in Brazil, since studies on these are still scarce and there is a considerable number of Annonaceae species that are unexplored in terms of their content, chemical composition, and the biological activities of their EOs. In addition, the ethnobotanical use of some plants of this family was demonstrated, and it was found that the most cited species in folk medicine belong to the *Annona* genus.

Author Contributions: Conceptualization, M.M.C., O.d.S.C., L.D.d.N., Â.A.B.d.M., M.S.d.O. and J.N.C.; methodology, M.M.C., O.d.S.C., L.D.d.N., Â.A.B.d.M., M.S.d.O. and J.N.C.; writing—original draft preparation, M.M.C.; writing—review and editing, O.d.S.C., L.D.d.N., Â.A.B.d.M., M.S.d.O. and J.N.C.; visualization, G.M.S.P.G. and E.H.d.A.A.; supervision, G.M.S.P.G. and E.H.d.A.A.; project administration, E.H.d.A.A. All authors have read and agreed to the published version of the manuscript.

Funding: Universidade Federal do Pará/Proposp/ PROGRAMA DE APOIO À PUBLICAÇÃO QUALIFICADA—PAPQ- EDITAL 06/2021.

Institutional Review Board Statement: Not applicable.

Informed Consent Statement: Not applicable.

Acknowledgments: The author Márcia Moraes Cascaes thanks CAPES for the Phd scholarship process number: [**88887.497476/2020-00**]. Ângelo Antônio Barbosa de Moraes thanks CNPq for the scientific initiation scholarship. The author Mozaniel Santana de Oliveira, thanks PCI-MCTIC/MPEG, as well as CNPq for the scholarship process number: [**302050/2021-3**]. The authors would like to thank the Universidade Federal do Pará.

Conflicts of Interest: The authors declare no conflict of interest.

References

1. Fechine, I.M.; Lima, M.A.; Navarro, V.R.; da Cunha, E.V.L.; Silva, M.S.; Barbosa-Filho, J.M.; Maia, J.G.S. Alcalóides de *Duguetia trunciflora* Maas (Annonaceae). *Rev. Bras. Farmacogn.* **2002**, *12*, 17–19. [CrossRef]
2. Tamokou, J.D.D.; Mbaveng, A.T.; Kuete, V. *Antimicrobial Activities of African Medicinal Spices and Vegetables*; Elsevier Inc.: Amsterdam, The Netherlands, 2017; ISBN 9780128094419.
3. De Souza Araújo, C.; De Oliveira, A.P.; Lima, R.N.; Alves, P.B.; Diniz, T.C.; Da Silva Almeida, J.R.G. Chemical constituents and antioxidant activity of the essential oil from leaves of *Annona vepretorum* Mart. (Annonaceae). *Pharmacogn. Mag.* **2015**, *11*, 615–618. [CrossRef]
4. Lobão, A.Q.; Lopes, J.C.; Erkens, R.H.J.; Mendes-Silva, I.; Pontes Pires, A.F.; Silva, L.V.; Oliveira, M.L.B.; Johnson, D.; Mello-Silva, R. Annonaceae in Flora do Brasil. 2020. Available online: http://reflora.jbrj.gov.br/reflora/floradobrasil/FB110760 (accessed on 9 September 2021).
5. De Lemos, E.E.P. A Produçâo de Anonáceas no Brasil. *Rev. Bras. Frutic.* **2014**, *36*, 86–93. [CrossRef]
6. Ferreira, P.; Martins, I.; Pereira, J.; Correia, A.; Sampaio, R.; Silva, M.; Costa, V.; Silva, M.; Cavalcante, F.; Silva, B. Tocolytic action of essential oil from *Annona leptopetala* R. E. Fries is mediated by oxytocin receptors and potassium channels. In Proceedings of the International Conference Series on Multidisciplinary Sciences, Puyo, Ecuador, 20 March–20 December 2019; MDPI: Basel, Switzerland, 2020; p. 6773. [CrossRef]
7. Rabelo, S.V.; Quintans, J.; de Sousa Siqueira Quintans, J.; Costa, E.V.; Guedes da Silva Almeida, J.R.; Quintans, L.J. *Annona species* (Annonaceae) oils. In *Essential Oils in Food Preservation, Flavor and Safety*; Elsevier Inc.: Amsterdam, The Netherlands, 2016; pp. 221–229, ISBN 9780124166448.
8. Rajca Ferreira, A.K.; Lourenço, F.R.; Young, M.C.M.; Lima, M.E.L.; Cordeiro, I.; Suffredini, I.B.; Lopes, P.S.; Moreno, P.R.H. Chemical composition and biological activities of *Guatteria elliptica* R. E. Fries (Annonaceae) essential oils. *J. Essent. Oil Res.* **2018**, *30*, 69–76. [CrossRef]
9. Shaaban, H.A.; El-Ghorab, A.H. Bioactivity of essential oils and their volatile aroma components: Review Bioactivity of essential oils and their volatile aroma components: Review. *J. Essent. Oil Res.* **2012**, *24*, 203–212. [CrossRef]
10. Fournier, G.; Leboeuf, M.; Cavé, A. Annonaceae essential oils: A review. *J. Essent. Oil Res.* **1999**, *11*, 131–142. [CrossRef]
11. Rodrigues, A.C.B.C.; Bomfim, L.M.; Neves, S.P.; Menezes, L.R.A.; Dias, R.B.; Soares, M.B.P.; Prata, A.P.N.; Rocha, C.A.G.; Costa, E.V.; Bezerra, D.P. Antitumor Properties of the Essential Oil from the Leaves of *Duguetia gardneriana*. *Planta Med.* **2015**, *81*, 798–803. [CrossRef]
12. Moura, A.P.G.; Beltrão, D.M.; Pita, J.C.L.R.; Xavier, A.L.; Brito, M.T.; de Sousa, T.K.G.; Batista, L.M.; de Carvalho, J.E.; Ruiz, A.L.T.G.; Della Torre, A.; et al. Essential oil from fruit of *Xylopia langsdorffiana*: Antitumour activity and toxicity. *Pharm. Biol.* **2016**, *54*, 3093–3102. [CrossRef] [PubMed]
13. Saldanha, A.A.; Vieira, L.; de Azambuja Ribeiro, R.I.M.; Thomé, R.G.; dos Santos, H.B.; Silva, D.B.; Carollo, C.A.; de Oliveira, F.M.; de Oliveira Lopes, D.; de Siqueira, J.M.; et al. Chemical composition and evaluation of the anti-inflammatory and antinociceptive activities of *Duguetia furfuracea* essential oil: Effect on edema, leukocyte recruitment, tumor necrosis factor alpha production, iNOS expression, and adenosinergic and opioidergic systems. *J. Ethnopharmacol.* **2019**, *231*, 325–336. [CrossRef]
14. De Oliveira, M.S.; Silva, S.G.; da Cruz, J.N.; Ortiz, E.; da Costa, W.A.; Bezerra, F.W.F.; Cunha, V.M.B.; Cordeiro, R.M.; de Jesus Chaves Neto, A.M.; de Aguiar Andrade, E.H.; et al. Supercritical CO_2 Application in Essential Oil Extraction. In *Industrial Applications of Green Solvents—Volume II*; Inamuddin, R.M., Asiri, A.M., Eds.; Materials Research Foundations: Millersville, PA, USA, 2019; pp. 1–28.
15. Turek, C.; Stintzing, F.C. Stability of Essential Oils: A Review. *Compr. Rev. Food Sci. Food Saf.* **2013**, *12*, 40–53. [CrossRef]

16. Hamid, A.A.; Aiyelaagbe, O.; Usman, L.A. Essential oils: Its medicinal and pharmacological uses. *Int. J. Curr. Res.* **2011**, *3*, 86–98.
17. De Oliveira, M.S.; Cruz, J.N.; Ferreira, O.O.; Pereira, D.S.; Pereira, N.S.; Oliveira, M.E.C.; Venturieri, G.C.; Guilhon, G.M.S.P.; da Silva Souza Filho, A.P.; de Aguiar Andrade, E.H. Chemical Composition of Volatile Compounds in *Apis mellifera* Propolis from the Northeast Region of Pará State, Brazil. *Molecules* **2021**, *26*, 3462. [CrossRef] [PubMed]
18. Santana de Oliveira, M.; Pereira da Silva, V.M.; Cantão Freitas, L.; Gomes Silva, S.; Nevez Cruz, J.; Aguiar Andrade, E.H. Extraction Yield, Chemical Composition, Preliminary Toxicity of *Bignonia nocturna* (Bignoniaceae) Essential Oil and in Silico Evaluation of the Interaction. *Chem. Biodivers.* **2021**, *18*, e2000982. [CrossRef] [PubMed]
19. Santana de Oliveira, M.; da Cruz, J.N.; Almeida da Costa, W.; Silva, S.G.; da Paz Brito, M.; de Menezes, S.A.F.; de Jesus Chaves Neto, A.M.; de Aguiar Andrade, E.H.; de Carvalho Junior, R.N. Chemical Composition, Antimicrobial Properties of *Siparuna guianensis* Essential Oil and a Molecular Docking and Dynamics Molecular Study of its Major Chemical Constituent. *Molecules* **2020**, *25*, 3852. [CrossRef]
20. Bezerra, F.W.F.; de Oliveira, M.S.; Bezerra, P.N.; Cunha, V.M.B.; Silva, M.P.; da Costa, W.A.; Pinto, R.H.H.; Cordeiro, R.M.; da Cruz, J.N.; Chaves Neto, A.M.J.; et al. Extraction of bioactive compounds. In *Green Sustainable Process for Chemical and Environmental Engineering and Science*; Elsevier: Amsterdam, The Netherlands, 2020; pp. 149–167.
21. Silva, S.G.; de Oliveira, M.S.; Cruz, J.N.; da Costa, W.A.; da Silva, S.H.M.; Barreto Maia, A.A.; de Sousa, R.L.; Carvalho Junior, R.N.; de Aguiar Andrade, E.H. Supercritical CO2 extraction to obtain *Lippia thymoides* Mart. & Schauer (Verbenaceae) essential oil rich in thymol and evaluation of its antimicrobial activity. *J. Supercrit. Fluids* **2021**, *168*, 105064. [CrossRef]
22. Burt, S. Essential oils: Their antibacterial properties and potential applications in foods—A review. *Int. J. Food. Macrobiol.* **2004**, *94*, 223–253. [CrossRef] [PubMed]
23. Perricone, M.; Arace, E.; Corbo, M.R.; Sinigaglia, M.; Bevilacqua, A. Bioactivity of essential oils: A review on their interaction with food components. *Front. Microbiol.* **2015**, *6*, 1–7. [CrossRef] [PubMed]
24. Silvestre, W.P.; Livinalli, N.F.; Baldasso, C.; Tessaro, I.C. Pervaporation in the separation of essential oil components: A review. *Trends Food Sci. Technol.* **2019**, *93*, 42–52. [CrossRef]
25. Kliszcz, A.; Danel, A.; Puła, J.; Barabasz-Krasny, B.; Mozdzen, K. Fleeting Beauty—The World of Plant Fragrances and Their Application. *Molecules* **2021**, *26*, 2473. [CrossRef]
26. Sharmeen, J.B.; Mahomoodally, F.M.; Zengin, G.; Maggi, F. Essential Oils as Natural Sources of Fragrance Compounds for Cosmetics and Cosmeceuticals. *Molecules* **2021**, *26*, 666. [CrossRef] [PubMed]
27. Ahmed, H.M. Ethnomedicinal, phytochemical and pharmacological investigations of *Perilla frutescens* (L.) Britt. *Molecules* **2019**, *24*, 102. [CrossRef] [PubMed]
28. Dos Santos Pereira, A.C.; das Graças Campolina Cunha, M. Medicina popular e saberes tradicionais sobre as propriedades medicinais da flora cerradeira. *Hygeia-Rev. Bras. Geogr. Médica Saúde* **2015**, *67*, 126–127.
29. Santos, J.J.F.; Coelho-Ferreira, M.; Lima, P.G.C. Etnobotânica de plantas medicinais em mercados públicos da Região Metropolitana de Belém do Pará, Brasil. *Biota Amaz.* **2018**, *8*, 1–9.
30. Vásquez, S.P.F.; de Mendonça, M.S.; do nascimento Noda, S. Etnobotânica de plantas medicinais em comunidades ribeirinhas do município de Manacapuru, Amazonas, Brasil. *Acta Amaz.* **2014**, *44*, 457–472. [CrossRef]
31. Xavier, M.N.; Alves, J.M.; Carneiro, N.S.; Souchie, E.L.; Da Silva, E.A.J.; Martins, C.H.G.; Ambrosio, M.A.L.V.; Egea, M.B.; Alves, C.C.F.; Miranda, M.L.D. Chemical composition from essential oil of *Cardiopetalum alophyllum* Schltdl. (Annonaceae) and their antioxidant, antibacterial and antifungal activities. *Rev. Virtual Quim.* **2016**, *8*, 1433–1448. [CrossRef]
32. Moghadamtousi, S.Z.; Fadaeinasab, M.; Nikzad, S.; Mohan, G.; Ali, H.M.; Kadir, H.A. *Annona muricata* (Annonaceae): A review of its traditional uses, isolated acetogenins and biological activities. *Int. J. Mol. Sci.* **2015**, *16*, 15625–15658. [CrossRef]
33. Ferraz, R.P.C.; Cardoso, G.M.B.; Da Silva, T.B.; Fontes, J.E.D.N.; Prata, A.P.D.N.; Carvalho, A.A.; Moraes, M.O.; Pessoa, C.; Costa, E.V.; Bezerra, D.P. Antitumour properties of the leaf essential oil of *Xylopia frutescens* Aubl. (Annonaceae). *Food Chem.* **2013**, *141*, 542–547. [CrossRef] [PubMed]
34. De Quintans, J.S.S. Chemical Constituents and Anticancer Effects of the Essential Oil from Leaves of *Xylopia laevigata*. *Planta Med.* **2013**, *79*, 123–130. [CrossRef] [PubMed]
35. Oliveira, V.B.; Yamada, L.T.; Fagg, C.W.; Brandão, M.G.L. Native foods from Brazilian biodiversity as a source of bioactive compounds. *Food Res. Int.* **2012**, *48*, 170–179. [CrossRef]
36. Maia, J.G.S.; Andrade, E.H.A.; Carreira, L.M.M.; Oliveira, J.; Araújo, J.S. Essential oils of the Amazon Guatteria and Guatteriopsis species. *Flavour Fragr. J.* **2005**, *20*, 478–480. [CrossRef]
37. Da Silva, D.B.; Tulli, E.C.O.; Garcez, W.S.; Nascimento, E.A.; De Siqueira, J.M. Chemical constituents of the underground stem bark of *Duguetia furfuracea* (Annonaceae). *J. Braz. Chem. Soc.* **2007**, *18*, 1560–1565. [CrossRef]
38. Sousa, O.V.; Del-Vechio-Vieira, G.; Alves, M.S.; Araújo, A.A.L.; Pinto, M.A.O.; Amaral, M.P.H.; Rodarte, M.P.; Kaplan, M.A.C. Chemical composition and biological activities of the essential oils from *Duguetia lanceolata* St. Hil. barks. *Molecules* **2012**, *17*, 11056–11066. [CrossRef]
39. Siqueira, C.A.T.; Oliani, J.; Sartoratto, A.; Queiroga, C.L.; Moreno, P.R.H.; Reimão, J.Q.; Tempone, A.G.; Fischer, D.C.H. Chemical constituents of the volatile oil from leaves of *Annona coriacea* and in vitro antiprotozoal activity. *Braz. J. Pharmacogn.* **2011**, *21*, 33–40. [CrossRef]
40. Rodrigues, V.E.G.; Carvalho, D.A. Etnobotanical survey of medicinal plants in the dominion of meadows in the region of the alto rio grande—MINAS GERAIS. *Cienc. Agrotec.* **2001**, *25*, 102–123.

41. Agra, M.D.F.; de Freitas, P.F.; Barbosa-Filho, J.M. Divulgação Synopsis of the plants known as medicinal and poisonous in Northeast of Brazil. *Rev. Bras. Farmacogn.* **2007**, *17*, 114–140. [CrossRef]
42. Brito, M.T.; Ferreira, R.C.; Beltrão, D.M.; Moura, A.P.G.; Xavier, A.L.; Pita, J.C.L.R.; Batista, T.M.; Longato, G.B.; Ruiz, A.L.T.G.; de Carvalho, J.E.; et al. Antitumor activity and toxicity of volatile oil from the leaves of *Annona leptopetala*. *Braz. J. Pharmacogn.* **2018**, *28*, 602–609. [CrossRef]
43. Costa, E.V.; Dutra, L.M.; Jesus, H.C.R.; Nogueira, P.C.L.; Moraes, V.R.S.; Salvador, M.J.; Cavalcanti, S.C.H.; Santos, R.C.; Prata, A.P.N. Chemical Composition and Antioxidant, Antimicrobial, and Larvicidal Activities of the Essential Oils of Annona salzmannii and A. pickelii (Annonaceae). *Nat. Prod. Commun.* **2011**, *6*, 907–912. [CrossRef]
44. Meira, C.S.; Guimarães, E.T.; MacEdo, T.S.; Da Silva, T.B.; Menezes, L.R.A.; Costa, E.V.; Soares, M.B.P. Chemical composition of essential oils from Annona vepretorum Mart. and *Annona squamosa* L. (Annonaceae) leaves and their antimalarial and trypanocidal activities. *J. Essent. Oil Res.* **2015**, *27*, 160–168. [CrossRef]
45. Formagio, A.S.N.; Vieira, M.D.C.; Dos Santos, L.A.C.; Cardoso, C.A.L.; Foglio, M.A.; De Carvalho, J.E.; Andrade-Silva, M.; Kassuya, C.A.L. Composition and evaluation of the anti-inflammatory and anticancer activities of the essential oil from *Annona sylvatica* A. St.-Hil. *J. Med. Food* **2013**, *16*, 20–25. [CrossRef] [PubMed]
46. Dutra, L.M.; Bomfim, L.M.; Rocha, S.L.A.; Nepel, A.; Soares, M.B.P.; Barison, A.; Costa, E.V.; Bezerra, D.P. Ent-Kaurane diterpenes from the stem bark of *Annona vepretorum* (Annonaceae) and cytotoxic evaluation. *Bioorganic Med. Chem. Lett.* **2014**, *24*, 3315–3320. [CrossRef] [PubMed]
47. Valter, J.L.; Alencar, K.M.C.; Sartori, Â.L.B.; Nascimento, E.A.; Chang, R.; De Morais, S.A.L.; Laura, V.A.; Yoshida, N.C.; Carollo, C.A.; Da Silva, D.B.; et al. Variação química no óleo essencial das folhas de seis indivíduos de *Duguetia furfuracea* (Annonaceae). *Braz. J. Pharmacogn.* **2008**, *18*, 373–378. [CrossRef]
48. Maia, J.G.S.; Andrade, E.H.A.; Carla, A.; Silva, M.; Oliveira, J.; Carreira, L.M.M.; Araújo, J.S. Leaf volatile oils from four Brazilian *Xylopia* species. *Flavour Fragr. J.* **2005**, *20*, 474–477. [CrossRef]
49. Alcântara, J.M.; De Lucena, J.M.V.M.; Facanali, R.; Marques, M.O.M.; Da Paz Lima, M. Chemical composition and bactericidal activity of the essential oils of four species of annonaceae growing in brazilian amazon. *Nat. Prod. Commun.* **2017**, *12*, 619–622. [CrossRef] [PubMed]
50. Mendes, R.d.F.; Pinto, d.C.C.; da Silva, J.M.; da Silva, J.B.; Hermisdorf, R.C.d.S.; Fabri, R.L.; Chedier, L.M.; Scio, E. The essential oil from the fruits of the Brazilian spice *Xylopia sericea* A. St.-Hil. presents expressive in-vitro antibacterial and antioxidant activity. *J. Pharm. Pharmacol.* **2017**, *69*, 341–348. [CrossRef]
51. De Alencar, D.C.; Pinheiro, M.L.B.; Pereira, J.L.D.S.; De Carvalho, J.E.; Campos, F.R.; Serain, A.F.; Tirico, R.B.; Hernández-Tasco, A.J.; Costa, E.V.; Salvador, M.J. Chemical composition of the essential oil from the leaves of *Anaxagorea brevipes* (Annonaceae) and evaluation of its bioactivity. *Nat. Prod. Res.* **2016**, *30*, 1088–1092. [CrossRef] [PubMed]
52. Cascaes, M.M.; Silva, S.G.; Cruz, J.N.; Santana de Oliveira, M.; Oliveira, J.; de Moraes, A.A.B.; da Costa, F.A.M.; da Costa, K.S.; Diniz do Nascimento, L.; de Aguiar Andrade, E.H. First report on the *Annona exsucca* DC. Essential oil and in silico identification of potential biological targets of its major compounds. *Nat. Prod. Res.* **2021**, *35*, 1–4. [CrossRef]
53. Costa, E.V.; Dutra, L.M.; Salvador, M.J.; Ribeiro, L.H.G.; Gadelha, F.R.; De Carvalho, J.E. Chemical composition of the essential oils of *Annona pickelii* and *Annona salzmannii* (Annonaceae), and their antitumour and trypanocidal activities. *Nat. Prod. Res.* **2013**, *27*, 997–1001. [CrossRef] [PubMed]
54. Bomfim, L.M.; Menezes, L.R.A.; Rodrigues, A.C.B.C.; Dias, R.B.; Gurgel Rocha, C.A.; Soares, M.B.P.; Neto, A.F.S.; Nascimento, M.P.; Campos, A.F.; Silva, L.C.R.C.E.; et al. Antitumour Activity of the Microencapsulation of *Annona vepretorum* Essential Oil. *Basic Clin. Pharmacol. Toxicol.* **2016**, *118*, 208–213. [CrossRef]
55. Diniz, T.C.; de Oliveira Júnior, R.G.; Miranda Bezerra Medeiros, M.A.; Gama e Silva, M.; de Andrade Teles, R.B.; dos Passos Menezes, P.; de Sousa, B.M.H.; Abrahão Frank, L.; de Souza Araújo, A.A.; Russo Serafini, M.; et al. Anticonvulsant, sedative, anxiolytic and antidepressant activities of the essential oil of *Annona vepretorum* in mice: Involvement of GABAergic and serotonergic systems. *Biomed. Pharmacother.* **2019**, *111*, 1074–1087. [CrossRef]
56. Costa, E.V.; Dutra, L.M.; Nogueira, P.C.L.; Moraes, V.R.S.; Salvador, M.J.; Ribeiro, L.H.G.; Gadelha, F.R. Essential oil from the leaves of *Annona vepretorum*: Chemical composition and bioactivity. *Nat. Prod. Commun.* **2012**, *7*, 265–266. [CrossRef] [PubMed]
57. Bay, M.; Souza de Oliveira, J.V.; Sales Junior, P.A.; Fonseca Murta, S.M.; Rogério dos Santos, A.; dos Santos Bastos, I.; Puccinelli Orlandi, P.; Teixeira de Sousa Junior, P. In Vitro Trypanocidal and Antibacterial Activities of Essential Oils from Four Species of the Family Annonaceae. *Chem. Biodivers.* **2019**, *16*, e1900359. [CrossRef] [PubMed]
58. Soares, E.R.; Da Silva, F.M.A.; De Almeida, R.A.; De Lima, B.R.; Koolen, H.H.F.; Lourenço, C.C.; Salvador, M.J.; Flach, A.; Da Costa, L.A.M.A.; De Souza, A.Q.L.; et al. Chemical composition and antimicrobial evaluation of the essential oils of Bocageopsis pleiosperma Maas. *Nat. Prod. Res.* **2015**, *29*, 1285–1288. [CrossRef] [PubMed]
59. Xavier, M.N.; Alves, C.C.F.; Cazal, C.d.M.; Santos, N.H. Chemical composition of the volatile oil of *Cardiopetalum calophyllum* collected in the Cerrado area. *Ciência Rural* **2016**, *46*, 937–942. [CrossRef]
60. Maia, D.S.; Lopes, C.F.; Saldanha, A.A.; Silva, N.L.; Sartori, Â.L.B.; Carollo, C.A.; Sobral, M.G.; Alves, S.N.; Silva, D.B.; de Siqueira, J.M. Larvicidal effect from different Annonaceae species on Culex quinquefasciatus. *Environ. Sci. Pollut. Res.* **2020**, *27*, 36983–36993. [CrossRef]

61. De Sousa Orlando, V.; Glauciemar, D.V.V.; Bruna, C.S.S.; Ceacutelia, H.Y.; Santos de Matos Arajo, A.L.; da Luz, A.; Aparecida de Oliveira Pinto, M.; Pereira Rodarte, M.; Alves, M.S. In-vivo and vitro bioactivities of the essential oil of *Duguetia lanceolata* branches. *Afr. J. Pharm. Pharmacol.* **2016**, *10*, 298–310. [CrossRef]
62. Ribeiro, L.P.; Domingues, V.C.; Gonçalves, G.L.P.; Fernandes, J.B.; Glória, E.M.; Vendramim, J.D. Essential oil from Duguetia lanceolata St.-Hil. (Annonaceae): Suppression of spoilers of stored-grain. *Food Biosci.* **2020**, *36*, 100653. [CrossRef]
63. Santos, A.R.; Benghi, T.G.S.; Nepel, A.; Marques, F.A.; Lobão, A.Q.; Duarte, M.C.T.; Ruiz, A.L.T.G.; Carvalho, J.E.; Maia, B.H.L.N.S. In vitro Antiproliferative and Antibacterial Activities of Essential Oils from Four Species of Guatteria. *Chem. Biodivers.* **2017**, *14*, e1700097. [CrossRef] [PubMed]
64. Siqueira, C.A.T.; Serain, A.F.; Pascoal, A.C.R.F.; Andreazza, N.L.; De Lourenço, C.C.; Góis Ruiz, A.L.T.; De carvalho, J.E.; De Souza, A.C.O.; Tonini Mesquita, J.; Tempone, A.G.; et al. Bioactivity and chemical composition of the essential oil from the leaves of *Guatteria australis* A.St.-Hil. *Nat. Prod. Res.* **2015**, *29*, 1966–1969. [CrossRef]
65. Aciole, S.D.G.; Piccoli, C.F.; Duque, J.E.L.; Costa, E.V.; Navarro-Silva, M.A.; Marques, F.A.; Maia, B.H.L.N.S.; Pinheiro, M.L.B.; Rebelo, M. Insecticidal activity of three species of Guatteria (Annonaceae) against *Aedes aegypti* (Diptera: Culicidae). *Rev. Colomb. Entomol.* **2011**, *37*, 262–268.
66. Meira, C.S.; Menezes, L.R.A.; dos Santos, T.B.; Macedo, T.S.; Fontes, J.E.N.; Costa, E.V.; Pinheiro, M.L.B.; da Silva, T.B.; Teixeira Guimarães, E.; Soares, M.B.P. Chemical composition and antiparasitic activity of essential oils from leaves of *Guatteria friesiana* and *Guatteria pogonopus* (Annonaceae). *J. Essent. Oil Res.* **2017**, *29*, 156–162. [CrossRef]
67. Costa, R.G.A.; Anunciação, T.A.D.; Araujo, M.d.S.; Souza, C.A.; Dias, R.B.; Sales, C.B.S.; Rocha, C.A.G.; Soares, M.B.P.; Silva, F.M.A.D.; Koolen, H.H.F.; et al. In vitro and in vivo growth inhibition of human acute promyelocytic leukemia HL-60 cells by *Guatteria megalophylla* Diels (Annonaceae) leaf essential oil. *Biomed. Pharmacother.* **2020**, *122*, 109713. [CrossRef] [PubMed]
68. Fontes, J.E.N.; Ferras, R.P.C.; Britto, A.C.S.; Carvalho, A.A.; Moraes, M.O.; Pessoa, C.; Costa, E.V.; Bezerra, D.P. Antitumor effect of the essential oil from leaves of *Guatteria pogonopus* (Annonaceae). *Chem. Biodivers.* **2013**, *10*, 722–729. [CrossRef] [PubMed]
69. De Lima, B.R.; da Silva, F.M.A.; Soares, E.R.; de Almeida, R.A.; da Silva Filho, F.A.; Pereira Junior, R.C.; Hernandez Tasco, Á.J.; Salvador, M.J.; Koolen, H.H.F.; de Souza, A.D.L.; et al. Chemical composition and antimicrobial activity of the essential oils of *Onychopetalum amazonicum* R.E.Fr. *Nat. Prod. Res.* **2016**, *30*, 2356–2359. [CrossRef] [PubMed]
70. De Lima, B.R.; da Silva, F.M.A.; Soares, E.R.; de Almeida, R.A.; Maciel, J.B.; Fernandes, C.C.; de Oliveira, A.C.; Tadei, W.P.; Koolen, H.H.F.; de Souza, A.D.L.; et al. Chemical composition and larvicidal activity of the essential oil from the leaves of *Onychopetalum periquino* (Rusby) D.M. Johnson & N.A. Murray. *Nat. Prod. Res.* **2019**, *35*, 1038–1041. [CrossRef] [PubMed]
71. Da Silva, E.B.P.; Soares, M.G.; Mariane, B.; Vallim, M.A.; Pascon, R.C.; Sartorelli, P.; Lago, J.H.G. The Seasonal variation of the chemical composition of essential oils from *Porcelia macrocarpa* r.e. fries (annonaceae) and their antimicrobial activity. *Molecules* **2013**, *18*, 13574–13587. [CrossRef] [PubMed]
72. Yoshida, N.C.; Saffran, F.P.; Lima, W.G.; Freire, T.V.; de Siqueira, J.M.; Garcez, W.S. Chemical characterization and bioherbicidal potential of the essential oil from the leaves of *Unonopsis guatterioides* (A.DC.) R.E.Fr. (Annonaceae). *Nat. Prod. Res.* **2019**, *33*, 3312–3316. [CrossRef]
73. Nascimento, M.N.G.; Junqueira, J.G.M.; Terezan, A.P.; Severino, R.P.; De Souza Silva, T.; Martins, C.H.G.; Severino, V.G.P. Chemical composition and antimicrobial activity of essential oils from *Xylopia aromatica* (Annonaceae) flowers and leaves. *Rev. Virtual Quim.* **2018**, *10*, 1578–1590. [CrossRef]
74. De Souza, I.L.L.; de Carvalho Correia, A.C.; da Cunha Araujo, L.C.; Vasconcelos, L.H.C.; da Conceicao Correia Silva, M.; de Oliveira Costa, V.C.; Tavares, J.F.; Paredes-Gamero, E.J.; de Andrade Cavalcante, F.; da Silva, B.A. Essential oil from *Xylopia frutescens* Aubl. reduces cytosolic calcium levels on guinea pig ileum: Mechanism underlying its spasmolytic potential. *BMC Complement. Altern. Med.* **2015**, *15*, 327. [CrossRef]
75. Nascimento, A.M.D.; Maia, T.D.S.; Soares, T.E.S.; Menezes, L.R.A.; Scher, R.; Costa, E.V.; Cavalcanti, S.C.H.; La Corte, R. Repellency and Larvicidal Activity of Essential oils from *Xylopia laevigata, Xylopia frutescens, Lippia pedunculosa*, and Their Individual Compounds against *Aedes aegypti* Linnaeus. *Neotrop. Entomol.* **2017**, *46*, 223–230. [CrossRef]
76. Pereira, T.S.; Machado Esquissato, G.N.; Costa, E.V.; de Lima Nogueira, P.C.; de Castro-Prado, M.A.A. Mutagenic and cytostatic activities of the *Xylopia laevigata* essential oil in human lymphocytes. *Nat. Prod. Res.* **2019**, *6419*. [CrossRef]
77. Costa, E.V.; Da Silva, T.B.; D'Souza Costa, C.O.; Soares, M.B.P.; Bezerra, D.P. Chemical composition of the essential oil from the fresh fruits of *Xylopia laevigata* and its cytotoxic evaluation. *Nat. Prod. Commun.* **2016**, *11*, 417–418. [CrossRef] [PubMed]
78. Gontijo, D.C.; do Nascimento, M.F.A.; Brandão, G.C.; de Oliveira, A.B. Phytochemistry and antiplasmodial activity of *Xylopia sericea* leaves. *Nat. Prod. Res.* **2019**, *6419*. [CrossRef]
79. Chandra, P.; Sharma, R.K.; Arora, D.S. Antioxidant compounds from microbial sources: A review. *Food Res. Int.* **2019**, *129*, 108849. [CrossRef] [PubMed]
80. Diniz Do Nascimento, L.; Antônio Barbosa De Moraes, A.; Santana Da Costa, K.; Marcos, J.; Galúcio, P.; Taube, P.S.; Leal Costa, M.; Neves Cruz, J.; Helena De Aguiar Andrade, E.; Guerreiro De Faria, L.J. Bioactive Natural Compounds and Antioxidant Activity of Essential Oils from Spice Plants: New Findings and Potential Applications. *Biomolecules* **2020**, *10*, 988. [CrossRef] [PubMed]
81. Caesar, L.K.; Cech, N.B. Synergy and antagonism in natural product extracts: When 1 + 1 does not equal 2. *Nat. Prod. Rep.* **2019**, *36*, 869–888. [CrossRef]

82. Alves, C.C.F.; Oliveira, J.D.; Estevam, E.B.B.; Xavier, M.N.; Nicolella, H.D.; Furtado, R.A.; Tavares, D.C.; Miranda, M.L.D. Antiproliferative activity of essential oils from three plants of the brazilian cerrado: *Campomanesia adamantium* (myrtaceae), *Protium ovatum* (burseraceae) and *Cardiopetalum calophyllum* (annonaceae). *Braz. J. Biol.* **2020**, *80*, 290–294. [CrossRef] [PubMed]
83. Cláudia, K.; Lidani, F.; Andrade, F.A.; Bavia, L. Chagas Disease: From Discovery to a Worldwide Health Problem. *Front. Public Health* **2019**, *7*, 166. [CrossRef]

International Journal of *Molecular Sciences*

MDPI

Review

The Effects of Antioxidants from Natural Products on Obesity, Dyslipidemia, Diabetes and Their Molecular Signaling Mechanism

Chindiana Khutami [1], Sri Adi Sumiwi [1], Nur Kusaira Khairul Ikram [2,3] and Muchtaridi Muchtaridi [4,*]

1 Department of Pharmacology and Clinical Pharmacy, Faculty of Pharmacy, Universitas Padjadjaran, Jl. Raya Bandung-Sumedang KM 21, Sumedang 45363, Indonesia; chindiana19001@mail.unpad.ac.id (C.K.); sri.adi@unpad.ac.id (S.A.S.)

2 Institute of Biological Sciences, Faculty of Science, Universiti Malaya, Kuala Lumpur 50603, Malaysia; nkusaira@um.edu.my

3 Centre for Research in Biotechnology for Agriculture (CEBAR), Kuala Lumpur 50603, Malaysia

4 Department of Pharmaceutical Analysis and Medicinal Chemistry, Faculty of Pharmacy, Universitas Padjadjaran, Jl. Raya Bandung-Sumedang KM 21, Sumedang 45363, Indonesia

* Correspondence: muchtaridi@unpad.ac.id

Abstract: Obesity is a risk factor that leads to the development of other diseases such as dyslipidemia and diabetes. These three metabolic disorders can occur simultaneously, hence, the treatment requires many drugs. Antioxidant compounds have been reported to have activities against obesity, dyslipidemia and diabetes via several mechanisms. This review aims to discuss the antioxidant compounds that have activity against obesity, dyslipidemia and diabetes together with their molecular signaling mechanism. The literature discussed in this review was obtained from the PUBMED database. Based on the collection of literature obtained, antioxidant compounds having activity against the three disorders (obesity, dyslipidemia and diabetes) were identified. The activity is supported by various molecular signaling pathways that are influenced by these antioxidant compounds, further study of which would be useful in predicting drug targets for a more optimal effect. This review provides insights on utilizing one of these antioxidant compounds as opposed to several drugs. It is hoped that in the future, the number of drugs in treating obesity, dyslipidemia and diabetes altogether can be minimized consequently reducing the risk of side effects.

Keywords: obesity; oxidative stress; dyslipidemia; diabetes; antioxidant

Citation: Khutami, C.; Sumiwi, S.A.; Khairul Ikram, N.K.; Muchtaridi, M. The Effects of Antioxidants from Natural Products on Obesity, Dyslipidemia, Diabetes and Their Molecular Signaling Mechanism. *Int. J. Mol. Sci.* **2022**, 23, 2056. https://doi.org/10.3390/ijms23042056

Academic Editors: Antonio Carrillo Vico and Ivan Cruz-Chamorro

Received: 7 January 2022
Accepted: 10 February 2022
Published: 12 February 2022

Publisher's Note: MDPI stays neutral with regard to jurisdictional claims in published maps and institutional affiliations.

1. Introduction

Obesity is a pathological condition of excessive fat accumulating in the tissues under the skin and spreading to the organs and tissues around the body. From a health perspective, obesity is malnutrition caused by long-term excessive consumption of unhealthy food. Obese patients have health problems, one of which is an increase in total cholesterol levels > 200 mg/dL. The World Health Organization (WHO) points out that obesity is a chronic disease and is one of the risk factors for degenerative diseases such as diabetes and dyslipidemia, as well as acute coronary disease, hypertension, hyperuricemia and polycystic ovary syndrome [1]. Obesity can trigger oxidative stress through various mechanisms such as oxidative phosphorylation, glyceraldehyde autoxidation and superoxide formation [2]. Oxidative stress is a condition when there is an increase in the number of free radicals and/or a decrease in antioxidant activity [3]. Oxidative stress plays a role in comorbid obesity such as diabetes, dyslipidemia, endothelial dysfunction and mitochondrial dysfunction [2].

Patients with obesity are often associated with lipid abnormalities [4]. Approximately 60–70% of obese patients have dyslipidemia. In addition, insulin resistance disorders also contribute to the development of dyslipidemia. In recent years, dyslipidemia caused by the combined action of insulin resistance and obesity has been referred to as "metabolic dyslipidemia" with the main characteristics of increasing levels of triglycerides (TG) and decreasing levels of high-density lipoprotein (HDL). Under this condition, there can also be an increase in low-density lipoprotein (LDL) levels [5].

Diabetes mellitus (DM) is one of the comorbidities of obesity caused by oxidative stress. DM is a chronic disease affecting the body's metabolism, characterized by increased blood sugar levels exceeding normal limits. In Southeast Asia, the number of diabetes cases in 2019 reached 88 million people and 90% of them were type 2 diabetes mellitus, half of which have complications that lead to death. The International Diabetes Federation (IDF) listed Indonesia as the 7th highest diabetic country with a prevalence of 8.5 million, and the number is predicted to increase to 14.1 million by 2035 [6]. The study by Mohieldein et al. (2015) reported that prediabetes was associated with obesity, development of dyslipidemia and decreasing total antioxidant status. Lifestyle changes such as weight loss, regular physical activity and a healthy diet should be encouraged to prevent progression to type 2 diabetes and its complications from prediabetes [7].

The hyperglycemia condition in DM has a significant impact on the vascular endothelium, which is caused by the auto-oxidation of glucose during the formation of free radicals, which in turn leads to macro- and microvascular dysfunction due to oxidative stress [3]. Oxidative stress conditions in DM are usually associated with an increase in endothelial cell apoptosis, which shows an increase in the free radical formation and a decrease in antioxidant capacity [8].

Metabolic disorders such as obesity, dyslipidemia and diabetes are the main causes of life-threatening ischemic heart disease [9]. Based on research by Vona et al. (2019) and Pechánová et al. (2015), it is reported that this metabolic disorder is accompanied by chronic inflammation mediated by oxidative stress. Increased oxidative stress in metabolic disorders plays a role in causing mitochondrial dysfunction, accumulation of protein and lipid oxidation products and disruption of the antioxidant system [10,11]. Clinical studies show that obesity co-occurring with metabolic disorders such as dyslipidemia and diabetes will increase the risk of death compared to obesity without metabolic disorders. However, when compared with lean individuals, obesity may increase the risk of death from various complications that accompany this condition [12].

At present, many people have adopted a healthy lifestyle, such as eating foods or taking medications derived from natural ingredients, especially those containing antioxidant compounds that can prevent and treat various diseases [13]. Compounds with antioxidant activity such as kahweol have been reported to have antidiabetic properties by suppressing pancreatic cell apoptosis and increasing insulin secretion in streptozotocin (STZ)-induced mice [14]. Another study by Pan et al. (2018) reported that flavonoids such as resveratrol (3,4′,5-trihydroxy-stilbene, RES) are widely present in vegetables and fruits with biological and pharmacological effects such as antiobesity, antioxidation activity and antidiabetic [15]. The combination of resveratrol and quercetin has also been reported to reduce hyperglycemia, serum glucose dysfunction and dyslipidemia in streptozotocin (STZ)-induced diabetic rats [16].

Of the many studies that discuss metabolic disorders and antioxidants from the literature, only two studies are obtained, which are closest to the discussion in this review. The two studies are conducted by Dal et al. (2016) and Shabbir et al. (2021). In a review article by Dal et al. (2016), the effect of consuming antioxidants from various sources such as functional foods, plants, fruits, vegetables, vitamins, supplements and other natural sources rich in polyphenols on diabetes and vascular complications based on in vivo, in vitro and clinical trials in humans were discussed [17]. Another review article by Shabbir et al. (2021) discusses the activity of polyphenol antioxidant compounds, namely, curcumin, quercetin and catechins, against metabolic disorders that focus on the role of the gut microbiota, which is affected by these antioxidant compounds to improve metabolic disorders [18].

In the two aforementioned reviews, there has been no detailed discussion on the effect of antioxidant compounds and their molecular signaling mechanisms against obesity, dyslipidemia and diabetes. The novelty of this review is the summary of information on antioxidant compounds derived from natural products based on the results of in vivo, in vitro and clinical trials that can treat obesity, dyslipidemia and diabetes. This could provide an insight on the antioxidant compounds that can simultaneously act as an antiobesity, antidyslipidemia and antidiabetic as compared to the current practices that require several drugs for the treatment of the three metabolic disorders, hence minimizing the use of multiple drugs and the risk of side effects and drug interactions. Identifying the plants containing these valuable compounds could also potentially yield a cheaper alternative treatment in the form of a herbal preparation to treat all three illnesses in the future.

In addition, this review also provides information on the molecular signaling pathways influenced by the antioxidant compounds from natural products that play a role in the development of obesity, dyslipidemia and diabetes. This would be useful for researchers to further investigate the activity of these antioxidant compounds to determine therapeutic targets.

2. Method

This review was made based on the results of the collection and review of journals obtained from the PUBMED database with several related keywords such as "antioxidant AND natural product AND obesity", "antioxidant AND natural product AND antidiabetes", "antioxidant AND natural product AND antidyslipidemia", "signaling pathways AND natural product", "obesity AND dyslipidemia AND natural product", "antioxidant AND oxidative stress AND obesity AND dyslipidemia AND diabetes", " resveratrol AND metabolis disorders AND clinical study", "quercetin AND clinical study AND antiobesity", "curcumin AND antiobesity AND clinical study", "anthocyanins AND clinical study AND anti obesity", "antioxidant AND metabolic disorders AND clinical study", "antioxidant AND antiobesity AND antidiabetes", "antioxidant AND antidislipidemia AND antidiabetes".

The inclusion criteria for the main article are articles published in ≥2016 and research articles that discuss pharmacological antioxidant activity against obesity, dyslipidemia and diabetes as well as the signaling pathways of these antioxidant activities. Inclusion criteria for supporting articles are articles that discuss the metabolic disorders of obesity, dyslipidemia, diabetes and oxidative stress including the mechanism of metabolic disorders and the relationship between these metabolic disorders. This supporting article is taken from articles published between 2000 and 2021 with most of the articles included being published after 2016. Exclusion criteria for the main articles were duplicate articles, review articles, research with crude extracts and unrelated articles/irrelevant articles that do not discuss in detail the activity of the chemical compounds contained.

Based on the search results using related keywords, 782 journals were obtained, which were then reduced after removing 60 duplicate articles, 155 review articles, 152 research articles with crude extracts and 270 unrelated/not specific articles. The results are 145 selected articles consisting of 49 supporting articles, 68 articles discussing antioxidant activity against obesity, dyslipidemia and diabetes and 28 articles discussing signaling pathways of antioxidant activity, which were used to be studied in this review. The article search flow can be seen in Figure 1.

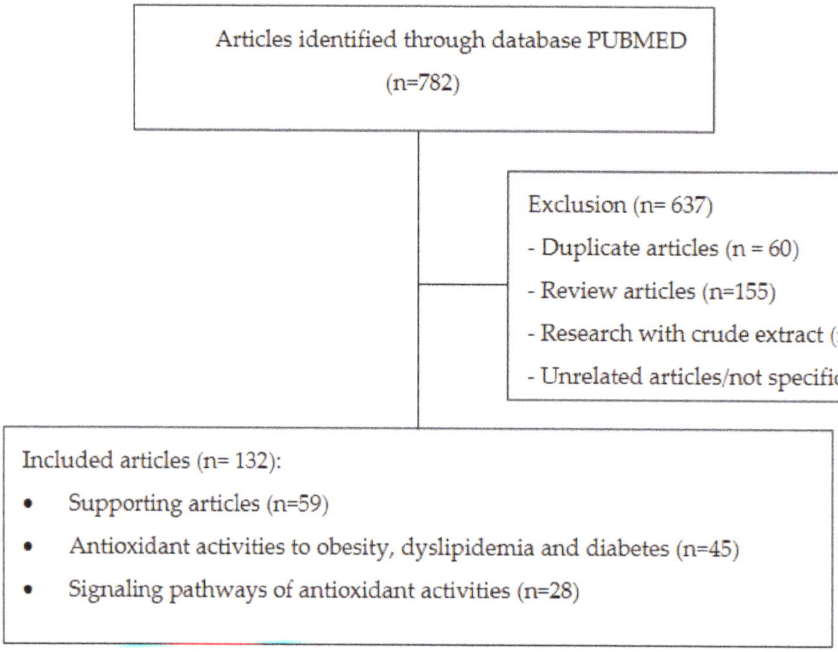

Figure 1. Article literature search flow chart.

3. Oxidative Stress and Its Relation to Metabolic Disorders (Obesity, Dyslipidemia and Diabetes)

Oxidative stress is a condition of an imbalance of production and accumulation of reactive oxygen species (ROS) in cells and tissues with the ability of biological systems to detoxify these reactive products [19]. Reactive oxygen species (ROS) in normal amounts contribute to various physiological processes such as hormone biosynthesis, defense systems, cellular signaling and fertilization. However, increased ROS production results in a condition known as oxidative stress, which has implications for various diseases such as diabetes, dyslipidemia, hypertension, atherosclerosis, heart failure, stroke and other chronic diseases [20,21]. Oxidative stress contributes to the development of obesity's comorbidities [2]. Possible contributors to oxidative stress in obesity include increased hyperglycemia in tissue, lipids, vitamin and mineral deficiencies, chronic inflammation, endothelial dysfunction and impaired mitochondrial function [19].

3.1. Obesity

Obesity is one of the metabolic disorders resulting from an imbalance between energy intake and consumption. Obesity is marked by the inflammation of many cells, including macrophages [22] and adipose tissue [23]. Macrophages produce cytokines such as IL-6 and tumor necrosis factor alpha (TNFα), which also play a role in causing insulin resistance. In addition, obesity can cause the body's resistance to insulin, mediated in part by free fatty acids (FFA) and adipokines such as retinol binding protein-4 (RBP4) and resistin, which can reduce insulin sensitivity [24]. Obesity is the main cause of metabolic syndromes such as type 2 diabetes mellitus, insulin resistance, dyslipidemia, hypertension and non-alcoholic fatty liver diseases (NAFLD) [1,25].

Obesity is known to have a relationship with the incidence of oxidative stress. Based on research done on several obese patients, it was reported that abdominal obesity may affect the occurrence of inflammation that triggers an increase in oxidative stress [26].

Studies show that obesity in visceral adipose tissue contributes to a state of oxidative stress that may lead to insulin resistance [27,28].

3.2. Dyslipidemia

Dyslipidemia is a serious problem because it is a major risk for coronary heart disease. Dyslipidemia is caused by several factors such as genes, diet, lifestyle, obesity, and many more [29]. Dyslipidemia in obesity consists of elevated levels of triglycerides (TG) and free fatty acids (FFA), decreased and dysfunctional high-density lipoprotein (HDL) and a slight increase of low-density lipoprotein-cholesterol (LDL) levels. In addition, the concentration of apolipoprotein B (apo B) may also increase due to excessive production of apo B in the liver containing lipoproteins [30].

Dyslipidemia in obesity is characterized by the occurrence of hypertriglyceridemia due to the accumulation of triglycerides in the liver. This leads to the inhibition of chylomicron lipolysis caused by increased synthesis of very low-density lipoprotein (VLDL) in the liver due to the competition for lipoprotein lipase (LPL). Hypertriglyceridemia will induce an increase in the exchange of cholesterol esters (CE) and triglycerides between VLDL, LDL and HDL via cholesteryl ester transfer protein (CETP) [25].

Based on previous research, hypercholesterolemia may induce apoptosis and autophagy caused by ROS activation [31]. Increased production of ROS affects the development of dyslipidemia and other cardiovascular diseases. Free radicals function physiologically as signal transducers and maintain homeostasis in cellular signaling, but when these pathways are disrupted, they may cause risk factors for atherosclerosis, one of which is dyslipidemia [32,33].

3.3. Diabetes Mellitus

Diabetes mellitus (DM) is a metabolic disorder characterized by increased blood sugar levels. DM can be divided into two types: DM type 1 and DM type 2. DM type 1 is caused by the unavailability of insulin produced by pancreatic beta cells, which can be caused by genetic or autoimmune disorders, while type 2 DM occurs due to lack of insulin secretion or insulin resistance, or both. DM type 2 can occur due to the influence of genetic, epigenetic or lifestyle factors [34].

Various inflammatory cytokines such as IL-1β produced by M1 macrophages can cause local and systemic inflammation, pancreatic cell dysfunction and insulin resistance in the liver, adipose and musculoskeletal tissues [35]. M1 macrophages are also associated with diabetes complications, such as kidney disease, neurological diseases, retinopathy, and cardiovascular diseases. However, to date, the underlying mechanism of M1 macrophage accumulation in diabetic patients is not fully known [22].

Hyperglycemia in prediabetes may trigger oxidative stress and increase inflammatory factors that affect vascular dysfunction. This oxidative stress may also cause interference with glucose uptake from muscle cells and fat cells and may reduce insulin sensitivity [27]. Another study also reported an increase in ROS concomitantly with suppression of the antioxidant enzyme superoxide dismutase (SOD) in rats induced by hyperglycemia [36].

4. Antioxidant Activities from Natural Products to Treat Obesity, Dyslipidemia and Diabetes Mellitus

Antioxidants are substances that can counteract free radicals and prevent free radicals from damaging cells. Free radicals are the root cause of health problems, such as cancer, premature aging, cardiovascular disease and digestive diseases. The body naturally produces antioxidants, but when free radicals are abundant, this process will not be efficient and its effectiveness also decreases with age. Increasing the intake of antioxidants can prevent various diseases and reduce health problems. Food such as fruits and vegetables contain important antioxidants such as vitamins A, C, E and beta-carotene, as well as essential minerals such as selenium and zinc [37]. Several antioxidant compounds from natural ingredients that have been widely studied and have antiobesity, antidyslipidemia and antidiabetic activity are resveratrol, curcumin, quercetin and anthocyanin as well as other antioxidants.

4.1. Resveratrol

Resveratrol (3,5,4′-trihydroxy-trans-stilbene) is a natural antioxidant compound that can be found in various plants, such as *Polygonum cuspidatum*, peanuts and fruits, such as grapes and berries. The use of resveratrol as a nutraceutical has also been widely reported based on testing in animal and human models as a treatment for obesity, metabolic disorders and cardiovascular disorders [38]. Resveratrol is a polyphenolic antioxidant compound that has various biological activities and has been used as a dietary supplement [38].

A clinical study was conducted on 13 patients with type 1 diabetes who were given resveratrol in 500 mg capsules for 60 days. From the results of this study, it is known that the administration of resveratrol can reduce fasting blood sugar (FBS) significantly ($p < 0.05$) compared to the initial value with FBS levels of 253.69 ± 49.67 vs. 174.38 ± 45.19 [39]. Another clinical study by Jorge et al. (2020) was conducted on 25 obese individuals (BMI 30 kg/m^2) aged 30–60 years who were randomly assigned to a placebo group and a group given resveratrol at a dose of 250 mg/day accompanied by the same program of physical activity and diet carried out for three months. After 3 months, it was reported that in the resveratrol group there was a significant decrease ($p < 0.05$) in body weight, BMI, waist circumference, total cholesterol (TC), VLDL and a significant increase in HDL levels, while in the placebo group significant reduction in body weight, BMI and waist circumference was also reported but no significant reduction in lipid profile [40].

Another study by Rabbani et al. (2021) reported that administration of oral capsules containing a combination of trans-resveratrol and hesperetin (90 mg tRES: 120 mg HESP) for 8 weeks tested on obese patients showed a decrease in inflammation which was characterized by a decrease in the expression of IL-8 and receptor for the advanced glycation end product (RAGE). In addition, this combination also improves insulin resistance and hyperglycemia [41].

In vivo testing has been carried out on rats induced with a high-fat diet for 12 weeks which were then given resveratrol at a dose of 20 mg/kg/day for 4 weeks, and the results showed a reduction in total cholesterol by 8.4% and LDL by 6.6%. compared to the HFD group (hyperlipidemia group) [42]. Another study was conducted by Campbell et al. (2019) on male C57BL/6J rats which were given a high-fat diet for 16 weeks accompanied by the administration of resveratrol doses of 50, 75 and 100 mg/kg body weight via drinking water. Based on the results, the administration of resveratrol doses of 75 and 100 mg/kgbw could significantly ($p < 0.05$) prevent weight gain in rats compared to the HFD group. In addition, the administration of resveratrol doses of 75 and 100 mg/kgbw was also reported to prevent chronic inflammation, which was characterized by a decrease in serum IL-1 and TNFα ($p < 0.05$), as well as oxidative stress in the liver and brain as indicated by an increase in activity of superoxide dismutase, catalase and glutathione peroxidase ($p < 0.05$) [43].

Research on the antiobesity activity of resveratrol compounds has also been carried out by Chang et al. (2016) in vivo and in vitro. In vivo testing activity of resveratrol was carried out on male C57BL/6C rats induced with high fat diet (HFD) accompanied by the administration of resveratrol at doses of 1, 10 and 30 mg/kgbw for 10 weeks with the results showing that the administration of resveratrol with these three doses may significantly attenuate dose-dependent HFD-induced weight gain compared to the HFD group without any treatment. Furthermore, in vitro testing on 3T3-L1 cells by administering resveratrol at a concentration of 0.03 to 100 μM for 24 h significantly inhibited dose-dependent adipose lipolysis [44]. Based on in vivo, in vitro and clinical trials discussed previously, resveratrol is a potent antioxidant compound in overcoming metabolic disorders of obesity, dyslipidemia and diabetes. However, there is very little information about the dosage and safety for long-term use of this compound, hence, further research is needed to provide an optimal effect of resveratrol and reduce the risk of side effects.

4.2. Curcumin

Curcumin found in turmeric root (*Curcuma longa* L.) has been reported to have physiological effects such as antioxidant, antiobesity, anti-inflammatory, antidyslipidemic and antidiabetic. Turmeric is widely consumed in Asian countries and used as a cooking spice with no reported toxicity [45]. In vitro assays were carried out by Zhao et al. (2021) on 3T3-L1 preadipocytes and the results showed that incubation of curcumin at doses of 10, 20 and 35 μM for 8 days could induce adipogenic differentiation and accumulation of intracellular fat droplets. These results also showed that there was a 55.0% and 74.7% decrease in preadipocyte viability compared to the control group on incubation with 50 μM and 75 μM curcumin, respectively ($p < 0.01$). Administration of curcumin has also been reported to enhance mitochondrial respiratory function, induce adipogenic differentiation and regulate peroxisome proliferation activated receptor γ (PPARγ) and peroxisome proliferator-activated receptor gamma coactivator 1-alpha (PGC1α) expression [46,47].

In vivo testing was carried out on male albino wistar rats induced with HFD for 4 weeks followed by administration of curcumin at a dose of 80 mg/kg body weight/day for the next 6 weeks. The results showed that there was a significant decrease in BMI ($p < 0.05$) in rats given curcumin, a BMI value of 0.78 g/cm^2 compared to the obese group with a BMI of 0.86 g/cm^2 [48].

Clinical trials have been carried out by Thota et al. (2019) on several individuals with a high risk of developing diabetes, having a BMI ranging from 25–45 kg/m^2, with fasting glucose levels of 6.1–6.9 mmol/L and HbA1c levels between 5.7–6.4% were given curcumin tablets at a dose of 2 × 500 mg curcumin/day taken every morning and night for 12 weeks. The results of this study indicate that curcumin can increase insulin sensitivity by 32.7 ± 10.3%, reduce serum triglycerides by 0.79%, compared to the placebo group, which had an increase of 26.89%, and significantly reduce insulin resistance (−0.3 ± 0.1 vs. 0.01 ± 0.05, $p = 0.0142$), as compared to the placebo group [49,50].

In vivo testing was carried out on male wistar rats by inducing rats with an intraperitoneal injection of nicotinamide (110 mg/kg) and streptozotocin (45 mg/kg) in a fasting state. The results of this study by Goushki et al. (2020) reported that the administration of curcumin (100 and 200 mg/kg/day) and nano curcumin (100 and 200 mg/kg/day) could significantly ($p < 0.001$) reduce fasting blood sugar (FBS) with FBS levels respectively also 158.13, 163.75, 173.38 and 158 mg/dl, compared to the FBS value in the diabetes group of 518.5 mg/dl [51]. In addition, a study by Roxo et al. (2019) was also carried out on male wistar rats induced with STZ at a dose of 40 mg/kg IV so that diabetic rats with blood sugar levels in the range of 380–510 mg/dl were then given a dose of curcumin, 30 mg/kg, 60 mg/kg and 90 mg/kg, and the results showed that none of these doses caused toxicity in rats. Furthermore, the study reported that administration of curcumin at a dose of 90 mg/kg could improve the lipid profile, which was indicated by a significant decrease ($p < 0.05$) in plasma triacylglycerol and cholesterol levels compared to diabetes controls, and inhibit the advanced glycation end products (AGE)/RAGE signaling pathway [52,53].

Tests in streptozotocin-induced diabetic mice show that tetrahydrocurcumin (THC) at a dose of 120 mg/kg/day for 12 weeks can relieve diabetic cardiomyopathy by attenuating oxidative stress due to hyperglycemia and activating the SIRT1 pathway [54]. Other tests by Lima et al. (2020) reported that administration of curcumin at a dose of 90 mg/kg for 45 days can significantly increase the activity of antioxidant enzymes such as superoxide dismutase, paraoxonase 1 and catalase in 40 mg/kg STZ-induced rats compared to negative controls [55].

Based on a study by Li et al. (2019), the administration of curcumin 20 μM for 24 h at 37 °C to INS-1 cells induced with high glucose/palmitate could effectively inhibit oxidative stress, cell proliferation, increase insulin levels and reduce nicotinamide adenine dinucleotide phosphate (NADPH) oxidase expression as compared to high palmitate (PH) [56], while an in vivo study conducted on C57BL/6J rats induced with a high-fat diet for 3 months and then given curcumin at a dose of 1.5 g/kg/day for 8 weeks showed that curcumin can protect islet cells of Langerhans from apoptosis by modulating the NADPH pathway [57].

Based on the results of curcumin testing in vivo, in vitro and based on clinical trials, it is known that curcumin compounds have good potential in the treatment of metabolic disorders of obesity, dyslipidemia and diabetes. However, further research is needed to obtain the optimal dose, the appropriate dosage form, safety testing for long-term use and the possible side effects so that it can provide optimal effects on metabolic disorders of obesity, dyslipidemia and diabetes in humans.

4.3. *Quercetin*

Quercetin belongs to the class of flavonols found in many fruits and vegetables such as apples, berries, cauliflower, cabbage and beans. Quercetin has also been widely studied as having antioxidant, antidyslipidemic, antidiabetic, anti-inflammatory and other activities [58]. In vitro tests were carried out on 3T3-L1 adipocytes, and it was reported that administration of pentamethylquercetin (PMQ) at concentrations of 1 and 10 M increased glucose consumption by 24.6% and 66.4% ($p < 0.05$ and $p < 0.01$ vs. vehicle). This suggests that PMQ can increase insulin activity in 3T3L1 adiposity. Furthermore, in vivo testing was carried out on wistar rats induced with HFD and given PMQ at a dose of 0.04% g/g for 17 weeks. The test results showed a significant decrease ($p < 0.05$) in serum glucose, TC, TG and LDL levels in rats given PMQ compared to the HFD group of rats [59]. Tests on rats induced using STZ reported that there was a significant decrease ($p < 0.05$) in body weight in diabetic rats, and administration of 75 mg/kg bw of quercetin for 28 days showed an increase in body weight of 17.83%, a decrease in blood glucose of 66.80%, triglycerides of 24%, VLDL of 48%, LDL of 31.75% and total cholesterol 26.43% and an increase in HDL of 78.88%, compared to the group of HFD mice [60].

In vivo testing was carried out on wistar rats with STZ induction of 55 mg/kg bw, and an excision wound of 2 cm × 2 cm (400 mm^2) was made. Then, after being declared diabetic on blood sugar measurements after 72 h, quercetin was given orally at a dose of 100 mg/kg body weight + quercetin ointment (1%) for 21 days. The results showed that quercetin can normalize changes in blood glucose levels comparable to the normal control group with blood sugar levels ranging from ±150 mg/dl, while the blood sugar levels of the HFD group were ±350 mg/dl and could heal wound areas in diabetic rats significantly greater than the diabetes group [58]. A study by Zhuang et al. (2018) reported that administration of quercetin isolated from *Edgeworthia gardneri* at a dose of 0.5 g/kg quercetin per day for 4 weeks in db/db mice with type 2 diabetes mellitus (T2DM) could induce insulin secretion at a concentration of 0.10 mol/L, inhibit palmitate-induced pancreatic cell apoptosis and ameliorate mitochondrial dysfunction [61].

Clinical testing was conducted by Lee et al. (2016) on obese male and female Korean individuals who were given quercetin-rich onion peel extract (OPE) capsules at a dose of 100 mg for 12 weeks. The results obtained were that quercetin-rich OPE supplementation significantly reduced body weight and BMI from 70.0 ± 11.4 to 69.2 ± 11.4 kg ($p = 0.02$) and BMI from 26.6 ± 3.3 to 26.3 ± 3.2 kg/m^2 ($p = 0.03$), whereas in the placebo group there was no significant change. Waist and hip circumference showed significant changes in both groups. The waist circumference of the control group decreased from 90.2 ± 6.5 to 89.5 ± 6.4 cm, while the OPE group decreased by 2 cm from 91.9 ± 7.6 to 89.9 ± 7.7 cm. The hip circumference of the control group decreased from 100.7 ± 5.2 to 99.9 ± 4.6 cm, while the OPE group decreased by 1.3 cm from 101.1 ± 5.9 cm before the experiment to 99.9 ± 6.3 cm after the experiment. In addition, skinfold thickness in the control group decreased significantly by 2.2 mm from 33.2 ± 5.5 to 31.1 ± 5.6 mm ($p < 0.001$), whereas in the OPE group it decreased significantly by 3.2 mm from 34.1 ± 7.1 to 30.9 ± 6.4 mm ($p < 0.001$). OPE also showed a significant reduction in arm fat percentage by 0.7% from 36.1% ± 8.8% to 35.5% ± 5.5% ($p = 0.03$) and total body fat by 0.6% from 38.2% ± 6.5% to 37.6% ± 6.4% ($p = 0.02$) [62].

4.4. Anthocyanin

Anthocyanins are polyphenolic compounds found in pigmented fruits and vegetables. It is reported that this compound has pharmacological activities such as antioxidant, anti-inflammatory and anti-obesity [63]. In vitro assays were carried out on 3T3-L1 cells by administering 5, 10, 15, 20, 25, 30, 50, 100 and 200 g/mL anthocyanin fraction (AnT Fr) for 24 h, and the results showed the inhibition of lipid accumulation via regulation of adipogenesis and lipogenesis-related genes and signaling proteins [64]. An in vitro study by Han et al. (2018) reported that administration of anthocyanins at a dose of 200 g/mL showed a lipid reduction of 60% [65] and the administration of 10 g/mL can decrease ROS and increase catalase (CAT) and superoxide dismutase (SOD) enzymes significantly [66]. A study by Suantawee et al. (2017) reported that administration of cyanidin, an anthocyanin, at a dose of 1–300 μM can increase insulin release from INS-1 cells and stimulate insulin secretion [67], whereas administration at 60, 100 and 300 μM increased insulin secretion six times higher than the control [68]. In addition, in vivo studies on obese rats reported that administration of blackberry anthocyanins (BLA) and blueberry anthocyanins (BBA) at a dose of 200 mg/kg food for 12 weeks could inhibit body weight gain by 40.5% and 55.4%, respectively [63].

A clinical trial was conducted by Zhang et al. (2020) on dyslipidemic patients who were given anthocyanins in the form of supplements to see the dose-response relationship of oxidative stress and inflammation in dyslipidemic patients. Based on these tests, it was found that anthocyanin supplementation (320 mg/day) for 6 weeks significantly increased total-SOD compared to the placebo ($p < 0.05$). Anthocyanins (80 mg/day) significantly reduced serum IL-6 (−20%), TNF-α (−11%) and urinary 8-iso-PGF2α (−27%) versus the placebo ($p < 0.05$). A dosage of 320 mg/day anthocyanin supplementation can significantly reduce serum IL-6 (−40%), TNF-α (−21%) and malondialdehyde (MDA) (−20%) [69].

4.5. Other Antioxidants

Obesity may cause a decrease in total antioxidant capacity (TAC) in obese compared to normal individuals. This condition may also lead to a decrease in HDL levels [70]. Various in vivo and in vitro studies have been carried out to investigate the activity of antioxidant compounds from natural ingredients towards obesity. Lemon is known to have antioxidant activity, and lemon fermented product (LFP) at 0.75 and 1 mg/mL for 10 days was reported to inhibit the accumulation of lipids by 8.3% in 3T3L1 adipocytes. In addition, based on in vivo studies using mice, LFP at a dose of 2.89 g/kg for 9 weeks can reduce the body weight of obese mice by 9.7%, decrease triglyceride levels (17.0%), glucose (29.3%) and free fatty acids (17.9%) and can increase serum HDL (17.6%) [71].

In other studies, by Liao et al. (2019), antioxidant polysaccharides okra (OP) derived from okra (*Abelmoschus esculentus* L.) at doses of 200 and 400 mg/kgbw showed a significant reduction of dyslipidemia in rats induced by a high-fat diet and streptozotocin 100 mg/kg. The lipid profiles such as total cholesterol, triglycerides and LDL were significantly reduced as compared to the negative control group. These studies also reported that there was an increase in antioxidant enzymes at a dose of 400 mg/kg of OP such as superoxide dismutase (sod), catalase (cat) and glutathione peroxidase (gsh-px) by 274.18 ± 24.1, 57.09 ± 6.91 and 530.08 ± 45.1 u/mg prot respectively [72].

Clinical trials were conducted on healthy individuals aged 30–75 years who consumed green tea in combination with glucosyl hesperidin (GT gH), which contained 178 mg glucosyl hesperidin and 146 mg epigallocatechin gallate (EGCG), for 12 weeks. The results showed that GT gH prevented the addition of body weight, and the antiobesity effect of GT gH is more pronounced in people < 50 years old [73].

In vivo testing was carried out on wistar rats fed with strawberry ellagitannins (ET), which showed that a level of 0.24% of the total diet for 4 weeks can be used effectively for the prevention and treatment of metabolic disorders associated with obesity, dyslipidemia, imbalanced redox status and inflammation [74]. In vivo testing was also carried out by Sousa et al. (2020) on male Sprague-Dawley rats induced on a high-fat diet then given α-terpineol at a dose of 50 mg/kg, which improved insulin sensitivity and reduced ($p < 0.05$) serum levels of the proinflammatory cytokines TNF-α and IL-1β, when compared with the control group [75].

5. Effect of Antioxidants on Metabolic Disorders of Obesity, Dyslipidemia and Diabetes

5.1. Relationship between Obesity, Dyslipidemia and Diabetes

In the obese population, there is a decrease in skeletal muscle strength and function as well as impaired skeletal muscle mitochondrial respiratory function that contributes to increased mitochondrial ROS production compared to normal-weight individuals [76]. It has been reported that the ratio of type II and type I skeletal muscle fibers is higher than that of normal individuals, with two to three-fold ROS production [77]. Tumor necrosis factor alpha (TNF-α) functions as a catalyst in oxidative stress which is only expressed by type II muscle fibers. Based on studies, systemic administration of TNF-α has been shown to reduce the production of skeletal muscle strength in test animals and can increase muscle protein loss through oxidative activation of the TNF-α/nuclear factor kappa B (NF-κB) signaling pathway. Skeletal muscle oxidative stress induced by TNF-α is preventable by the administration of antioxidants, suggesting that TNF-α may provide an important target for confirming obesity-associated oxidative stress [78].

Obesity can trigger various complications, as shown in Figure 2. Obesity leads to increased and dysfunctional adipocytes [79]. Adipocytes produce adipokines and hormones whose rate and effect of secretion are influenced by the distribution and amount of available adipose tissue. High secretion of pro-inflammatory adipokines by adipocytes and macrophages may cause systemic inflammation in some obese patients [80]. The accumulation of excess lipid intermediates (such as ceramides) triggers lipotoxicity with cell dysfunction and apoptosis in some non-obese tissues. Inflammatory cytokines that are elevated in non-adipose tissue cause impaired signaling and insulin resistance, especially in obese patients, leading to type 2 DM [81].

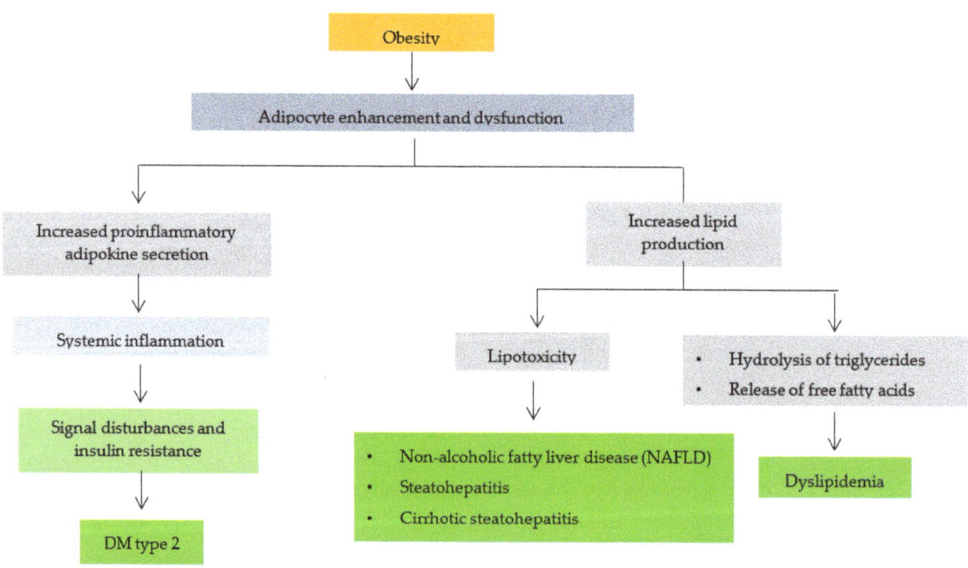

Figure 2. The relationship of obesity with other disease complications (modification of B. Heymsfield et al., 2017) [81].

Increased adipocytes can also trigger increased lipid production, causing the triglycerides in adipocytes to hydrolyze and release free fatty acids (FFA). Dilation of adipose tissue causes high plasma FFA levels in some patients. Apart from adipose tissue, lipids can also be found in liposomes [82]. Too many fat cells can cause liposomes (steatosis) in liver cells to expand and form large vacuoles associated with diseases, including non-alcoholic fatty liver disease (NAFLD), steatohepatitis and cirrhotic steatohepatitis [83]. This finding is one of many pathophysiological mechanisms of obesity-induced dyslipidemia (increased triglyceride levels, LDL, decreased HDL), type 2 DM, obesity-associated liver disease and osteoarthritis [84–86].

In obese patients, there is an increase in reactive oxygen species (ROS) and a decrease in antioxidant defense. Increased oxidative stress in obesity can cause inflammation. In addition, the hormone leptin secreted by adipocytes also plays a role in inducing oxidative stress [87]. Oxidative stress plays a role in causing insulin resistance leading to type 2 diabetes and dyslipidemia [88].

Compounds with antioxidant activity in the treatment of type 2 diabetes can activate the 5′adenosine monophosphate-activated protein kinase (AMPK) pathways, downregulate the expression of cyclooxygenase-2 (COX2) related genes to release pro-inflammatory mediators, increase glucose tolerance and insulin sensitivity, reduce inflammatory cells and reduce cytokines levels. Pro-inflammatory agents in the serum, such as IL-1B, IL-6 and TNF-α, can inhibit the activation of NF-κB and inhibit the expression of macrophage chemotactic protein (MCP1) [89]. It is reported that the use of antioxidants in patients with type 2 diabetes can effectively prevent complications, which is supported by various studies on antioxidants and the pathological process of diabetes caused by increased oxidative stress [13].

5.2. Antioxidant Mechanisms Associated with Obesity, Dyslipidemia and Diabetes

Several antioxidant compounds have been studied and have activity against obesity, oxidative stress, dyslipidemia and diabetes. Kukoamine B compounds are known to have activity in preventing inflammation and reducing lipid accumulation and oxidative stress [90]. Another antioxidant compound, salidroside, has also been investigated to have activity in inhibiting the formation of ROS that leads to oxidative stress [91,92]. In addition, salidroside can also protect cells from apoptosis caused by H_2O_2 induction [93]. The antioxidant compound polysaccharide okra (OP) is known to prevent an increase in levels of free fatty acids (FFA), triglycerides and LDL and can prevent a decrease in HDL levels [72].

Kahweol is an antioxidant diterpene compound derived from coffee. Based on research, kahweol can inhibit adipogenesis and lipid accumulation while lowering blood glucose levels in rats induced by hyperglycemia [14,94]. Besides playing a role in inhibiting ROS and suppressing lipid accumulation, antioxidant compounds also have activity against insulin resistance. An antioxidant compound known to affect insulin resistance is asphalathin. Asphalathin can improve insulin resistance in in vitro testing with palmitate induction [95,96], while other studies have also shown that asphalathin can treat hyperglycemia accompanied by inflammation and apoptosis [97].

Another antioxidant compound that also affects insulin resistance is isothiocyanate. This compound reduced lipid accumulation and inflammation in the palmitate-induced test [98]. In another study, it was also reported that isothiocyanate compounds can suppress inflammation caused by an increase in pro-inflammatory cytokines and can reduce oxidative stress levels [99]. The activities of some of these antioxidant compounds against obesity, oxidative stress, dyslipidemia and diabetes are summarized in Figure 3.

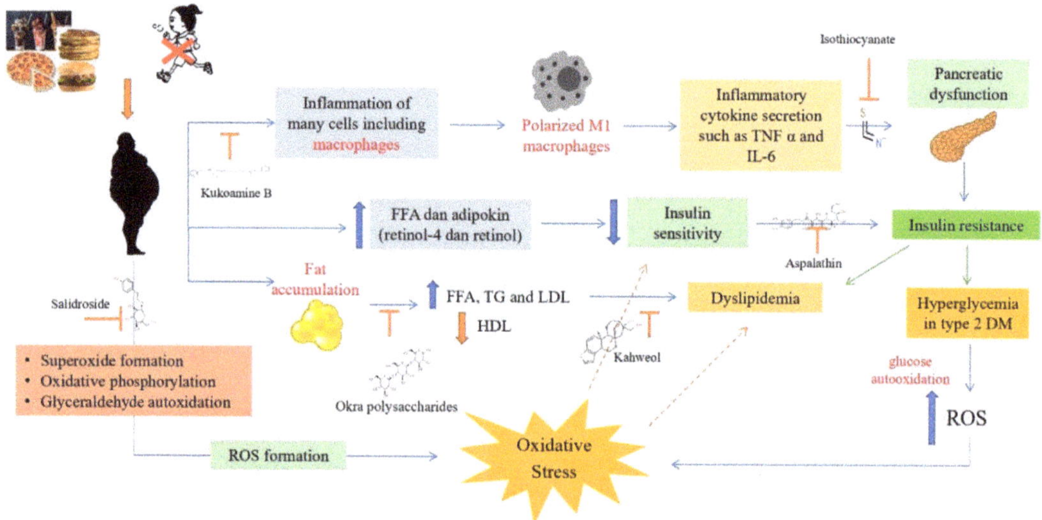

Figure 3. Antioxidant mechanisms in obesity, dyslipidemia and diabetes [100]. Obesity can cause inflammation of macrophage cells so macrophages will be polarized into M1 macrophages due to inflammation. M1 polarized macrophages will secrete inflammatory cytokines such as TNFα and IL-6, which can cause pancreatic dysfunction that leads to insulin resistance. Insulin resistance will cause type 2 DM with hyperglycemia, which can increase ROS, causing oxidative stress. Obesity can form ROS through the formation of superoxide, oxidative phosphorylation and auto-oxidation of glyceraldehyde, causing oxidative stress. Obesity can also cause an increase in FFA and adipokines, which can reduce insulin sensitivity and lead to type 2 diabetes and dyslipidemia. In addition, obesity also affects fat accumulation, which can cause an increase in FFA, TG and LDL and a decrease in HDL, which can cause dyslipidemia.

6. Antioxidant Compound Signaling Pathways

Based on the previous discussion, it is known that several antioxidant compounds have antiobesity, antidyslipidemic and antidiabetic activities both in vitro and in vivo and clinically. Furthermore, this review discusses the signaling pathways of antioxidant compounds derived from natural products against obesity, dyslipidemia and diabetes as well as oxidative stress that can trigger the emergence of these metabolic disorders (Table 1).

Table 1. Antioxidant signaling pathways of natural product.

Compounds	Sources	Experimental Models	Mechanisms	Ref.
Anthocyanin (100 and 400 mg/kg for 5 weeks)	*Vaccinium corymbosum*	Streptozotocin-induced diabetic rats and HepG2 cells.	Hyperglycemia and hyperlipidemia are inhibited by reducing the expression of enzymes involved in gluconeogenesis, lipogenesis, and lipolysis via the adenosine monophosphate (AMPK)-activated kinase signaling pathway in HepG2 cells.	[101]

Table 1. *Cont.*

Compounds	Sources	Experimental Models	Mechanisms	Ref.
Aspalathin (10 g/mL for 3 h)	*Aspalathus linearis* (Green rooibos)	C2C12 skeletal muscle cells and 3T3-L1 fat cells induced with palmitate.	Aspalathin modulates the major insulin signaling PI3K/AKT and AMPK effectors to ameliorate insulin resistance by increasing glucose transporter expression.	[96]
Bouchardatine (50 mg/kg/days)	*Bouchardatia neurococca*	Male C57BL/6J mice induced with HFD.	Bou may have therapeutic potential for obesity-related metabolic diseases by increasing the capacity of energy expenditure in adipose tissues and liver through a mechanism involving the SIRT1–LKB1–AMPK axis.	[102]
Ginseng oligopeptides (GOPs) (0.125, 0.5 and 2.0 g/kg bw for 7, 24 and 52 weeks)	*Panax ginseng*	Mice that were induced with a high-fat diet for 4 weeks.	Oligopeptides increase the normal content of insulin and protect pancreatic cells from apoptosis associated with type 2 diabetes mellitus by inhibiting NF-κB activity to protect against inflammation due to diabetes.	[103]
Gossypol in vivo (1 and 2.5 mg/kg at 0, 30, 60, 90, 120, 150 and 180 min on glucose tolerance test) and in vitro (25, 50 and 75 μg/mL for 24 h)	*Gossypium* sp.	Mouse myoblast cells (C2C12) and streptozotocin-induced (STZ) mouse myoblasts.	Gossypol (GSP) can activate the insulin receptor substrate 1 (IRS-1)/protein kinase B (Akt) signaling pathway and can translocate glucose transporter 4 (GLUT 4) into the plasma membrane at C2C12 myotube, thereby increasing glucose uptake.	[104]
Hyperoside (200, 100 and 50 mg/kg for 4 weeks)	*Zanthoxylum bungeanum*	Mice that were induced with alloxan and a high-fat diet.	Hyperoside inhibits the phosphorylation of p65/NF-κB, MAPK (including p38, JNK and ERK1/2).	[105]
Isothiocyanate (*Moringa isothiocyanate*/MIC-1) (5 μM for 24 h)	*Moringa oleifera*	HK-2 cells were given high glucose to induce oxidative stress.	Nrf2-ARE is activated by MIC1 to suppress inflammation and reduce oxidative stress.	[99]
Kahweol (2.5 and 5 μM for 24 h)	*Coffea* sp.	INS-1 cells tonal clonal induced with streptozotocin (STZ).	Kahweol downregulates NF-κB, antioxidant proteins, inhibitors of DNA binding and cell differentiation.	[14]
Kukoamine B (50mg/kg/day for 9 weeks)	*Lycium chinense*	Diabetic mouse model (dB/dB) using metabolomics approach (Biocrates p180)	Kukoamine B regulates the NF-κB/PPAR transcriptional pathway to reduce inflammation in diabetes.	[106]
Lycium barbarum Polysaccharide (LBPS) (100, 250, and 500mg/kg for 4 weeks)	*Lycium barbarum*	HFD and streptozotocin-induced mice.	It inhibits serum levels of inflammatory factors (IL-2, IL-6, TNF-α, and IFN-α), protects kidney damage and inhibits NF-κB expression.	[107]

Table 1. *Cont.*

Compounds	Sources	Experimental Models	Mechanisms	Ref.
Mangiferin (40 mg/kg for 28 days)	*Mangifera indica*	Research on myocardial ischemia-reperfusion (IR) in diabetic rats.	Mangiferin can reduce IR injury in diabetic rats through inhibition of the AGE-RAGE/MAPK pathway thereby preventing oxidative stress, apoptosis and inflammation.	[108]
Morroniside (6.25, 12.5, 25, 50 and 100 μmol/L for 24 h)	*Cornus officinalis* Sieb.	In vitro study using rat renal tubular epithelial cells (mRETCs) induced with palmitate and glucose.	Morroniside increases cholesterol reduction via the PGC1a/LXR pathway and and it also downregulates RAGE, p38MAPK and NF-κB expression via the AGEs/RAGE signaling pathway.	[109]
Nodakenin (NK) (10 and 20 mg/kg for 5 weeks)	*Angelicae gigas*	Male C57BL/6N mice with a high-fat diet.	Administration of NK can improve the phosphorylation level of AMPK, indicating that NK exerts anti-adipogenic and antioxidant effects.	[110]
Onopordopicrin (0.125, 0.25 and 0.5 μg/mL for 24 h)	*Arctium lappa*	A model of human muscle cells exposed to H2O2 oxidative stress.	Onopordopicrin has antioxidant activity by limiting the production of free radicals and DNA damage and through activation of the Nrf2/HO-1 signaling pathway in muscle cells.	[111]
Pectic bee pollen polysaccharide (RBPP-P) in vitro (0.1 mg/mL for 24 h and in vivo (20 mg/kg for 8 weeks)	*Rosa rugosa*	HepG2 cells treated with high-glucose and high-fatty acids and obese mice with a high-fat diet (HFD) inducer.	This polysaccharide is able to decrease hepatic steatosis and insulin resistance by promoting autophagy through AMPK/mTOR-mediated signaling pathways.	[112]
Phanginin A (250 mg/kg for 26 days)	*Caesalpinia sappan*	Male ob/ob mice.	Phanginin A activates SIK1 and causes inhibition of gluconeogenesis with increased PDE4 and inhibition of the cAMP/PKA/CREB pathway in the liver.	[113]
Polyphenol (125–500 mg GP/mL for 8 days)	*Vitis vinivera*	Preadiposit 3T3-F442A cells.	It induces adiposity differentiation through upregulation of GLUT-4, PI3K and adipogenic genes.	[114]
Polysaccharide (200 and 400 mg/kg bw for 8 weeks)	Okra (*Abelmoschus esculentus* (L.) Moench).	Rats that were given a high-fat diet (HFD) combined with injection of 100 mg/kg streptozotocin (STZ) intraperitoneally (ip).	Okra polysaccharide (OP) exert their type 2 antidiabetic effects in part by modulating oxidative stress via Nrf2 transport in the PI3K/AKT/GSK3β pathway.	[72]

Table 1. *Cont.*

Compounds	Sources	Experimental Models	Mechanisms	Ref.
Polysaccharide (80, 160 and 320 mg/kg/day for 4 weeks)	*Angelica sinensis*	BALB/C mice induced with a high-fat diet were used.	Angelica sinensis polysaccharide (ASP) is reported to lower blood glucose and improve insulin resistance through regulation of metabolic enzymes and activation of the PI3K/Akt pathway in HFD mice. It can also decrease lipid accumulation and fatty liver by increasing PPARγ expression and activation of the adiponectin signaling pathway SIRT1 and AMPK.	[115]
Polysaccharide (0.1, 1.0, 10 and 100 μg/mL for 0, 12, 24, 48 and 72 h)	*Astragalus mongholicus*	AGE-induced DCM cell model.	Astragalus polysaccharides can decrease intracellular ROS levels, increase SOD activity and GSH-Px and lower MDA and NO levels.	[116]
Procyanidin (25, 50 and 75 μg/mL for 24 h)	*Rubus amabilis*	MIN6 cells were given 0.5 mM palmitate (PA) for 24 h to induce cell apoptosis.	Procyanidin can activate the PI3K/Akt/FoxO1 signal to protect MIN6 cells from apoptosis induced by palmitate induction.	[117]
Puerarin (25, 50 and 100 mg/kg for 12 weeks)	*Pueraria lobata*	Mice induced with streptozotocin.	Puerarin significantly lowers blood sugar levels and prevents cataracts as well as lowers the level of expression of retinal vascular endothelial growth factor and interleukin 1β and increases the expression of Nrf2 and Ho-1 mRNA so that it can reduce oxidative stress in diabetic rats.	[118]
Pyrogallol-phloroglucinol-6,6-bieckol (PPB) (2 mg/kg for 4 weeks)	*Ecklonia cava*	C57BL/6N mice induced with HFD for 8 weeks.	It inhibits RAGE ligands, reduces RAGE expression and binding of RAGE and RAGE ligands and reduces proinflammatory cytokines that cause obesity.	[119]
Resveratrol (1 mg/kg/day for 8 weeks)	*Polygonum cuspidatum*	Goto-Kakizaki (GK) type 2 diabetic female rats.	Resveratrol increases adenine nucleotide and citrate synthase activity by increasing the expression of eNOS-SIRT1 and P-AKT.	[120]
Salidroside (100 mg/kg/day for 5 weeks)	*Rhodiola rosea*	Mice induced by high-fat diet (HFD).	Salidroside suppresses ROS production and inhibits the JNK-caspase apoptotic cascade, inhibiting FOXO-1 by activating AMPK-AKT.	[121]
Saponins (40 mg/kg)	*Momordica carantia* L.	Mice that were induced with a high-fat diet and streptozotocin.	Saponins exhibit hypoglycemic activity possibly via the AMPK/NF-κB signaling pathway by activating AMPK phosphorylation and energy metabolism of the body.	[122]

Table 1. *Cont.*

Compounds	Sources	Experimental Models	Mechanisms	Ref.
Simmondsin (10, 20, 40, 80 and 150 μg/mL for simmondsin for 24 h)	*Simmondsia Chinensis*	Fructose-induced oxidative stress in RIN5f beta cells.	Simmondsin is reported to reduce ROS by 69%, activate caspase-3, increase antioxidant defense, inhibit p22phox and increase Nrf2 factor.	[123]
Toosendanin (TSN) in vitro (12.5 nM, 25 and 50 nM for 6 days) in vivo (0.1 mg/kg/day for one month)	*Melia toosendan*	3T3L1 preadipocytes and mice induced with a high-fat diet.	TSN can inhibit adipocyte differentiation and lipid accumulation by activating Wnt/β-catenin signaling, inhibiting mRNA and protein levels of PPAR-γ and C/EBP-α, which proves that TSN can inhibit adipogenesis via its mechanism in inhibiting transcription factor cascades.	[124]

6.1. *The Phosphoinositide 3-Kinase/Protein Kinase B (PI3K/AKT)*

The phosphoinositide 3-kinase/protein kinase B (PI3K/AKT) signaling pathway is a regulator of physiological processes associated with type 2 diabetes mellitus. Most studies reported that the PI3K/AKT pathway not only promotes insulin signal transduction but can also stimulate glucose uptake in adipose and liver [72]. Phosphorylated protein kinases can activate glycogen synthase kinase 3 beta (GSK3β), which then triggers NF-E2-related factor (Nrf2) from the binding of Keap1 to the nucleus. Then, target genes are transactivated through antioxidant response elements (AREs) to inhibit oxidative stress. Some of the compounds in Table 1 that can induce vasodilation through the PI3K/AKT signaling pathway are polysaccharides [72], anthocyanin [101], resveratrol [120], gossypol [104], procyanidins [117] and polyphenol [114].

6.2. *The Nuclear Factor Erythroid 2-Related Factor 2 (Nrf2) Signaling*

The NFE2 system associated with kelch-like ECH-associated protein 1 (Keap1) factor 2 (Nrf2) is a defense system for cellular homeostasis. The interaction between Nrf2 and Keap1 can trigger the expression of the B globin gene known as a key marker of oxidative stress in cells [125]. Levels of oxidative stress and inflammation in cells are common in most tissues. Nrf2 and NF-κB (nuclear factor kappa-light-chain-enhancer of activated B cells) are the two main transcription factors that play a role in regulating cellular responses to oxidative stress and inflammation. There is functional crosstalk between these two pathways based on pharmacological and genetic studies which stated that NF-κB activity will be disrupted with the absence of Nrf2, causing an increase in cytokine production. In addition, NF-B also plays a role in modulating the activity and transcription of Nrf2 [126]. Based on Table 1, several natural compounds were reported to act on the Nrf2 cell homeostasis system such as isothiocyanates [99], ocra polysaccharides [72], simmondsine [123] and puerarine [118].

Moringa isothiocyanate (MIC1) is the main isothiocyanate found in *Moringa oleifera*. MIC-1 can activate Nrf2-ARE at levels similar to sulforaphane (SFN), suppress pro-inflammatory cytokines, reduce ROS and inhibit high glucose (HG)-induced transforming growth factor beta 1 (TGFβ1) [99]. Another antioxidant compound, namely, okra polysaccharide (OP), significantly reduces the increase in blood sugar, cholesterol, triglycerides and LDL. OP also decreases ROS and mitochondrial dysfunction by inhibiting activation of NADPH oksidase 2 (Nox2). In summary, OP has activity against type 2 DM via Nrf2 transport of the PI3K/AKT pathway [72]. Simmondsin and puerarin are also reported to reduce oxidative stress via the same mechanism, namely, by activating the Nrf2 pathway [118,123].

6.3. The Peroxisome Proliferation Activated Receptor γ (PPARγ)

The peroxisome proliferation activated receptor γ (PPARγ) is a transmembrane transcription factor. When activated by the ligand, PPARγ inhibits the transcription of NF-κB and reduces the expression of the cytokine gene in inflammation, which may decrease the inflammatory response. In vivo studies have shown that the induction of a high-fat diet can lead to an increase in glucose levels and insulin resistance accompanied by a decrease in PPAR activation [127]. Compounds reported in Table 1 with activity to increase PPARγ expression are kahweol [14], *Angelica sinensis* polysaccharide (ASP) [115] and toosendanin [128].

6.4. The Nuclear Factor Kappa-Light-Chain-Enhancer of Activated B Cells (NF-κB) Signaling

NF-κB is a transcription factor consisting of seven transcription factors that are structurally related and play a role in regulating the expression of many genes. NF-κB usually represents the p50-p65 heterodimer, which is the major Rel/NF-κB complex in most cells. The NF-κB subunit is expressed in many places but its induction and expression depend on the stimulus and cell type. NF-κB can be activated by cytokines, ROS, viral infection, vasopressors and DNA damage. NF-κB and Nrf2 play a role in cellular homeostasis and responses to stress and inflammation, whose molecular mechanisms depend on cell type and tissue context [129]. Based on Table 1, several natural compounds are reported to act on the NFκB cell homeostasis system such as kahweol [14], kukoamine B [106], saponins [122], oligopeptides [103] and hyperoside [105]. These compounds can prevent cell apoptosis due to the induction of hyperglycemia and downregulation of NFκB to protect against inflammation caused by diabetes [14,103].

6.5. 5′AMP-Activated Protein Kinase (AMPK) Signaling

AMP-activated protein kinase (AMPK) plays a role in regulating energy metabolism through the inhibition of anabolic pathways and stimulation of catabolic pathways. In addition, AMPK also plays a role in enzyme regulation through phosphorylation and regulation of transcription factors and coactivators [130]. Based on Table 1, several natural compounds can increase AMPK levels, such as salidroside [121], aspalathin [102], nodakenine [110], anthocyanins [101], saponin [122], peptic bee pollen polysaccharide [112] and bouchardatine [102]. Salidroside (Figure 4) and other antioxidant compounds that act in the AMPK pathways can decrease ROS production, improve mitochondrial function by reducing NADPH oxidase-2 (NOX2) expression and inhibit the JNK-caspase 3 apoptotic cascade by activating AMPK [121].

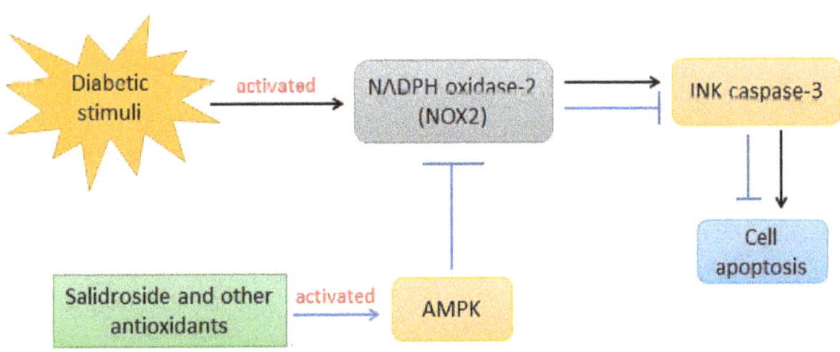

Figure 4. The salidroside mechanism in preventing oxidative diabetes (modification from Ju L et al., 2017) [121].

6.6. AGE/RAGE

In conditions of persistent hyperglycemia of uncontrolled diabetes mellitus, the end product is advanced glycation end (AGE). Glycation is one of the mechanisms that contribute to diabetes complications such as cardiomyopathy, nephropathy, retinopathy and neuropathy [131]. Advanced glycation ends (AGEs) play a role in the disruption of cellular functions including denaturation of target proteins and reduced function of AGEs. The accumulation of AGEs in tissues can damage organs and trigger chemical oxidative stress. Another mechanism of AGEs is the receptor associated with the receptor for the advanced glycation end product (RAGE). The increase in RAGE causes an increase in ROS synthesis and oxidative stress. One of the oxidative stresses is the phosphorylation of the primary signal transduction cascade, which is the molecularly activated protein kinase (MAPK) that activates NF-κB [132]. In Table 1, the compounds with a AGE/RAGE inhibitory mechanism are mangiferin [108], morroniside [109] and pyrogallol-phloroglucinol-6,6-bieckol (PPB) [119].

6.7. SIRT (Sirtuin)

Sirtuin 1 (SIRT1) and sirtuin 3 (SIRT3) proteins play an important role in counteracting oxidative stress. In degenerative diseases such as type 2 diabetes, especially in women, there is a high risk of death from myocardial infarction, even with drug therapy for diabetes. Compounds that can increase SIRT1 are resveratrol and bouchardatine [102] (Table 1). Resveratrol (RSV) is a natural polyphenol that has antioxidant activity and can improve mitochondrial dysfunction. RSV has been shown to increase NO production, increase NOS expression and activity, prevent eNOS release and increase NO bioavailability [120].

7. Conclusions

Various antioxidant compounds have been reported to have beneficial activities against obesity, dyslipidemia and diabetes in the literature. The molecular signaling mechanism of the reported compounds associated with obesity, dyslipidemia and diabetes has also been discussed. However, further research is needed to determine the optimal dose of these antioxidant compounds so as to provide optimal effects on these metabolic disorders. Furthermore, the review also provides insights into antioxidant compounds that act simultaneously against obesity, dyslipidemia and diabetes to minimize the use of drugs and the risk of side effects. Further research on the side effects of long-term use of these antioxidant compounds should also be explored, thereby increasing the safety of long-term use of these compounds.

Author Contributions: C.K. conceived and drafted the manuscript. S.A.S., N.K.K.I. and M.M. were involved in the editing process. All authors have read and agreed to the published version of the manuscript.

Funding: This work is supported by Universitas Padjadjaran.

Institutional Review Board Statement: Not applicable.

Informed Consent Statement: Not applicable.

Data Availability Statement: All data included in the article.

Acknowledgments: We thank Rector of Universitas Padjadjaran for supporting this study through Academic Leadership Grant No. 1959/UN6.3.1/PT.00/2021.

Conflicts of Interest: The authors declare no conflict of interest.

References

1. Leibowitz, K.L.; Moore, R.H.; Ahima, R.S.; Stunkard, A.J.; Stallings, V.A.; Berkowitz, R.I.; Chittams, J.L.; Faith, M.S.; Stettler, N. Maternal obesity associated with inflammation in their children. *World J. Pediatr.* **2012**, *8*, 76–79. [CrossRef] [PubMed]
2. Manna, P.; Jain, S.K. Obesity, Oxidative Stress, Adipose Tissue Dysfunction, and the Associated Health Risks: Causes and Therapeutic Strategies. *Metab. Syndr. Relat. Disord.* **2015**, *13*, 423–444. [CrossRef] [PubMed]
3. Oyenihi, A.B.; Ayeleso, A.O.; Mukwevho, E.; Masola, B. Antioxidant strategies in the management of diabetic neuropathy. *Biomed Res. Int.* **2015**, *2015*, 515042. [CrossRef]
4. Samavat, H.; Newman, A.R.; Wang, R.; Yuan, J.M.; Wu, A.H.; Kurzer, M.S. Effects of green tea catechin extract on serum lipids in postmenopausal women: A randomized, placebo-controlled clinical trial. *Am. J. Clin. Nutr.* **2016**, *104*, 1671–1682. [CrossRef] [PubMed]

5. Vekic, J.; Zeljkovic, A.; Stefanovic, A.; Jelic-Ivanovic, Z.; Spasojevic-Kalimanovska, V. Obesity and dyslipidemia. *Metabolism* **2019**, *92*, 71–81. [CrossRef]
6. Atlas, D. *International Diabetes Federation*, 9th ed.; International Diabetes Federation: Brussels, Belgium, 2019.
7. Mohieldein, A.H.; Hasan, M.; Al-Harbi, K.K.; Alodailah, S.S.; Azahrani, R.M.; Al-Mushawwah, S.A. Dyslipidemia and reduced total antioxidant status in young adult Saudis with prediabetes. *Diabetes Metab. Syndr. Clin. Res. Rev.* **2015**, *9*, 287–291. [CrossRef]
8. Rambhade, S.; Chakraborty, A.K.; Patil, U.K.; Rambhade, A. Diabetes Mellitus-Its complications, factors influencing complications and prevention-An Overview. *J. Chem. Pharm. Res.* **2010**, *2*, 7–25.
9. Kashiyama, K.; Sonoda, S.; Otsuji, Y. Reconsideration of Secondary Risk Management Strategies in Patients with Ischemic Heart Disease. *J. UOEH* **2017**, *39*, 11–24. [CrossRef] [PubMed]
10. Vona, R.; Gambardella, L.; Cittadini, C.; Straface, E.; Pietraforte, D. Biomarkers of oxidative stress in metabolic syndrome and associated diseases. *Oxid. Med. Cell. Longev.* **2019**, *2019*. [CrossRef]
11. Pechánová, O.; Varga, Z.V.; Cebová, M.; Giricz, Z.; Pacher, P.; Ferdinandy, P. Cardiac NO signalling in the metabolic syndrome. *Br. J. Pharmacol.* **2015**, *172*, 1415–1433. [CrossRef]
12. Kuk, J.L.; Rotondi, M.; Sui, X.; Blair, S.N.; Ardern, C.I. Individuals with obesity but no other metabolic risk factors are not at significantly elevated all-cause mortality risk in men and women. *Clin. Obes.* **2018**, *8*, 305–312. [CrossRef] [PubMed]
13. Prawitasari, D.S. Diabetes Melitus dan Antioksidan. *Keluwih J. Kesehat. Kedokt.* **2019**, *1*, 48–52. [CrossRef]
14. El-Huneidi, W.; Anjum, S.; Bajbouj, K.; Abu-Gharbieh, E.; Taneera, J. The coffee diterpene, kahweol, ameliorates pancreatic β-cell function in streptozotocin (Stz)-treated rat ins-1 cells through nf-kb and p-akt/bcl-2 pathways. *Molecules* **2021**, *26*, 5167. [CrossRef] [PubMed]
15. Pan, B.; Ge, L.; Xun, Y.Q.; Chen, Y.J.; Gao, C.Y.; Han, X.; Zuo, L.Q.; Shan, H.Q.; Yang, K.H.; Ding, G.W.; et al. Exercise training modalities in patients with type 2 diabetes mellitus: A systematic review and network meta-analysis. *Int. J. Behav. Nutr. Phys. Act.* **2018**, *15*, 1–14. [CrossRef]
16. Yang, D.K.; Kang, H.S. Anti-diabetic effect of cotreatment with quercetin and resveratrol in streptozotocin-induced diabetic rats. *Biomol. Ther.* **2018**, *26*, 130–138. [CrossRef]
17. Dal, S.; Sigrist, S. The Protective Effect of Antioxidants Consumption on Diabetes and Vascular Complications. *Diseases* **2016**, *4*, 24. [CrossRef] [PubMed]
18. Shabbir, U.; Rubab, M.; Daliri, E.B.; Chelliah, R.; Javed, A.; Oh, D. Curcumin, Quercetin, Catechins and Metabolic Diseases : The Role of Gut Microbiota. *Nutrients* **2021**, *13*, 206. [CrossRef]
19. Pizzino, G.; Irrera, N.; Cucinotta, M.; Pallio, G.; Mannino, F.; Arcoraci, V.; Squadrito, F.; Altavilla, D.; Bitto, A. Oxidative Stress: Harms and Benefits for Human Health. *Oxid. Med. Cell. Longev.* **2017**, *2017*. [CrossRef]
20. Paravicini, T.M.; Touyz, R.M. NADPH oxidases, reactive oxygen species, and hypertension: Clinical implications and therapeutic possibilities. *Diabetes Care* **2008**, *31* (Suppl. 2), S170–S180. [CrossRef]
21. Corbi, S.C.T.; Bastos, A.S.; Orrico, S.R.P.; Secolin, R.; Dos Santos, R.A.; Takahashi, C.S.; Scarel-Caminaga, R.M. Elevated micronucleus frequency in patients with type 2 diabetes, dyslipidemia and periodontitis. *Mutagenesis* **2014**, *29*, 433–439. [CrossRef]
22. Ren, W.; Xia, Y.; Chen, S.; Wu, G.; Bazer, F.W.; Zhou, B.; Tan, B.; Zhu, G.; Deng, J.; Yin, Y. Glutamine Metabolism in Macrophages: A Novel Target for Obesity/Type 2 Diabetes. *Adv. Nutr.* **2019**, *10*, 221–230. [CrossRef] [PubMed]
23. Wu, H.; Ballantyne, C.M. Metabolic Inflammation and Insulin Resistance in Obesity. *Circ. Res.* **2020**, *126*, 1549–1564. [CrossRef]
24. Flock, M.R.; Green, M.H.; Kris-Etherton, P.M. Effects of adiposity on plasma lipid response to reductions in dietary saturated fatty acids and cholesterol. *Adv. Nutr.* **2011**, *2*, 261–274. [CrossRef] [PubMed]
25. Klop, B.; Elte, J.W.F.; Cabezas, M.C. Dyslipidemia in Obesity: Mechanisms and Potential Targets. *Nutrients* **2013**, *5*, 1218–1240. [CrossRef] [PubMed]
26. Na, I.J.; Park, J.S.; Park, S.B. Association between abdominal obesity and oxidative stress in Korean adults. *Korean J. Fam. Med.* **2019**, *40*, 395–398. [CrossRef]
27. Luc, K.; Schramm-Luc, A.; Guzik, T.J.; Mikolajczyk, T.P. Oxidative stress and inflammatory markers in prediabetes and diabetes. *J. Physiol. Pharmacol.* **2019**, *70*, 809–824. [CrossRef]
28. Guzik, T.J.; Skiba, D.S.; Touyz, R.M.; Harrison, D.G. The role of infiltrating immune cells in dysfunctional adipose tissue. *Cardiovasc. Res.* **2017**, *113*, 1009–1023. [CrossRef]
29. Niemann, B.; Rohrbach, S.; Miller, M.R.; Newby, D.E.; Fuster, V.; Kovacic, J.C. Oxidative Stress and Cardiovascular Risk: Obesity, Diabetes, Smoking, and Pollution: Part 3 of a 3-Part Series. *J. Am. Coll. Cardiol.* **2017**, *70*, 230–251. [CrossRef]
30. Wang, H.; Peng, D.Q. New insights into the mechanism of low high-density lipoprotein cholesterol in obesity. *Lipids Health Dis.* **2011**, *10*, 176. [CrossRef]
31. Li, K.; Deng, Y.; Deng, G.; Chen, P.; Wang, Y.; Wu, H.; Ji, Z.; Yao, Z.; Zhang, X.; Yu, B.; et al. High cholesterol induces apoptosis and autophagy through the ROS-activated AKT/FOXO1 pathway in tendon-derived stem cells. *Stem Cell Res. Ther.* **2020**, *11*, 1–16. [CrossRef]
32. Pignatelli, P.; Menichelli, D.; Pastori, D.; Violi, F. Oxidative stress and cardiovascular disease: New insights. *Kardiol. Pol.* **2018**, *76*, 713–722. [CrossRef]
33. Czaja, A.J. Autoimmune Hepatitis: Focusing on Treatments other than Steroids. *Can. J. Gastroenterol.* **2012**, *26*, 615–620. [CrossRef] [PubMed]
34. Khan, R.M.M.; Chua, Z.J.Y.; Tan, J.C.; Yang, Y.; Liao, Z.; Zhao, Y. From pre-diabetes to diab. *Medicine* **2019**, *55*, 1–30.

35. Makki, K.; Froguel, P.; Wolowczuk, I. Adipose Tissue in Obesity-Related Inflammation and Insulin Resistance: Cells, Cytokines, and Chemokines. *ISRN Inflamm.* **2013**, *2013*, 1–12. [CrossRef] [PubMed]
36. Zhou, X.; Li, M.; Xiao, M.; Ruan, Q.; Chu, Z.; Ye, Z.; Zhong, L.; Zhang, H.; Huang, X.; Xie, W.; et al. ERβ Accelerates Diabetic Wound Healing by Ameliorating Hyperglycemia-Induced Persistent Oxidative Stress. *Front. Endocrinol.* **2019**, *10*, 1–11. [CrossRef]
37. Sen, S.; Chakraborty, R.; Sridhar, C.; Reddy, Y.S.R.; De, B. Free radicals, antioxidants, diseases and phytomedicines: Current status and future prospect. *Int. J. Pharm. Sci. Rev. Res.* **2010**, *3*, 91–100.
38. Kan, N.W.; Lee, M.C.; Tung, Y.T.; Chiu, C.C.; Huang, C.C.; Huang, W.C. The synergistic effects of resveratrol combined with resistant training on exercise performance and physiological adaption. *Nutrients* **2018**, *10*, 1360. [CrossRef] [PubMed]
39. Movahed, A.; Raj, P.; Nabipour, I.; Mahmoodi, M.; Ostovar, A. Efficacy and Safety of Resveratrol in Type 1 Diabetes. *Nutrients* **2020**, *12*, 161. [CrossRef]
40. Batista-Jorge, G.C.; Barcala-Jorge, A.S.; Silveira, M.F.; Lelis, D.F.; Andrade, J.M.O.; de Paula, A.M.B.; Guimarães, A.L.S.; Santos, S.H.S. Oral resveratrol supplementation improves Metabolic Syndrome features in obese patients submitted to a lifestyle-changing program. *Life Sci.* **2020**, *256*, 117962. [CrossRef]
41. Rabbani, N.; Xue, M.; Martin, O.W.; Thornalley, P.J. Subjects by trans -Resveratrol and Hesperetin. *Nutrients* **2021**, *13*, 2374. [CrossRef]
42. Zhu, X.; Yang, J.; Zhu, W.; Yin, X.; Yang, B.; Wei, Y.; Guo, X. Combination of berberine with resveratrol improves the lipid-lowering efficacy. *Int. J. Mol. Sci.* **2018**, *19*, 3903. [CrossRef]
43. Campbell, L.; Yu, R.; Li, F.; Zhou, Q.; Chen, D.; Qi, C.; Yin, Y.; Sun, J. Diabetes, Metabolic Syndrome and Obesity: Targets and Therapy Dovepress Modulation of fat metabolism and gut microbiota by resveratrol on high-fat diet-induced obese mice. *Diabetes Metab. Syndr. Obes. Targets Ther.* **2019**, *12*, 97–107. [CrossRef]
44. Chang, C.C.; Lin, K.Y.; Peng, K.Y.; Day, Y.J.; Hung, L.M. Resveratrol exerts anti-obesity effects in high-fat diet obese mice and displays differential dosage effects on cytotoxicity, differentiation, and lipolysis in 3T3-L1 cells. *Endocr. J.* **2016**, *63*, 169–178. [CrossRef] [PubMed]
45. Song, W.Y.; Choi, J.H. Korean Curcuma longa L. induces lipolysis and regulates leptin in adipocyte cells and rats. *Nutr. Res. Pract.* **2016**, *10*, 487–493. [CrossRef]
46. Ferguson, B.S.; Nam, H.; Morrison, R.F. Curcumin inhibits 3T3-L1 preadipocyte proliferation by mechanisms involving post-transcriptional p27 regulation. *Biochem. Biophys. Rep.* **2016**, *5*, 16–21. [CrossRef] [PubMed]
47. Zhao, D.; Pan, Y.; Yu, N.; Bai, Y.; Ma, R.; Mo, F.; Zuo, J.; Chen, B.; Jia, Q.; Zhang, D.; et al. Curcumin improves adipocytes browning and mitochondrial function in 3T3-L1 cells and obese rodent model. *R. Soc. Open Sci.* **2021**, *8*, 200974. [CrossRef] [PubMed]
48. Labban, R.S.M.; Alfawaz, H.A.; Almnaizel, A.T.; Al-Muammar, M.N.; Bhat, R.S.; El-Ansary, A. Garcinia mangostana extract and curcumin ameliorate oxidative stress, dyslipidemia, and hyperglycemia in high fat diet-induced obese Wistar albino rats. *Sci. Rep.* **2021**, *11*, 1–11. [CrossRef]
49. Thota, R.N.; Acharya, S.H.; Garg, M.L. Curcumin and/or omega-3 polyunsaturated fatty acids supplementation reduces insulin resistance and blood lipids in individuals with high risk of type 2 diabetes: A randomised controlled trial. *Lipids Health Dis.* **2019**, *18*, 1–11. [CrossRef] [PubMed]
50. Thota, R.N.; Rosato, J.I.; Dias, C.B.; Burrows, T.L.; Martins, R.N.; Garg, M.L. Dietary supplementation with curcumin reduce circulating levels of glycogen synthase kinase-3B and islet amyloid polypeptide in adults with high risk of type 2 diabetes and Alzheimer's disease. *Nutrients* **2020**, *12*, 1032. [CrossRef] [PubMed]
51. Shamsi-Goushki, A.; Mortazavi, Z.; Mirshekar, M.A.; Mohammadi, M.; Moradi-Kor, N.; Jafari-Maskouni, S.; Shahraki, M. Comparative effects of curcumin versus nano-curcumin on insulin resistance, serum levels of apelin and lipid profile in type 2 diabetic rats. *Diabetes Metab. Syndr. Obes. Targets Ther.* **2020**, *13*, 2337–2346. [CrossRef] [PubMed]
52. Roxo, D.F.; Arcaro, C.A.; Gutierres, V.O.; Costa, M.C.; Oliveira, J.O.; Lima, T.F.O.; Assis, R.P.; Brunetti, I.L.; Baviera, A.M. Curcumin combined with metformin decreases glycemia and dyslipidemia, and increases paraoxonase activity in diabetic rats. *Diabetol. Metab. Syndr.* **2019**, *11*, 1–8. [CrossRef]
53. Xie, T.; Chen, X.; Chen, W.; Huang, S.; Peng, X.; Tian, L.; Wu, X.; Huang, Y. Curcumin is a Potential Adjuvant to Alleviates Diabetic Retinal Injury via Reducing Oxidative Stress and Maintaining Nrf2 Pathway Homeostasis. *Front. Pharmacol.* **2021**, *12*, 1–15. [CrossRef] [PubMed]
54. Li, K.; Zhai, M.; Jiang, L.; Song, F.; Zhang, B.; Li, J.; Li, H.; Li, B.; Xia, L.; Xu, L.; et al. Tetrahydrocurcumin ameliorates diabetic cardiomyopathy by attenuating high glucose-induced oxidative stress and fibrosis via activating the SIRT1 pathway. *Oxid. Med. Cell. Longev.* **2019**, *2019*, 6746907. [CrossRef]
55. Lima, T.F.O.; Costa, M.C.; Figueiredo, I.D.; Inácio, M.D.; Rodrigues, M.R.; Assis, R.P.; Baviera, A.M.; Brunetti, I.L. Curcumin, Alone or in Combination with Aminoguanidine, Increases Antioxidant Defenses and Glycation Product Detoxification in Streptozotocin-Diabetic Rats: A Therapeutic Strategy to Mitigate Glycoxidative Stress. *Oxid. Med. Cell. Longev.* **2020**, *2020*, 1036360. [CrossRef]
56. Li, J.; Wu, N.; Chen, X.; Chen, H.; Yang, X.; Liu, C. Curcumin protects islet cells from glucolipotoxicity by inhibiting oxidative stress and NADPH oxidase activity both in vitro and in vivo. *Islets* **2019**, *11*, 152–164. [CrossRef] [PubMed]
57. Weisberg, S.; Leibel, L.; Tortoriello, D. Proteasome inhibitors, including curcumin, improve pancreatic β-cell function and insulin sensitivity in diabetic mice. *Nutr. Diabetes* **2016**, *48*, 1–8. [CrossRef] [PubMed]

58. Ahmad, M.; Sultana, M.; Raina, R.; Pankaj, N.K.; Verma, P.K.; Prawez, S. Hypoglycemic, Hypolipidemic, and Wound Healing Potential of Quercetin in Streptozotocin-Induced Diabetic Rats. *Pharmacogn. Mag.* **2017**, *13*, 633–639. [CrossRef]
59. Han, Y.; Wu, J.Z.; Shen, J.Z.; Chen, L.; He, T.; Jin, M.W.; Liu, H. Pentamethylquercetin induces adipose browning and exerts beneficial effects in 3T3-L1 adipocytes and high-fat diet-fed mice. *Sci. Rep.* **2017**, *7*, 1–13. [CrossRef]
60. Srinivasan, P.; Vijayakumar, S.; Kothandaraman, S.; Palani, M. Anti-diabetic activity of quercetin extracted from Phyllanthus emblica L. fruit: In silico and in vivo approaches. *J. Pharm. Anal.* **2018**, *8*, 109–118. [CrossRef]
61. Zhuang, M.; Qiu, H.; Li, P.; Hu, L.; Wang, Y.; Rao, L. Islet protection and amelioration of type 2 diabetes mellitus by treatment with quercetin from the flowers of Edgeworthia gardneri. *Drug Des. Devel. Ther.* **2018**, *12*, 955–966. [CrossRef]
62. Lee, J.S.; Cha, Y.J.; Lee, K.H.; Yim, J.E. Onion peel extract reduces the percentage of body fat in overweight and obese subjects: A 12-week, randomized, double-blind, placebo-controlled study. *Nutr. Res. Pract.* **2016**, *10*, 175–181. [CrossRef]
63. Wu, T.; Gao, Y.; Guo, X.; Zhang, M.; Gong, L. Blackberry and blueberry anthocyanin supplementation counteract high-fat-diet-induced obesity by alleviating oxidative stress and inflammation and accelerating energy expenditure. *Oxid. Med. Cell. Longev.* **2018**, *2018*, 4051232. [CrossRef]
64. Khan, M.I.; Shin, J.H.; Shin, T.S.; Kim, M.Y.; Cho, N.J.; Kim, J.D. Anthocyanins from Cornus kousa ethanolic extract attenuate obesity in association with anti-angiogenic activities in 3T3-L1 cells by down-regulating adipogeneses and lipogenesis. *PLoS ONE* **2018**, *13*, e0208556. [CrossRef] [PubMed]
65. Han, M.H.; Kim, H.J.; Jeong, J.W.; Park, C.; Kim, B.W.; Choi, Y.H. Inhibition of adipocyte differentiation by anthocyanins isolated from the fruit of Vitis coignetiae Pulliat is associated with the activation of AMPK signaling pathway. *Toxicol. Res.* **2018**, *34*, 13–21. [CrossRef] [PubMed]
66. Huang, W.; Yan, Z.; Li, D.; Ma, Y.; Zhou, J.; Sui, Z. Antioxidant and anti-inflammatory effects of blueberry anthocyanins on high glucose-induced human retinal capillary endothelial cells. *Oxid. Med. Cell. Longev.* **2018**, *2018*, 1862462. [CrossRef] [PubMed]
67. Suantawee, T.; Elazab, S.T.; Hsu, W.H.; Yao, S.; Cheng, H.; Adisakwattana, S. Cyanidin stimulates insulin secretion and pancreatic β-cell gene expression through activation of L-type voltage-dependent ca2+ channels. *Nutrients* **2017**, *9*, 814. [CrossRef] [PubMed]
68. Kongthitilerd, P.; Thilavech, T.; Marnpae, M.; Rong, W.; Yao, S.; Adisakwattana, S.; Cheng, H.; Suantawee, T. Cyanidin-3-rutinoside stimulated insulin secretion through activation of L-type voltage-dependent Ca2+ channels and the PLC-IP3 pathway in pancreatic β-cells. *Biomed. Pharmacother.* **2022**, *146*, 112494. [CrossRef]
69. Zhang, H.; Xu, Z.; Zhao, H.; Wang, X.; Pang, J.; Li, Q.; Yang, Y.; Ling, W. Anthocyanin supplementation improves anti-oxidative and anti-inflammatory capacity in a dose–response manner in subjects with dyslipidemia. *Redox Biol.* **2020**, *32*, 101474. [CrossRef] [PubMed]
70. Rowicka, G.; Dyląg, H.; Ambroszkiewicz, J.; Riahi, A.; Weker, H.; Chełchowska, M. Total Oxidant and Antioxidant Status in Prepubertal Children with Obesity. *Oxid. Med. Cell. Longev.* **2017**, *2017*, 5621989. [CrossRef]
71. Wu, C.C.; Huang, Y.W.; Hou, C.Y.; Chen, Y.T.; Di Dong, C.; Chen, C.W.; Singhania, R.R.; Leang, J.Y.; Hsieh, S.L. The anti-obesity effects of lemon fermented products in 3t3-l1 preadipocytes and in a rat model with high-calorie diet-induced obesity. *Nutrients* **2021**, *13*, 2809. [CrossRef] [PubMed]
72. Liao, Z.; Zhang, J.; Liu, B.; Yan, T.; Xu, F.; Xiao, F. Polysaccharide from Okra (*Abelmoschus esculentus*). *Molecules* **2019**, *24*, 1906. [CrossRef]
73. Yoshitomi, R.; Yamamoto, M.; Kumazoe, M.; Fujimura, Y.; Yonekura, M.; Shimamoto, Y.; Nakasone, A.; Kondo, S.; Hattori, H.; Haseda, A.; et al. The combined effect of green tea and α-glucosyl hesperidin in preventing obesity: A randomized placebo-controlled clinical trial. *Sci. Rep.* **2021**, *11*, 1–8. [CrossRef] [PubMed]
74. Fotschki, B.; Jurgo, A.; Kosmala, M.; Majewski, M.; Ognik, K. Extract against Pro-Oxidative and Pro-Inflammatory Rat Model. *Molecules* **2020**, *25*, 5874. [CrossRef]
75. de Sousa, G.M.; Cazarin, C.B.B.; Maróstica Junior, M.R.; Lamas, C.d.A.; Quitete, V.H.A.C.; Pastore, G.M.; Bicas, J.L. The effect of α-terpineol enantiomers on biomarkers of rats fed a high-fat diet. *Heliyon* **2020**, *6*, e03752. [CrossRef] [PubMed]
76. Hey-Mogensen, M.; Højlund, K.; Vind, B.F.; Wang, L.; Dela, F.; Beck-Nielsen, H.; Fernström, M.; Sahlin, K. Effect of physical training on mitochondrial respiration and reactive oxygen species release in skeletal muscle in patients with obesity and type 2 diabetes. *Diabetologia* **2010**, *53*, 1976–1985. [CrossRef]
77. Anderson, J.C.; Clarke, E.J.; Arkin, A.P.; Voigt, C.A. Environmentally controlled invasion of cancer cells by engineered bacteria. *J. Mol. Biol.* **2006**, *355*, 619–627. [CrossRef]
78. Li, Y.P.; Reid, M.B. NF-κB mediates the protein loss induced by TNF-α in differentiated skeletal muscle myotubes. *Am. J. Physiol.-Regul. Integr. Comp. Physiol.* **2000**, *279*, 1165–1170. [CrossRef]
79. Heymsfield, S.B.; Wadden, T.A. Mechanisms, Pathophysiology, and Management of Obesity. *N. Engl. J. Med.* **2017**, *376*, 254–266. [CrossRef]
80. Jung, U.J.; Choi, M.S. Obesity and its metabolic complications: The role of adipokines and the relationship between obesity, inflammation, insulin resistance, dyslipidemia and nonalcoholic fatty liver disease. *Int. J. Mol. Sci.* **2014**, *15*, 6184–6223. [CrossRef]
81. Tchkonia, T.; Thomou, T.; Zhu, Y.I.; Karagiannides, I.; Pothoulakis, C.; Jensen, M.D.; Kirkland, J.L. Mechanisms and metabolic implications of regional differences among fat depots. *Cell Metab.* **2013**, *17*, 644–656. [CrossRef]
82. Heymsfield, S.B.; Hu, H.H.; Shen, W.; Carmichael, O. Emerging Technologies and Their Applications in Lipid Compartment Measurement Lipid Compartment Measurement Advances HHS Public Access. *Trends Endocrinol Metab* **2015**, *26*, 688–698. [CrossRef]

83. McCullough, A.J. The clinical features, diagnosis and natural history of nonalcoholic fatty liver disease. *Clin. Liver Dis.* **2004**, *8*, 521–533. [CrossRef]
84. Calle, E.E.; Thun, M.J. Obesity and cancer. *Oncogene* **2004**, *23*, 6365–6378. [CrossRef] [PubMed]
85. Furukawa, S.; Fujita, T.; Shimabukuro, M.; Iwaki, M.; Yamada, Y.; Nakajima, Y.; Nakayama, O.; Makishima, M.; Matsuda, M.; Shimomura, I. Increased oxidative stress in obesity and its impact on metabolic syndrome. *J. Clin. Invest.* **2004**, *114*, 1752–1761. [CrossRef]
86. Zhang, Y.; Li, Y.; Ma, P.; Chen, J.; Xie, W. Ficus carica leaves extract inhibited pancreatic β-cell apoptosis by inhibiting AMPK/JNK/caspase-3 signaling pathway and antioxidation. *Biomed. Pharmacother.* **2020**, *122*, 109689. [CrossRef] [PubMed]
87. Huang, C.J.; McAllister, M.J.; Slusher, A.L.; Webb, H.E.; Mock, J.T.; Acevedo, E.O. Obesity-Related Oxidative Stress: The Impact of Physical Activity and Diet Manipulation. *Sport. Med.-Open* **2015**, *1*, 1–12. [CrossRef] [PubMed]
88. Tangvarasittichai, S. Oxidative stress, insulin resistance, dyslipidemia and type 2 diabetes mellitus. *World J. Diabetes* **2015**, *6*, 456. [CrossRef] [PubMed]
89. Drago, D.; Manea, M.M.; Timofte, D.; Ionescu, D. Mechanisms of Herbal Nephroprotection in diabetes mellitus. *J. Diabetes Res.* **2020**, *2020*. [CrossRef]
90. Zhao, Q.; Li, L.; Zhu, Y.; Hou, D.; Li, Y.; Guo, X.; Wang, Y.; Olatunji, O.J.; Wan, P.; Gong, K. Kukoamine B ameliorate insulin resistance, oxidative stress, inflammation and other metabolic abnormalities in high-fat/high-fructose-fed rats. *Diabetes Metab. Syndr. Obes. Targets Ther.* **2020**, *13*, 1843–1853. [CrossRef]
91. Chen, L.; Liu, P.; Feng, X.; Ma, C. Salidroside suppressing LPS-induced myocardial injury by inhibiting ROS-mediated PI3K/Akt/mTOR pathway in vitro and in vivo. *J. Cell. Mol. Med.* **2017**, *21*, 3178–3189. [CrossRef]
92. Zheng, T.; Yang, X.; Li, W.; Wang, Q.; Chen, L.; Wu, D.; Bian, F.; Xing, S.; Jin, S. Salidroside attenuates high-fat diet-induced nonalcoholic fatty liver disease via AMPK-dependent TXNIP/NLRP3 pathway. *Oxid. Med. Cell. Longev.* **2018**, *2018*, 8597897. [CrossRef]
93. Qi, Z.L.; Liu, Y.H.; Qi, S.M.; Ling, L.F.; Feng, Z.Y.; Li, Q. Salidroside protects PC12 cells from H2O2-induced apoptosis via suppressing NOX2-ROS-MAPKs signaling pathway. *Nan Fang Yi Ke Da Xue Xue Bao* **2016**, *37*, 178–183. [CrossRef] [PubMed]
94. Baek, J.H.; Kim, N.J.; Song, J.K.; Chun, K.H. Kahweol inhibits lipid accumulation and induces Glucose-uptake through activation of AMP-activated protein kinase (AMPK). *BMB Rep.* **2017**, *50*, 566–571. [CrossRef] [PubMed]
95. Mazibuko-Mbeje, S.E.; Dludla, P.V.; Johnson, R.; Joubert, E.; Louw, J.; Ziqubu, K.; Tiano, L.; Silvestri, S.; Orlando, P.; Opoku, A.R.; et al. Aspalathin, a natural product with the potential to reverse hepatic insulin resistance by improving energy metabolism and mitochondrial respiration. *PLoS ONE* **2019**, *14*, e0216172. [CrossRef] [PubMed]
96. Mazibuko-Mbeje, S.E.; Dludla, P.V.; Roux, C.; Johnson, R.; Ghoor, S.; Joubert, E.; Louw, J.; Opoku, A.R.; Muller, C.J.F. Aspalathin-enriched green rooibos extract reduces hepatic insulin resistance by modulating PI3K/AKT and AMPK pathways. *Int. J. Mol. Sci.* **2019**, *20*, 633. [CrossRef]
97. Johnson, R.; Dludla, P.V.; Muller, C.J.F.; Huisamen, B.; Essop, M.F.; Louw, J. The Transcription Profile Unveils the Cardioprotective Effect of Aspalathin against Lipid Toxicity in an in Vitro H9c2 Model. *Molecules* **2017**, *22*, 219. [CrossRef]
98. Li, C.X.; Gao, J.G.; Wan, X.Y.; Chen, Y.; Xu, C.F.; Feng, Z.M.; Zeng, H.; Lin, Y.M.; Ma, H.; Xu, P.; et al. Allyl isothiocyanate ameliorates lipid accumulation and inflammation in nonalcoholic fatty liver disease via the Sirt1/AMPK and NF-κB signaling pathways. *World J. Gastroenterol.* **2019**, *25*, 5120–5133. [CrossRef]
99. Cheng, D.; Gao, L.; Su, S.; Sargsyan, D.; Wu, R.; Raskin, I.; Kong, A.N. Moringa Isothiocyanate Activates Nrf2: Potential Role in Diabetic Nephropathy. *AAPS J.* **2019**, *21*, 31. [CrossRef] [PubMed]
100. Khutami, C.; Sumiwi, S.A.; Ikram, N.K.K.; Muchtaridi, M. *The Effect of Antioxidants from Natural Products on Obesity, Dyslipidemia and Diabetes and Their Molecular Signaling Mechanism*; Universitas Padjadjaran: Bandung, Indonesia, 2021.
101. Herrera-Balandrano, D.D.; Chai, Z.; Hutabarat, R.P.; Beta, T.; Feng, J.; Ma, K.; Li, D.; Huang, W. Hypoglycemic and hypolipidemic effects of blueberry anthocyanins by AMPK activation: In vitro and in vivo studies. *Redox Biol.* **2021**, *46*, 102100. [CrossRef]
102. Rao, Y.; Yu, H.; Gao, L.; Lu, Y.T.; Xu, Z.; Liu, H.; Gu, L.Q.; Ye, J.M.; Huang, Z.S. Natural alkaloid bouchardatine ameliorates metabolic disorders in high-fat diet-fed mice by stimulating the sirtuin 1/liver kinase B-1/AMPK axis. *Br. J. Pharmacol.* **2017**, *174*, 2457–2470. [CrossRef]
103. Xu, M.; Sun, B.; Li, D.; Mao, R.; Li, H.; Li, Y.; Wang, J. Beneficial effects of small molecule oligopeptides isolated from Panax ginseng meyer on pancreatic beta-cell dysfunction and death in diabetic rats. *Nutrients* **2017**, *9*, 1061. [CrossRef]
104. Alam, M.B.; An, H.; Ra, J.S.; Lim, J.Y.; Lee, S.H.; Yoo, C.Y.; Lee, S.H. Gossypol from cottonseeds ameliorates glucose uptake by mimicking insulin signaling and improves glucose homeostasis in mice with streptozotocin-induced diabetes. *Oxid. Med. Cell. Longev.* **2018**, *2018*, 5796102. [CrossRef]
105. Zhang, Y.; Wang, M.; Dong, H.; Yu, X.; Zhang, J. Anti-hypoglycemic and hepatocyte-protective effects of hyperoside from Zanthoxylum bungeanum leaves in mice with high-carbohydrate/high-fat diet and alloxan-induced diabetes. *Int. J. Mol. Med.* **2018**, *41*, 77–86. [CrossRef] [PubMed]
106. Li, Y.Y.; Stewart, D.A.; Ye, X.M.; Yin, L.H.; Pathmasiri, W.W.; McRitchie, S.L.; Fennell, T.R.; Cheung, H.Y.; Sumner, S.J. A metabolomics approach to investigate kukoamine B—A potent natural product with anti-diabetic properties. *Front. Pharmacol.* **2019**, *9*, 1–16. [CrossRef] [PubMed]

107. Du, M.; Hu, X.; Kou, L.; Zhang, B.; Zhang, C. Lycium barbarum Polysaccharide Mediated the Antidiabetic and Antinephritic Effects in Diet-Streptozotocin-Induced Diabetic Sprague Dawley Rats via Regulation of NF- B. *Biomed Res. Int.* **2016**, *2016*, 3140290. [CrossRef] [PubMed]
108. Suchal, K.; Malik, S.; Khan, S.I.; Malhotra, R.K.; Goyal, S.N.; Bhatia, J.; Kumari, S.; Ojha, S.; Arya, D.S. Protective effect of mangiferin on myocardial ischemia-reperfusion injury in streptozotocin-induced diabetic rats: Role of AGE-RAGE/MAPK pathways. *Sci. Rep.* **2017**, *7*, 1–11. [CrossRef]
109. Gao, J.; Liu, P.; Shen, Z.; Xu, K.; Wu, C.; Tian, F.; Chen, M.; Wang, L.; Li, P. Morroniside Promotes PGC-1 α -Mediated Cholesterol Efflux in Sodium Palmitate or High Glucose-Induced Mouse Renal Tubular Epithelial Cells. *Biomed Res. Int.* **2021**, *2021*, 9942152. [CrossRef]
110. Jin, B.R.; Lee, M.; An, H.J. Nodakenin represses obesity and its complications via the inhibition of the VLDLR signalling pathway in vivo and in vitro. *Cell Prolif.* **2021**, *54*, 1–15. [CrossRef]
111. El Khatib, N.; Morel, S.; Hugon, G.; Rapior, S.; Carnac, G.; Saint, N. Identification of a sesquiterpene lactone from arctium lappa leaves with antioxidant activity in primary human muscle cells. *Molecules* **2021**, *26*, 1328. [CrossRef]
112. Li, X.; Gong, H.; Yang, S.; Yang, L.; Fan, Y.; Zhou, Y. Pectic Bee Pollen Polysaccharide from Rosa rugosa Alleviates Diet-Induced Hepatic Steatosis and Insulin Resistance via Induction of AMPK/mTOR-Mediated Autophagy. *Molecules* **2017**, *22*, 699. [CrossRef]
113. Liu, S.; Huang, S.; Wu, X.; Feng, Y.; Shen, Y.; Zhao, Q.S.; Leng, Y. Activation of SIK1 by phanginin A inhibits hepatic gluconeogenesis by increasing PDE4 activity and suppressing the cAMP signaling pathway. *Mol. Metab.* **2020**, *41*, 101045. [CrossRef]
114. Torabi, S.; DiMarco, N.M. Original Research: Polyphenols extracted from grape powder induce lipogenesis and glucose uptake during differentiation of murine preadipocytes. *Exp. Biol. Med.* **2016**, *241*, 1776–1785. [CrossRef] [PubMed]
115. Wang, K.; Cao, P.; Wang, H.; Tang, Z.; Wang, N.; Wang, J.; Zhang, Y. Chronic administration of Angelica sinensis polysaccharide effectively improves fatty liver and glucose homeostasis in high-fat diet-fed mice. *Sci. Rep.* **2016**, *6*, 1–11. [CrossRef] [PubMed]
116. Chang, X.; Lu, K.; Wang, L.; Lv, M.; Fu, W. Astraglaus polysaccharide protects diabetic cardiomyopathy by activating NRG1/ErbB pathway. *Biosci. Trends* **2018**, *12*, 149–156. [CrossRef] [PubMed]
117. Sun, M.; Zhu, T.; Tong, J.; Caidan, R.; Wang, K.; Kai, G.; Zhang, W.; Ru, L.; Pengcuo, J.; Tong, L. Screening active components from Rubus amabilis for pancreatic β-cells protection. *Pharm. Biol.* **2020**, *58*, 674–685. [CrossRef]
118. Zhang, D.; Li, M. Puerarin prevents cataract development and progression in diabetic rats through Nrf2/HO-1 signaling. *Mol. Med. Rep.* **2019**, *20*, 1017–1024. [CrossRef]
119. Choi, J.; Oh, S.; Son, M.; Byun, K. Pyrogallol-phloroglucinol-6,6-bieckol alleviates obesity and systemic inflammation in a mouse model by reducing expression of RAGE and RAGE ligands. *Mar. Drugs* **2019**, *17*, 612. [CrossRef]
120. Fourny, N.; Lan, C.; Sérée, E.; Bernard, M.; Desrois, M. Protective effect of resveratrol against ischemia-reperfusion injury via enhanced high energy compounds and eNOS-SIRT1 expression in type 2 diabetic female rat heart. *Nutrients* **2019**, *11*, 105. [CrossRef]
121. Ju, L.; Wen, X.; Wang, C.; Wei, Y.; Peng, Y.; Ding, Y.; Feng, L.; Shu, L. Salidroside, a natural antioxidant, improves β-cell survival and function via activating AMPK pathway. *Front. Pharmacol.* **2017**, *8*, 1–12. [CrossRef]
122. Wang, Q.; Wu, X.; Shi, F.; Liu, Y. Comparison of antidiabetic effects of saponins and polysaccharides from Momordica charantia L. in STZ-induced type 2 diabetic mice. *Biomed. Pharmacother.* **2019**, *109*, 744–750. [CrossRef]
123. Belhadj, S.; Hentati, O.; Hamdaoui, G.; Fakhreddine, K.; Maillard, E.; Dal, S.; Sigrist, S. Beneficial effect of Jojoba seed extracts on hyperglycemia-induced oxidative stress in RINm5f beta cells. *Nutrients* **2018**, *10*, 384. [CrossRef] [PubMed]
124. Chen, T.X.; Cheng, X.Y.; Wang, Y.; Yin, W. Toosendanin inhibits adipogenesis by activating Wnt/β-catenin signaling. *Sci. Rep.* **2018**, *8*, 1–12. [CrossRef] [PubMed]
125. Suzuki, T.; Yamamoto, M. Molecular basis of the Keap1–Nrf2 system. *Free Radic. Biol. Med.* **2015**, *88*, 93–100. [CrossRef] [PubMed]
126. Wardyn, J.D.; Ponsford, A.H.; Sanderson, C.M. Dissecting molecular cross-talk between Nrf2 and NF-κB response pathways. *Biochem. Soc. Trans.* **2015**, *43*, 621–626. [CrossRef]
127. Albrahim, T.; Alonazi, M.A. Lycopene corrects metabolic syndrome and liver injury induced by high fat diet in obese rats through antioxidant, anti-inflammatory, antifibrotic pathways. *Biomed. Pharmacother.* **2021**, *141*, 111831. [CrossRef]
128. Chen, H.; Jiang, Y.; Yang, Z.; Hu, W.; Xiong, L.; Wang, N.; Liu, X.; Zheng, G.; Ouyang, K.; Wang, W. Effects of Chimonanthus nitens Oliv. Leaf Extract on Glycolipid Metabolism and Antioxidant Capacity in Diabetic Model Mice. *Oxid. Med. Cell. Longev.* **2017**, *2017*, 7648505. [CrossRef] [PubMed]
129. Brasier, A.R. The NF-κB regulatory network. *Cardiovasc. Toxicol.* **2006**, *6*, 111–130. [CrossRef] [PubMed]
130. Jiang, W.L.; Zhao, K.C.; Yuan, W.; Zhou, F.; Song, H.Y.; Liu, G.L.; Huang, J.; Zou, J.J.; Zhao, B.; Xie, S.P. MicroRNA-31-5p Exacerbates Lipopolysaccharide-Induced Acute Lung Injury via Inactivating Cab39/AMPK α Pathway. *Oxid. Med. Cell. Longev.* **2020**, *2020*, 8822361. [CrossRef] [PubMed]
131. Tayanloo-Beik, A.; Roudsari, P.P.; Rezaei-Tavirani, M.; Biglar, M.; Tabatabaei-Malazy, O.; Arjmand, B.; Larijani, B. Diabetes and Heart Failure: Multi-Omics Approaches. *Front. Physiol.* **2021**, *12*, 1–12. [CrossRef]
132. Kay, A.M.; Simpson, C.L.; Stewart, J.A. The Role of AGE/RAGE Signaling in Diabetes-Mediated Vascular Calcification. *J. Diabetes Res.* **2016**, *2016*, 6809703. [CrossRef]

International Journal of
Molecular Sciences

Article

Studies on the Accumulation of Secondary Metabolites and Evaluation of Biological Activity of In Vitro Cultures of *Ruta montana* L. in Temporary Immersion Bioreactors

Agnieszka Szewczyk [1,*], Andreana Marino [2], Maria Fernanda Taviano [2], Lucia Cambria [2], Federica Davì [2,3], Monika Trepa [1], Mariusz Grabowski [1] and Natalizia Miceli [2]

1 Department of Pharmaceutical Botany, Faculty of Pharmacy, Jagiellonian University Medical College, 30-688 Krakow, Poland
2 Department of Chemical, Biological, Pharmaceutical and Environmental Sciences, University of Messina, Viale F. Stagno d'Alcontres, 31, 98166 Messina, Italy
3 Foundation "Prof. Antonio Imbesi", University of Messina, Piazza Pugliatti 1, 98122 Messina, Italy
* Correspondence: agnieszka.szewczyk@uj.edu.pl

Abstract: The present work focuses on in vitro cultures of *Ruta montana* L. in temporary immersion Plantform^{TM} bioreactors. The main aim of the study was to evaluate the effects of cultivation time (5 and 6 weeks) and different concentrations (0.1–1.0 mg/L) of plant growth and development regulators (NAA and BAP) on the increase in biomass and the accumulation of secondary metabolites. Consequently, the antioxidant, antibacterial, and antibiofilm potentials of methanol extracts obtained from the in vitro-cultured biomass of *R. montana* were evaluated. High-performance liquid chromatography analysis was performed to characterize furanocoumarins, furoquinoline alkaloids, phenolic acids, and catechins. The major secondary metabolites in *R. montana* cultures were coumarins (maximum total content of 1824.3 mg/100 g DM), and the dominant compounds among them were xanthotoxin and bergapten. The maximum content of alkaloids was 561.7 mg/100 g DM. Concerning the antioxidant activity, the extract obtained from the biomass grown on the 0.1/0.1 LS medium variant, with an IC_{50} 0.90 ± 0.03 mg/mL, showed the best chelating ability among the extracts, while the 0.1/0.1 and 0.5/1.0 LS media variants showed the best antibacterial (MIC range 125–500 µg/mL) and antibiofilm activity against resistant *Staphylococcus aureus* strains.

Keywords: in vitro cultures; *Ruta montana*; coumarins; alkaloids; phenolic compounds; antibiofilm formation; antibacterial activity; antioxidant activity

Citation: Szewczyk, A.; Marino, A.; Taviano, M.F.; Cambria, L.; Davì, F.; Trepa, M.; Grabowski, M.; Miceli, N. Studies on the Accumulation of Secondary Metabolites and Evaluation of Biological Activity of In Vitro Cultures of *Ruta montana* L. in Temporary Immersion Bioreactors. *Int. J. Mol. Sci.* **2023**, *24*, 7045. https://doi.org/10.3390/ijms24087045

Academic Editors: Gea Guerriero and David Della-Morte

Received: 11 January 2023
Revised: 5 April 2023
Accepted: 7 April 2023
Published: 11 April 2023

1. Introduction

Plants of the *Ruta* genus have a rich and varied chemical composition. They mainly contain compounds from the group of coumarins, alkaloids, flavonoids, phenolic acids, and essential oil [1,2]. *Ruta montana* is widespread in the Mediterranean region since it grows in sandy and dry areas, specifically in countries such as Algeria, Morocco, and Tunisia. These plants also grow in Greece, Portugal, and Turkey in Europe [3–6]. The diverse chemical composition of the species includes compounds from groups of alkaloids, flavonoids (flavanols, including catechins and flavones), coumarins (hydroxycoumarins and furanocoumarins), organic acids, tannins, among other essential oils (sesquiterpenes and monoterpenes). The genus *R. montana* also contains derivatives of leucoanthocyanins, sterols, triterpenes, mucus, monosaccharides, and anthracene [1–3,7–10]. Additionally, the diverse chemical composition of *R. montana* favors multidirectional therapeutic uses. The species further contains furanocoumarins, such as bergapten and xanthotoxin, which are used in dermatology to treat vitiligo and psoriasis [11,12]. Although these compounds have a therapeutic effect, they also have some toxic effects, such as photodermatitis as well as kidney and liver damage [6]. Therefore, their application should be controlled.

The raw materials of *R. montana* have antidiabetic properties, which cause a decrease in the blood glucose concentration and improve glucose tolerance. Further, they may also have a positive effect on the architecture of the pancreatic islets [2,13]. The extracts of *R. montana* also show antioxidant activity due to the high content of phenolic compounds. The compounds have the ability to scavenge free radicals [2,6]. The essential oil obtained from *R. montana* has antifungal, insecticidal, and larvicidal properties, which showed an inhibitory effect on the growth of the *Candida albicans* and *Aspergillus niger* species [2,3]. Additionally, the species have proven antimicrobial properties. Rue extracts from the species inhibited the growth of pathogens, such as *Staphylococcus aureus*, *Bacillus subtilis*, *Listeria innocua*, *Proteus mirabilis*, and *Clavibacter michiganensis*. The ketone components of the essential oil are the substances that are responsible for this activity [7]. The extracts from *R. montana* are also considered to be a potential antihypertensive drug. Studies conducted on rats suffering from hypertension revealed that this agent lowered systolic and diastolic blood pressure, mean blood pressure, and heart rate. Moreover, it caused dose-dependent relaxation of the aorta. It is likely that the prostaglandin pathway mediates these processes [14].

In vitro plant cultures have great potential for the commercial production of secondary metabolites. This method has many advantages. It allows the synthesis of bioactive compounds under controlled conditions, regardless of climate and soil conditions. In vitro plant cultures can be carried out continuously using selected cell lines, which makes them a valuable and reliable source of secondary metabolites [15].

A modern method of cultivation, a temporary immersion system (Plantform^{TM}), was selected for the in vitro cultures. This type of bioreactor allows for obtaining a large amount of biomass in a short period of time [16]. Plantform^{TM} bioreactors have recently been used to produce secondary metabolites. They are very useful for cultivating shoot cultures, the enlargement of which on a larger scale has been a major technical problem so far. This method allowed the production of secondary metabolites from various groups of compounds on a larger scale. The authors of studies on *Schizandra chinensis* cultures conducted in various types of bioreactors found the highest production of lignans in cultures carried out in Plantform^{TM} bioreactors [17]. Subsequently, *Centella asiatica* cultures in Plantform^{TM} bioreactors turned out to be a valuable source for obtaining asiaticoside, phenolic acids, and flavonoids [18]. Cultures of *Scutellaria lateriflora* maintained in this type of bioreactor showed a high ability to accumulate verbascoside and flavonoids [19]. The in vitro plant cultures are considered an alternative high-quality plant material for pharmaceutical or cosmetic purposes. Currently, there are no studies on the in vitro cultures of *R. montana*. Therefore, the present study aimed to investigate the biosynthetic, antioxidant, antibacterial, and antibiofilm potential of the in vitro cultures of *R. montana* using a temporary immersion system (Plantform^{TM}).

2. Results and Discussion

2.1. Biomass Growth

The *R. montana* cultures grown in bioreactors mainly in the form of shoots (Figure 1). The cultures were characterized by an 8-fold increase in biomass. The content of dry biomass obtained from the *R. montana* bioreactor cultures ranged between 8.042 (LS medium containing 0.5/1.0 mg/L NAA/BAP, 5-week growth cycle) and 7.247 g (LS medium containing 0.1/0.1 mg/L NAA/BAP, 6-week growth cycle). The maximum weight was obtained after a 5-week cultivation period (growth cycle) on the LS medium containing 0.5/1.0 mg/L NAA/BAP. There were no significant differences in biomass growth after 5 and 6 weeks of the culture cycle for most of the LS medium. Only in case of the LS 0.5/0.5 medium variant was the dry biomass after 5 weeks significantly higher compared to those for 6 weeks. This was due to the gradual withering away of cultures. Macroscopic observations of the tissue showed the cultures had a tendency to die out after 6 weeks. An intense green color was observed on the 5-week-old cultures, while the 6-week-old cultures were partially covered in brown. The content of obtained dry biomass is summarized in Table 1.

Figure 1. *Ruta montana* bioreactor culture (LS NAA/BAP 0.1/0.1 mg/L, 5-week growth cycle).

Table 1. The dry weight (DW) [g] obtained from *R. montana* bioreactor cultures for different growth cycle and LS medium variants.

Growth Cycle	LS Medium Variant NAA/BAP (mg/L)				
	LS 0.1/0.1	LS 0.1/0.5	LS 0.5/0.5	LS 0.5/1.0	LS 1.0/1.0
5 weeks	7.734 ± 0.31 [ab]	7.775 ± 0.36 [ab]	8.030 ± 0.47 [b]	8.042 ± 0.31 [b]	7.738 ± 0.59 [ab]
6 weeks	7.247 ± 0.86 [a]	7.916 ± 0.02 [ab]	7.265 ± 0.18 [a]	7.787 ± 0.25 [ab]	7.946 ± 0.21 [ab]

Mean of three replications ± SD. Different letters (a, b) indicate significant differences ($p < 0.05$).

The increase in dry biomass varied from 8.9-fold (LS 0.1/0.1, 6 weeks) to 9.9-fold (LS 0.5/1.0, 5 weeks). Previous studies on the increase in biomass in in vitro cultures of another species of the *Ruta* genus indicated a wide variation, depending on the type of culture used, the duration of its cultivation, and the medium variant used. In the studies on the accumulation of furanocoumarins in *R. graveolens* shoot cultures, which were conducted for 6 weeks in liquid stationary cultures on medium with NAA/BAP 2.0/2.0 mg/L, a 5-fold growth of dry biomass was observed. The highest increase occurred between the 7th and 21st day of culture [20]. In the next studies, the in vitro cultures of *Ruta graveolens* ssp. *divaricata* were grown for 4 weeks on the LS medium, supplemented with NAA/BAP at various concentrations, ranging between 0.1 and 3.0 mg/L. The fold increase in the dry biomass was 6.2–6.3 [21]. Another research was carried out on agitated cultures on the species *Ruta corsica*, *Ruta chalepensis*, and *Ruta graveolens*, on the LS medium with the addition of NAA/BAP at a concentration of 0.1/0.1 mg/L for a period of 3, 4, 5, 6, and 7 weeks. The highest increase in biomass for *R. graveolens* was observed for the culture grown for 4 weeks (33.3-fold increase), while for *R. chalepensis* and *R. corsica*, the highest increase occurred after 5 weeks (33.7-fold and 31.3-fold increase, respectively) [22]. The increase in biomass in cultures of other plant species cultivated in temporary immersion bioreactors differ depending on the duration of the culture and the media variants used. The microshoot cultures of *Scutellaria lateriflora* were maintained in Plantform^{TM} bioreactors on two media (LS and MS—Murashige—Skoog, both contained 0.5/1.0 mg/L NAA/BAP) for 4 weeks. The increase in dry biomass was 6.3-fold on MS medium and 6.9-fold on the LS medium [19]. Shoot cultures of *Schizandra chinensis* conducted in Plantform^{TM} bioreactors on MS medium containing 1.0/3.0 mg/L NAA/BAP showed good increase in biomass (from 2.5-fold after 30 days of cultivation to 9-fold after 60 days of cultivation) [17].

In conclusion the *R. montana* cultures showed good biomass growth. There are no statistically significant differences between the media used.

2.2. Phytochemical Investigation

2.2.1. HPLC Analysis

The HPLC analysis of methanol extracts obtained from the biomass of *R. montana* bioreactor cultures identified various secondary metabolites related to alkaloids and phenolic compounds (coumarins and catechins) (Table 2). There were only small amounts of phenolic acids accumulated. Spectrophotometric determinations indicated the presence of polyphenolic and flavonoid compounds in the extracts (Table 3). However, according to the HPLC analysis, there were no flavonoids corresponding to the reference substances used. The absence of the main flavonoid—rutoside, which is typically present in the parent plant, was particularly significant. The absence can be explained due to the fact that in vitro cultures generally tend to have different metabolic pathways compared to the parent plants. Some of the enzymes can also become inactive or complex glycosidic bonds, which are not present in the parent plant, often form due to the glycosylation process [15]. The HPLC analysis confirmed the accumulation of linear furanocoumarins (bergapten retention time (RT) = 63.72 min, isoimperatorin RT = 70.55 min, isopimpinellin RT = 60.91 min, psoralen RT = 53.38 min, xanthotoxin RT = 54.18 min), furoquinoline alkaloids (γ-fagarine RT = 64.68 min, 7-isopentenyloxy-γ-fagarine RT = 71.92 min, skimmianine RT = 62.36 min), and catechins (catechin RT = 7.3 min). Sample chromatograms of the extract from *R. montana* in vitro cultures are included in Supplementary Materials (Figures S1 and S2).

The dynamics of the accumulation of individual metabolites varied according to the time of culture and the LS medium variant. Overall, the production of the metabolites was higher after the 5-week growth cycle. The exceptions are isopimpinellin and 7-isopentenyloxy-γ-fagarine, which, on most variants of the LS medium, reached a higher content after a 6-week growth cycle. Table 2 shows the mean contents [mg/100 g DW] of individual metabolites depending on the LS medium variant and the growth cycle. A comparison of homogeneous groups marked with the letters a-g was used, where "a" means the group of lowest mean, followed by "b" for the next average, and so on to the highest averages. The results of detailed statistical analyzes and the list of homogeneous groups are included in the Supplementary Materials, for each metabolite separately.

Among the coumarins, the highest accumulated individual metabolite was xanthotoxin (maximum content 885.9 mg/100 g DW, LS 0.5/0.5 medium, 5-week growth cycle). The second highest accumulated metabolite was the coumarin bergapten, with the maximum content obtained after the 5-week growth cycle on the 1.0/1.0 LS medium (446 mg/100 g DW). The quantity of psoralen was also high, with the maximum content of 340.1 mg/100 g DW (5-week growth cycle, LS 0.1/0.1). The amounts of isopimpinellin and isoimperatorin were lower, with the maximum content of 223.2 (6-week growth cycle, LS 1.0/1.0) and 105.8 mg/100 g DM (5-week growth cycle, LS 1.0/1.0), respectively.

Among the confirmed furoquinoline alkaloids, γ-fagarine was accumulated in the highest amount (305.4 mg/100 g DW) on the 0.1/0.1 LS medium, after the 5-week growth cycle. The maximum skimmianine content was 233.7 mg/100 g DW (5-week growth cycle, LS 0.5/1.0). Isopentenyloxy-γ-fagarine was accumulated in the lowest amount, with the maximum content of 42.2 mg/100 g DW (6-week growth cycle, LS 0.5/0.5).

In the analyzed catechins, catechin was accumulated after the 5-week growth cycle, in the maximum amounts of 89.6 mg/100 g DW on the 0.1/0.1 LS medium; this result is in agreement with those obtained with the spectrophotometric determinations of the condensed tannins (Tables 2 and 3).

Table 2. Average content of metabolites [mg/100 g DW] in methanol extracts obtained from biomass of *R. montana* bioreactor cultures depending on the duration of the culture (5 and 6 weeks) growth cycle and LS medium variant.

	LS Medium Variant	Growth Cycle	
Accumulated Compounds	**NAA/BAP (mg/L)**	**5 Weeks**	**6 Weeks**
Bergapten	0.1/0.1	435.32 ± 50.26 ^{bcd}	403.75 ± 35.98 ^{abcd}
	0.1/0.5	375.70 ± 27.51 ^{a}	364.99 ± 17.31 ^{a}
	0.5/0.5	440.91 ± 21.31 ^{cd}	395.18 ± 34.86 ^{abcd}
	0.5/1.0	386.32 ± 30.63 ^{ab}	378.85 ± 21.30 ^{a}
	1.0/1.0	445.99 ± 18.08 ^{d}	388.82 ± 42.73 ^{abc}
Isoimperatorin	0.1/0.1	104.01 ± 9.72 ^{de}	86.44 ± 3.95 ^{c}
	0.1/0.5	79.11 ± 5.75 ^{c}	56.35 ± 14.32 ^{b}
	0.5/0.5	87.65 ± 7.76 ^{cd}	35.23 ± 2.37 ^{a}
	0.5/1.0	84.00 ± 9.72 ^{c}	82.81 ± 18.34 ^{c}
	1.0/1.0	105.78 ± 7.29 ^{e}	82.33 ± 8.48 ^{c}
Isopimpinellin	0.1/0.1	81.09 ± 8.77 ^{a}	79.50 ± 7.52 ^{a}
	0.1/0.5	100.75 ± 13.77 ^{ab}	148.58 ± 6.99 ^{c}
	0.5/0.5	92.89 ± 0.61 ^{ab}	154.44 ± 20.64 ^{c}
	0.5/1.0	90.25 ± 14.70 ^{ab}	168.38 ± 12.19 ^{c}
	1.0/1.0	113.97 ± 16.58 ^{b}	223.21 ± 36.68 ^{d}
Psoralen	0.1/0.1	340.07 ± 14.03 ^{e}	236.59 ± 40.05 ^{d}
	0.1/0.5	182.13 ± 15.72 ^{bc}	185.22 ± 22.54 ^{c}
	0.5/0.5	150.21 ± 11.24 ^{a}	181.77 ± 9.57 ^{bc}
	0.5/1.0	152.67 ± 6.37 ^{ab}	169.00 ± 12.13 ^{abc}
	1.0/1.0	168.45 ± 14.68 ^{abc}	150.23 ± 1.91 ^{a}
Xanthotoxin	0.1/0.1	863.78 ± 57.12 ^{f}	772.92 ± 45.05 ^{cde}
	0.1/0.5	682.89 ± 26.80 ^{ab}	825.65 ± 64.82 ^{def}
	0.5/0.5	885.92 ± 34.83 ^{f}	846.06 ± 18.84 ^{ef}
	0.5/1.0	731.96 ± 50.07 ^{abc}	771.51 ± 47.02 ^{cde}
	1.0/1.0	748.75 ± 38.92 ^{bcd}	665.85 ± 57.15 ^{a}
Total coumarins	0.1/0.1	1824.26 ± 98.07 ^{d}	1579.20 ± 50.29 ^{bc}
	0.1/0.5	1420.58 ± 35.81 ^{a}	1580.80 ± 68.30 ^{bc}
	0.5/0.5	1657.57 ± 45.79 ^{c}	1612.67 ± 49.17 ^{bc}
	0.5/1.0	1445.20 ± 86.25 ^{ab}	1570.55 ± 18.55 ^{bc}
	1.0/1.0	1582.94 ± 33.95 ^{bc}	1510.44 ± 104.38 ^{ab}
γ-Fagarine	0.1/0.1	305.42 ± 19.28 ^{e}	215.82 ± 12.41 ^{d}
	0.1/0.5	172.27 ± 36.64 ^{c}	159.87 ± 15.20 ^{bc}
	0.5/0.5	146.64 ± 6.45 ^{abc}	151.05 ± 13.10 ^{abc}
	0.5/1.0	133.85 ± 8.53 ^{ab}	128.08 ± 18.70 ^{a}
	1.0/1.0	133.61 ± 13.47 ^{ab}	130.70 ± 13.75 ^{ab}
Isopentenyloxy-γ-fagarine	0.1/0.1	30.79 ± 3.61 ^{d}	39.42 ± 3.45 ^{e}
	0.1/0.5	18.59 ± 2.40 ^{a}	30.66 ± 1.57 ^{cd}
	0.5/0.5	26.17 ± 2.02 ^{bc}	42.20 ± 4.30 ^{e}
	0.5/1.0	21.93 ± 1.64 ^{ab}	32.88 ± 0.72 ^{d}
	1.0/1.0	25.90 ± 2.26 ^{b}	22.34 ± 3.02 ^{ab}
Skimmianine	0.1/0.1	225.46 ± 14.16 ^{fg}	126.58 ± 12.12 ^{c}
	0.1/0.5	197.64 ± 18.26 ^{ef}	89.38 ± 6.45 ^{b}
	0.5/0.5	195.22 ± 23.06 ^{ef}	51.95 ± 14.28 ^{a}
	0.5/1.0	233.73 ± 27.88 ^{g}	199.69 ± 30.19 ^{ef}
	1.0/1.0	179.84 ± 24.17 ^{de}	158.27 ± 4.47 ^{cd}
Total alkaloids	0.1/0.1	561.66 ± 29.70 ^{f}	381.82 ± 15.50 ^{e}
	0.1/0.5	388.50 ± 25.29 ^{e}	279.91 ± 22.69 ^{ab}
	0.5/0.5	368.04 ± 31.01 ^{de}	245.20 ± 2.97 ^{a}
	0.5/1.0	389.51 ± 22.39 ^{e}	360.66 ± 21.15 ^{de}
	1.0/1.0	339.35 ± 17.92 ^{cd}	311.31 ± 12.33 ^{bc}
Catechin	0.1/0.1	89.62 ±0.97 ^{f}	50.80 ±6.41 ^{bc}
	0.1/0.5	76.12 ±11.16 ^{e}	59.25 ±3.91 ^{cd}
	0.5/0.5	59.70 ±5.98 ^{d}	46.38 ±3.68 ^{b}
	0.5/1.0	61.16 ±2.69 ^{d}	22.01 ±1.83 ^{a}
	1.0/1.0	73.56 ±3.51 ^{e}	60.82 ±1.91 ^{d}

Means of three measurements ± SD. Different letters $^{(a–g)}$ indicate significant differences between homogenous groups ($p < 0.05$) between LS medium variants for each compound after a certain growth cycle.

Table 3. Total phenolic, flavonoid, and condensed tannin content, free radical scavenging activity (DPPH) test, reducing power, and ferrous ion chelating activity of methanol extracts obtained from biomass of *R. montana* bioreactor cultures grown on LS medium variant supplemented with different concentrations of NAA/BAP (0.5/0.5, 0.1/0.5, 1.0/1.0, 0.1/0.1 mg/L), after 5-week growth cycle.

LS Medium Variant NAA/BAP (mg/L)	Total Polyphenols (mg GAE/g)	Total Flavonoids (mg QE/g)	Condensed Tannins (mg CE/g)	DPPH IC_{50} (mg/mL)	Reducing Power ASE/mL	Fe^{2+} Chelating Activity IC_{50} (mg/mL)
0.1/0.1	26.94 ± 0.67 [a]	24.60 ± 0.58 [b]	10.97 ± 0.5 [c]	>2 mg/mL	28.95 ± 2.37 [b]	0.90 ± 0.03 [a]
0.1/0.5	41.61 ± 0.77 [e]	45.65 ± 0.33 [e]	9.42 ± 0.57 [b]	>2 mg/mL	26.62 ± 3.25 [b]	2.47 ± 0.01 [c]
0.5/0.5	29.64 ± 0.76 [b]	16.96 ± 0.01 [a]	4.68 ± 0.32 [a]	>2 mg/mL	19.54 ± 0.64 [a]	0.94 a ± 0.02 [a]
0.5/1.0	31.10 ± 0.88 [c]	28.44 ± 0.20 [c]	5.40 ± 0.2 [a]	>2 mg/mL	39.75 ± 0.50 [c]	0.93 ± 0.03 [a]
1.0/1.0	34.32 ± 0.22 [d]	33.00 ± 0.70 [d]	6.21 ± 0.44 [a]	>2 mg/mL	18.28 ± 4.30 [a]	1.68 ± 0.02 [b]
Reference standard				BHT 0.07 ± 0.01	BHT 1.44 ± 0.02 [d]	EDTA 0.01 ± 0.001 [d]

Values are expressed as mean ± SD (n = 3). [a–e] Different letters within the same column indicate significant differences between mean values ($p < 0.05$).

All the analyzed extracts had very high furanocoumarin total content. The maximum content of 1824.3 mg/100 g DW was obtained on the 0.1/0.1 LS medium after the 5-week growth cycle. High contents of furanocoumarins were also noted after the 5-week growth cycle on LS 0.5/0.5 and LS 1.0/1.0 media 1657.6 and 1582.9 mg/100 g DW, respectively.

The maximum content of the analyzed furoquinoline alkaloids (561.6 mg/100 g DW) was also obtained on the 0.1/0.1 LS medium after the 5-week growth cycle.

The obtained results suggest that *R. montana* bioreactor cultures can be proposed as an alternative and rich in vitro controlled source of linear furanocoumarins and furoquinoline alkaloids. Xanthotoxin and bergapten have photosensitizing effects on human skin and their pigmentation-stimulating and antiproliferative properties are utilized in the symptomatic treatment of vitiligo, psoriasis, and mycosis fungoides. Therefore, Bergapten is better tolerated by patients [11,12]. The obtained results showed very high levels of xanthotoxin (above 800 mg/100 g DW) and bergapten (above 400 mg/100 g DW). For comparison, the content of these compounds in plants cultivated in the field is 410 and 110 mg/100 g DW, respectively [23]. The biosynthetic potential for the production of furoquinoline alkaloids is also an interesting area of study. Furoquinoline alkaloids show a number of biological activities, including antifungal and antibacterial properties, inhibitory activity toward AchE (acetylcholinesterase), and 5-HT2 receptor-inhibiting properties [24]. The content of the above-mentioned groups of metabolites in cultures of another species of rue—*R. graveolens* varies depending on the type of culture, culture conditions, and the strategies used to increase the production of bioactive metabolites. Study on the level of accumulation of furanocoumarins in stationary liquid shoot cultures of *R. graveolens* confirmed the presence of the following coumarins: psoralen, xanthotoxin, isopimpinellin, bergapten, imperatorin, and umbelliferon. The maximum content of coumarins (966 mg/100 g DW) was determined after a 4-week growth cycle in the LS medium containing 2/2 mg/L NAA/BAP, and the dominant metabolites were found to be xanthotoxin (330 mg/100 g DW) and bergapten (320 mg/100 g DM) [20]. In subsequent studies performed in stationary liquid cultures maintained under various light conditions on the LS medium containing 2/2 mg/L NAA/BAP (6-week breeding cycle), the highest total content of coumarins (1022 mg/100 g DM) was observed in the cultures grown under white constant artificial light. The maximum content of the main furanocoumarins (xanthotoxin and bergapten) was 433.4 and 219.5 mg/100 g DM, respectively [25]. Similar to previous studies, in the study on *R. graveolens* agitated cultures, it was observed that the most dominant coumarins were xanthotoxin (428.3 mg/100 g DW) and bergapten (186.6 mg/100 g DW). The highest total content (917.2 mg/100 g DW) of linear furanocoumarins was reached after a 5-week growth cycle on the LS medium containing 0.1/0.1 mg/L NAA/BAP. The dominant furoquinolic alkaloids were skimmianine (94.6 mg/100 g DW) and γ-fagarine (54.5 mg/100 g DW). The highest total content

(155.9 mg/100 g DW) of these alkaloids was noted after a 5-week growth cycle on the LS medium containing 0.1/0.1 mg/L NAA/BAP [22].

Use of abiotic and biotic elicitors in agitated shoot cultures of *R. graveolens* caused an increase in the production of furanocoumarins and furoquinoline alkaloids. The highest obtained contents of the main coumarins were as follows: xanthotoxin (288.36 mg/100 g DW, 8.5-fold increase, compared to control cultures), bergapten (153.78 mg/100 g DW, 3.7-fold increase) The highest obtained contents of main furoquinoline alkaloids were as follows: γ-fagarine (68.0 mg/100 g DW), skimmianine (48.0 mg/100 g DW) [26–28].

2.2.2. Total Phenolic, Flavonoid, and Condensed Tannin Content

Total content of phenolic, flavonoid, and condensed tannins was measured in extracts from biomass obtained after 5-week growth cycle. The results of Folin–Ciocâlteu assay indicated that the total phenolic content was highest in the extract of *R. montana* biomass grown on the LS medium variant supplemented with NAA/BAP at the concentration of 0.1/0.5 mg/L, followed by the in vitro culture maintained on the LS medium variant 1.0/1.0 (Table 3).

Similar trend was observed for the total flavonoid content that, in the different extracts, varied from 16.96 to 45.65 mg QE/g extract. The extract obtained from in vitro culture maintained on the LS medium variant 0.1/0.5 contained the highest content of flavonoids, followed by the extract maintained on the LS medium variant 1/1 (Table 3).

The condensed tannin content, as evaluated by the vanillin assay, was found to be low in all the extracts. Among the extracts, that obtained from *R. montana* in vitro culture maintained on the LS medium variant 0.1/0.1 resulted as the richest (10.97 ± 0.50 mg QE/g extract) (Table 3). The content of condensed tannins determined by spectrophotometric method followed the same trend observed for these compounds determined by HPLC analysis; in fact, the content of condensed tannins as well as of catechins detected by HPLC resulted higher in the extract obtained from biomass grown on the variant LS 0.1/0.1 and decreased in the same order (Tables 2 and 3).

Recently, the content of total polyphenols was determined spectrophotometrically in a methanolic extract obtained from the aerial parts of *R. montana* collected in Algeria and in Morocco [29,30]. Interestingly, our extracts, obtained from biomass cultured in vitro, were found much richer both in phenolic compounds and in flavonoids than those prepared from the same species growing in field, demonstrating that the in vitro plant cultures can be considered an alternative high-quality plant material for pharmaceutical or cosmetic purposes. The content of phenolic compounds in cultures of other species of rue has not been extensively studied. This is due to the generally low content of these compounds. In the study on *R. graveolens* shoots grown in stationary liquid cultures on four different variants of the LS medium (with NAA and BAP added at different concentrations, in the range from 0.1 to 3.0 mg/L), the total content of phenolic acids was determined in the range from 85.04 to 108.28 mg/100 g DW (depending on the medium variant). The highest content of the tested phenolic acids was observed on the LS medium containing 2.0/2.0 mg/L NAA/BAP [31]. In the study on effect of monochromatic light conditions on the production of free phenolic acids in stationary liquid shoot cultures of R. graveolens, the highest total content (103.4 mg/100 g DW) of phenolic acids was observed in the biomass from the cultures cultivated on the LS medium, containing 3.0/1.0 mg/L NAA/BAP under white light [32]. An attempt was also made to increase the production of phenolic compounds in agitated shoot cultures of *R. graveolens* using the phenylalanine feeding method. The addition of this precursor increased 1.5-fold the production of phenolic compounds (phenolic acids and catechins). The highest obtained contents were, respectively, 109 mg/100 g DW (total content of phenolic acids) and 65.9 mg/100 g DW (catechin content) [33].

2.3. *Antioxidant Activity*

Reactive oxygen species (ROS) are widely regarded as etiologic factors for several diseases, including cancer, inflammation, and organ injuries. Evidence suggests that the antioxidants, scavenging ROS in pathological condition, may reduce cell damage and control the pathological process [34]. Antioxidants can be divided into primary (or chain-breaking) and secondary (or preventive) categories, and the primary antioxidant reactions can be classified into hydrogen-atom transfer (HAT) and single-electron transfer (SET). The HAT mechanism takes place when an antioxidant scavenges free radicals donating hydrogen atoms; an antioxidant acting by SET mechanism transfers a single electron to reduce any compound. Antioxidants can act through various mechanisms, and therefore it is vital to use methodologies with distinct mechanisms to evaluate the antioxidant capacity of plant-derived phytocomplexes or isolated compounds.

Based on the different mechanisms of determination of the antioxidant capacity, three in vitro tests were used to determine the in vitro antioxidant effectiveness of *R. montana* extracts: the DPPH assay (involving both HAT and SET mechanisms), and the reducing power assay (based on SET mechanism); the ferrous ions chelating activity was determined to assess the secondary antioxidant ability.

The DPPH test is the most common spectrophotometric method used to evaluate the free scavenging properties of a phytocomplexes or pure compounds due to its simple, rapid, sensitive, and reproducible procedures. Figure 2A shows the results of the DPPH test; it is evident from the comparison of all extracts of the species of *R. montana* and the reference standard BHT that, in the range of concentrations assayed, all of them display a weak activity. This result is confirmed by the IC_{50} values as well, which were higher than 2 mg/mL for all the extracts (Table 3).

The Fe^{3+}–Fe^{2+} transformation method was used to evaluate the reducing power of the *R. montana* extracts, which show that all extracts displayed mild activity compared to the standard BHT (Figure 2B). The calculated ASE/mL values indicated that the best reducing efficacy was highlighted for the extracts obtained from *R. montana* biomass grown on 1/1 and 0.5/0.5 LS medium variants (18.28 $\pm$ 4.30 and 19.54 $\pm$ 0.64 ASE/mL, respectively) (Table 3).

The *R. montana* extracts exhibited good, dose-dependent, and chelating properties in the Fe^{2+} chelating activity assay (Figure 2C). Comparing the IC_{50} values calculated for all the extracts, the values obtained from biomass in vitro cultured on the 0.1/0.1 LS medium variant, with an IC_{50} of 0.90 $\pm$ 0.03 mg/mL, with about 84% activity at the maximum concentration assayed, resulted as the most active, followed by those from in vitro culture grown on the 0.5/0.5 and 0.5/1 LS medium variants (Table 3).

It is evident from the obtained results that *R. montana* extracts act as weak primary antioxidants and possess good secondary antioxidant properties.

Ruta genus is rich in bioactive phytochemicals, such as coumarins, phenolic acids, flavonoids, alkaloids, and tannins [6]. The phytochemical investigations carried out by HPLC have highlighted a different metabolic profile of the in vitro cultures of *R. montana* than that of the parent plant. The flavonoids in all the extracts were undetectable, and phenolic acids were only present in small amounts. The main secondary metabolites in *R. montana* cultures were linear furanocoumarins (the dominant compound among them was xanthotoxin), followed by furoquinoline alkaloids (γ-fagarine, 7-isopentenyloxy-γ-fagarine, skimmianine) and catechins. It was reported that coumarins imperatorin, xanthotoxin, and bergapten did not possess antiradical activity against DPPH, though they exhibited a moderate level of reducing power and were effective in the ferrous ion-chelation test [35]. Coumarins reduce the level of oxidative stress via chelation of redox-active Cu and Fe, thus suppressing the ROS formation via the Fenton reaction [36]. It was reported that catechin can act as free radical scavengers as well as a metal chelator [37].

It is evident from the result of the chelating activity assay that the extract obtained on the 0.1/0.1 LS medium variant is the most effective; this extract is also the richest in coumarins and catechin. Therefore, the good secondary antioxidant properties highlighted for all *R. montana* extracts could mainly be related to the great amount of coumarins and catechin.

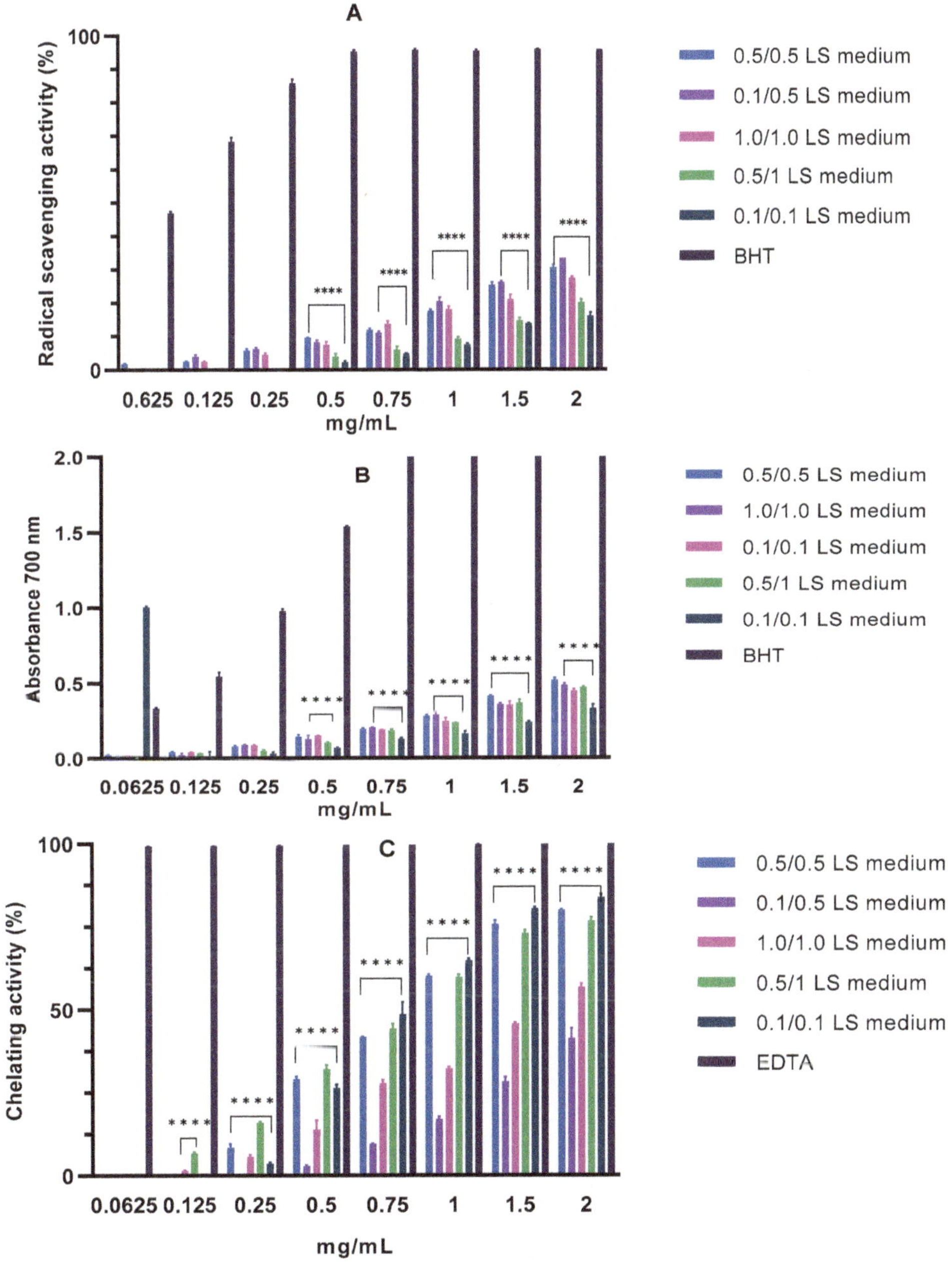

Figure 2. Free radical scavenging activity (DPPH assay) (**A**), reducing power (**B**), and ferrous ion chelating activity (**C**) of methanol extracts obtained from biomass of *R. montana* bioreactor cultures 2 grown on LS medium variant supplemented with different concentrations of NAA/BAP mg/L 293 (0.5/0.5, 0.1/0.5, 1.0/1.0, 0.1/0.1), after 5-week growth cycle. Reference standard: BHT (**A**,**B**), EDTA (**C**). Values are expressed as the mean ± SD (n = 3). Statistically significant differences between different variant are indicated as **** $p < 0.0001$.

2.4. Antibacterial Screening

There has been an increase in the spread of resistant bacterial strains due to the continued inappropriate usage of antibiotics. Many of these produce biofilms, which are a complex bacterial community encased in an extracellular polymeric matrix notoriously difficult to eradicate once established [38]. The biofilm network possesses the ability to evade environmental threats, such as antimicrobials and host defense mechanisms [39]. Due to this global health issue, there has been extensive research on finding alternative therapeutics. Plants rich in natural secondary metabolites are one of the go-to reservoirs in discovering potential resources to alleviate this problem [40]. In this study, the antibacterial screening of *R. montana* extracts was performed against a representative set of Gram-positive and Gram-negative bacterial strains. All the extracts obtained from in vitro culture of *R. montana* grown on the LS medium variants supplemented with NAA/BAP at different concentrations showed selective antibacterial activity. Table 4 shows the results of the antibacterial activity screening of extracts of the *R. montana* biomass grown on the different LS medium variants. The extract the 0.1/0.1 LS medium showed the highest activity against all the *Staphylococcus* strains and *E. coli* ATCC 10536 with the best effect against *S. aureus* ATCC 6538, *S. aureus* ATCC 43300, and *S. aureus* 74CCH (MIC = 125 µg/mL; MBC = 250–500 µg/mL), followed by the 0.5/0.5 LS medium active against *S. aureus* ATCC 43300 (MIC = 125 µg/mL) and *S. epidermidis* ATCC 35984 (MIC = 250 µg/mL), the 0.5/1.0 LS medium active against *S. aureus* 74CCH (MIC = 125 µg/mL) and *S. epidermidis* ATCC 35984 (MIC = 250 µg/mL), and the 0.1/0.5 LS medium active against *S. aureus* ATCC 43300 (MIC = 125 µg/mL). The extracts showed no activity against *E. coli* DSM 105388, *P. aeruginosa* ATCC 9027, and *P. aeruginosa* DSM 102273. The presence of a high content of xanthotoxin, followed by bergapten, psoralen, γ -fagarine, and catechins, likely justifies the efficacy of these extracts [41,42]. The good activity of the 0.5/0.5 LS medium against all the *Staphylococcus* strains can be due to the high content of xanthotoxin, followed by bergapten and skimmianine [41–44]. Coumarins have the ability to bind to the B subunit of DNA gyrase in bacteria and can inhibit DNA supercoiling by blocking the ATPase activity [45]. Alkaloids, such as γ-fagarine, have been shown to exhibit antimicrobial activity by inhibiting enzyme activity or other mechanisms, such as bacterial membrane disruption of Gram-negative bacteria [43]. Catechins demonstrated bactericidal effects on both Gram-positive and Gram-negative bacteria, including multidrug-resistant strains through membrane disruption and damage in DNA and protein oxidation [46].

Table 4. MIC and MBC values (µg/mL) of methanol extracts obtained from biomass of *R. montana* bioreactor cultures grown on LS medium variant supplemented with different concentrations of NAA/BAP, 5-week growth cycle.

Strains	LS Medium Variant NAA/BAP									
	0.1./0.1		0.1/0.5		0.5/0.5		0.5/1.0		1.0/1.0	
	MIC	MBC	MIC	MBC	MIC	MBC	MIC	MBC	MIC	MBC
S. aureus ATCC 6538	125	500	1000	n.a	500	1000	n.a.	n.a.	1000	1000
S. aureus ATCC 43300	125	250	125	n.a	125	n.a.	n.a.	n.a.	500	1000
S. aureus 815	500	n.a.	1000	n.a	1000	n.a.	500	1000	n.a.	n.a.
S. aureus 74CCH	125	250	1000	n.a	500	n.a.	125	n.a.	500	n.a.
S. epidermidis ATCC 35984	500	500	500	n.a	250	n.a.	250	n.a.	1000	n.a.
E. coli ATCC 10536	500	1000	n.a.	n.a	n.a.	n.a.	n.a.	n.a.	n.a.	n.a.

n.a. not active at concentrations assayed.

In a previous study, the extracts obtained from in vitro-cultured biomass of different species of *Ruta*, such as *R. chalepensis*, *R. corsica*, or *R. graveolens*, showed a good bacteriostatic activity against *S. aureus* [22]. The results obtained in this study highlight that the biomass of *R. montana* produced in vitro by bioreactors can also represent a source of compounds with antibacterial activity, which are effective against resistant *S. aureus* strains as well.

2.5. Effect on Biofilm Formation

The *R. montana* extracts demonstrated a good capacity to reduce biofilm formation of methicillin-resistant *S. aureus* (MRSA) strains and biofilm producers (Figure 3). Compared to the demonstration on *S. aureus* 74CCH (53–27%) at 1/2 MIC, there was a higher activity on *S. aureus* 815 with a biofilm biomass reduction range from 78 to 37% and on *S. aureus* ATCC 43,300 with reduction from 71 to 28%. The extract 0.5./1.0 LS medium, particularly, reduced the biofilm formation of *S. aureus* 815 (78%) and *S. aureus* 43,300 (69%) followed by the 0.5/05 LS medium on *S. aureus* 43,300 (71%) and the 0.1/0.1 LS medium on *S. aureus* 815 (69%). The extract of the 0.5/0.5 LS medium showed a high activity against *S. aureus* ATCC 43,300 with a reduction in biofilm formation of 72% at 1/4 MIC. According to a study by Lemos et al. (2016), the action of the extracts was classified as highly effective in the range from ≤60% to ≤90% of the biofilm reduction. Statistically significant differences are indicated as * $p < 0.05$ vs. each control group [47]. The activity of extracts on strong biofilm producer strains is interesting. LS 0.5/1.0, followed by the LS 0.5/0.5 and 0.1/0.1 LS media variants among all the extracts, showed the best activity. Coumarins and alkaloids can interfere in biofilm production by repressing curli genes and motility genes, while catechins disrupt glycocalyx [48–50].

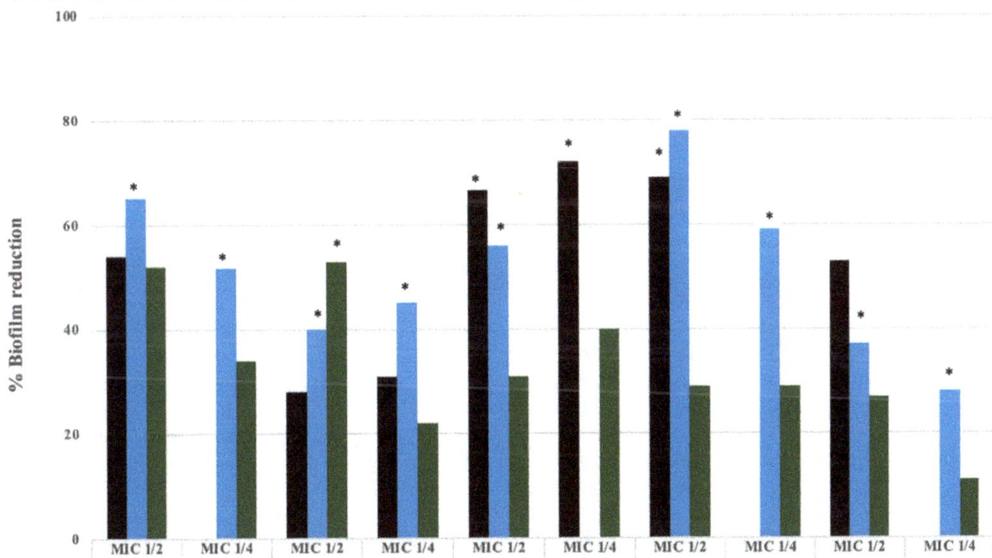

	MIC 1/2	MIC 1/4	MIC 1/2	MIC 1/4	MIC 1/2	MIC 1/4	MIC 1/2	MIC 1/4	MIC 1/2	MIC 1/4
	0.1/0.1		0.1/0.5		0.5/0.5		0.5/1.0		1.0/1.0	
■S. a. ATCC 43300	54	0	28	31	71	72	69	0	53	0
■S. a. 815	69	55	40	45	56	0	78	59	37	28
■S. a. 74CCH	52	34	53	22	31	40	29	29	27	11

Figure 3. The effect of methanol extracts obtained from the biomass of *R. montana* bioreactor cultures grown on LS medium variant supplemented with different concentrations of NAA/BAP mg/L (0.1/0.1, 0.1/0.5, 0.5/0.5, 0.5/1.0, 1.0/1.0), 5-week growth cycle, on *S. aureus* strains biofilm formation reduction. The reduction percentage of biofilm formation was calculated using the following formula: [(OD_{492} nm with extract/OD_{492} nm without extract) × 100]. Statistically significant differences are indicated as * $p < 0.05$ vs. each control group.

3. Materials and Methods

3.1. *In Vitro Cultures*

The starting material was the in vitro cultures of *R. montana*, established in 2018 from the seeds obtained from the Botanical Garden in Vácrátót, Hungary. The initial cultures were carried out in the form of liquid stationary cultures on Linsmaier and Skoog (LS) medium [51] and plant growth and development regulators: naphthyl-1-acetic acid (NAA) and 6-benzylaminopurine (BAP) at the concentration of 1.0/1.0 mg/L.

Ruta montana cultures were conducted in Plantform^{TM} bioreactors (Plant Form AB, Lomma, Sweden). For this, 15.0 g of previously grown plant biomass was placed in the bioreactors, which was then poured over with 500 mL of LS medium. Growth and development regulators in five concentration variants were added to it (Table 5). Immersion frequency was 5 min every 90 min. The cultures were maintained at 25 ± 2 °C under constant artificial light (16 $\mu M \cdot m^{-2} \cdot s^{-2}$). The cultivation was carried out for a period of 5 and 6 weeks. After this time, fresh biomass was collected and dried at approximately 38 °C.

Table 5. Concentration of plant growth regulator [mg/L] in LS medium variants.

Growth Regulator	LS Medium Variant NAA/BAP (mg/L)				
	0.1/0.1	**0.1/0.5**	**0.5/0.5**	**0.5/1.0**	**1.0/1.0**
NAA	0.1	0.1	0.5	0.5	1.0
BAP	0.1	0.5	0.5	1.0	1.0

3.2. *Extraction*

Briefly, 1.0 g of micronized dry biomass was weighed into 250 mL round bottom flasks and 50 mL of methanol was extracted from the flasks for 2 h at the solvent boiling point (64.7 °C). Following this, the extracts were evaporated. They were then dissolved in 4.0 mL high-performance liquid chromatography (HPLC)-grade methanol and filtered through Millipore membrane filters with a pore size of 0.22 µm for the HPLC analysis. Micronized dry biomass, 6–10 g of DW obtained after 5-week growth cycle, was used to prepare the extracts for the biological assays following the above-mentioned procedure.

3.3. *Phytochemical Investigation*

3.3.1. Total Phenolic, Flavonoid, and Condensed Tannin Content

The spectrophotometric methods were used to determine the total phenolic, flavonoid, and condensed tannin content of *R. montana* extracts. The Folin–Ciocâlteu method was used to measure the total phenolic content with gallic acid used as a standard phenolic compound [52]. An aliquot of 0.1 mL of each sample solution was mixed with 0.2 mL Folin–Ciocâlteu reagent, 2 mL of distilled water, and 1 mL of 15% Na_2CO_3. A linear calibration curve of gallic acid in the range from 125 to 500 µg/mL was constructed. After a 2 h incubation at room temperature, the absorbance was measured at 765 nm, using a UV-1601 spectrophotometer (Shimadzu, Milan, Italy). The total phenolic content was expressed as mg gallic acid equivalents (GAE)/g of extract (dw) ± standard deviation (SD).

The aluminum chloride spectrophotometric assay was used to measure the total flavonoid content of the extracts [53]. Appropriately diluted 250 µL of each sample solution was mixed with 750 µL MeOH, 50 µL µL of 10% aluminum chloride, 50 µL of 1 M potassium acetate, and 1.4 mL of distilled water. The samples were incubated at room temperature in the dark for 30 min, and the reaction absorbance was measured at 415 nm using a UV-1601 spectrophotometer (Shimadzu, Milan, Italy). Quercetin was used to make the calibration curve, and the total flavonoid content was expressed as mg quercetin equivalents (QE)/g extract (dw) ± SD.

The vanillin method was used to determine the condensed tannin content of the extracts [54]. Briefly, 25 µL of each sample solution was mixed with 750 µL of 4% vanillin in

MeOH and 375 μL of concentrated hydrochloric acid. After incubation at room temperature in the dark for 20 min, the absorbance of the reaction mixture was measured at 500 nm. The (+)-catechin was used as the reference standard, and the results were estimated as catechin equivalents (CE) and mg CE/g extract (dw) ± SD.

The results of the spectrophotometric determinations were obtained from the average of three independent experiments.

3.3.2. HPLC Analyses

Reversed-phase HPLC analysis was performed as described elsewhere [55] on a Merck–Hitachi liquid chromatograph (LaChrom Elite, Hitachi, Tokyo, Japan) equipped with a DAD detector L-2455 and a Purospher ® RP-18e 250 × 4 mm/5 mm column (Merck, Darmstadt, Germany). The analysis was conducted at 25 °C, with a mobile phase consisting of A—methanol, B—methanol:0.5% acetic acid 1:4 (*v/v*). The gradient elution, at the flow rate of 1 mL min^{-1}, was as follows: 100% B for 0–20 min; 100–80% B for 20–35 min; 80–60% B for 35–55 min; 60–0% B for 55–70 min; 0% B for 70–75 min; 0–100% B for 75–80 min; 100% B for 80–90 min; wavelength range 200–400 nm. The quantification was performed at λ = 254 and 330 nm (phenolic acids, catechins, coumarins, and alkaloids) and at 330 and 370 nm (flavonoids).

The standards were purchased from the following companies: bergapten, imperatorin, xanthotoxin, and psoralen from Roth (Karlsruhe, Germany); caffeic acid, chlorogenic acid, cinnamic acid, ellagic acid, gallic acid, gentizic acid, isoferulic acid, neochlorogenic acid, *o*-coumaric acid, protocatechuic acid, rosmarinic acid, salicylic acid, sinapic acid, syringic acid, apigenin, apigetrin (apigenin 7-glucoside), hyperoside (quercetin 3-*O*-galactoside), isoquercetin (quercetin 3-*O*-glucoside), isorhamnetin, kaempferol, luteolin, myricetin, populnin (kaempferol 7-*O*-glucoside), robinin (kaempferol 3-*O*-robinoside-7-*O*-rhamnoside), quercetin, quercitrin (quercetin 3-*O*-rhamnoside), rhamnetin, rutoside, vitexin, 5,7-dimethoxycoumarin, 4-hydroxy 6 methylcoumarin, 6-methylcoumarin, osthole, and umbelliferone from Sigma-Aldrich (St Louis, MO, USA); *p*-coumaric acid, vanillic acid, ferulic acid, *p*-hydroxybenzoic acid, coumarin, and scopoletin from Fluka (Bucha, Switzerland); caftaric acid, cryptochlorogenic acid, isochlorogenic acid, catechin, epigallocatechin, epicatechin gallate, epicatechin, epigallocatechin gallate, cinaroside (luteolin 7-*O*-glucoside), osthenol, 4-methylumbelliferone, 4,6-dimethoxy-2H-1-benzopyran-2-one, and skimmianine from ChromaDex (Irvine, CA, USA); 4-O-feruloylquinic acid, apigetrin (apigenin 7-*O*-glucoside), apigenin 7-*O*-glucuronide, astragalin (kaempferol 3-*O*-glucoside), avicularin (quercetin 3-*O*-α-L-arabinofuranoside), trifolin (kaempferol 3-*O*-galactoside), isopimpinellin, isoimperatorin, daphnetin 7-methyl ether, rutaretin, daphnetin, osthenol, bergaptol, daphnetin dimethyl ether, γ-fagarine, and 7-isopentenyloxy-γ-fagarine from ChemFaces (Wuhan, China).

3.4. *Antioxidant Activity*

3.4.1. DPPH Assay

The DPPH (1,1-diphenyl-2-picrylhydrazyl) assay was carried out to evaluate the free radical scavenging activity of *R. montana* extracts, using the protocol reported in a previous study by the authors [56]. The extracts were tested in the range from 0.0625 to 2 mg/mL, and butylated hydroxytoluene (BHT) was utilized as positive control. A volume of 0.5 mL of each sample was mixed with 3 mL of methanol DPPH solution (0.1 mM) and incubated in the dark for 20 min at room temperature. Then, the color change of the solutions was estimated by measuring the absorbance using a spectrophotometer (UV-1601, Shimadzu) at the wavelength of 517 nm. The scavenging activity was measured as the decrease in absorbance of the samples vs. DPPH control solution. The radical scavenging activity percentage (%) was calculated by the formula $[(A_o - A_c)/A_o] \times 100$, where A_o is the absorbance of the control and A_c is the absorbance in the presence of the sample or standard. Three independent experiments were carried out, and the results are reported as the mean radical scavenging activity percentage (%) ± SD and mean 50% inhibitory concentration

(IC_{50}) ± SD defined as the concentration of the essential oil necessary to reduce or inhibit 50% of DPPH radical solution. The best activity against the DPPH radical was obtained with the lowest value of IC_{50}. IC_{50} values were estimated by a nonlinear regression curve with the use of Prism Graphpad Prism version 9.0 for Windows, GraphPad Software, San Diego, CA, USA (www.graphpad.com (accessed on 10 January 2023)). The dose–response curve was obtained by plotting the percentage of inhibition *versus* the concentrations.

3.4.2. Reducing Power Assay

The Fe^{3+}–Fe^{2+} transformation method was used to estimate the reducing power of *R. montana* extracts [57]. The extracts were tested in the range from 0.0625 to 2 mg/mL, and BHT and ascorbic acid were utilized as positive controls. One mL of each sample was added after mixing 2.5 mL of phosphate buffer (0.2 M, pH 6.6) and 2.5 mL of 1% potassium ferricyanide. Following the incubation at 50 °C for 20 min and rapid cooling, 2.5 mL of 10% trichloroacetic acid was added, and the mixture was centrifuged (3000 rpm, 10 min). Then, 2.5 mL of the supernatant was transferred into another test tube and mixed with 2.5 mL of distilled water and 0.5 mL of 0.1% ferric chloride ($FeCl_3$). After a 10-min incubation in the dark at room temperature, the color change of the samples was estimated by measuring absorbance at 700 nm. Three independent experiments were carried out, and the results are expressed as the mean absorbance values ± SD and ascorbic acid equivalents (ASE/mL) ± SD.

3.4.3. Ferrous Ion (Fe^{2+}) Chelating Activity Assay

The spectrophotometric measurement of the Fe^{2+}-ferrozine complex was used following the protocol of Deker and Welch to determine the Fe^{2+} chelating activity of the *R. montana* extracts [58]. The extracts were tested in the range from 0.0625 to 2 mg/mL, and ethylenediaminetetraacetic acid (EDTA) was used as positive control. After mixing 1 mL of each sample with 0.5 mL of methanol and 0.05 mL of 2 mM $FeCl_2$, 0.1 mL of 5 mM ferrozine was added to initiate the reaction. The mixture was incubated in the dark at room temperature for 10 min, and the color change of the solutions was estimated by measuring absorbance spectrophotometrically at 562 nm. Three independent experiments were carried out, and the results are reported as the mean inhibition of the ferrozine–(Fe^{2+}) complex formation (%) ± SD and IC_{50} ± SD.

3.5. Antibacterial Bioassays

3.5.1. Bacterial Strains

The antibacterial activity of *R. montana* extracts was tested against the following strains: *S. aureus* ATCC 6538, *S. aureus* ATCC 43300, *S. aureus* 815, *S. aureus* 74CCH, *S. epidermidis* ATCC 35984, *Escherichia coli* ATCC 25922, *E. coli* ATCC 10536, *E. coli* DSM 105388, *Pseudomonas aeruginosa* ATCC 9027, and *P. aeruginosa* DSM 102273 [59]. The strains were stored in the private collection of the Department of Chemical, Biological, Pharmaceutical and Environmental Sciences, University of Messina (Italy). They were stored at −70 °C in Microbanks™ (Pro-lab Diagnostics, Neston, UK). All reagents were purchased from Sigma-Aldrich (Milan, Italy), unless otherwise specified in the text.

3.5.2. Antibacterial Screening

The minimum inhibitory concentration (MIC) and the minimum bactericidal concentration (MBC) of *R. montana* extracts were established according to the Clinical and Laboratory Standards Institute, with a few modifications [60]. Overnight cultures of the bacterial strains were grown at 37 °C in Mueller–Hinton Broth (MHB; Oxoid, Milan, Italy). The methanol extracts were dissolved in dimethyl sulfoxide (DMSO) and further diluted using MHB to obtain a final concentration of 2 mg/mL. Two-fold serial dilutions were prepared in a 96-well plate. The tested concentrations ranged from 1000 to 7.8 µg/mL. Working bacterial cultures were adjusted to the required inoculum of 1×10^5 CFU/mL. Positive controls (medium with inocula, but without the extracts) and vehicle controls (medium with inocula and DMSO) were included. The concentration of solvent (DMSO) did not exceed 1%. Tetracycline was tested against all bacteria at concentrations ranging from 32 to 0.016 μg/mL. To determine MBC, aliquots (10 μL) were taken from each well and inoculated in Mueller–Hinton Agar (MHA, Oxoid, Basingstoke, UK). The cultures were incubated for 24 h at 37 °C. The bacterial growth was indicated visually and by a developer of the enzymatic activity (triphenyl tetrazolium chloride 0.05%), which reveals bacterial growth of purple color after 15 min of heating at 37 °C. MIC was defined as the lowest concentration of the extracts that completely inhibited growth compared to the growth controls. MBC was defined as the lowest concentration of extracts that did not allow visible growth when the aliquots of the well contents were plated on MHA and grown for 24 h at 37 °C. All experiments were repeated thrice in duplicate.

3.5.3. Effect on Biofilm Formation

The effect of *R. montana* extracts on the biofilm formation by *Staphylococcus* strains was assayed based on the results obtained from the antibacterial screening. *Staphylococcus aureus* 815 and *S. aureus* 74CCH clinical isolates and *S. epidermidis* ATCC 35984 reference strain, among all the strains, are known as slime producers. They have been selected for their icaA/icaD gene presence, slime production, capability of forming biofilm on polystyrene surface, hemolytic activity, and agar typing [61]. The antibiofilm effect was assessed according to the description by Cramton et al., with a few modifications [54]. Overnight culture in 10 mL TSB with 1% glucose (TSBG) was diluted to standardize the *Staphylococcus* strain suspensions (1×10^6 CFU mL). Aliquots of 100 μL were dispensed into each well of the sterile flat-bottom 96-well polystyrene microtiter plates (Corning Inc., Corning, NY) in the presence of 100 μL subinhibitory concentration (1/2 and 1/4 MIC) of each extract or 100 μL medium (control). The microtiter plates were incubated for 24 h at 37 °C. Positive biofilm controls (cells + TSBG) and negative controls (TSBG) were included. Planktonic growth was determined by spectrophotometric values (OD_{492} nm) using the microplate reader. The medium was then aspirated and the wells, rinsed twice with phosphate-buffered saline, were fixed by drying for 1 h. Once the wells were fully dry, 200 μL of 0.1% safranin was added for 2 min. The content of the wells was then aspirated, and 200 μL of 30% acetic acid (*v/v*) was added to the wells after they were rinsed with water, for the spectrophotometric analysis (OD_{492} nm). The results were derived from three separate experiments. The OD_{492} nm value obtained for each *Staphylococcus* strain without the extracts was used as the control. The ratio between the values of OD_{492} nm with and without the extracts was used to calculate the reduction percentage of biofilm formation in the presence of different extracts, adopting the following formula: [(OD_{492} nm with extract/OD_{492} nm without extract) × 100].

3.6. Statistical Analysis

Statistical comparison of the HPLC data was performed using the two-way analysis of variance (ANOVA), followed by the NIR post hoc test (STATISTICA v. 13.3 software, StatSoft, Inc., Tulsa, OK, USA). Three replicates were performed for each treatment. The comparison of the data obtained from the spectrophotometric determinations and antioxidant and antibiofilm tests was made using the ANOVA, followed by Tukey–Kramer multiple comparisons test (GraphPAD Prism Software for Science). *p*-values less than 0.05 were considered to be statistically significant. Detailed results of the statistical analysis are included in the Supplementary Materials.

4. Conclusions

The study proved the ability of *R. montana* bioreactor cultures to accumulate secondary metabolites from various groups of compounds such as alkaloids, coumarins, flavonoids, and catechins. Furanocoumarins, xanthotoxin and bergapten in particular, and furoquinoline alkaloids were produced in the highest amounts. Thus, *R. montana* cultures can be proposed as a biotechnological source to obtain these valuable compounds. Further,

R. montana bioreactor cultures showed good chelating properties as well as antibacterial and antibiofilm efficacy against resistant *Staphylococcus* strains. The use of a modern temporary immersion system (Plantform^{TM}) allows for obtaining a large amount of biomass containing secondary metabolites with therapeutical potential in a short time.

Supplementary Materials: The following supporting information can be downloaded at: https://www.mdpi.com/article/10.3390/ijms24087045/s1, Figure S1: Sample chromatogram of the extract from *Ruta montana* in vitro cultures (0.1/0.1 LS medium, 5-week growth cycle). Figure S2: Enlarged fragment of a sample chromatogram of the extract from *Ruta montana* in vitro cultures (0.1/0.1 LS medium, 5-week growth cycle) 1. psoralen, 2. xanthotoxin, 3. isopimpinellin, 4. skimmianine, 5. bergapten, 6. γ-fagarine, 7. 7-isopentenyloxy-γ-fagarine,8. Isoimperatorin.

Author Contributions: Conceptualization, A.S.; investigation, A.M., A.S., F.D., L.C., M.F.T., M.G., M.T. and N.M.; data curation, A.M., A.S. and N.M.; writing—original draft preparation, A.M., A.S. and N.M.; writing—review and editing, A.S. and N.M. All authors have read and agreed to the published version of the manuscript.

Funding: The research was realized as a part of the research project N42/DBS/000272 supported by the Polish Ministry of Science and Higher Education; this publication was created with the use of equipment co-financed by the qLIFE Priority Research Area under the program "Excellence Initiative—Research University" at Jagiellonian University.

Institutional Review Board Statement: Not applicable.

Informed Consent Statement: Not applicable.

Data Availability Statement: Not applicable.

Acknowledgments: The authors wish to thank the Foundation "Antonio Imbesi" for financial support.

Conflicts of Interest: The authors declare no conflict of interest.

References

1. Akalin, E.; Ertug, F. An Anatomical and Ethnobotanical Study on Ruta Species in Turkey. *J. Pharm. Istanb. Univ.* **2013**, *35*, 121–132.
2. Coimbra, A.T.; Ferreira, S.; Duarte, A.P. Genus Ruta: A natural source of high value products with biological and pharmacological properties. *J. Ethnopharmacol.* **2020**, *260*, 113076. [CrossRef] [PubMed]
3. Mohammedi, H.; Mecherara-Idjeri, S.; Hassani, A. Variability in essential oil composition, antioxidant and antimicrobial activities of *Ruta montana* L. collected from different geographical regions in Algeria. *J. Essent. Oil Res.* **2019**, *32*, 23–36. [CrossRef]
4. Nahar, L.; El-Seedi, H.R.; Khalifa, S.A.M.; Mohammadhosseini, M.; Sarker, S.D. *Ruta* Essential Oils: Composition and Bioactivities. *Molecules* **2021**, *26*, 4766. [CrossRef]
5. Hammami, I.; Smaoui, S.; Hsouna, A.B.; Hamdi, N.; Triki, M.A. *Ruta montana* L. leaf essential oil and extracts: Characterization of bioactive compounds and suppression of crown gall disease. *EXCLI J.* **2015**, *14*, 83–94.
6. Benkhaira, N.; Koraichi, S.I.; Fikri-Benbrahim, K. *Ruta montana* (L.) L.: An insight into its medicinal value, phytochemistry, biological properties, and toxicity. *J. Herbmed Pharmacol.* **2022**, *11*, 305–319. [CrossRef]
7. Fekhar, N.; Boutoumi, H.; Krea, M.; Moulay, S.; Asma, D.; Benmaamar, Z. Thionation of essential oils from Algerian Artemisia herba-alba L. and Ruta montana L. Impact on their antimicrobial and insecticidal activities. *Chem. J. Mold.* **2017**, *12*, 50–57. [CrossRef]
8. Daoudi, A.; Hrouk, H.; Belaidi, R.; Slimani, R.; Ibijbijen, J.; Nassiri, L. Valorisation de Ruta montana et Ruta chalepensis: Etude ethnobotanique, Screening phytochimique et pouvoir antibactérien. Valorization of Ruta montana and Ruta chalepensis: Ethnobotanical study, phytochemical screening and Antibacterial activity. *J. Mater. Environ. Sci.* **2015**, *7*, 926–935.
9. Kabouche, Z.; Benkiki, N.; Seguin, E.; Bruneau, C. A new dicoumarinyl ether and two rare furocoumarins from *Ruta montana*. *Fitoterapia* **2003**, *74*, 194–196. [CrossRef]
10. Barbouchi, M.; Benzidia, B.; Choukrad, M. Chemical variability in essential oils isolated from roots, stems, leaves and flowers of three Ruta species growing in Morocco. *J. King Saud Univ.—Sci.* **2021**, *33*, 101634. [CrossRef]
11. Roelandts, R. Photo(chemo) therapy for vitiligo. *Photodermatol. Photoimmunol. Photomed.* **2003**, *19*, 261–277. [CrossRef]
12. Wolf, P. Psoralen-ultraviolet A endures as one of the most powerful treatments in dermatology: Reinforcement of this 'triple-product therapy' by the 2016 British guidelines. *Br. J. Dermatol.* **2016**, *174*, 11–14. [CrossRef]
13. Farid, O.; Hebi, M.; Ajebli, M.; El Hidani, A.; Eddouks, M. Antidiabetic effect of *Ruta montana* L. in streptozotocin-induced diabetic rats. *J. Basic Clin. Physiol. Pharmacol.* **2017**, *28*, 275–282. [CrossRef]
14. El-Ouady, F.; Eddouks, M. Ruta Montana Evokes Antihypertensive Activity Through an Increase of Prostaglandins Release in L-NAME-Induced Hypertensive Rats. *Endocr. Metab. Immune Disord.-Drug Targets* **2021**, *21*, 305–314. [CrossRef]

15. Karuppusamy, S. A review on trends in production of secondary metabolites from higher plants by in vitro tissue, organ and cell cultures. *J. Med. Plants Res.* **2009**, *3*, 1222–1239.
16. Benelli, C.; De Carlo, A. In vitro multiplication and growth improvement of *Olea europaea* L. *cv Canino* with temporary immersion system (Planform^{TM}). *3 Biotech* **2018**, *8*, 317.
17. Szopa, A.; Kokotkiewicz, A.; Luczkiewicz, M.; Ekiert, H. Schisandra lignans production regulated by different bioreactor type. *J. Biotechnol.* **2017**, *247*, 11–17. [CrossRef]
18. Skrzypczak-Pietraszek, E.; Urbańska, A.; Żmudzki, P.; Pietraszek, J. Elicitation with methyl jasmonate combined with cultivation in the Plantform™ temporary immersion bioreactor highly increases the accumulation of selected centellosides and phenolics in Centella asiatica (L.) Urban shoot culture. *Eng. Life Sci.* **2019**, *19*, 931–943. [CrossRef]
19. Kwiecień, I.; Miceli, N.; D'Arrigo, M.; Marino, A.; Ekiert, H. Antioxidant Potential and Enhancement of Bioactive Metabolite Production in In Vitro Cultures of Scutellaria lateriflora L. by Biotechnological Methods. *Molecules* **2022**, *27*, 1140. [CrossRef]
20. Ekiert, H.; Chołoniewska, M.; Gomółka, E. Accumulation of furanocoumarins in *Ruta graveolens* L. shoot culture. *Biotechnol. Lett.* **2001**, *23*, 543–545. [CrossRef]
21. Ekiert, H.; Piekoszewska, A.; Muszyńska, B.; Baczyńska, S. Accumulation of p-coumaric acid and other bioactive phenolic acids in in vitro culture of *Ruta graveolens* ssp. *divaricata (Tenore) Gams. MIR* **2014**, *26*, 24–30.
22. Szewczyk, A.; Marino, A.; Molinari, J.; Ekiert, H.; Miceli, N. Phytochemical Characterization, and Antioxidant and Antimicrobial Properties of Agitated Cultures of Three Rue Species: *Ruta chalepensis*, *Ruta corsica*, and *Ruta graveolens*. *Antioxidants* **2022**, *11*, 592. [CrossRef] [PubMed]
23. Milesi, S.; Massot, B.; Gontier, E.; Bourgaud, F.; Guckert, A. *Ruta graveolens* L.: A promising species for the production of furanocoumarins. *Plant Sci.* **2001**, *161*, 189–199. [CrossRef]
24. Adamska-Szewczyk, A.; Głowniak, K.; Baj, T. Furochinoline alkaloids in plants from *Rutaceae* family—A review. *Curr. Issues Pharm. Med. Sci.* **2016**, *29*, 33–38. [CrossRef]
25. Ekiert, H.; Gomółka, E. Effect of light on contents of coumarin compounds in shoots of *Ruta graveolens* L. cultivated in vitro. *Acta Soc. Bot. Pol.* **1999**, *68*, 197–200. [CrossRef]
26. Orlita, A.; Sidwa-Gorycka, M.; Kumirska, J. Identification of *Ruta graveolens* L. metabolites accumulated in the presence of abiotic elicitors. *Biotechnol. Prog.* **2008**, *24*, 128–133. [CrossRef]
27. Orlita, A.; Sidwa-Gorycka, M.; Paszkiewicz, M.; Maliński, E.; Kumirska, J.; Siedlecka, E.M.; Łojkowska, E.; Stepnowski, P. Applicationof chitin and chitosan as elicitors of coumarins and furoquinolone alkaloids in *Ruta graveolens* L. (common rue). *Biotechnol. Appl. Biochem.* **2008**, *51*, 91–96. [CrossRef]
28. Orlita, A.; Sidwa-Gorycka, M.; Malinski, E. Effective biotic elicitation of *Ruta graveolens* L. shoot cultures by lysates from *Pectobacterium atrosepticum* and *Bacillus* sp. *Biotechnol. Lett.* **2007**, *30*, 541–545. [CrossRef]
29. Merghem, M.; Dahamna, D. In-Vitro Antioxidant Activity and Total Phenolic Content of Ruta montana L. Extracts. *J. Drug Deliv. Ther.* **2020**, *10*, 69–75. [CrossRef]
30. Drioiche, A.; Amine, S.; Boutahiri, S.; Saidi, S.; Ailli, A.; Rhafouri, R.; Mahjoubi, M.; El Hilali, F.; Mouradi, A.; Eto, B.; et al. Antioxidant and Antimicrobial Activity of Essential Oils and Phenolic Extracts from the Aerial Parts of *Ruta montana* L. of the Middle Atlas Mountains-Morocco. *J. Essent. Oil Bear. Plants* **2020**, *23*, 902–917. [CrossRef]
31. Ekiert, H.; Szewczyk, A.; Kuś, A. Free phenolic acids in *Ruta graveolens* L. in vitro culture. *Pharmazie* **2009**, *64*, 694–696.
32. Szopa, A.; Ekiert, H.; Szewczyk, A.; Fugas, E. Production of bioactive phenolic acids and furanocoumarins in in vitro cultures of *Ruta graveolens* L. and *Ruta graveolens* ssp. *divaricata* (Tenore) Gams. under different light conditions. *Plant Cell Tissue Organ Cult.* **2012**, *110*, 329–336. [CrossRef]
33. Szewczyk, A.; Paździora, W.; Ekiert, H. The Influence of Exogenous Phenylalanine on the Accumulation of Secondary Metabolites in Agitated Shoot Cultures of *Ruta graveolens* L. *Molecules* **2023**, *28*, 727. [CrossRef]
34. Yang, S.; Lian, G. ROS and diseases: Role in metabolism and energy supply. *Mol. Cell. Biochem.* **2020**, *467*, 1–12, Erratum in: Mol. Cell. Biochem. 2020, 467, 13. [CrossRef]
35. Senol, F.S.; Woźniak, K.S.; Khan, M.T.H.; Orhan, I.E.; Sener, B.; Głowniak, K. An in vitro and in silico approach to cholinesterase inhibitory and antioxidant effects of the methanol extract, furanocoumarin fraction, and major coumarins of *Angelica officinalis* L. fruits. *Phytochem. Lett.* **2011**, *4*, 462–467. [CrossRef]
36. Simunkova, M.; Alwasel, S.H.; Alhazza, I.M.; Jomova, K.; Kollar, V.; Rusko, M.; Valko, M. Management of oxidative stress and other pathologies in Alzheimer's disease. *Arch. Toxicol.* **2019**, *93*, 2491–2513. [CrossRef]
37. Bernatoniene, J.; Kopustinskiene, D.M. The Role of Catechins in Cellular Responses to Oxidative Stress. *Molecules* **2018**, *23*, 965. [CrossRef]
38. Listerand, J.L.; Horswill, A.R. *Staphylococcus aureus* biofilms: Recent developments in biofilm dispersal. *Front. Cell. Infect. Microbiol.* **2014**, *4*, 178. [CrossRef]
39. Magana, M.; Sereti, C.; Ioannidis, A.; Mitchell, C.A.; Ball, A.R.; Magiorkinis, E.; Chatzipanagiotou, S.; Hamblin, M.R.; Hadjifrangiskou, M.; Tegos, G.P. Options and Limitations in Clinical Investigation of Bacterial Biofilms. *Clin. Microbiol. Rev.* **2018**, *31*, e00084-16. [CrossRef]
40. Mahizan, N.A.; Yang, S.K.; Moo, C.L.; Song, A.A.; Chong, C.M.; Chong, C.W.; Abushelaibi, A.; Lim, S.E.; Lai, K.S. Terpene Derivatives as a Potential Agent against Antimicrobial Resistance (AMR) Pathogens. *Molecules* **2019**, *24*, 2631. [CrossRef]

41. Ozçelik, B.; Kusmenoglu, S.; Turkoz, S.; Abbasoglu, U. Antimicrobial Activities of Plants from the Apicaceae. *Pharm. Biol.* **2004**, *42*, 526–528. [CrossRef]
42. Kuete, V.; Metuno, R.; Ngameni, B.; Tsafack, A.M.; Ngandeu, F.; Fotso, G.W. Antimicrobial activity of the methanolic extracts and compounds from Treculia obovoidea (Moraceae). *J. Ethnopharmacol.* **2007**, *112*, 531–536. [CrossRef] [PubMed]
43. Othman, L.; Sleiman, A.; Abdel-Massih, R.M. Antimicrobial Activity of Polyphenols and Alkaloids in Middle Eastern Plants. *Front. Microbiol.* **2019**, *10*, 911. [CrossRef] [PubMed]
44. Murugan, N.; Srinivasan, R.; Murugan, A.; Kim, M.; Natarajan, D. Glycosmis pentaphylla (Rutaceae): A Natural Candidate for the Isolation of Potential Bioactive Arborine and Skimmianine Compounds for Controlling Multidrug-Resistant Staphylococcus aureus. *Front. Public Health* **2020**, *8*, 176. [CrossRef]
45. Feng, D.; Zhang, A.; Yang, Y.; Yang, P. Coumarin-containing hybrids and their antibacterial activities. *Arch. Pharm.* **2020**, *353*, 1900380. [CrossRef]
46. Wu, M.; Brown, A.C. Applications of Catechins in the Treatment of Bacterial Infections. *Pathogens* **2021**, *10*, 546. [CrossRef]
47. Lemos, M.; Wang, S.; Alia, A.; Simõesb, M.; Wilson, D.I. A fluid dynamic gauging device for measuring biofilm thickness on cylindrical surfaces. *Biochem. Eng. J.* **2016**, *106*, 48–60. [CrossRef]
48. Jain, A.; Parihar, D.K. Antibacterial, biofilm dispersal and antibiofilm potential of alkaloids and flavonoids of Curcuma. *Biocatal. Agric. Biotechnol.* **2018**, *16*, 677–682. [CrossRef]
49. da Cunha, M.G.; de Cássia, O.; Sardi, J.; Freires, I.A.; Franchin, M.; Rosalen, P.L. Antimicrobial, anti-adherence and antibiofilm activity against Staphylococcus aureus of a 4-phenyl coumarin derivative isolated from Brazilian geopropolis. *Microb. Pathog.* **2020**, *139*, 103855. [CrossRef]
50. Blanco, A.R.; Sudano-Roccaro, A.; Spoto, G.C.; Nostro, A.; Rusciano, D. Epigallocatechin Gallate Inhibits Biofilm Formation by Ocular Staphylococcal Isolates. *Antimicrob. Agents Chemother.* **2005**, *49*, 4339–4343. [CrossRef]
51. Linsmaier, E.; Skoog, F. Organic growth factor requirements of tobacco tissue cultures. *Physiol. Plant.* **1965**, *18*, 100–127. [CrossRef]
52. Gao, X.; Ohlander, M.; Jeppsson, N.; Björk, L.; Trajkovski, V. Changes in antioxidant effects and their relationship to phytonutrients in fruits of sea buckthorn (Hippophae rhamnoides L.) during maturation. *J. Agric. Food Chem.* **2000**, *48*, 1485–1490. [CrossRef]
53. Chang, C.C.; Yang, M.H.; Wen, H.M.; Chern, J.C. Estimation of total flavonoid content in propolis by two complementary colorimetric methods. *J. Food Drug Anal.* **2002**, *10*, 178–182.
54. Julkunen-Titto, R. Phenolic constituents in the leaves of northern willows: Methods for the analysis of certain phenolics. *J. Agric. Food Chem.* **1985**, *33*, 213–217. [CrossRef]
55. Sułkowska-Ziaja, K.; Maślanka, A.; Szewczyk, A.; Muszyńska, B. Physiologically active compounds in four species of *Phellinus*. *Nat. Prod. Commun.* **2017**, *12*, 363–366. [CrossRef]
56. Ohnishi, M.; Morishita, H.; Iwahashi, H.; Shitzuo, T.; Yoshiaki, S.; Kimura, M.; Kido, R. Inhibitory effects of chlorogenic acid on linoleic acid peroxidation and haemolysis. *Phytochemistry* **1994**, *36*, 579–583. [CrossRef]
57. Oyaizu, M. Studies on products of browning reaction: Antioxidative activities of products of browning reaction prepared from glucosamine. *Jpn. J. Nutr. Diet.* **1986**, *44*, 307–315. [CrossRef]
58. Decker, E.A.; Welch, B. Role of ferritin as a lipid oxidation catalyst in muscle food. *J. Agric. Food Chem.* **1990**, *38*, 674–677. [CrossRef]
59. Marino, A.; Nostro, A.; Mandras, N.; Roana, J.; Ginestra, G.; Miceli, N.; Taviano, M.F.; Gelmini, F.; Beretta, G.; Tullio, V. Evaluation of antimicrobial activity of the hydrolate of Coridothymus capitatus (L.) Reichenb. fil. (Lamiaceae) alone and in combination with antimicrobial agents. *BMC Complement. Med. Ther.* **2020**, *20*, 89. [CrossRef]
60. *Methods for Dilution Antimicrobial Susceptibility Tests for Bacteria That Grow Aerobically*; CLSI: Wayne, PA, USA, 2018.
61. Cramton, S.E.; Gerke, C.; Schnell, N.F.; Nichols, W.W.; Götz, F. The intercellular adhesion (ica) locus in *Staphylococcus aureus* and is required for biofilm formation. *Infect. Immun.* **1999**, *67*, 5427–5433. [CrossRef]

International Journal of *Molecular Sciences*

MDPI

Review

Phytosterols: Potential Metabolic Modulators in Neurodegenerative Diseases

Niti Sharma [1,†], Mario A. Tan [2,†] and Seong Soo A. An [1,*]

1 Bionano Research Institute, Gachon University, 1342 Seongnam-daero, Sujeong-gu, Seongnam-si 461-701, Gyeonggi-do, Korea; nitisharma@gachon.ac.kr

2 Research Center for the Natural and Applied Sciences, College of Science, University of Santo Tomas, Manila 1015, Philippines; matan@ust.edu.ph

* Correspondence: seongaan@gachon.ac.kr; Tel.: +82-31-750-8755

† These authors contributed equally to work.

Abstract: Phytosterols constitute a class of natural products that are an important component of diet and have vast applications in foods, cosmetics, and herbal medicines. With many and diverse isolated structures in nature, they exhibit a broad range of biological and pharmacological activities. Among over 200 types of phytosterols, stigmasterol and β-sitosterol were ubiquitous in many plant species, exhibiting important aspects of activities related to neurodegenerative diseases. Hence, this mini-review presented an overview of the reported studies on selected phytosterols related to neurodegenerative diseases. It covered the major phytosterols based on biosynthetic considerations, including other phytosterols with significant in vitro and in vivo biological activities.

Keywords: phytosterols; neurodegeneration; Alzheimer's disease; natural products; blood brain barrier

Citation: Sharma, N.; Tan, M.A.; An, S.S.A. Phytosterols: Potential Metabolic Modulators in Neurodegenerative Diseases. *Int. J. Mol. Sci.* **2021**, *22*, 12255. https://doi.org/10.3390/ijms222212255

Academic Editors: Antonio Carrillo Vico and Iván Cruz-Chamorro

Received: 15 October 2021
Accepted: 9 November 2021
Published: 12 November 2021

1. Introduction

Neurodegenerative disease (ND) is a group of disorders (Alzheimer's, Parkinson's, multiple sclerosis, amyotrophic lateral sclerosis, Huntington's disease, prion disease, etc.) that result in progressive degeneration of structure or function of the neurons. Neurodegeneration may progress differently with various levels and locations of neuronal circuitry in the brain, ranging from molecular to systemic pathways.

Alzheimer's disease (AD) is the most common form of dementia, characterized by progressive memory loss, cognitive impairment, and behavioral complications. Pathophysiologically, AD is characterized by excessive amyloid-*beta* (Aβ) peptide aggregation, intracellular neurofibrillary tangles, highly phosphorylated tau protein, deficiency of essential neurotransmitters, and oxidative stress-induced neuronal damage [1–3]. To date, the FDA approved drugs have included medicines that may help minimize or alleviate symptoms by regulating chemicals implicated in neurotransmission, such as acetyl cholinesterase inhibitors (donepezil, galantamine, and rivastigmine) and glutamate receptor antagonists (memantine). Aducanumab (marketed as Aduhelm, an anti-amyloid antibody intravenous (IV) infusion therapy), a new drug recently approved by the FDA, may delay clinical degeneration with benefits to both cognitive and motor function in patients with AD. However, all of these treatments suffer from side effects ranging from headache, nausea, confusion, dizziness, falls, and amyloid-related imaging abnormalities (ARIA; including swelling in the brain and micro hemorrhaging/superficial siderosis). This justified the screening of new and safe therapeutic agents from natural sources. In this context, several plants and their bioactive components were reported to inhibit Aβ formation, cholinesterases, tau proteins aggregation, and free radicals [4–7]. Various phytosterols are anti-inflammatory in action and are known to abrogate production and activation of nitric oxide (NO), tumor necrosis factor-α (TNF-α), cyclooxygenase-2 (COX-2), inducible nitric oxide synthase (iNOS), and phosphorylated extracellular signal-regulated protein kinase (p-ERK) [8]. Hence, phytosterols may possess therapeutic implications in neurodegenerative diseases.

Phytosterol is a collective term for plant sterols and their saturated forms (stanols). They are present in dietary sources such as unrefined plant oils, nuts, seeds, and legumes [9]. Phytosterols can be called 'plant cholesterol' due to their structural similarity to cholesterol. An additional methyl group (campesterol) or ethyl group (sitosterol) at the C-24 position or an additional double bond at the C-22 position (brassicasterol or stigmasterol) differentiate them from cholesterol. Depending on the nature of the diet, sitosterol (65%), campesterol (23%), and stigmasterol (10%) belong to the major group among over 200 types of phytosterols [10]. Other important phytosterols such as brassicasterol, avenasterol, and fucosterol, along with the saturated forms of phytosterols (sitostanol, campestanol, and stigmastanol), are present in minor quantities in the diet. The dietary intake of phytosterols varies from 150–450 mg/day [11], but their intestinal absorption is less than 2% (phytosterols) and 0.2% (phytostanols) in comparison to cholesterol, which is nearly 50% [12,13]. As a result of low absorption and rapid biliary elimination, physiological concentrations of phytosterols in the plasma are on the order of 10^{-3} in comparison to cholesterol [14].

Phytosterols have recently gained the attention of researchers due to cholesterol and lipid lowering, anti-atherogenic and immunomodulating properties, especially with a report on the changes in cholesterol metabolism of patients with AD [15–17]. Essentially, there is rarely any information in association with their toxicity apart from the indistinct anabolic effect for some [18], and they have already been declared safe (up to 3 g/day) by the Food and Drug Administration (FDA) and the European Union Scientific Committee (EUSC) [19].

Phytosterols have the ability to cross the blood–brain barrier (BBB) and accumulate in the brain [20]. Hence, they might have a significant role in modulating various pathways in the brain linked to neurodegeneration. As the research or review reports on this important aspect were limited, herein, the influences of phytosterols in neurodegenerative diseases were reviewed and discussed. The most prevalent plant sterols, stigmasterol, β-sitosterol, and campesterol [21], were mainly discussed with several minor sterols— brassicasterol, lanosterol, 24(S)-saringosterol, 4,4-dimethyl phytosterols, and ergosterol—for their action mechanisms in potentially treating AD.

2. Biosynthesis of the Phytosterols in Plants

The biosynthesis of phytosterols in plants consists of three major steps. As highlighted in Zhang et al. [22], step one involves the mevalonate or isoprenoid pathway. With acetyl-CoA as the starting point, the process lead to the formation of mevalonate, which serves as an intermediate in the biosynthesis of isoprenoids. Step two is described as the cyclization of squalene and its subsequent conversion to 2,3-oxidosqualene, generating various triterpenes and the phytosterols (lanosterol and cycloartenol). In step three, cycloartenol is converted to 24-methylene cycloartenol by the action of C-24-sterol methyltransferase 1 (SMT1). As shown in Figure 1, the formation of the 24-methylene cycloartenol would facilitate the biosynthesis of the major plant sterols β-sitosterol, stigmasterol, and campesterol.

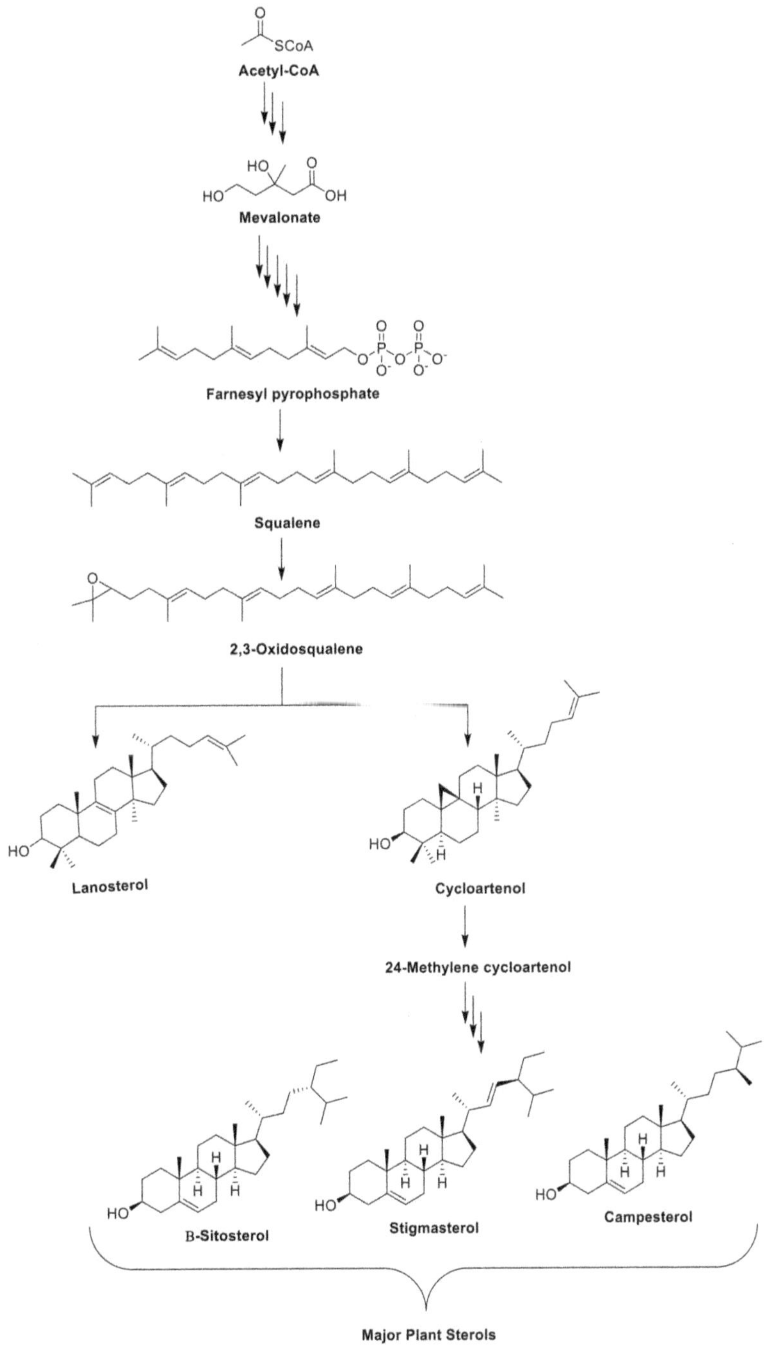

Figure 1. A simplified overview of the biosynthetic pathway of the major phytosterols in plants.

3. Plant Sterols in the Brain

As the exact mechanism through which phytosterols cross the BBB is still not fully understood, it was speculated that high density lipoprotein (HDL)-like particles carrying plant sterols and apolipoprotein E (ApoE) assist this process in brain [18,23,24]. The HDL-like particles are transported through the scavenger receptor class B type 1 (SR-B1) HDL receptors, present on the apical side of the BBB and released into the brain through ATP-binding cassette (ABCA/ABCG1) transporters, on the basolateral side of the BBB and on the astrocytes membrane. Low density and very low density lipoprotein (LDL/VLDL) receptors could transfer these HDL-like particles loaded with plant sterols on the surface of microglia and oligodendrocytes (Figure 2). Even though the brain accounts for 2.1% of the total body weight, it contains nearly 23% of all free cholesterol [25], which is synthesized by the brain itself (in situ), because the cholesterol in the circulation will not cross the BBB. A steady cholesterol turnover plays a critical role in synapse formation, cell–cell interactions, and intracellular signaling [25,26]. The level of cholesterol is maintained in the brain by removal of excess cholesterol out of the BBB in the form of polar 24-(S)-OH-cholesterol [27]. On the other hand, the metabolism of phytosterols is entirely different in the brain. Due to the presence of alkyl groups at C-24, they cannot be converted to more polarized derivatives. Therefore, after entering the brain circulation, phytosterols are irretrievably accumulated and incorporated into the cell membranes [28], where they display a positive role in reducing inflammation, Aβ levels, and β-secretase activity [29]. Again, the mechanism is still not confirmed, but is expected to occur through peroxisome proliferator-activated receptors (PPAR) signaling, involving the activation of liver X receptor/retinoid X receptor (LXR/RXR) and the regulation of apolipoprotein E (ApoE) [18]. The sterols from Aloe vera were reported to act through PPAR receptors and increased the expression of fatty acid transporter (FATP1), acyl-CoA- oxidase 1 (ACOX1), and carnitine palmitoyl transferase 1 (CPT1) in a dose-dependent manner, as well as by increasing glutathione and decreasing IL-18 expression [30]. Cholesterol is highly amyloidogenic; however, stigmasterol (having an additional double-bond at C-22/C-23 and an ethyl-group at C-24 in cholesterol skeleton) significantly decreased Aβ levels [31]. Evidence linked raised cholesterol levels to increased Aβ generation in cellular and most animal models of AD [32], and studies have indicated that drugs that inhibit cholesterol synthesis also lower Aβ in these models [33]; however, some other results are contradictory [34]. Studies showed that changes in cholesterol homeostasis, distribution within neurons, and compartmentation affected the processing of APP and Aβ generation [35]. The identification of a variant of the apolipoprotein E (*APOE*) gene as a major genetic risk factor for AD is also consistent with a role for cholesterol in the pathogenesis of AD [36]. The activity of an enzyme responsible for clearing cholesterol from peripheral cells, lecithin cholesterol acyltransferase (LCAT), is significantly decreased in atherosclerosis and AD [37].

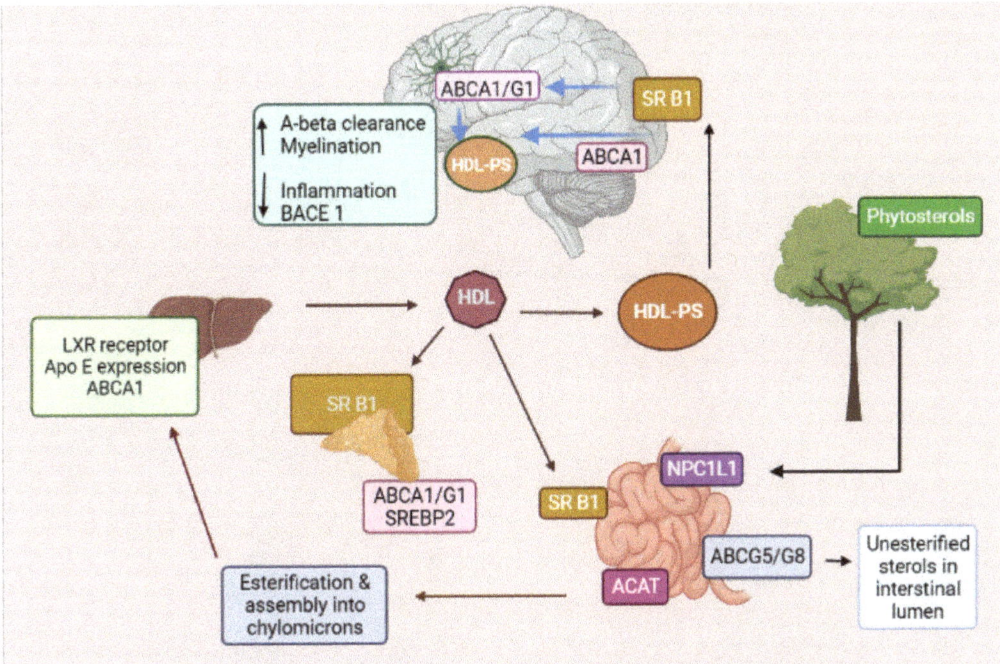

Figure 2. Pathway of dietary phytosterol assimilation in humans. Phytosterols and dietary cholesterol are mainly absorbed in the intestine through NPC1L1 transporter. After absorption, sterols are esterified by ACAT and transported to the liver in the form of chylomicrons. The unesterified sterols are pumped out by ABCG5/ABCG8 transporters. The plasma efflux of sterols is regulated by ABCA1, which is involved in the assembly of HDL-like particles with the assimilated sterols. In liver, phytosterols stimulate LXR receptors regulating ApoE expression required for HDL and LDL assembly and uptake. Stimulated LXR receptors upregulate ABCG5/G8 transporters and augment sterol absorption. The exported HDL-like particles containing phytosterols are taken up by SR-IB receptors expressed on the liver, adrenal, and brain surface. SR-B1 has a crucial role in the flux of phytosterols across the BBB. ABCA1/ABCG1 transporters present on astrocytes and the basolateral side of the cerebral endothelium also assist in the transfer of phytosterols. Inside the brain, PS exert a positive effect on brain function by escalating A-β plaque clearance, increasing re-myelination besides reducing neuro inflammation and BACE 1 activity. Abbreviations: ABC, ATP-binding cassette transporters; A-β, Amyloid-β; ACAT, Acyl-coenzyme A cholesterol acyl transferase; ApoE, Apolipoprotein E; BACE1, β-Secretase 1; HDL, High-density lipoprotein; SR-B1, Scavenger receptor class B type 1; LDL, Low-density lipoprotein; LXR, Liver X receptor; NPC1L1, Niemann-Pick C1 like 1 protein; PS, Phytosterols.

In particular, binding of cholesterol to the C-terminal transmembrane domain (C99, also known as the β-CTF) of the amyloid precursor protein (APP) seems to favor the amyloidogenic pathway in cells by endorsing localization of C99 in lipid rafts where γ-secretase and possibly β-secretase are present [38]. The products formed (amyloid-β and the intracellular domain of the APP (AICD)) might possibly down-regulate ApoE-mediated cholesterol uptake and cholesterol biosynthesis. Another possible mechanism by which cholesterol/C99 promote amyloidogenesis is via directly modifying substrate binding, catalysis, or product dissociation by β- or γ-secretase. The activity of these enzymes is much higher in the presence of cholesterol than in its absence [39]. The APP appears as a cellular cholesterol sensor; when membrane cholesterol levels are high, APP binds cholesterol to form a complex and promote the amyloidogenic pathway [40]. In a study using HT22 mouse hippocampal cells, the effect of substituting cholesterol with β- sitosterol was evaluated on APP metabolism. The results showed that the substitution stimulated non-amyloidogenic APP processing possibly by redistribution of APP from lipid rafts toward non-raft regions, without affecting membrane fluidity (Figure 3) [41].

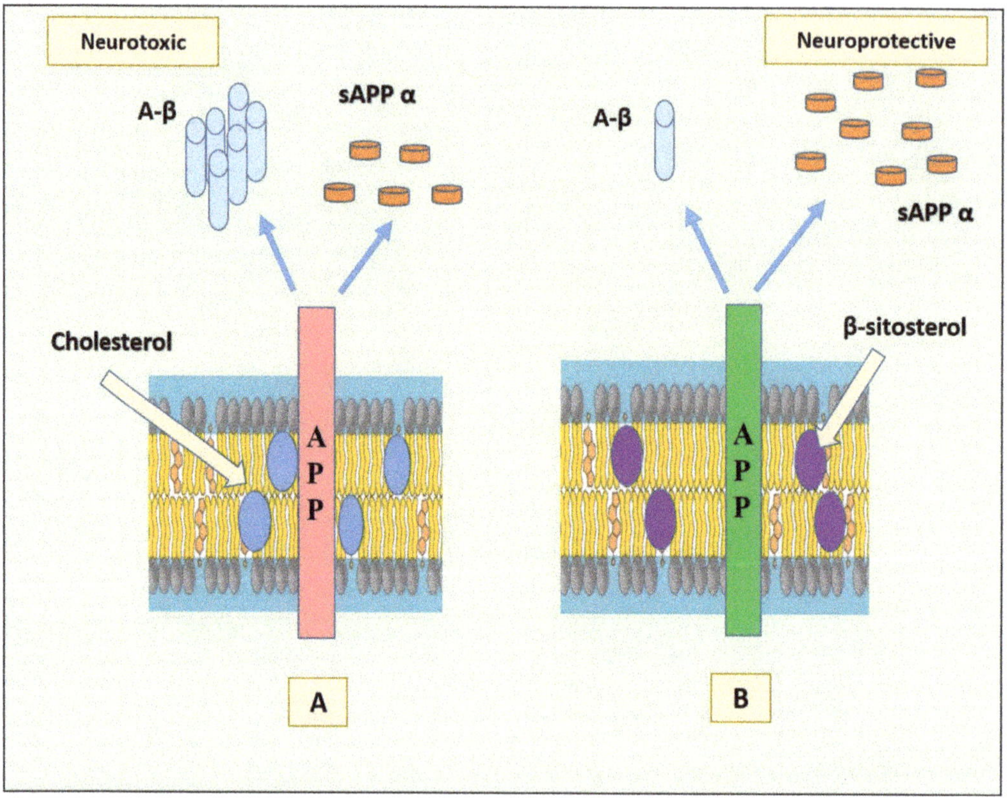

Figure 3. (**A**) Membrane cholesterol favors β-secretase cleavage of APP by direct binding to the C-terminal transmembrane domain of APP, generating neurotoxic Aβ and less sAPP α. (**B**) Substitution of membrane cholesterol with β-sitosterol promotes the re-distribution of APP in non-raft region and non-amyloidogenic processing, generating less Aβ and more of neuroprotective sAPP α. Abbreviations: Aβ: beta amyloid; APP: amyloid precursor protein; sAPP α: soluble alpha-amyloid precursor protein.

In fact, they significantly reduced brain Aβ levels and β- and γ-secretase activities in vivo, suggesting that a diet enriched in plant sterols might be beneficial for AD. Additionally, the administration of a blend of sitosterol (60%), campesterol (25%), and stigmasterol (15%) has been shown to have an anti-inflammatory effect on central nervous system (CNS) demyelination in an autoimmune encephalomyelitis (EAE) model of multiple sclerosis (MS) [42]. The similar outcomes were expected for AD, as both AD and MS were tightly related in lipid metabolism. Phytosterols are also known to prevent disease progression and to improve motor defects by increasing myelin content by facilitating incorporation of proteolipid protein (PLP) into myelin membranes, improving density of oligodendrocytes and reducing inflammation in animal models [43,44]. RT-PCR results indicated the role of phytosterols in modulating the action of a variety of growth factors from their involvement in the differentiation and survival of oligodendrocyte precursor cells [43]. Phytosterol supplementation elevated the expression levels of fibroblast growth factor 1 (FGF1) and Sonic hedgehog (SHH), without affecting insulin-like growth factor 1 (IGF1), epidermal growth factor (EGF), and ciliary neurotrophic factor (CNTF).

4. Neuroprotective Effects of the Selected Phytosterols

This section highlights various biological activities in vitro or in vivo by the major phytosterols, such as β-sitosterol, stigmasterol, and campesterol, and other minor sterols brassicasterol, lanosterol, 24(*S*)-saringosterol, and the 4,4-dimethyl sterols. The structures of the selected phytosterols included in this review are shown in Figure 4, and a summary of their implicated neuroprotective activities is shown in Table 1 and Figure 5.

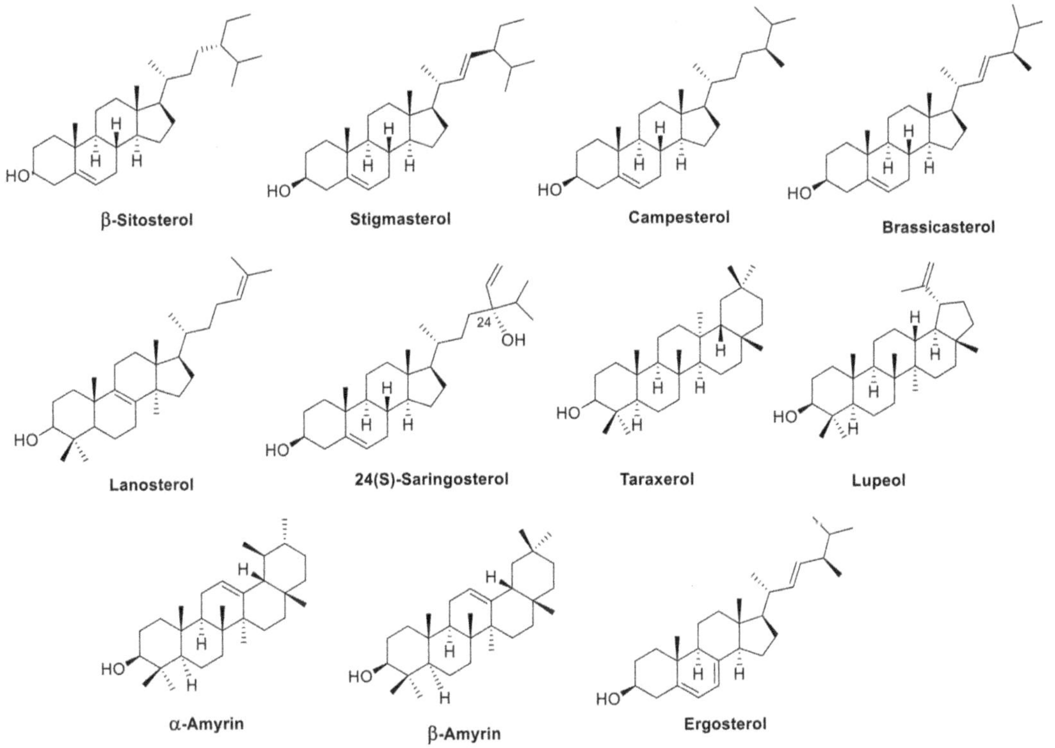

Figure 4. Structure of the phytosterols with potential activities against neurodegenerative diseases.

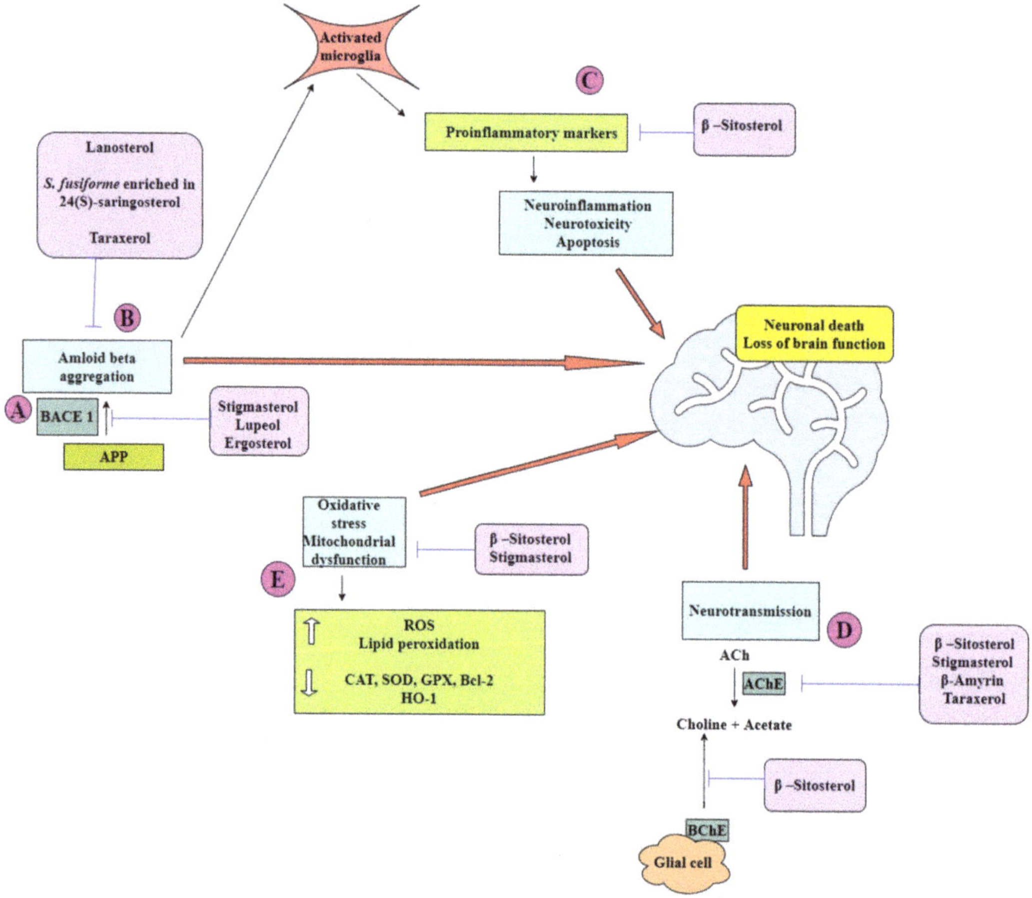

Figure 5. Phytosterols exert a multitarget approach to ameliorate symptoms of AD. (**A**) Some prevent amyloid-*beta* aggregation by inhibiting cleavage of the amyloid precursor protein (APP) by β-secretase (BACE-I). This causes a shift in the non-amyloidogenic pathway and reduces the levels of Aβ produced. (**B**) Aβ can self-aggregate to form oligomers and eventually amyloid plaques. Some phytosterols are able to inhibit the formation of amyloid plaques by binding to Aβ, inhibiting aggregation, and thereby promoting the formation of nontoxic oligomers. Toxic Aβ monomers and oligomers have been shown to induce microglial activation and proliferation. Activated microglia secrete pro-inflammatory cytokines such as IL-1β and IL-6. (**C**) Some phytosterols have been shown to reduce the levels of these cytokines. Some phytosterols reduce oxidative stress by increasing the levels of antioxidant enzymes and reducing lipid peroxidation. (**D**) Acetylcholine (ACh), a neurotransmitter essential for processing memory and learning, is decreased in both concentration and function in AD. Decreased levels of ACh can be restored by anticholinesterase activity of various phytosterols. (**E**) ROS irreversibly oxidize DNA and are important mediators of Aβ-induced neuronal cell death in the development of AD. Abbreviations: APP: Amyloid Precursor Protein; AChE: Acetyl Cholinesterase Enzyme; BACE 1: Beta-Secretase 1; BChE: Butyl Cholinesterase Enzyme; Bcl-2: B-Cell Lymphoma 2; CAT: Catalase; GPX: Glutathione Peroxidase; HO 1: Heme Oxygenase; ROS: Reactive Oxygen Species; SOD: Superoxide Dismutase.

4.1. β-Sitosterol

β–Sitosterol is one of the main dietary phytosterols belonging to the class of organic compounds known as stigmastanes. These are sterol lipids with a structure based on the stigmastane skeleton, which consists of a cholestane moiety bearing an ethyl group at the C-24 position. In a comprehensive study, the effects of β-sitosterol in behavioral studies were observed in transgenic mice using a shallow water maze (SWM), Y-maze, and balance beam tests. The in vivo (in frontal cortex and hippocampus) and in vitro anti-acetyl choline esterase (AChE), anti-butyl choline esterase (BChE) inhibitory potentials, antioxidant activity, and molecular docking were also analyzed [45,46]. β-sitosterol revealed strong Ache properties under both in vitro and in vivo conditions. An IC_{50} value of 55 and 50 μg/mL against AChE and BChE, respectively [45], and 62 μg/mL against both AChE and BChE [46] were reported in vitro, while the activity was significantly lower in brain tissue homogenates. The AChE inhibition was further confirmed by in silico studies in which β-sitosterol was strongly bound to the active sites of AChE and BChE through the *para*-hydroxyl group of the phenolic moiety networked with the active site water molecule and the side chain carbonyl residues through H-bonding. The rest of the active compound was packed in a shallow pocket through H-bonding [46]. In the antioxidant assays, the IC_{50} values were observed as 140, 120, and 280 μg/mL in the 2,2-diphenyl-1-picryl-hydrazyl-hydrate (DPPH), 2,2 -azino-bis (3-ethylbenzothiazoline-6-sulfonic acid) (ABTS), and hydrogen peroxide (H_2O_2) assays, respectively. The treatment group had lower oxidative stress in comparison to the disease control. β-Sitosterol was reported to increase levels of antioxidant enzymes by activating the estrogen receptor/PI3-kinase pathway. It also regulated glutathione levels, suggesting its role as an effective free radical scavenger [47,48]. Furthermore, glucose oxidase-mediated oxidative stress and lipid peroxidation might be inhibited from the incorporation of β-sitosterol into cell membrane, which showed valuable effects of this compound in neurodegenerative disorders including AD [47]. β-sitosterol also restored behavioral deficits, working memory, and motor coordination in transgenic animals [45]. It also hastened neuron degeneration in mice deficient for LXRβ [49], suggesting that LXRβ activation by non-glucosylated sitosterol is neuroprotective.

In another study, the effect and mechanism of β-sitosterol on deficits in learning and memory in amyloid protein precursor/presenilin 1 (APP/PS1) double transgenic mice was investigated. APP/PS1 mice were treated with β-sitosterol for 4 weeks, from the age of 7 months. Brain Aβ metabolism was evaluated using ELISA and Western blotting. β-sitosterol treatment ameliorated spatial learning and recognition memory capacity, along with reduction in plaque deposition and renovation of the excitatory postsynaptic current frequency in the hippocampus [50]. Substitution of cholesterol by β-sitosterol promoted non-amyloidogenic AP processing by migrating it to non-lipid raft region [41]. However, contradictory results were obtained in another study [31], where β-sitosterol increased the secretion of amyloid by 115.2% ± 2.2%.

Reports have suggested the role of cholesterol in amyloid precursor protein (APP) processing [40,51–53]; thus, cholesterol may be considered as a target for developing drugs to treat AD. Phytosterols are structurally similar to cholesterol and have been widely used to reduce blood cholesterol [54]. It was found that β-sitosterol efficiently repressed the release of high cholesterol-driven platelet Aβ. In addition, β-sitosterol also prevented high cholesterol-induced increases in β- and γ-secretase activity [55].

Mitochondrial dysfunction has a critical role in neuronal degeneration [56]. Therefore, augmentation of mitochondrial function by increasing the mitochondrial membrane potential (ΔΨm) and mitochondrial adenosine triphosphate (ATP) may be valuable in treating AD. Previously, it was suggested that an enhancement in mitochondrial ATP levels may be beneficial for neurodegenerative diseases [57]. Integration of β-sitosterol into mitochondrial membrane augmented the mitochondrial function by endorsing inner mitochondrial membrane fluidity, thereby raising the ΔΨm and ATP concentrations. Hence, β-sitosterol could be an effective dietary therapy for neurodegenerative diseases such as AD [58].

Continual neuroinflammation in neurodegenerative diseases damages the neurons due to the release of toxic factors such as nitric oxide. In a study, β-sitosterol displayed anti-inflammatory action in BV2 cells upon exposure to LPS by reducing the expression of pro-inflammatory markers, such as interleukin-6 (IL-6), inducible nitric oxide (iNOS), tumor necrosis factor-α (TNF-α), and cyclooxygenase-2 (COX-2). It also suppressed the phosphorylation and degradation of inhibitor of nuclear factor kappa B (IκB) and inhibited the phosphorylation of nuclear factor kappa B (NF-κB) and extracellular signal-regulated kinase (ERK), which regulated various cytokines in inflammatory pathway [59].

All of these studies confirmed the multi-target potential of β-sitosterol in management of memory deficit disorders such as AD.

4.2. Stigmasterol

Stigmasterol is a steroid derivative having a hydroxyl group in position C-3 of the steroid skeleton, unsaturated bonds in position C-5/C-6 of the B ring, and alkyl substituents in position C-22/C-23. Stigmasterol is shown to be involved in neuroprotection through a multi-target approach. Stigmasterol isolated from *Rhazya stricta* fruits displayed in vitro AChE inhibitory activity with an IC_{50} of 644 ± 11.75 μM [60]. It was also reported to reduce amyloid plaques by decreasing the β-secretase cleavage of APP [31]. Stigmasterol displayed neuroprotective activity against glutamate-induced toxicity by inhibition of reactive oxygen species (ROS) and Ca^{2+} production in in vitro studies [61]. To investigate the possible neuroprotective mechanisms of stigmasterol, an H_2O_2-induced oxidative stress model in SH-SY5Y neuroblastoma cells was established [62]. H_2O_2 exposure raised the levels of ROS significantly within the cells, inducing apoptosis. However, pre-incubation with stigmasterol prevented oxidative stress-induced cell death by reducing the level of ROS in the cells. It also upregulated catalase, forkhead box O (FoxO) 3a, and anti-apoptotic protein B-cell lymphoma 2 (Bcl-2) in the neurons. Additionally, it also increased the expression levels of sirtuin 1 (SIRT1) and decreased the levels of acetylated lysine [62]. Sirtuins are highly conserved NAD (+)-dependent enzymes that have positive effects in age-related neurodegenerative diseases. In vitro and in vivo studies revealed that the increased SIRT1 protein ameliorated AD-like symptoms by reducing the memory decline [63]. Similarly, reduction of Aβ aggregations was reported through activation of SIRT1 causing upregulation of APP metabolism by α-secretase [64]. Moreover, overexpression of SIRT1 protected SH-SY5Y cells from toxicity-induced cell death [65]. Using docking studies, the possible interaction between stigmasterol and the activator binding site of SIRT1 was also inspected. According to the binding analysis, SIRT1 interacted with stigmasterol and resveratrol; both compounds bind to Thr^{209} and Pro^{212} by van der Waals and hydrophobic interactions, respectively. Moreover, they also shared binding interactions with Phe^{414}, Asp^{292}, Gln^{294}, and Ala^{295} [62]. Stigmasterol was also reported to inhibit several pro-inflammatory cytokines in IL-1β-treated cells without affecting IL-6 levels, suggesting its role in the IL-1β-induced NF-κB inflammatory pathway [66].

Transcriptomic analysis implicated the role of stigmasterol in upregulating genes involved in neuritogenesis (*Map2, Dcx, Reln*) and synaptogenesis (*Arc, Egr1, Nr4a1*), thereby stimulating neuronal architecture in primary hippocampal neurons for processing memory and learning. Stigmasterol also decreased the expression of K^+ transport genes to sustain neuronal excitability under adverse conditions [67]. Thus, the broad spectrum of action has made stigmasterol a potential therapeutic candidate for the prevention and treatment of brain disorders, especially AD. Mice fed with stigmasterol-enriched diets exhibited the reduction of Aβ generation by diminishing β-secretase activity, decreasing expression of all γ-secretase components, reducing cholesterol and presenilin distribution in lipid rafts involved in amyloidogenic APP cleavage, and by decreasing β-secretase 1 (BACE1) internalization to endosomal compartments [31].

4.3. Campesterol

Campesterol is characterized by the presence of a hydroxyl group at C-3 of the steroid skeleton and saturated bonds throughout the sterol structure, with the exception of the C-5/C-6 double bond in the B ring. Campesterol only marginally affected Aβ secretion in APP695-expressing SH-SY5Y cells, mainly due to increased γ-secretase activity. The β-secretase activity was also significantly increased in purified membranes of mouse brains (104.8% ± 1.4), whereas no effect was observed in SH-SY5Y wild type (wt) membranes. Both gene expression and protein levels of BACE1 were unchanged. However, campesterol significantly increased γ-secretase activity in the purified membranes of SH-SY5Y wt cells and mouse brains (115.5% ± 3.0 and 106.9%, respectively). Gene expression of all components of the γ-secretase complex remained unchanged in the presence of campesterol, reflected by unchanged PS1 protein level [31].

In a clinical trial of AD patients, sitosterol and campesterol levels in the brain were almost similar in comparison to the healthy control group (6.3 ± 0.8 ng/mg and 6.2 ± 0.8 ng/mg versus 5.0 ± 0.8 ng/mg and 5.0 ± 0.8 ng/mg, $p < 0.05$, in the assayed and control groups, respectively). However, the levels of 27-hydroxycholesterol were decreased due to oxidative damage, where the BBB was not disturbed [68].

4.4. Brassicasterol

Brassicasterol is a 3β-sterol, that is, (22*E*)-ergosta-5,22-diene substituted by a hydroxy group at position 3β. It is found in marine algae, fish, and rapeseed oil. In the early stages of AD, the functions of the BBB and choroid plexus are impaired, resulting in reduced concentrations of phytosterols in cerebrospinal fluid (CSF). Both sitosterol and brassicasterol were significantly reduced, but when corrected for CSF cholesterol concentrations, only brassicasterol was reduced significantly in CSF of AD patients. Brassicasterol also amended the predictive value, when added to the biomarkers pTau and Aβ-42. Therefore, brassicasterol may be a novel additional CSF biomarker of AD patients [69]. Although their study was not designed to determine whether brassicasterol levels in CSF may be prognostic of AD progression, their observations nonetheless generated interesting hypotheses about the utility of plant sterol metabolism as a potential biomarker in AD.

However, in another study [31], brassicasterol directly increased β-secretase activity in purified membranes of SH-SY5Y wt cells and mouse brains (108.5% ± 1.9 and 106.6% ± 1.7, respectively) without affecting gene expression and protein level of BACE1. Similarly, γ-secretase activity was also slightly, but significantly, increased in purified membranes of SH-SY5Y wt cells and mouse brains (108.2% ± 1.0 and 111.4% ± 1.6%, respectively). Considerable changes in gene expression were only found for the γ-secretase components presenilin 2(PS2)/anterior-pharynx-defective protein 1 (APH1B), but not for the homologues PS1 and APH1A, presenilin enhancer 2 (PEN2) [31].

4.5. Lanosterol

Lanosterol is a $\Delta^{8,24}$ phytosterol formed from the cyclization of 2,3-oxidosqualene catalyzed by lanosterol synthetase [22]. It was shown to be a potential candidate in the treatment of cataracts by disrupting the aggregations of intracrystalline proteins [70]. Several studies suggested it as a promising modulator of neurodegenerative diseases [71–74]. In silico studies revealed that lanosterol disrupted the aggregations of the amyloidogenic KLVFFA peptide chain in $A\beta_{42}$ by hydrophobic interaction with Phe-19 and Phe-20 and interfering with the β-sheet-β-sheet packing interactions. This was further correlated with the thioflavin T (ThT) fluorescence assay and atomic force microscope (AFM) imaging. Furthermore, lanosterol also revealed neuroprotective effects with PC12 cells upon $A\beta_{42}$ treatment [71].

The aggregation of misfolded proteins in neuronal cells is interconnected with neurodegeneration. In vitro report utilizing HeLa and HEK-293A cells revealed that lanosterol and lanosterol synthase facilitated the renewal of the cell's trapped misfolded proteins in intracellular proteostasis. Lanosterol effectively minimized the number and size of sequestosomes/aggresomes in HeLa and HEK-293A cells. These results suggested a cytoprotective property of lanosterol in facilitating cell proliferation and survival by disaggregating and refolding ubiquitinated misfolded proteins [72]. Lanosterol treatment reduced the accumulations and cytotoxicity of misfolded protein aggregations through induction of co-chaperone (CHIP) and by promoting autophagy [73].

The neuroprotective effects of lanosterol was also demonstrated in vivo [74]. Administration of lanosterol to 1-methyl-4-phenyl-1,2,3,6-tetrahydropyridine-treated mice showed lanosterol reduction in the nigrostriatal region, suggesting an altered lanosterol metabolism involvement in Parkinson's disease pathogenesis. Treatment with lanosterol also indicated the increased cell viability in dopaminergic neurons, when cells were treated with 1-methyl-4-phenylridium, by regulating and inducing mitochondrial function and depolarization, and promoting autophagy.

4.6. 24(S)-Saringosterol

24(*S*)-Saringosterol or (3β)-stigmasta-5,28-diene-3,24-diol is a marine phytosterol commonly isolated from the edible seaweed *Sargassum fusiforme*. 24(*S*)-saringosterol was shown to be a potential anti-atherosclerotic natural agent by lowering cholesterol and a selective LXR-β agonist [75,76]. An in vivo study of APPswePS1DE9 mice treated with 24(*S*)-saringosterol for 10 weeks revealed a slowing of cognitive decline based on enhanced spatial and object memory assessments. It also reduced the in vivo expression of ionized calcium-binding adapter molecule 1 (Iba1), a marker for microglial activation and inflammation. No effect was detected on the amount of Aβ plaques, hence suggesting the neuroinflammatory effects of 24(*S*)-saringosterol in the deterrence of cognitive decline or limited penetration through BBB [77].

Dietary supplementation of *S. fusiforme* enriched in 24(*S*)-saringosterol in an AD mouse model indicated a reduction in hippocampal Aβ plaques and improvement in short-term memory. In vitro treatment with the same extract on mouse neuroblastoma (N2a) cells also exhibited a reduced secretion of Aβ plaques [78]. The reduction in the levels of Aβ plaques by dietary supplementation of *S. fusiforme* extracts enriched with phytosterols including 24(*S*)-saringosterol may indicate a synergistic model of the phytosterols in the prevention of neurodegenerative diseases.

4.7. 4,4-Dimethyl Phytosterols

4,4-Dimethyl phytosterols are a class of bioactive compounds having two methyl groups at the C-4 position of the aliphatic A-ring. Significant amounts of α-amyrin, β-amyrin, taraxerol, and lupeol are present in plants and vegetable oils. In scopolamine-induced cognitive impairment in mice, elevated levels of memory-related proteins in hippocampus were reported in the presence of α-amyrin and β-amyrin. Improvement in cognitive function was found to be induced through the activation of extracellular signal-regulated kinase (ERK) and glycogen synthase kinase-3β (GSK-3β). Additionally, β-amyrin and not α-amyrin displayed anti-AChE activity [79,80]. In scopolamine- and streptozotocin-induced memory deficit studies in mice, taraxerol also displayed anti-AChE activity through activation of the AChE receptor system [81]. Molecular docking studies confirmed high affinity of taraxerol for fibrils and Aβ [82]. Lupeol was found to inhibit BACE1 in both enzymatic and docking studies and is therefore a promising candidate for AD treatment [83,84]. The reduced levels of LDL-C were typically related to the prevalence of AD and PD [85]. Consequently, the anti-Alzheimer's and anti-Parkinsonian potential of 4,4-dimethy sterols may also be linked to their cholesterol metabolism regulating activity [86].

In addition, the elimination of the misfolded proteins through autophagy is an important mechanism in preventing neurodegenerative diseases. It was found that β-amyrin participated in the LGG-1 (ubiquitin-like modifier involved in the formation of autophagosomes) related autophagy pathway by improving LGG-1 expression and exhibiting a protective effect on dopaminergic neurons by decreasing cell damage, and α-synuclein aggregation, which improves PD symptoms [87].

4.8. Ergosterol

Ergosterol is the most abundant fungal phytosterol bearing the $\Delta^{5,7}$ diene oxysterol skeleton [88]. In in vitro β- and γ-secretase assays utilizing N2a cells, ergosterol slightly decreased the β-secretase activity at 20–80 μM concentrations, while strongly inhibiting the γ-secretase activity at 40 μM [53].

These data suggest the promising potentials of the phytosterols as modulators of neurodegenerative diseases. The neuroprotective effects exhibited by the phytosterols are summarized in Table 1.

Table 1. Neuroprotective mechanisms of the reported phytosterols.

Phytosterol	Mode of Action	Study	References
β–Sitosterol	AChE and BChE inhibitory activity	in vivo (mice), in vitro, in silico	[45,46]
	Increased levels of antioxidant enzymes by activating estrogen receptor/PI3-kinase pathway	in vitro (RAW 264.7; HT22)	[47,48]
	Anti-inflammatory	in vitro (BV12)	[59]
	Increase mitochondrial potential	in vitro (HT22)	[58]
Stigmasterol	AChE inhibitory activity	in vitro	[60]
	Reduced the β-secretase activity. Reduced the expression of all γ-secretase components Reduced the cholesterol and presenilin distribution in lipid rafts implicated in amyloidogenic APP cleavage. Decreased the BACE1 internalization to endosomal compartments, involved in APP β-secretase cleavage	in vivo (mice), in vitro (HT22)	[31,55]
	Decrease ROS	in vitro (SH-SY5Y)	[61,62]
	Anti-inflammatory	in vitro (mouse chondrocytes and human osteoarthritis chondrocytes)	[66]
Brassicasterol	Marker in CSF of AD patients	cerebrospinal fluid (CSF)	[69]
	Minor or no effect on Aβ secretion	in vivo (mice)	[31]
Campesterol	Minor or no effect on Aβ secretion	in vivo (mice)	[31]
Lanosterol	Reduced the accumulations and cytotoxicity of Aβ aggregation through induction of co-chaperone and by promoting autophagy	in silico, in vitro (HeLa, PC12, HEK-293A), in vivo (mice)	[71–74]
24(*S*)-Saringosterol	Reduced the in vivo expressions of Iba1	in vivo (mice)	[77]
Dietary supplementation of *S. fusiforme* enriched in 24(*S*)-saringosterol	Reduced secretion of Aβ plaques. Improves memory in AD mice model	in vivo (mice), in vitro (N2a)	[78]
α-Amyrin	Elevated levels of memory related proteins through the activation of ERKGSK-3β	in vivo (mice)	[79]
β-Amyrin	Elevated levels of memory related proteins through the activation of ERKGSK-3β. AChE inhibitory activity	in vivo (mice)	[79,80]
Taraxerol	AChE inhibitory activity	in vivo (mice)	[83]
	High affinity of taraxerol for fibrils and amyloid- β	in silico	[84]
Lupeol	BACE-I inhibitory activity	in vitro, in silico	[85,86]
Ergosterol	Reduced the β- and γ-secretase activity	in vitro	[44]

5. Physiological versus Pathological Features of Phytosterols

In diet, phytosterols primarily reduce blood cholesterol concentrations [89] by lowering serum LDL-cholesterol concentrations, with no effect on either serum HDL-cholesterol concentrations or triacylglycerol levels [90]. They are also anti-inflammatory and have physiological functions such as growth regulation and the promotion of protein synthesis, immune regulation, and hormone-like effects [91]. The hypocholesterolemic effect of 4-desmethyl sterols is well known [92], whereas 4,4-dimethylsterols (lupeol, α-amyrin, cycloartenol) have limited action on cholesterol reduction [93]. As phytosterols reduce the solubility of cholesterol, some other lipophilic compounds such as lipophilic antioxidant nutrients may also be displaced. Randomized trials have shown that phytosterols lower the blood concentrations of β-carotene (by about 25%), α-carotene (by 10%), and vitamin E (by 8%) [94,95], which protect the oxidation of LDL cholesterol. However, after the action of phytosterols, β-carotene was found to be reduced by 8–19%, while fat-soluble vitamins (A, D, E, K) remained unchanged [94]. Therefore, intake of food rich in carotenes may balance this side effect.

The most adversarial effect of phytosterols is seen in a rare inherited disease, sitosterolemia, described by tendon xanthoma and premature coronary disease [96]. These patients have mutated ABCG5 and ABCG8 transporters that cause reduced transport of phytosterols from enterocytes back into the intestinal lumen and reduced secretion into bile. Hence, these patients have increased phytosterol absorption (3–4 fold) and low biliary excretion, resulting in build-up of these compounds in plasma and tissues [97]. A relationship between increased plasma phytosterols (7–16%) and increased risk of coronary heart disease (CHD) has been reported [98]. However, various contradictory reports rule out any relation between phytosterols and the risk of incident CHD. Additionally, plasma concentrations of the main phytosterols (sitosterol and campesterol) can be used as bio markers for cardiometabolic risk, as moderately elevated plasma sitosterol, but not campesterol, may possibly indicate decreased risk for CHD [14]. Thus, the available data are contradictory and more in-depth studies are required to confirm whether phytosterols are friend or foe.

6. Safety Concerns for the Phytosterols

Even though intake of phytosterols (up to 3 g/day) is considered safe by the FDA, information on their toxicity is limited, including their indistinct anabolic effect [18]. No adverse effect of phytosterols on mental or cognitive activity has been reported yet. However, oxidized phytosterols were reported to exhibit neurotoxicity with glutamate excitotoxicity and ROS generation [87,99]. Several unfavorable effects of phytosterols on endothelium-dependent vasorelaxation in *wt* mice have been reported [100–103]. The types and concentrations of phytosterols or their structure may determine and influence the production of superoxides by endothelial cells [101–103]. Collectively, these studies indicated that phytosterols could modulate various endothelium-dependent processes, such as vasorelaxation, oxidative stress, ischemia–reperfusion, and neuroinflammation, which are key biological processes in the progression of CNS disorders. Therefore, depending on the nature, concentrations, and target cells, the phytosterols may have critical influences on neurodegenerative disorders [104].

7. Future Perspectives and Conclusions

Phytosterols are an important component of diet. They have wide applications in foods and cosmetics. To date, our knowledge has been limited to their cholesterol-lowering properties, protective effects mostly against cardiovascular diseases, and their anti-cancer and anti-inflammatory potential. Extensive research on the use of phytosterols for the preventive and therapeutic management of other diseases should be highly explored. The impact of phytosterols on the central nervous system is another exciting avenue of investigation. Phytosterols could cross the BBB through a less-known mechanism and become accumulated in the brain. The BBB is one of the most important and largest barriers among the three CNS barriers for the exchange of constituents between the blood and the CNS [105]. Any impairment in the brain endothelial cells may alter the normal function of adhesion molecules, chemotactic proteins, and angiogenic factors; thus, increased ROS production, infiltration of immune cells, and neuroinflammation could be the consequences in CNS disorders [106–109]. Moreover, normal functioning of various signaling pathways occurring at lipid rafts (cholesterol-rich domains), such as γ-aminobutyric acid and glutamate signaling, may perturb the synaptic vesicle turnover, protection of motor neurons by brain-derived neurotrophic factor (BDNF), functioning of the calcium channel, and calcium-dependent neurotransmitter release [110–113].

AD patients are shown to have compromised cholesterol turnover [114,115], where suppressed cholesterol biosynthesis may reduce the production of Aβ both in vitro and in vivo [68]. Likewise, a cholesterol-rich diet increased the generation of Aβ in mice [116–118]. The increased cellular cholesterol turnovers were in association with increased expression of the LXR-activated genes, which significantly amended cognitive performance in AD animal models [119–121].

Because the early stage of AD is linked to BBB dysfunction, the decreased concentrations of phytosterols in CSF could be used as a promising prognostic biomarker. Importantly, cholesterol is highly amyloidogenic, whereas the phytosterols are not. In fact, phytosterols significantly reduced brain Aβ levels and β- and γ-secretase activities in vivo, suggesting that a diet enriched in plant sterols might be beneficial for neurodegenerative diseases. Research on the optimal doses of phytosterols through oral administration would be essential. Because the consumed phytosterols would have limited intestinal absorption and crossing the BBB may take longer periods of ~6 weeks, higher doses may be required to capture the benefits [122]. Researchers have resorted to nanoencapsulation for improving the bioavailability of phytosterols in the food industry [123–125], but only for research purposes as a methodology of drug delivery. Thus, future studies should be carried out to explore the therapeutic and disease-specific mechanisms of phytosterols for their neuroprotective role in neurodegenerative diseases.

Author Contributions: Conceptualization, M.A.T. and S.S.A.A.; writing—original draft preparation, N.S. and M.A.T.; writing—review and editing, N.S., M.A.T. and S.S.A.A.; funding acquisition, S.S.A.A. All authors have read and agreed to the published version of the manuscript.

Funding: This research was funded by the National Research Foundation of Korea and by the Korean Government (2020R1A2B5B01002463 and 2021R1A6A1A03038996).

Institutional Review Board Statement: Not applicable.

Informed Consent Statement: Not applicable.

Data Availability Statement: Not applicable.

Conflicts of Interest: The authors declare no conflict of interest.

References

1. LaFerla, F.M.; Green, K.N.; Oddo, S. Intracellular amyloid-β in Alzheimer's disease. *Nat. Rev. Neurosci.* **2007**, *8*, 499–509. [CrossRef]
2. Minati, L.; Edginton, T.; Bruzzone, M.G.; Giaccone, G. Current concepts in Alzheimer's disease: A multidisciplinary review. *Am. J. Alzheimers Dis. Other Dement.* **2009**, *24*, 95–121. [CrossRef] [PubMed]
3. Selkoe, D.J. Alzheimer's disease: Genes, proteins, and therapy. *Physiol. Rev.* **2001**, *81*, 741–766. [CrossRef] [PubMed]
4. Armand, E.F.; Shantaram, M.; Nico, N.F.; Simon, F.N.; Paul, M.F. Potential of medicinal plant compounds to targeting Tau protein in the therapy of Alzheimer's disease—A review. *Biomedicine* **2019**, *39*, 217–227. [CrossRef]
5. Cui, X.; Lin, Q.; Liang, Y. Plant-derived antioxidants protect the nervous system from aging by inhibiting oxidative stress. *Front. Aging Neurosci.* **2020**, *12*, 209. [CrossRef] [PubMed]
6. Murray, A.P.; Faraoni, M.B.; Castro, M.J.; Alza, N.P.; Cavallaro, V. Natural AChE inhibitors from plants and their contribution to Alzheimer's disease therapy. *Curr. Neuropharmacol.* **2013**, *11*, 388–413. [CrossRef]
7. Witter, S.; Witter, R.; Vilu, R.; Samoson, A. Medical plants and nutraceuticals for amyloid-β fibrillation inhibition. *J. Alzheimers Dis. Rep.* **2018**, *2*, 239–252. [CrossRef] [PubMed]
8. Yuan, L.; Zhang, F.; Shen, M.; Jia, S.; Xie, J. Phytosterols suppress phagocytosis and inhibit inflammatory mediators via ERK pathway on LPS-triggered inflammatory responses in RAW264. 7 macrophages and the correlation with their structure. *Foods* **2019**, *8*, 582. [CrossRef] [PubMed]
9. Piironen, V.; Lindsay, D.G.; Miettinen, T.A.; Toivo, J.; Lampi, A.M. Plant sterols: Biosynthesis, biological function and their importance to human nutrition. *J. Sci. Food Agric.* **2000**, *80*, 939–966. [CrossRef]
10. Weihrauch, J.L.; Gardner, J.M. Sterol content of foods of plant origin. *J. Am. Diet. Assoc.* **1978**, *73*, 39–47. [CrossRef]
11. Othman, R.A.; Myrie, S.B.; Jones, P.J. Non-cholesterol sterols and cholesterol metabolism in sitosterolemia. *Atherosclerosis* **2013**, *231*, 291–299. [CrossRef] [PubMed]
12. Ostlund Jr, R.E.; McGill, J.B.; Zeng, C.M.; Covey, D.F.; Stearns, J.; Stenson, W.F.; Spilburg, C.A. Gastrointestinal absorption and plasma kinetics of soy Δ5-phytosterols and phytostanols in humans. *Am. J. Physiol. Endocrinol. Metab.* **2002**, *282*, E911–E916. [CrossRef]

13. Von Bergmann, K.; Sudhop, T.; Lütjohann, D. Cholesterol and plant sterol absorption: Recent insights. *Am. J. Cardiol.* **2005**, *96*, 10–14. [CrossRef] [PubMed]
14. Escurriol, V.; Cofán, M.; Moreno-Iribas, C.; Larrañaga, N.; Martínez, C.; Navarro, C.; Rodriguez, L.; Gonzalez, C.A.; Corella, D.; Ros, E. Phytosterol plasma concentrations and coronary heart disease in the prospective Spanish EPIC cohort. *J. Lipid Res.* **2010**, *51*, 618–624. [CrossRef] [PubMed]
15. Morand, C.; Tomás Barberán, F.A. Contribution of plant food bioactives in promoting health effects of plant foods: Why look at interindividual variability? *Eur. J. Nutr.* **2019**, *58*, 13–19. [CrossRef]
16. Nashed, B.; Yeganeh, B.; HayGlass, K.T.; Moghadasian, M.H. Antiatherogenic effects of dietary plant sterols are associated with inhibition of proinflammatory cytokine production in Apo E-KO mice. *J. Nutr.* **2005**, *135*, 2438–2444. [CrossRef]
17. Trautwein, E.A.; Vermeer, M.A.; Hiemstra, H.; Ras, R.T. LDL-cholesterol lowering of plant sterols and stanols—Which factors influence their efficacy? *Nutrients* **2018**, *10*, 1262. [CrossRef] [PubMed]
18. Kopylov, A.T.; Malsagova, K.A.; Stepanov, A.A.; Kaysheva, A.L. Diversity of plant sterols metabolism: The impact on human health, sport, and accumulation of contaminating sterols. *Nutrients* **2021**, *13*, 1623. [CrossRef] [PubMed]
19. Dutta, P.C. *Phytosterols as Functional Food Components and Nutraceuticals*, 1st ed.; CRC Press: Boca Raton, FL, USA, 2003; p. 96.
20. Vanmierlo, T.; Weingärtner, O.; van der Pol, S.; Husche, C.; Kerksiek, A.; Friedrichs, S.; Sijbrands, S.; Steinbusch, H.; Grimm, M.; Hartmann, T.; et al. Dietary intake of plant sterols stably increases plant sterol levels in the murine brain. *J. Lipid Res.* **2012**, *53*, 726–735. [CrossRef] [PubMed]
21. Moreau, R.A.; Nystrom, L.; Whitaker, B.D.; Winkler-Moser, J.K.; Baer, D.J.; Gebauer, S.K.; Hicks, K.B. Phytosterols and their derivatives: Structural diversity, distribution, metabolism, analysis, and health-promoting uses. *Prog. Lipid Res.* **2018**, *70*, 35–61. [CrossRef] [PubMed]
22. Zhang, X.; Lin, K.; Li, Y. Highlights to phytosterols accumulation and equilibrium in plants: Biosynthetic pathway and feedback regulation. *Plant. Physiol. Biochem.* **2020**, *155*, 637–649. [CrossRef] [PubMed]
23. Fricke, C.B.; Schrøder, M.; Poulsen, M.; von Bergmann, K.; Wester, I.; Knudsen, I.; Mortensen, A.; Lütjohann, D. Increased plant sterol and stanol levels in brain of Watanabe rabbits fed rapeseed oil derived plant sterol or stanol esters. *Br. J. Nutr.* **2007**, *98*, 890–899. [CrossRef] [PubMed]
24. Jansen, P.J.; Lutjohann, D.; Abildayeva, K.; Vanmierlo, T.; Plosch, T.; Plat, J.; von Bergmann, K.; Groen, A.; Ramaekers, F.; Kuipers, F.; et al. Dietary plant sterols accumulate in the brain. *Biochim. Biophys. Acta* **2006**, *1761*, 445–453. [CrossRef] [PubMed]
25. Dietschy, J.M.; Turley, S.D. Thematic review series: Brain lipids. Cholesterol metabolism in the central nervous system during early development and in the mature animal. *J. Lipid Res.* **2004**, *45*, 1375–1397. [CrossRef]
26. Björkhem, I. Crossing the barrier: Oxysterols as cholesterol transporters and metabolic modulators in the brain. *J. Intern. Med.* **2006**, *260*, 493–508. [CrossRef] [PubMed]
27. Wang, Y.; Muneton, S.; Sjövall, J.; Jovanovic, J.N.; Griffiths, W.J. The effect of 24S-hydroxycholesterol on cholesterol homeostasis in neurons: Quantitative changes to the cortical neuron proteome. *J. Proteome Res.* **2008**, *7*, 1606–1614. [CrossRef]
28. Van Kampen, J.M.; Robertson, H.A. The BSSG rat model of Parkinson's disease: Progressing towards a valid, predictive model of disease. *EPMA J.* **2017**, *8*, 261–271. [CrossRef] [PubMed]
29. Mongrand, S.; Stanislas, T.; Bayer, E.M.; Lherminier, J.; Simon-Plas, F. Membrane rafts in plant cells. *Trends Plant. Sci.* **2010**, *15*, 656–663. [CrossRef] [PubMed]
30. Klaikeaw, N.; Wongphoom, J.; Werawatganon, D.; Chayanupatkul, M.; Siriviriyakul, P. Anti-inflammatory and anti-oxidant effects of Aloe vera in rats with non-alcoholic steatohepatitis. *World J. Hepatol.* **2020**, *12*, 363–377. [CrossRef] [PubMed]
31. Burg, V.K.; Grimm, H.S.; Rothhaar, T.L.; Grösgen, S.; Hundsdörfer, B.; Haupenthal, V.J.; Zimmer, V.; Mett, J.; Weingartner, O.; Laufs, U.; et al. Plant sterols the better cholesterol in Alzheimer's disease? A mechanistical study. *J. Neurosci.* **2013**, *33*, 16072–16087. [CrossRef] [PubMed]
32. Panchal, M.; Loeper, J.; Cossec, J.; Perruchini, C.; Lazar, A.; Pompon, D.; Duyckaerts, C. Enrichment of cholesterol in microdissected Alzheimer's disease senile plaques as assessed by mass spectrometry. *J. Lipid Res.* **2010**, *51*, 598–605. [CrossRef] [PubMed]
33. Kandiah, N.; Feldman, H. Therapeutic potential of statins in Alzheimer's disease. *J. Neurol. Sci.* **2009**, *283*, 230–234. [CrossRef]
34. Fonseca, A.; Resende, R.; Oliveira, C.; Pereira, C. Cholesterol and statins in Alzheimer's disease: Current controversies. *Exp. Neurol.* **2010**, *223*, 282–293. [CrossRef] [PubMed]
35. Puglielli, L.; Tanzi, R.; Kovacs, D. Alzheimer's disease: The cholesterol connection. *Nat. Neurosci.* **2003**, *6*, 345–351. [CrossRef] [PubMed]
36. Kuo, Y.; Emmerling, M.; Bisgaier, C.; Essenburg, A.; Lampert, H.; Drumm, D.; Roher, A. Elevated low-density lipoprotein in Alzheimer's disease correlates with brain Aβ1-42 levels. *Biochem. Biophys. Res. Commun.* **1998**, *252*, 711–715. [CrossRef] [PubMed]
37. Knebl, J.; DeFazio, P.; Clearfield, M.; Little, L.; McConathy, W.; McPherson, R.; Lacko, A. Plasma lipids and cholesterol esterification in Alzheimer's disease. *Mech. Ageing Dev.* **1994**, *73*, 69–77. [CrossRef]
38. Tang, B.L. Neuronal protein trafficking associated with Alzheimer disease: From APP and BACE1 to glutamate receptors. *Cell Adh. Migr.* **2009**, *3*, 118–128. [CrossRef] [PubMed]
39. Kalvodova, L.; Kahya, N.; Schwille, P.; Ehehalt, R.; Verkade, P.; Drechsel, D.; Simons, K. Lipids as modulators of proteolytic activity of BACE: Involvement of cholesterol, glycosphingolipids, and anionic phospholipids in vitro. *J. Biol. Chem.* **2005**, *280*, 36815–36823. [CrossRef] [PubMed]

40. Beel, A.J.; Sakakura, M.; Barrett, P.J.; Sanders, C.R. Direct binding of cholesterol to the amyloid precursor protein: An important interaction in lipid–Alzheimer's disease relationships? *Biochim. Biophys. Acta Mol. Cell Biol. Lipids* **2010**, *1801*, 975–982. [CrossRef] [PubMed]
41. Wang, J.; Wu, F.; Shi, C. Substitution of membrane cholesterol with β-sitosterol promotes nonamyloidogenic cleavage of endogenous amyloid precursor protein. *Neuroscience* **2013**, *247*, 227–233. [CrossRef] [PubMed]
42. Valerio, M.; Liu, H.B.; Heffner, R.; Zivadinov, R.; Ramanathan, M.; Weinstock-Guttman, B.; Awad, A.B. Phytosterols ameliorate clinical manifestations and inflammation in experimental autoimmune encephalomyelitis. *Inflamm. Res.* **2011**, *60*, 457–465. [CrossRef] [PubMed]
43. Berghoff, S.A.; Gerndt, N.; Winchenbach, J.; Stumpf, S.K.; Hosang, L.; Odoardi, F. Dietary cholesterol promotes repair of demy-elinated lesions in the adult brain. *Nat. Commun.* **2017**, *8*, 14241. [CrossRef] [PubMed]
44. Saher, G.; Rudolphi, F.; Corthals, K.; Ruhwedel, T.; Schmidt, K.F.; Lowel, S.; Dibaj, P.; Barrette, B.; Mobius, W.; Nave, K. Therapy of Pelizaeus-Merzbacher disease in mice by feeding a cholesterol-enriched diet. *Nat. Med.* **2012**, *18*, 1130–1135. [CrossRef] [PubMed]
45. Ayaz, M.; Junaid, M.; Ullah, F.; Subhan, F.; Sadiq, A.; Ali, G.; Ovais, M.; Shahid, M.; Ahmad, A.; Wadood, A.; et al. Anti-Alzheimer's studies on β-sitosterol isolated from *Polygonum hydropiper* L. *Front. Pharmacol.* **2017**, *8*, 697. [CrossRef]
46. Bari, W.U.; Zahoor, M.; Zeb, A.; Khan, I.; Nazir, Y.; Khan, A.; Ur Rehman, N.; Ullah, R.; Shahat, A.; Mahmood, H.M. Anti-cholinesterase, antioxidant potentials, and molecular docking studies of isolated bioactive compounds from *Grewia optiva*. *Int. J. Food Prop.* **2019**, *22*, 1386–1396. [CrossRef]
47. Shi, C.; Wu, F.; Zhu, X.; Xu, J. Incorporation of β-sitosterol into the membrane increases resistance to oxidative stress and lipid peroxidation via estrogen receptor-mediated PI3K/GSK3β signalling. *Biochim. Biophys. Acta* **2013**, *1830*, 2538–2544. [CrossRef] [PubMed]
48. Vivancos, M.; Moreno, J.J. β-Sitosterol modulates antioxidant enzyme response in RAW 264.7 macrophages. *Free Radic. Biol. Med.* **2005**, *39*, 91–97. [CrossRef] [PubMed]
49. Kim, H.J.; Fan, X.; Gabbi, C.; Yakimchuk, K.; Parini, P.; Warner, M.; Gustafsson, J.A. Liver X receptor β (LXRβ): A link between β-sitosterol and amyotrophic lateral sclerosis-Parkinson's dementia. *Proc. Natl. Acad. Sci. USA* **2008**, *105*, 2094–2099. [CrossRef]
50. Ye, J.Y.; Li, L.; Hao, Q.M.; Qin, Y.; Ma, C.S. β-Sitosterol treatment attenuates cognitive deficits and prevents amyloid plaque deposition in amyloid protein precursor/presenilin 1 mice. *Korean J. Physiol. Pharmacol.* **2020**, *24*, 39–46. [CrossRef] [PubMed]
51. Runz, H.; Rietdorf, J.; Tomic, I.; de Bernard, M.; Beyreuther, K.; Pepperkok, R.; Hartmann, T. Inhibition of intracellular cholesterol transport alters presenilin localization and amyloid precursor protein processing in neuronal cells. *J. Neurosci.* **2002**, *22*, 1679–1689. [CrossRef] [PubMed]
52. Wang, H.; Kulas, J.A.; Wang, C.; Holtzman, D.M.; Ferris, H.A.; Hansen, S.B. Regulation of beta-amyloid production in neurons by astrocyte-derived cholesterol. *Proc. Nat. Acad. Sci. USA* **2021**, *118*, e2102191118. [CrossRef] [PubMed]
53. Xiong, H.; Callaghan, D.; Jones, A.; Walker, D.G.; Lue, L.F.; Beach, T.G.; Sue, L.; Woulfe, J.; Xu, H.; Stanimirovic, D.; et al. Cholesterol retention in Alzheimer's brain is responsible for high β-and γ-secretase activities and Aβ production. *Neurobiol. Dis.* **2008**, *29*, 422–437. [CrossRef]
54. Clifton, P. Lowering cholesterol: A review on the role of plant sterols. *Aust. Fam. Physician* **2009**, *38*, 218–221. [PubMed]
55. Shi, C.; Liu, J.; Wu, F.; Zhu, X.; Yew, D.T.; Xu, J. β-sitosterol inhibits high cholesterol-induced platelet β-amyloid release. *J. Bioenerg. Biomembr.* **2011**, *43*, 691–697. [CrossRef] [PubMed]
56. Schon, E.A.; Manfredi, G. Neuronal degeneration and mitochondrial dysfunction. *J. Clin. Investig.* **2003**, *111*, 303–312. [CrossRef]
57. Reddy, P.H. Role of mitochondria in neurodegenerative diseases: Mitochondria as a therapeutic target in Alzheimer's disease. *CNS Spect.* **2009**, *8* (Suppl. 7), 8–13.
58. Shi, C.; Wu, F.; Xu, J. Incorporation of β-sitosterol into mitochondrial membrane enhances mitochondrial function by promoting inner mitochondrial membrane fluidity. *J. Bioenerg. Biomembr.* **2013**, *45*, 301–305. [CrossRef] [PubMed]
59. Sun, Y.; Gao, L.; Hou, W.; Wu, J. β-Sitosterol alleviates inflammatory response via inhibiting the activation of ERK/p38 and NF-κB pathways in LPS-exposed BV2 cells. *BioMed Res. Int.* **2020**, *2020*, 7532306. [CrossRef]
60. Sultana, N.; Khalid, A. Phytochemical and enzyme inhibitory studies on indigenous medicinal plant *Rhazya stricta*. *Nat. Prod. Res.* **2010**, *24*, 305–314. [CrossRef] [PubMed]
61. Lee, J.; Weon, J.B.; Ma, C.J. Neuroprotective activity of phytosterols isolated from *Artemisia apiacea*. *Korean J. Pharmacogn.* **2014**, *45*, 214–219.
62. Pratiwi, R.; Nantasenamat, C.; Ruankham, W.; Suwanjang, W.; Prachayasittikul, V.; Prachayasittikul, S.; Phopin, K. Mechanisms and neuroprotective activities of stigmasterol against oxidative stress-induced neuronal cell death via sirtuin family. *Front. Nutr.* **2021**, *8*, 648995. [CrossRef]
63. Donmez, G. The neurobiology of sirtuins and their role in neurodegeneration. *Trends Pharmacol. Sci.* **2012**, *33*, 494–501. [CrossRef] [PubMed]
64. Braidy, N.; Jayasena, T.; Poljak, A.; Sachdev, P.S. Sirtuins in cognitive ageing and Alzheimer's disease. *Curr. Opin. Psychiatry* **2012**, *25*, 226–230. [CrossRef] [PubMed]
65. Singh, P.; Hanson, P.S.; Morris, C.M. SIRT1 ameliorates oxidative stress induced neural cell death and is down-regulated in Parkinson's disease. *BMC Neurosci.* **2017**, *18*, 46. [CrossRef] [PubMed]
66. Gabay, O.; Sanchez, C.; Salvat, C.; Chevy, F.; Breton, M.; Nourissat, G.; Wolf, C.; Jacques, C.; Berenbaum, F. Stigmasterol: A phytosterol with potential anti-osteoarthritic properties. *Osteoarthr. Cartil.* **2010**, *18*, 106–116. [CrossRef]

67. Haque, M.N.; Moon, I.S. Stigmasterol upregulates immediate early genes and promotes neuronal cytoarchitecture in primary hippocampal neurons as revealed by transcriptome analysis. *Phytomedicine* **2018**, *46*, 164–175. [CrossRef]
68. Shafaati, M.; Marutle, A.; Pettersson, H.; Lövgren-Sandblom, A.; Olin, M.; Pikuleva, I. Marked accumulation of 27- hydroxycholesterol in the brains of Alzheimer's patients with the Swedish APP 670/671 mutation. *J. Lipid Res.* **2011**, *52*, 1004–1010. [CrossRef] [PubMed]
69. Vanmierlo, T.; Popp, J.; Kölsch, H.; Friedrichs, S.; Jessen, F.; Stoffel-Wagner, B.; Bertsch, T.; Hartmann, T.; Maier, W.; von Bergmann, K.; et al. The plant sterol brassicasterol as additional CSF biomarker in Alzheimer's disease. *Acta Psychiatr. Scand.* **2011**, *124*, 184–192. [CrossRef] [PubMed]
70. Pei, H.; Ma, X.; Pan, Y.; Han, T.; Lu, Z.; Wu, R.; Cao, X.; Zheng, J. Separation and purification of lanosterol, dihydrolanosterol, and cholesterol from lanolin by high-performance counter-current chromatography dual-mode elution method. *J. Sep. Sci.* **2019**, *42*, 2171–2178. [CrossRef] [PubMed]
71. Zhou, H.; Yang, Z.; Tian, X.; Chen, L.; Lee, S.; Huynh, T.; Ge, C.; Zhou, R. Lanosterol disrupts the aggregation of amyloid-β peptides. *ACS Chem. Neurosci.* **2019**, *10*, 4051–4060. [CrossRef] [PubMed]
72. Hu, L.-D.; Wang, J.; Chen, X.-J.; Yan, Y.-B. Lanosterol modulates proteostasis via dissolving cytosolic sequestosomes/aggresome-like induced structures. *Biochim. Biophys. Acta Mol. Cell Res.* **2020**, *1867*, 118617. [CrossRef] [PubMed]
73. Upadhyay, A.; Amanullah, A.; Mishra, R.; Kumar, A.; Mishra, A. Lanosterol suppresses the aggregation and cytotoxicity of misfolded proteins linked with neurodegenerative diseases. *Mol. Neurobiol.* **2018**, *55*, 1169–1182. [CrossRef] [PubMed]
74. Lim, L.; Jackson-Lewis, V.; Wong, L.C.; Shui, G.H.; Goh, A.X.; Kesavapany, S.; Jenner, A.M.; Fivaz, M.; Przedborski, S.; Wenk, M.R. Lanosterol induces mitochondrial uncoupling and protects dopaminergic neurons from cell death in a model for Parkinson's disease. *Cell Death Differ.* **2012**, *19*, 416–427. [CrossRef] [PubMed]
75. Chen, Z.; Liu, J.; Fu, Z.; Ye, C.; Zhang, R.; Song, Y.; Zhang, Y.; Li, H.; Ying, H.; Liu, H. 24(S)-Saringosterol from dibble marine seaweed *Sargassum fusiforme* is a novel selectve LXRβ agonist. *J. Agric. Food Chem.* **2014**, *62*, 6130–6137. [CrossRef] [PubMed]
76. Yan, Y.; Niu, Z.; Wang, B.; Zhao, S.; Sun, C.; Wu, Y.; Li, Y.; Ying, H.; Liu, H. Saringosterol from sargassum fusiforme modulates cholesterol metabolism and alleviates atherosclerosis in ApoE-deficient mice. *Mar. Drugs* **2021**, *19*, 485. [CrossRef] [PubMed]
77. Martens, N.; Schepers, M.; Zhan, N.; Leijten, F.; Voortman, G.; Tiane, A.; Rombaut, B.; Poisquet, J.; van de Sande, N.; Kerksiek, A.; et al. 24(S)-Saringosterol prevents cognitive decline in a mouse model for Alzheimer's disease. *Mar. Drugs* **2021**, *19*, 190. [CrossRef]
78. Bogie, J.; Hoeks, C.; Schepers, M.; Tiane, A.; Cuypers, A.; Leijten, F.; Chintapakorn, Y.; Suttiyut, T.; Pornpakakul, S.; Struik, D.; et al. Dietary *Sargassum fusiforme* imprives memory and reduces amyloid plaque load in an Alzheimer's disease mouse model. *Sci. Rep.* **2019**, *9*, 4908.
79. Park, S.J.; Ahn, J.Y.; Oh, S.R.; Lee, Y.; Kwon, G.; Woo, H.; Lee, H.E.; Jang, D.S.; Ryu, J.H. Amyrin attenuates scopolamine-induced cognitive impairment in mice. *Biol. Pharm. Bull.* **2014**, *37*, 1207–1213. [CrossRef] [PubMed]
80. Schwarz, S.; Loesche, A.; Lucas, S.D.; Sommerwerk, S.; Serbian, I.; Siewert, B.; Pianowski, E.; Csuk, R. Converting maslinic acid into an effective inhibitor of acetylcholinesterases. *Eur. J. Med. Chem.* **2015**, *103*, 438–445. [CrossRef]
81. Berté, T.E.; Dalmagro, A.P.; Zimath, P.L.; Gonçalves, A.E.; Meyre-Silva, C.; Bürger, C.; Weber, C.J.; Dos Santos, D.A.; Cechinel-Filho, V.; de Souza, M.M. Taraxerol as a possible therapeutic agent on memory impairments and Alzheimer's disease: Effects against scopolamine and streptozotocin-induced cognitive dysfunctions. *Steroids* **2018**, *132*, 5–11. [CrossRef] [PubMed]
82. Ngo, S.T.; Li, M.S. Top-leads from natural products for treatment of Alzheimer's disease: Docking and molecular dynamics study. *Mol. Simul.* **2013**, *39*, 279. [CrossRef]
83. Hosen, S.M.; Rubayed, M.; Dash, R.; Junaid, M.; Mitra, S.; Alam, M.; Dey, R. Prospecting and structural insight into the binding of novel plant-derived molecules of *Leea indica* as inhibitors of BACE1. *Curr. Pharm. Des.* **2018**, *24*, 3972–3979. [CrossRef]
84. Koirala, P.; Su, H.S.; Jung, H.A.; Choi, J.S. Comparative molecular docking studies of lupeol and lupenone isolated from *Pueraria lobata* that inhibits BACE1: Probable remedies for Alzheimer's disease. *Asian Pac. J. Trop. Med.* **2017**, *10*, 1117–1122. [CrossRef] [PubMed]
85. Huang, X.; Chen, H.; Miller, W.C.; Mailman, R.B.; Woodard, J.L.; Chen, P.C.; Dong, X.; Murrow, R.W.; Wang, Y.Z.; Poole, C. Lower low density lipid cholesterol levels are associated with Parkinson's disease. *Mov. Disord.* **2007**, *22*, 377–381. [CrossRef] [PubMed]
86. Zhang, T.; Liu, R.; Chang, M.; Jin, Q.; Zhang, H.; Wang, X. Health benefits of 4, 4-dimethyl phytosterols: An exploration beyond 4-desmethyl phytosterols. *Food Funct.* **2020**, *11*, 93–110. [CrossRef] [PubMed]
87. Wei, C.-C.; Chang, C.-H.; Liao, H.-C. Anti-Parkinsonian effects of β-amyrin are regulated via LGG-1 involved autophagy pathway in *Caenorhabditis elegans*. *Phytomedicine* **2017**, *36*, 118–125. [CrossRef] [PubMed]
88. Rodrigues, M.L. The multifunctional fungal ergosterol. *mBio* **2018**, *9*, e01755-18. [CrossRef]
89. Brufau, G.; Canela, M.A.; Rafecas, M. Phytosterols: Physiologic and metabolic aspects related to cholesterol-lowering properties. *Nutr. Res.* **2008**, *28*, 217–225. [CrossRef]
90. Miettinen, T.A.; Vuoristo, M.; Nissinen, M.; Jarvinen, H.J.; Gylling, H. Serum, biliary, and fecal cholesterol and plant sterols in colectomized patients before and during consumption of stanol ester margarine. *Am. J. Clin. Nutr.* **2000**, *71*, 1095–1102. [CrossRef]
91. Xinmei, X.; Ning, X.; Jianbin, H. Physiological function of phytosterol and its application. *Anim. Husb. Feed Sci.* **2015**, *7*, 67–69.
92. Katan, M.B.; Grundy, S.M.; Jones, P.; Law, M.; Miettinen, T.A.; Paoletti, R.; Stresa Workshop Participants. Efficacy and safety of plant stanols and sterols in the management of blood cholesterol levels. *Mayo Clin. Proc.* **2003**, *78*, 965–978. [CrossRef]

93. Sierksma, A.; Weststrate, J.A.; Meijer, G.W. Spreads enriched with plant sterols, either esterified 4,4 dimethyl-sterols or free 4-desmethylsterols, and plasma total- and LDL-cholesterol concentrations. *Br. J. Nutr.* **1999**, *82*, 273–282. [CrossRef] [PubMed]
94. Hendriks, H.F.; Weststrate, J.A.; van Vliet, T.; Meijer, G.W. Spreads enriched with three different levels of vegetable oil sterols and the degree of cholesterol lowering in normocholesterolaemic and mildly hypercholesterolaemic subjects. *Eur. J. Clin. Nutr.* **1999**, *53*, 319–327. [CrossRef] [PubMed]
95. Hallikainen, M.A.; Sarkkinen, E.S.; Uusitupa, M.I. Effects of low-fat stanol ester enriched margarine on concentrations of serum carotenoids in subjects with elevated serum cholesterol concentrations. *Eur. J. Clin. Nutr.* **1999**, *53*, 966–969. [CrossRef]
96. Salen, G.; Shefer, S.; Nguyen, L.; Ness, G.C.; Tint, G.S.; Shore, V. Sitosterolemia. *J. Lipid Res.* **1992**, *33*, 945–955. [CrossRef]
97. Lütjohann, D.; Björkhem, I.; Ose, L. Phytosterolemia in a Norwegian family: Diagnosis and characterization of the first Scandinavian case. *Scand. J. Clin. Lab. Investig.* **1996**, *56*, 229–240. [CrossRef] [PubMed]
98. Assmann, G.; Cullen, P.; Erbey, J.; Ramey, D.R.; Kannenberg, F.; Schulte, H. Plasma sitosterol elevations are associated with an increased incidence of coronary events in men: Results of a nested case control analysis of the prospective cardiovascular munster (PROCAM) study. *Nutr. Metab. Cardiovasc. Dis.* **2006**, *16*, 13–21. [CrossRef] [PubMed]
99. Panov, A.; Kubalik, N.; Brooks, B.R.; Shaw, C.A. In vitro effects of cholesterol β-D-glucoside, cholesterol and cycad phytosterol glucosides on respiration and reactive oxygen species generation in brain mitochondria. *J. Membr. Biol.* **2010**, *237*, 71–77. [CrossRef] [PubMed]
100. Tabata, R.C.; Wilson, J.M.; Ly, P.; Zwiegers, P.; Kwok, D.; Van Kampen, J.M.; Cashman, N.; Shaw, C. Chronic exposure to dietary sterol glucosides is neurotoxic to motor neurons and induces an ALS-PDC phenotype. *Neuromolecular Med.* **2008**, *10*, 24–39. [CrossRef] [PubMed]
101. Weingartner, O.; Lutjohann, D.; Ji, S.; Weisshoff, N.; List, F.; Sudhop, T.; von Bergmann, K.; Gertz, K.; Konig, J.; Schafers, H.J.; et al. Vascular effects of diet supplementation with plant sterols. *J. Am. Coll. Cardiol.* **2008**, *51*, 1553–1561. [CrossRef]
102. Weingartner, O.; Ulrich, C.; Lutjohann, D.; Ismail, K.; Schirmer, S.H.; Vanmierlo, T.; Bohm, M.; Laufs, U. Differential effects on inhibition of cholesterol absorption by plant stanol and plant sterol esters in apoE-/- mice. *Cardiovasc. Res.* **2011**, *90*, 484–492. [CrossRef]
103. Nam, Y.; Lee, D. Ameliorating effects of constituents from Cortex Acanthopanacis Radicis on memory impairment in mice induced by scopolamine. *J. Tradit. Chin. Med.* **2014**, *34*, 57–62. [CrossRef]
104. Yang, C.; Chen, Z.-Y.; Wong, S.-L.; Liu, J.; Liang, Y.-T.; Lau, C.-W.; Lee, H.-K.; Huang, Y.; Tsang, S.-Y. β-Sitosterol oxidation products attenuate vasorelaxation by increasing reactive oxygen species and cyclooxygenase-2. *Cardiovasc. Res.* **2013**, *97*, 520–532. [CrossRef] [PubMed]
105. Vanmierlo, T.; Bogie, J.F.; Mailleux, J.; Vanmol, J.; Lütjohann, D.; Mulder, M.; Hendriks, J.J. Plant sterols: Friend or foe in CNS disorders? *Prog. Lipid Res.* **2015**, *58*, 26–39. [CrossRef]
106. Abbott, N.J.; Patabendige, A.A.; Dolman, D.E.; Yusof, S.R.; Begley, D.J. Structure and function of the blood–brain barrier. *Neurobiol. Dis.* **2010**, *37*, 13–25. [CrossRef] [PubMed]
107. Zlokovic, B.V. The blood–brain barrier in health and chronic neurodegenerative disorders. *Neuron* **2008**, *57*, 178–201. [CrossRef]
108. Greenberg, D.A.; Jin, K. From angiogenesis to neuropathology. *Nature* **2005**, *438*, 954–959. [CrossRef] [PubMed]
109. Uttara, B.; Singh, A.V.; Zamboni, P.; Mahajan, R.T. Oxidative stress and neurodegenerative diseases: A review of upstream and downstream antioxidant therapeutic options. *Curr. Neuropharmacol.* **2009**, *7*, 65–74. [CrossRef] [PubMed]
110. Grammas, P.; Martinez, J.; Miller, B. Cerebral microvascular endothelium and the pathogenesis of neurodegenerative diseases. *Expert Rev. Mol. Med.* **2011**, *13*, e19. [CrossRef]
111. Davies, A.; Douglas, L.; Hendrich, J.; Wratten, J.; Tran Van Minh, A.; Foucault, I.; Koch, D.; Pratt, W.; Saibil, H.; Dolphin, A. The calcium channel $\alpha_2\delta$-2 subunit partitions with CaV2.1 into lipid rafts in cerebellum: Implications for localization and function. *J. Neurosci.* **2006**, *26*, 8748–8757. [CrossRef] [PubMed]
112. Mojsilovic-Petrovic, J.; Jeong, G.B.; Crocker, A.; Arneja, A.; David, S., Russell, D.; Kalb, R. Protecting motor neurons from toxic insult by antagonism of adenosine A2a and Trk receptors. *J. Neurosci.* **2006**, *26*, 9250–9263. [CrossRef] [PubMed]
113. Li, X.; Serwanski, D.R.; Miralles, C.P.; Bahr, B.A.; De Blas, A.L. Two pools of Triton X-100-insoluble GABA(A) receptors are present in the brain, one associated to lipid rafts and another one to the postsynaptic GABAergic complex. *J. Neurochem.* **2007**, *102*, 1329–1345. [CrossRef] [PubMed]
114. Wasser, C.R.; Ertunc, M.; Liu, X.; Kavalali, E.T. Cholesterol-dependent balance between evoked and spontaneous synaptic vesicle recycling. *J. Physiol.* **2007**, *579*, 413–429. [CrossRef] [PubMed]
115. Kölsch, H.; Heun, R.; Jessen, F.; Popp, J.; Hentschel, F.; Maier, W.; Lütjohann, D. Alterations of cholesterol precursor levels in Alzheimer's disease. *Biochim. Biophys. Acta Mol. Cell Biol. Lipids* **2010**, *1801*, 945–950. [CrossRef]
116. Fassbender, K.; Simons, M.; Bergmann, C.; Stroick, M.; Lutjohann, D.; Keller, P.; Runz, H.; Kuhl, S.; Bertsch, T.; von Bergmann, K.; et al. Simvastatin strongly reduces levels of Alzheimer's disease beta-amyloid peptides Abeta 42 and Abeta 40 in vitro and in vivo. *Proc. Natl. Acad. Sci. USA* **2001**, *98*, 5856–5861. [CrossRef]
117. Thirumangalakudi, L.; Prakasam, A.; Zhang, R.; Bimonte-Nelson, H.; Sambamurti, K.; Kindy, M.S.; Bhat, N.R. High cholesterol-induced neuroinflammation and amyloid precursor protein processing correlate with loss of working memory in mice. *J. Neurochem.* **2008**, *106*, 475–485. [CrossRef] [PubMed]
118. Julien, C.; Tremblay, C.; Phivilay, A.; Berthiaume, L.; Emond, V.; Julien, P.; Calon, F. Highfat diet aggravates amyloid-beta and tau pathologies in the 3xTg-AD mouse model. *Neurobiol. Aging* **2010**, *31*, 1516–1531. [CrossRef]

119. Knight, E.M.; Martins, I.V.; Gumusgoz, S.; Allan, S.M.; Lawrence, C.B. High-fat diet induced memory impairment in triple-transgenic Alzheimer's disease (3xTgAD) mice is independent of changes in amyloid and tau pathology. *Neurobiol. Aging* **2014**, *35*, 1821–1832. [CrossRef]
120. Riddell, D.R.; Zhou, H.; Comery, T.A.; Kouranova, E.; Lo, C.F.; Warwick, H.K.; Ring, R.; Kirksey, Y.; Aschmies, S.; Xu, J.; et al. The LXR agonist TO901317 selectively lowers hippocampal Abeta42 and improves memory in the Tg2576 mouse model of Alzheimer's disease. *Mol. Cell. Neurosci.* **2007**, *34*, 621–628. [CrossRef]
121. Jiang, Q.; Lee, C.Y.; Mandrekar, S.; Wilkinson, B.; Cramer, P.; Zelcer, N.; Mann, K.; Lamb, B.; Wilson, T.; Collins, J.; et al. ApoE promotes the proteolytic degradation of Abeta. *Neuron* **2008**, *58*, 681–693. [CrossRef]
122. Vanmierlo, T.; Rutten, K.; Dederen, J.; Bloks, V.W.; van Vark-van der Zee, L.C.; Kuipers, F.; Kiliaan, A.; Blokland, A.; Sijbrands, E.; Steinbusch, H.; et al. Liver X receptor activation restores memory in aged AD mice without reducing amyloid. *Neurobiol. Aging* **2011**, *32*, 1262–1272. [CrossRef]
123. Mohammadi, M.; Jafari, S.M.; Hamishehkar, H.; Ghanbarzadeh, B. Phytosterols as the core or stabilizing agent in different nanocarriers. *Trends Food Sci. Technol.* **2020**, *101*, 73–88. [CrossRef]
124. Tolve, R.; Cela, N.; Condelli, N.; Di Cairano, M.; Caruso, M.C.; Galgano, F. Microencapsulation as a tool for the formulation of functional foods: The phytosterols' case study. *Foods* **2020**, *9*, 470. [CrossRef] [PubMed]
125. Joye, I.J.; Davidov-Pardo, G.; McClements, D.J. Nanotechnology for increased micronutrient bioavailability. *Trends Food Sci. Technol.* **2014**, *40*, 168–182. [CrossRef]

International Journal of
Molecular Sciences

MDPI

Review

Natural Products for the Prevention and Treatment of Oral Mucositis—A Review

Ana Sofia Ferreira [1,†], Catarina Macedo [1,†], Ana Margarida Silva [1], Cristina Delerue-Matos [1], Paulo Costa [2,3] and Francisca Rodrigues [1,*]

1 REQUIMTE/LAQV—Instituto Superior de Engenharia do Porto, Rua Dr. António Bernardino de Almeida 431, 4249-015 Porto, Portugal; asofiamferreira.95@gmail.com (A.S.F.); catpmacedo@gmail.com (C.M.); ana.silva@graq.isep.ipp.pt (A.M.S.); cmm@isep.ipp.pt (C.D.-M.)
2 UCIBIO—Applied Molecular Biosciences Unit, MedTech-Laboratory of Pharmaceutical Technology, Faculty of Pharmacy, University of Porto, 4050-313 Porto, Portugal; pccosta@ff.up.pt
3 Associate Laboratory i4HB—Institute for Health and Bioeconomy, Faculty of Pharmacy, University of Porto, 4050-313 Porto, Portugal
* Correspondence: francisca.rodrigues@graq.isep.ipp.pt; Tel.: +351-22-83-40-500
† These authors contributed equally to this work.

Abstract: Cancer, a major world public health problem, is associated with chemotherapy treatments whose administration leads to secondary concerns, such as oral mucositis (OM). The OM disorder is characterized by the presence of ulcers in the oral mucosa that cause pain, bleeding, and difficulty in ingesting fluids and solids, or speaking. Bioactive compounds from natural sources have arisen as an effective approach for OM. This review aims to summarize the new potential application of different natural products in the prevention and treatment of OM in comparison to conventional ones, also providing a deep insight into the most recent clinical studies. Natural products, such as *Aloe vera*, *Glycyrrhiza glabra*, *Camellia sinensis*, *Calendula officinalis*, or honeybee crops, constitute examples of sources of bioactive compounds with pharmacological interest due to their well-reported activities (e.g., antimicrobial, antiviral, anti-inflammatory, analgesic, or wound healing). These activities are associated with the bioactive compounds present in their matrix (such as flavonoids), which are associated with in vivo biological activities and minimal or absent toxicity. Finally, encapsulation has arisen as a future opportunity to preserve the chemical stability and the drug bioa vailability of bioactive compounds and, most importantly, to improve the buccal retention period and the therapeutic effects.

Keywords: cancer; drug delivery; natural products; oral mucositis; treatment

Citation: Ferreira, A.S.; Macedo, C.; Silva, A.M.; Delerue-Matos, C.; Costa, P.; Rodrigues, F. Natural Products for the Prevention and Treatment of Oral Mucositis—A Review. *Int. J. Mol. Sci.* **2022**, 23, 4385. https://doi.org/10.3390/ijms23084385

Academic Editor: Lorenzo Lo Muzio

Received: 14 March 2022
Accepted: 14 April 2022
Published: 15 April 2022

Publisher's Note: MDPI stays neutral with regard to jurisdictional claims in published maps and institutional affiliations.

1. Introduction

Cancer is currently a major public health problem all over the world. In 2020, almost 19.3 millions new cases were diagnosed worldwide [1]. Treatment of malignancies with cytotoxic chemotherapy (CT), radiation (RT), or a combination of the two is becoming more effective, as it is associated with short- and long-term adverse effects, including mucositis [2,3]. This secondary reaction may occur in any area of the gastrointestinal tract's mucosal layer, from the mouth to the anus, with the oral cavity being the most prevalent location. The cytotoxicity is caused by a variety of mechanisms, including inhibition of DNA replication and repair, cell-cycle arrest, DNA damage, and cell death [4]. However, the precise and complex molecular pathways underlying the oral epithelial damage are not completely known [5,6].

Oral mucositis (OM) is a painful inflammatory and frequently ulcerative disorder of the oral mucosa that severely reduces the patient's quality of life [3,7,8]. OM occurs in 20 to 40% of the patients submitted to conventional CT, 80% of patients on high-dose CT, 75 to 100% of patients receiving hematopoietic cell transplants, and practically all patients with head and neck squamous carcinoma (HNSC) undergoing RT [4,5,9–11]. Common symptoms of OM include pain, bleeding, ulcers, and difficulty ingesting fluids or solids and speaking, as well as severe complications, such as secondary infections and significant weight loss, which may compromise the treatment of the primary disease and its outcome [2,12,13]. In addition, OM may result in the need for enteral or parenteral nutrition [14,15] and systemic analgesics [16–18], thus increasing hospitalizations [13,19], the use of resources and higher costs [19,20], and, in some cases, the risk of sepsis [8,21]. However, when mucositis progresses, topical analgesics become less effective and systemic opioids may be required [22–24]. Different strategies have been used to attempt the prevention or amelioration of this condition, and some clinical trials were effective [8,16,17,25]. For example, cryotherapy [11,26] and keratinocyte growth factor [11,27] demonstrated some benefits in preventing mucositis. Zinc [28,29] and vitamin E [28,30,31] were effective in reducing the severity of OM, but *Aloe vera* [32], amifostine [4,33], glutamine [28,30,34], honey [32,35–38], photobiomodulation (PBM) therapy [39–41], and antibiotics [21] demonstrated lower evidence of benefits. The studies reviewed were evaluated in patients with different types of cancer who underwent different treatment approaches.

While there are a growing number of innovative anticancer agents, few therapeutic alternatives for the prevention or treatment of oral mucositis have been reported. Most important, the scarce alternatives that have been successfully achieved are still unsatisfactory [6,16]. Therefore, the search for alternative compounds obtained from natural sources could be an option and a challenge for this research field. Natural compounds, in contrast to synthetic ones, are often thought to have fewer side effects, are easy to access, and present beneficial bioactive properties (e.g., anti-inflammatory, antioxidant, and antimicrobial properties), making them interesting solutions as promising therapeutics. Aside from the protective results of natural products against toxicity induced by radiation or antineoplastic drugs, one of the most promising preventive measures in patients during therapy may be the employment of natural products. The aim of this review is to provide an overview of the use of natural compounds for the prevention and eventual treatment of OM in cancer patients and their potential applications in drug delivery systems to overcome the specific limitations of the oral cavity environment.

2. Oral Mucositis

As previously stated, mucositis is an inflammatory response condition of the oral mucous membrane that is frequently observed in malignant neoplastic patients undergoing CT, RT, or both. This condition develops due to interactions among an oral tissue injury, the oral cavity environment, bone marrow suppression, and innate predisposing factors in the patient [18,42,43]. The symptoms of OM, such as oral mucosal atrophy, swelling, erythema and subsequent pain, bleeding, ulceration, difficulty in feeding and even swallowing saliva, or a combination thereof, may be diverse [2,11,44]. Difficulties with eating reduce the nutritional intake, resulting in a decline in the patient's nutritional status. This can also seriously affect their speech due to an uncomfortably dry mouth and a decrease or increase in salivation [11,16,44]. OM may also be aggravated by injuries induced by sharpened teeth, bruxism, food, and microorganisms [44,45]. Naturally, additional ulcers provide an easy access point for microorganisms, including bacteria, fungi, and viruses, to enter the bloodstream because of the loss of mucosal integrity, culminating in systemic infections that may cause the treatment for fighting the primary disease to be discontinued or even threaten the patient's survival. Moreover, the dysphagia, xerostomia, and changes in taste caused by OM can increase the systemic symptoms, such as lethargy and anorexia, as well as psychological issues. Consequently, OM is associated with increased resource needs and potentially major economic impacts—depending on its severity—due to the more frequent and prolonged hospitalizations for support and nutritional care and analgesic treatments.

Three tools are available for assessing the severity of OM. The most extensively used is the World Health Organization's Oral Mucositis Grading Scale (WHO-OMGS), which incorporates clinical criteria to evaluate the OM lesion and eating capacity [46]. On the other hand, the Common Terminology Criteria for Adverse Events in its fifth revision (CTCAE v5.0) considers the following factors when assessing the impact of OM: pain intensity, ability to eat, and need for intervention [46]. Finally, Radiation Therapy Oncology Group (RTOG) defines the severity of RT-induced OM using a four-point scale [46]. OM is classified according to these three criteria, as summarized in Table 1.

Table 1. Available clinical scales for oral mucositis assessment. Adapted from [16]. NA—Not applicable.

Scale	Grade 0	Grade 1 (Mild)	Grade 2 (Moderate)	Grade 3 (Severe)	Grade 4 (Life-Threatening)	Grade 5 (Death)
WHO	No findings	Oral erythema and soreness; no ulcers	Oral erythema, ulcers; solid diet tolerated	Oral ulcers; liquid diet only	Oral alimentation impossible	NA
CTCAE	None	Asymptomatic or mild symptoms; intervention not indicated	Moderate pain or ulcer that does not interfere with oral intake; modified diet indicated	Severe pain, interfering with oral intake	Life-threatening consequences; urgent intervention indicated	Death
RTOG	No change over baseline	Irritation; may experience mild pain, not requiring analgesics	Patchy mucositis that may produce an inflammatory serosanguinous discharge; may experience moderate pain requiring analgesia	Confluent, fibrinous mucositis; may include severe pain requiring narcotics	Ulceration, hemorrhage, or necrosis	NA

2.1. Physiopathology of OM

In the last decades, substantial evolution has taken place in the understanding of the complex mechanism behind the development of mucositis [6]. A five-phase model that begins with an (i) initiation involving cell injury, (ii) elevation of inflammatory cytokines, a (iii) primary damage response, and (iv) signaling and amplification of the inflammatory cascade, followed by (v) ulceration and mucosal repair through epithelial proliferation, has been reported by different authors [2,3,16,47]. Thus, OM is characterized by a cascade of events that occur simultaneously and are mechanistically related (Figure 1). Therefore, each factor that drives each phase may constitute a possible therapeutic target [16].

The mucositis initiation phase—initiation—corresponds to the injury of oral mucosal cells caused by CT and/or RT. This phase begins instantaneously as the antineoplastic treatment is being administered [5,6,48,49]. The second phase—upregulation with messenger generation—involves the cytotoxic effect, resulting in the generation of reactive oxygen and nitrogen species (ROS and RNS, respectively) and DNA damage, leading to basal and suprabasal epithelial cell death [2,3,6]. Particularly, when DNA strands breaks, the apoptotic process is activated, with p53 and nuclear factor κB (NF- κB) playing major roles [50,51]. At this point, inflammatory cytokines, chemokines, and adhesion molecules are generated when NF- κB, the key mediator of pro-inflammatory gene expression, is activated, which is clinically manifested as mucosal damage. The release of pro-inflammatory cytokines, such as tumor necrosis factor (TNF-α), interleukin-1β (IL-1β), and interleukin-6 (IL-6), is mediated through transcription factor activation, and this promotes connective tissue and endothelial damage, limiting the tissue oxygenation and stimulating epithelial basal cell death [2,18,50–53]. The third phase—signaling and amplification—is the consequence of tissue damage, apoptosis, enzyme activation, and vascular permeability, which amplify the molecules of the innate immune response as pro-inflammatory cytokines in a positive feedback mechanism, leading to more tissue damage [18,54]. In the fourth phase—ulceration—clinical signs of mucositis become visible, as the integrity of the mucosa and submucosa is disrupted, causing pain control to be required [3,5,16]. In neutropenic patients, the immune cells cannot respond properly, and the ulcerative lesions allow several microorganisms to penetrate into the connective tissue, triggering the production of more pro-inflammatory cytokines and increasing the tissue damage [50,51]. Bacteremia and sepsis are mostly caused by herpes simplex virus, *Candida albicans*, or other fungal genera, such as *Aspergillus* [10]. Healing usually occurs naturally after the cancer treatment is ceased, and it is marked by epithelial proliferation, migration, and differentiation promoted by the extracellular matrix [2,3,49]. The oral mucosa recovers, but the patient remains at risk for recurrent episodes due to residual angiogenesis [16,18,49,55].

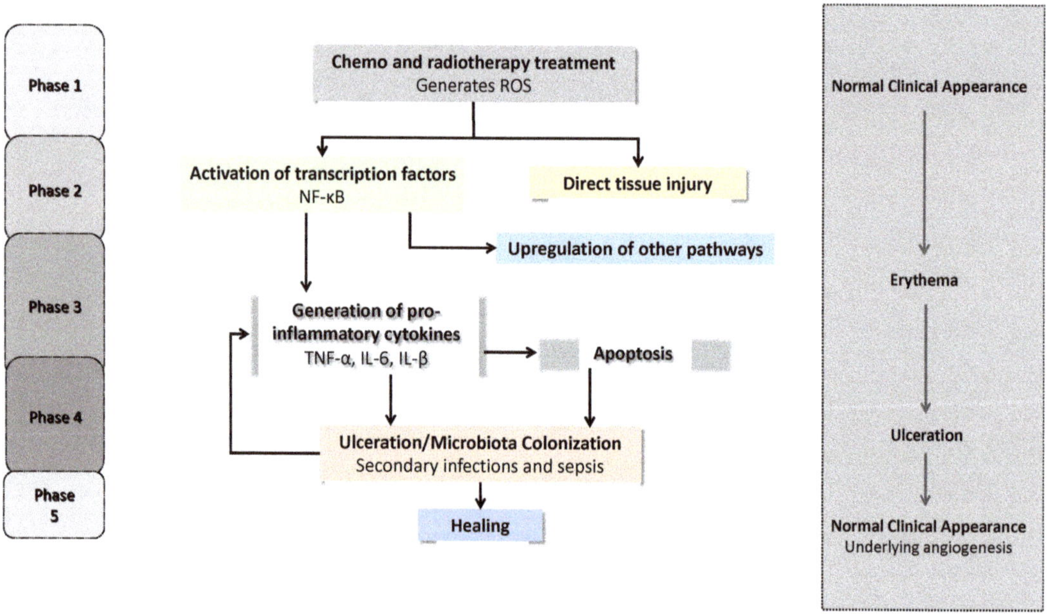

Figure 1. Diagram representing the mucosal cells and clinical manifestations of oral mucositis.

CT patients often experience acute symptoms 3–5 days following its administration, with ulcerative lesions appearing a few days later and resolving within 2 weeks [3,44,51]. On the other hand, RT mucositis is a chronic condition that lasts up to 7 weeks. The radiation doses range from 2 to 70 Gy per day and cause ulcerations that remain for 3–4 weeks after the treatment is ceased [9,11,18]. The lack of taste develops because the oral mucosa is exposed to radiation after few weeks, compromising nutrition and psychological status, while the recovery begins 6–8 weeks after the completion of the treatment [5,9].

2.2. Risk Factors

The risk factors of OM can be classified as patient-related, tumor-related, and treatment-related variables, as summarized in Table 2 [16,45].

Table 2. Risk factors related to patients, tumors, and treatments in the development of oral mucositis.

Risk Factor	Criteria	References
Related to patient		
Age	Extremities	[9,12,25,45,56–58]
Gender	Female	[2,3,10,23]
Body mass index (BMI)	Low and high body mass index	[2,3,10,23]
Dental prosthesis	Orthodontics and prosthesis	[9,11]
Education	Lack of health literacy	[7,55,59,60]
Oral hygiene	Oral hygiene less than 2 times/day Periodontal disease	[2,3,10,23]
Comorbidities	Diabetes mellitus, renal and hepatic dysfunction	[16,54,61]
Leucocytes	Neutropenic patients are immunocompromised	[2,18,45]
Alcohol	Use of alcohol prior to and during treatment	[2,3,18,45]
Smoking	Smoking prior to and during treatment increases the severity	[5,45]
Genetics	Genetic polymorphisms (e.g., TNF-α)	[10,45]
Mucosal trauma	Sharpened teeth	[5]
Related to tumor		
Types of cancer	Solid tumors have higher risk, mainly those located near oral cavity	[12,45,47]
Related to treatment		
Type of treatment	5-fluorouracil, Doxorubicin, Methotrexate, Cisplatin, Vinblastine, Mitomycin, Transtuzumabe, Docetaxel, Melphalan	[10,16,54]
Dose	High doses over short periods and their extension	[10,16,54]
Type of administration	Intravenous	[2,10,16,45,54]
Microbiota		

In the patient-related factors, gender has been linked to mucositis, since women are associated with a higher risk, which could be due to dosimetric considerations [12,25,58,62]. However, other studies reported the absence of evidence that gender and OM are correlated [16,45,58,63]. Although age is frequently reported as a mucositis risk factor, there are few consistent reports that link younger and older patients and mucositis severity [45,58]. Likewise, the effect of body mass index (BMI) on mucositis risk is inconsistent, with data suggesting that a low BMI and a BMI higher than 25 are related with a superior risk, as body composition can affect drug metabolism, as can smoking and poor oral hygiene [18,45,58]. Genetic variants, previous treatment, and comorbidities (such as renal dysfunction and diabetes mellitus) have been indicated as possible factors for chronic OM associated with RT [16,61].

In what concerns the tumor's nature, its location, size, and stadium may also influence the grade of OM [45]. For instance, in HNSC patients, the standard protocol includes RT with a specific area and prescription dose, which influences the exposure to radiation and the subsequent mucosal damage [16,45]. However, in recent years, there was an increased investment in intraoral medical devices that enable the minimization of excessive irradiation of normal tissues [64].

Although the risk factors of OM are not completely understood, the characteristics of anticancer therapeutics (mechanism of action, dose, planning, and number of cycles) are closely associated with the prevalence and severity of the lesions, as their effects accumulate [12,18,45]. It is well known that female patients using methotrexate and melphalan have a greater chance of developing this local inflammatory condition [16].

Along with investigating intrinsic patient characteristics, such as pre-existing medical conditions, altered oral dynamics, and general health, age, oral health (hygiene prior to treatment), nutritional status, and liver and kidney function are critical, as they are parameters that a medical team must consider [2,18]. Aside from that, it is necessary to emphasize that OM is frequently documented only in its advanced phases owing to the requirements for clinical therapy and assistance [13,14,19,65]. Therefore, the search for new active ingredients that could be used in the prevention (and even treatment) of OM is of utmost importance.

2.3. Prevention and Management of OM

OM management strategies include either preventive or symptom control strategies [8,18,23,25]. The primary key measure in preventing OM is the preservation of tissue during RT treatment planning and the use of RT procedures that conserve the uninvolved oral mucosal surface [17,66]. Some strategies are addressed in the evidence-based guidelines developed by the Multinational Association of Supportive Care in Cancer and the International Society of Oral Oncology (MASCC/ISOO), which present three categories: a recommendation, a suggestion, and a situation where no guideline is possible [54,63]. These guidelines can be adjusted at any time to compensate for possible restraints in the clinic and patient choices [63]. Table 3 summarizes the recommended or suggested strategies for most of the groups of cancer patients.

Table 3. Management of Oral Mucositis Guidelines created by the Multinational Association for Supportive Care in Cancer and the International Society of Oral Oncology.

Intervention	Aim	MASCC/ISOO Guideline Category	Results	References
Oral care	Prevention	Suggestion	Increases patient's awareness and enhances their compliance with treatment	[5,37,46,60,63,67–69]
Oral cryotherapy	Prevention	Recommendation	Local vasoconstriction that minimizes drug absorption	[11,46,70–73]
Photobiomodulation therapy	Prevention	Recommendation	Promotes wound healing and has an anti-inflammatory effect	[23,26,39–41,46]
Benzydamine mouthwash	Prevention	Recommendation	Anti-inflammatory properties by inhibiting the production of pro-inflammatory cytokines	[46,53,54,74–76]
Keratinocyte growth factor-1 (palifermin)	Prevention	Recommendation	Proliferation and restoration of epithelial cells	[26,27,46]
Glutamine	Prevention	Suggestion	It is used by cells of the immune system	[28,30,34,46]
Honey	Prevention	Suggestion	Inhibits bacterial growth and enhances healing rate	[32,35,37,38,46,77]
Patient-controlled analgesia (e.g., 0.2% morphine mouthwash)	Treatment	Recommendation	Pain management	[46,78,79]
Zinc supplements	Prevention	Suggestion	Prevents lipids peroxidation and replaces redox-reactive metals	[28,30,46]
Doxepin mouthwash	Treatment	Suggestion	In topical application, it has analgesic and anesthetic properties	[46,78,80]
Vitamin E	Prevention	Suggestion	Antioxidant that may protect tissue damage from free oxygen radicals	[28,30,31,46,73]
Amifostine	Prevention	Suggestion	Reduces DNA strand breaks, recruits ROS scavengers, and preserves salivary glands, endothelium, and connective tissue integrity	[4,33,46]

Proper oral health and hygiene are essential for mitigating the risk and severity of OM [11,60]. Before initiating CT or RT, all potential causes of mucosal irritation should be removed, as they may worsen and prolong the development of oral mucositis [60]. Teeth with sharp surfaces must be restored, orthodontics and protheses should be removed, and the maintenance of a stable oral microbiome is also an important aspect. The presence of a balanced nutrition is another variable that may help in the relief of discomfort from mucositis [9,11,60]. Alcohol, smoking, and foods that are crunchy, acidic, spicy, or sweetened should be limited or eliminated [81].

As previously stated, OM can make the ingestion process a challenge, as it is normally unpleasant and, in extreme cases, impossible due to painful symptoms; therefore, a liquid diet is the only solution [2,18]. Therefore, soft and liquid diets may be necessary, and, in the case of patients that cannot tolerate a liquid diet, the solution is parenteral nutrition [17]. The patients' complaints can be reduced with the use of specific mouthwashes with topical analgesics, anesthetics, antibiotics, and steroids [67,74,80,82], as topical analgesics and anesthetics are intended to relieve localized pain [23].

According to the MASCC/ISOO guidelines (Table 3), a benzydamine mouthwash may be useful due to its anti-inflammatory properties, which inhibit the production of TNF-α and IL-1β [46,53,63,74]. However, the use of saline, sodium bicarbonate, and antimicrobial (e.g., chlorhexidine 0.12%) rinses can ameliorate the symptoms of moderate mucositis [53,83]. In clinical practice, topical analgesics (e.g., morphine, benzocaine, and menthol) are applied to provide temporary relief in some patients, but their concentrations are not well established [22,84].

Currently, palifermin was the only agent that has been approved by the European Medical Agency (EMA) and the American Food and Drug Administration (FDA) for the prevention of OM in HSCT patients receiving CT and RT. However, on 1 April 2016, the European Commission withdrew the marketing authorization for this drug in the European Union (EU). The withdrawal was at the request of the marketing authorisation holder, which notified the European Commission of its decision to permanently discontinue the marketing of the product for commercial reasons. It has also been tested in HNSC patients in terms of its reduction of the state of pathogenic severity [10]. The MASSC/ISOO guidelines also indicate cryotherapy and photobiomodulation (PBM) protocols for more advanced phases. In particular, cryotherapy has been reported to reduce the symptoms of oral mucositis in patients undergoing CT as a result of its vasoconstriction, decrease in the blood flow, and reduction of the local distribution of the chemotherapeutic agent (e.g., fluorouracil (5-FU) and melphalan) [23,71]. Thirty minutes of ice chips used prior to the administration CT are the recommended and tolerable period [71,73].

PBM is another method employed to stabilize and inhibit the development of OM [39,85]. It has anti-inflammatory effects, diminishes the pain, and improves the healing rate of the basal wound. The energy applied to the specific area must be adapted according to the patient's lesion. To relieve the most common complaints, PBM can be used both prophylactically and therapeutically, that is, it can be used before and after an antineoplastic treatment [40,86,87]. Mucositis may also be treated with supplementary vitamins and minerals. For instance, vitamin E, a potent antioxidant, may reduce the grade of mucositis by preventing the damage caused by ROS [2]. A blood test performed in severe OM patients demonstrated a lack of some vitamins (such as vitamins E, A, and D), which inhibited the pro-inflammatory pathways [34]. Different studies also showed that oral zinc supplements may be applied as a prophylactic treatment [18,29,30].

Therefore, for most of the strategies recommended or suggested in Table 3, the research in the literature displays minimal evidence or even contradictory results, thus invalidating the definitions of the guidelines [7,8,54,63]. Consequently, the search for new active ingredients with potential therapeutic effects for preventing or treating OM is a challenge. Natural compounds, the majority of which are rich in polyphenols, are an option that should be explored.

3. Natural Compounds and Their Properties for Preventing/Treating OM

Currently, the protocols and therapeutical agents available from the different authorities have the purpose of ameliorating the OM grade, as mentioned in the previous section, but no treatments with reasonable results have been established [42,53,88]. Aside from that, many of these compounds have been associated with adverse effects and high costs [8,19]. Thus, natural products, such as honey, *Aloe vera*, curcumin, or propolis, are of huge interest for the nutraceutical and pharmaceutical industries, as they are easily accessible and allow more cost-effective treatments with minimal or no toxicity when compared to conventional strategies [89,90]. Their richness in bioactive compounds with anti-inflammatory, antioxidant, antiseptic, analgesic, and wound-healing properties that may interfere with many cellular signaling pathways could play an important role in the progression of OM and the activity of carcinogenic cells (e.g., HNSC) [10].

3.1. Bee Products

Honey is a natural product generated by bees and has been used since ancient times in traditional medicine. The huge diversity of studies has shown the multiplicity of beneficial applications of honey based on its antioxidant, anti-inflammatory, antibacterial, antiviral, antifungal, antitumoral, antimutagenic, and wound-healing properties [32,35,37,38,77]. The composition of honey is difficult to exactly define, as the components and relative amounts are conditioned by the flora of the geographical area from which honeybees collect pollen [91]. In a general way, honey is a heterogeneous mixture of water, nectar sugars, and glandular secretions produced by honeybees that contain proteins, vitamins, and enzymes [92]. One of the enzymes present is glucose oxidase, which, when in contact with body tissue, may stimulate the production of hydrogen peroxide, which acts as a messenger and promote wound healing and rapid epithelization at low concentrations by stimulating the proliferation of fibroblasts and epithelial cells [92,93]. It is also suggested that matrix metalloproteases of connective tissue and neutrophil serine proteases may be activated by hydrogen peroxide [94]. Furthermore, the levulose and fructose present in honey may improve local nutrition and promote epithelialization [93,94]. Honey also has immunomodulatory effects, as it influences the activation of macrophages and the proliferation of B-lymphocytes and T-lymphocytes [95], in addition to decreasing the inflammatory process by inhibiting cyclooxygenase pathway and reducing prostaglandin synthesis [96]. The beneficial effects of honey may also be due to its moisturizing effect, low pH, and viscosity which inhibit the proliferation of bacteria [35].

Charalambous et al. conducted a randomized, controlled trial to evaluate the potential effect of thyme honey rinses on HNSC patients [35]. In this study that involved 72 participants, a solution of 20 mL of thyme honey diluted in 100 mL of purified water was given to the patients to gargle in the oral cavity three times per day (15 min before and after the RT session and 6 h later) for 7 weeks, starting from the first day of the fourth week of RT. The results showed a significant improvement ($p < 0.001$) in the patients' quality of life, leading to fewer symptoms and maintenance of the body weight ($p = 0.001$) when compared to saline rinses [35]. Honey mouthwash also proved to be effective in a randomized, single-blind controlled trial that enrolled 53 patients [97]. The honey solution (honey-to-water ratio of 1:20) at 37 °C was gargled and kept in the mouth before and after each meal and before sleeping for 30 s by the treatment group, while the patients in the control group received routine care, such as ingestion of fluconazole capsules, nursing care, and mouth hygiene training [97]. According to the authors, the solution reduced or eliminated weight loss, leading to some weight gain and preventing and reducing the severity of OM in the acute myeloid leukemia patients receiving CT ($p < 0.001$ at the fourth week of treatment) [97]. In another randomized, controlled trial with a parallel design involving 150 children, Sener et al. treated 25 OM patients with honey (with vitamin E as the most effective compound) for 21 days, applying the amount of 1–1.5 g of honey per weight (kg) of the child twice per day (every 12 h) [31]. Honey was found to be more effective in the management of OM ($p < 0.05$) when compared to chlorhexidine, a wide-spectrum antifungal and bactericidal antiseptic solution that is frequently used in oral care [31]. Motallbnejad et al. also conducted a randomized single-blind (examiner-blind) clinical trial to evaluate the effect of pure honey on radiation-induced mucositis in a total of 40 patients with head and neck cancer receiving RT [95]. Twenty patients were instructed to rinse and gradually swallow 20 mL of pure honey 15 min before radiation, then again at intervals of 15 min and six hours after radiation, while the control group was advised to rinse with 20 mL of saline before and after radiation. This procedure was repeated weekly from the beginning of the treatment until the end of the RT. The honey-receiving patients exhibited a significant reduction in OM ($p < 0.001$) when compared to the control group [95]. In a unicenter randomized, controlled clinical human study involving 82 patients with head and neck cancer treated with RT over 4–6 weeks, the treatment group was instructed to take 20 mL of Ziziphus honey 15 min before and after the radiation and before sleeping at night, while the control group repeated the process using 20 mL of 0.9% saline [92]. The results showed that the proportion of mucositis (Grades 3 and 4) was lower in the honey-treated group ($p = 0.016$ and $p = 0.032$ for Grades 3 and 4 of mucositis, respectively) than in the control group at the end of 6 weeks of RT [92]. In 2010, Khanal et al. conducted a single-blinded, randomized, controlled clinical trial over 6 weeks on 40 oral carcinoma patients receiving RT [91]. Radiation was given once per day for 5 days a week, and the application was performed 15 min before and after radiation and once before going to bed. Honey extracted from beehives of the Western Ghats forests or lignocaine gel 2% (control group) was swished around the oral cavity for 2 min and expectorated. Only one of the 20 patients of the treatment group developed intolerable mucositis ($p < 0.0001$) compared to 15 of the 20 patients of the lignocaine group [91]. Caffeine, a natural alkaloid with hypoalgesic, antioxidant, and anti-inflammatory effects, has also been screened as a potential ingredient to work against oral mucositis [98–101]. In a double-blinded randomized clinical trial involving 75 patients (that randomly fell into three treatment groups) presenting OM after CT, the therapeutic effects of coffee plus honey were compared with those of topical steroids that are usually used in the treatment of OM after CT [98]. A syrup-like solution was prepared for each treatment group: 300 g of honey plus 20 g of instant coffee for the honey-plus-coffee group; 300 mg of honey for the honey group; the control group was treated with 20 eight-milligram ampoules of betamethasone solution. All groups were instructed to sip 10 mL of the prescribed product and swallow every 3 h for 1 week. While all treatment regimens decreased the severity of the lesions, the best result was achieved in the honey–coffee group ($p < 0.05$), followed by the honey-and-steroid groups [98].

3.1.1. Propolis

Propolis is a resinous material produced by bees and is frequently used as natural nutritional supplement [102]. It is composed of a mix of plant buds and exudates, bee enzymes, pollen, and wax, and it has been widely used by different civilizations to treat colds, wounds, and ulcers due to its anesthetic, antimicrobial, anti-inflammatory, antitumor, immunomodulatory, and antioxidant properties [102]. Similarly to honey, the chemical composition of propolis is highly dependent on the diversity of the flora and bee species [103,104]. It is mainly composed of proteins, amino acids, vitamins (A, B1, B2, B3, and B7), minerals, essential oils, phenolic acids, alcohols, fatty acids, and flavonoids [102,105–108]. Regarding OM, the bioactivity of propolis is mainly associated with flavonoids, as these molecules are capable of sequestering or inhibiting the formation of free radicals, and they promote immunomodulatory, antioxidant, wound-healing, and anti-inflammatory activities [109]. The anti-inflammatory properties are directly related with the inhibition of the synthesis of prostaglandins and promotion of phagocytic activity [110]. In addition, propolis promotes healing effects in epithelial tissues, while the presence of iron and zinc improves the synthesis of collagen [108].

Akhavan-Karbassi et al. conducted a randomized double-blind placebo-controlled trial to evaluate the potential effect of propolis mouthwash on head and neck tumor patients undergoing CT [111]. In the treatment group ($n = 20$), 5 mL of propolis mouth rinse (30%

extract) was administered every 8 h for 7 consecutive days. The solution was swished in the patients' mouths for 60 s, gargled, and expectorated. In the control group ($n = 20$), the process was repeated with a placebo mouth rinse. OM, erythema, and eating and drink ability were evaluated. When compared to the control group, the treatment group presented significant improvement in OM, wound healing, and erythema at day 7 ($p = 0.006$), but no significant differences in eating and drinking ability were observed ($p = 0.21$). Moreover, 65% of the patients in the propolis group were completely healed by day 7 of the trial [111].

3.1.2. Royal Jelly

Royal jelly is a secretory product of the cephalic glands of nurse bees that serves as the diet of honeybee larvae in their first 2–3 days, while for the queen, it is the specific food for her whole life period [112]. It is widely used in folk and mainstream medicines and as a dietary supplement due to its antioxidant, anti-inflammatory, hypoglycemic, antibiotic, antitumor, antiallergic, antiaging, immunomodulatory, neurotrophic, hypocholesterolemic, hepatoprotective, hypotensive, and blood pressure regulatory activities [112–120].

Similarly to the aforementioned bee products, the composition of royal jelly is dependent on the geography and climate [121]. It is a complex substance with a unique combination of sugars (mainly glucose and fructose, as well as traces of sucrose, maltose, trehalose, melibiose, ribose, and erlose), proteins (which represent >50% of the dry weight of royal jelly), amino acids, nucleotides, ascorbic acid, phenols, waxes, fatty acids, steroids, and phospholipids [121]. The impact that royal jelly has on OM may be closely related to its anti-inflammatory and wound-healing activities. However, the active compounds of royal jelly and the mechanisms underlying these activities are still largely unknown. In vitro studies performed on mice revealed that supernatants of royal jelly suspensions added to a mouse peritoneal macrophage culture stimulated with lipopolysaccharides and IFN-γ efficiently suppressed the secretion of pro-inflammatory cytokines TNF-α, IL-6, and IL-1, which was probably due to protein factors such as Major Royal Jelly Protein 3 (MRJP3) [119]. MRJP2, MRJP3, and MRJP7 are thought to be responsible for the wound-healing bioactivity of royal jelly, as they stimulate cell migration and proliferation [122], along with the antioxidant compounds present in royal jelly, which, when taken orally, lowered the levels of 8-hydroxy-2-deoxyguanosine, a marker of oxidative stress in mouse kidney DNA and serum [123].

Suemaru et al. evaluated the effects of royal jelly, honey, and propolis on OM induced with 5-fluorouracil and mild abrasions made on the cheek pouch in hamsters [124]. The bee products were topically administered to the oral mucosa. Royal jelly ointments at 3%, 10%, and 30% improved the recovery from 5-fluorouracil-induced damage in a dose-dependent manner, while the results of ointments of honey at 1%, 10%, and 100% and propolis at 0.3%, 1%, and 3% were not statically different from those of the Vaseline-treated control group [124]. In a more in-depth trial in Golden Syrian hamsters, the influence of royal jelly on 5-fluorouracil-induced OM was assessed using oral mucosal adhesive films containing royal jelly [125]. The 5-fluorouracil was administered through intraperitoneal injections on days 0 and 2, and the left cheek pouches of hamsters ($n = 12$ per group) were everted and scratched with a small wire brush on days 1 and 2. Royal-jelly-containing sodium alginate–chitosan films (10% or 30%) were applied to the cheek pouches every day from day 3. Royal-jelly-containing films (both 10% and 30%) improved the recovery from 5-fluorouracil-induced OM, which presented lower erythema and absence of ulceration and abscesses on day 8. They also reduced the myelo-peroxidase (MPO) activity and the expression of pro-inflammatory cytokines. The data suggest that these effects were caused by the anti-inflammatory or antioxidative properties of royal jelly [125]. In humans, the effect of royal jelly on OM in patients with different types of malignancies undergoing RT and CT was evaluated by Erdem et al. in a randomized, controlled trial [126]. In this clinical trial that involved 103 patients, all patients received a mouthwash therapy with benzydamine hydrochloride and nystatin rinses. In addition, patients in the experimental group received royal jelly two times per day for a total daily dose of 1 g. Royal jelly was orally swished for 30 s and swallowed. The treatment group showed a mean resolution time of OM that was significantly shorter than that of the control group (OM Grade 1: $p = 0.0001$; OM Grade 2: $p = 0.0001$; OM Grade 3: $p = 0.05$) [126]. In a single-blind clinical trial that involved 13 patients with head and neck cancer receiving CT, 1 g of royal jelly was given three times per day to the treatment group during the RT period [127]. Royal jelly was shown to have a preventive effect on the progression of CT-induced OM from the early phase ($p < 0.001$) [127].

3.2. *Spondias Mombin*

The leaves of *Spondias mombin*, commonly known as the cashew tree, are a rich source of interesting bioactive compounds, with particular emphasis on tannins, saponins, triterpenes, and flavonoids [128]. Traditionally, the leaves have been used to treat inflammatory pathologies, making them a promising source for the development of new therapeutic agents for OM [128]. Gomes et al. assessed the effects of a hydroethanolic extract of *S. mombin* leaves on 5-fluorouracil-induced OM in Golden Syrian male hamsters [128]. The animals were orally pre-treated with the hydroethanolic extract of *S. mombin* leaves (50, 100, or 200 mg/kg) for ten days [128]. The treatment with the highest dose of the extract (200 mg/kg) showed the best healing effect, with hamsters displaying reduced oxidative stress and inflammation and no evidence of ulceration. Further analysis showed re-epithelialization, absence of hemorrhage, discrete mononuclear inflammatory infiltration, and lower expression levels of different molecules involved in the modulation of inflammation, such as MMP-2, COX-2, TNF-α, NF-κB p50 NLS, iNOS, and IL-1β, as well as an increase in glutathione (GSH) levels [128]. Although the mechanisms behind these effects remain under investigation, the hydroethanolic extract of *S. mombin* leaves is rich in potent antioxidant phenolic phytochemicals, such as ellagic acid (12 mg/g) and chlorogenic acid (19.4 mg/g), which could justify these activities [129]. Studies have demonstrated that chlorogenic acid acts on the reduction of COX-2 expression in macrophages, as well as in the inhibition of the production of pro-inflammatory cytokines, such as IL-1β and TNF-α, and of NF-κB activation [129]. On the other hand, chlorogenic acid was proven to promote wound healing in rats [130], while ellagic acid acted by down-regulating MMP-2 expression and inhibiting NF-κB-mediated transcriptional activation [129]. These activities may justify the results achieved in the previously detailed trial.

3.3. *Camellia sinensis*

Camellia sinensis (green tea) is one of the most popular drinks in the world and is widely known for its antimicrobial, antitumoral, antioxidant, and anti-inflammatory activities [67]. Different compounds with therapeutic effects have been discovered in this plant. The majority of the health-promoting properties are associated with polyphenols [131], which represent almost 30% of the fresh-leaf dry weight, including flavandiols, flavonols, flavonoids, and phenolic acids [132]. However, most of the polyphenols present in leaves of *C. sinensis* are catechins, namely, (+)-catechin, (−)-epicatechin, (+)-gallocatechin, (−)-epigallocatechin (EGC), (−)-epicatechin gallate, and (−)-epigallocatechin gallate (EGCG) [133]. Catechins are mainly responsible for the ROS scavenging and antioxidant activities of *C. sinensis* [134,135]. EGCG, in particular, efficiently inhibits the transcription of NF-κB, resulting in a decrease in the expression of different pro-inflammatory genes [136]. The anti-inflammatory effect of catechins may be due to the activation of endothelial nitric oxide synthase (eNOS) [137,138].

The effect of green tea on OM was evaluated in oral cancer patients [67]. For that, a single-blind randomized, controlled trial was made with 63 participants. For 6 months, after the tooth-brushing procedure, the intervention group rinsed the mouth with a solution of 5 g of green tea dissolved in 100 mL of water for 60 s, and the control group rinsed the mouth with 100 mL of tap water for the same period. The results demonstrated an improvement in oral health status and the preservation of the oral mucosa at the end of the follow-up period (6 months), with a higher reduction of the oral health status score in the intervention group than in the control group ($p = 0.008$) [67]. In another randomized study, the effect of Baxidil Onco® mouthwash (Sanitas Farmaceutici Srl, Tortona, Italy), composed of C. *Sinensis* leaf extract and palmitoyil hydrolyzed wheat protein, was tested in 60 hematologic patients undergoing hematopoietic stem cell transplantation (HCST) [139]. Twenty mL of Baxidil Onco® was used to rinse the mouths of 28 patients four times per day for at least one minute without swallowing, while the remaining 32 patients were treated with standard prophylactic schedules and served as control. The results demonstrated that the incidence, severity, and duration of OM were significantly reduced ($p = 0.022$) by the oral rinsing with Baxidil Onco® [139].

3.4. *Plantago Major*

In traditional Persian medicine, *Plantago major* was used as a wound-healing herb, as it possesses a wide range of bioactive properties, such as anti-inflammatory, antiulcerogenic, antioxidant, antimicrobial, analgesic, wound-healing, and immunomodulatory effects [140].

Soltani et al. conducted a randomized, double-blind, placebo-controlled clinical trial to assess the effects of *P. major* syrup as a natural agent against OM for 7 weeks [141]. The participants were HNSC patients who were going to receive RT. The 23 patients of the intervention group received 7.5 cc of *P. major* syrup three times per day, starting from three days before the start of RT until the end of it, while the placebo group received 7.5 cc of placebo syrup. The *P. major* syrup was shown to be effective in the reduction of the mucositis and the severity of pain caused by RT ($p < 0.001$) [141]. A multicenter randomized, controlled trial developed by Cabrera-Jaime et al. evaluated the efficacy of *P. major* extract vs. chlorhexidine vs. sodium bicarbonate in the treatment of CT-induced OM in solid-tumor cancer patients with grade II–III mucositis [140]. A total of 45 patients were randomized for one of the treatments, consisting of a 5% aqueous solution of sodium bicarbonate together with (i) an additional dose of 5% sodium bicarbonate, (ii) *P. major extract*, or (iii) 0.12% chlorhexidine. The solutions were applied over 14 days. The differences in healing time and the lower pain levels among the three groups were not statistically significant ($p = 0.702$) [140].

The properties of *P. major* leaves are dependent on the different compounds present. The leaves are rich in different bioactive molecules, such as aucubin, a glycoside with antitoxin activity, and ursolic, oleanolic, and α-linoleic acids, which inhibit COX-2-catalyzed prostaglandin production [142–144]. Extracts of *P. major* leaves have remarkable antioxidant and antiradical capacities due to the presence of baicalein, lutolin, salicylic acid, citric acid, ascorbic acid, apigenin, ferulic acid, benzoic acid, chlorogenic acid, oleanolic acid, and ursolic acids [145,146]. According to different studies, the bioactivity of *P. major* is due to the decrease in the inflammatory reaction through the modulation of NF-κB, NO, COX-2, and B4 leukotriene (LB4) levels [140].

3.5. *Aloe vera*

A. vera is a plant that has been used for medical purposes for thousands of years. It is widely employed for the treatment of various medical conditions, such as oral ulcers, psoriasis, skin burns, and frostbite, since it presents analgesic, liver-protection, antifungal, antidiabetic, anti-inflammatory, antiproliferative, anticarcinogenic, antiaging, and immunomodulatory properties [147–149]. In addition, it can scavenge free radicals, improve wound oxygenation, promote wound healing, increase collagen formation, and inhibit metalloproteinase and collagenase activity [150–154]. Different studies have shown the potent free-radical and superoxide anion activity of three derivatives from *A. vera*, namely, isorabaichromone, feruoylaloesin, and *p*-coumaroylaloesin [150,151]. The beneficial effects, assumed to be exerted in the oral cavity, may also be due to its moisturizing effect, which is provided by the polysaccharide components (principally mannose, glucose, xylose, arabinose, galactose, and rhamnose), which provide and sustain moisture in tissues [155]. One of the sugars present in a higher quantity, mannose-6-phosphate, acted as an active-growth substance and anti-inflammatory agent in in vivo studies on mice [156]. The anti-inflammatory effects of *A. vera* extracts are attributable to the inhibitory action on the arachidonic acid pathway via COX-2 inhibition [150,157], as well as the reduction of leukocyte adhesion molecules and TNF-α levels [158]. In vitro and animal assays suggest that *A. vera* promotes wound healing through the reduction of the vasoconstriction and the platelet aggregation at the wound site [152].

An initial assessment of *A. vera*'s potential in preventing RT-induced OM did not yield promising results in a single-institution, double-blind, prospective, randomized trial that involved 58 head and neck cancer patients [159]. The patients were instructed to take a 20 mL swish (*A. vera* solution or placebo) and swallow four times daily, beginning on the first day and continuing throughout the course of RT. However, no significant differences were observed between treatments ($p = 0.07$) [159]. Better results were achieved in other studies. Mansouri et al. evaluated the effect of *A. vera* on CT-induced OM in patients with acute lymphocytic leukemia and acute myeloid leukemia [160]. In this randomized, controlled clinical trial, 64 patients were divided into an intervention group and a control group. The first group was instructed to wash their mouths with 5 mL of *A. vera* solution for 2 min three times per day for 14 days. The control group repeated the procedure using mouthwashes that are typically recommended by hematologic centers, including normal saline, nystatin, and chlorhexidine. An evaluation of the patients' mouths was performed on days 1, 3, 5, 7, and 14. Even though, regarding the intensity of stomatitis and pain, no significant differences were found between the two groups on the first day, a significant difference was observed in this regard on the other days ($p < 0.001$) [160]. In a similar study, an assessment of the effect of *A. vera* mouthwash on CT-induced OM was performed in a double-blinded randomized clinical trial on 120 patients, who were divided into three groups [161]. Until 2 weeks after the CT sessions, group 1 received tablets with 10 mg of atorvastatin daily plus a placebo mouthwash, group 2 received placebo tablets and *A. vera* mouthwash, and group 3 received placebo tablets and placebo mouthwash. The analysis of the results showed that 50% of the placebo patients (group 3) experienced mucositis, while that value decreased to 2.5% in group 2 ($p < 0.042$), with no significant differences between groups 1 and 3 ($p < 0.674$) [161]. Likewise, the efficacy of *A. vera* use for prevention of CT-induced OM was evaluated in a randomized, controlled clinical trial in 26 children with acute lymphoblastic leukemia [162]. Depending on the treatment group, a 70% *A. vera* solution or a 5% sodium bicarbonate solution was applied twice per day to oral tissues with spongeous sticks. The application started 3 days before the CT therapy. The application of *A. vera* solution showed to be effective in the prevention and reduction of OM severity ($p < 0.001$) [162]. A triple-blind randomized and controlled interventional quality-of-life clinical trial on the efficacy of *A. vera* and a benzydamine mouthwash in the alleviation of RT-induced OM was performed by Sahebjamee et al. in a study with 26 head and neck cancer patients [163]. The intervention group rinsed the mouth three times per day with 5 mL of an *A. vera* mouthwash, while the control group repeated the procedure with benzydamine mouthwash. The protocol was applied from the first day of RT until the end of the treatment, demonstrating that *A. vera* mouthwash was as efficient as benzydamine at reducing the severity of RT-induced OM, without differences between them ($p < 0.09$) [163].

3.6. Curcuma Longa

Curcuma longa, also known as turmeric, is an herb that is extensively grown in Asia [164] and is often used culinarily as a spice and in traditional Asian medical treatments for depression, stress, infection, and dermatological diseases [165,166]. Various compounds were identified in this plant, including polyphenols, sesquiterpenes, diterpenes, triterpenoids, sterols, and alkaloids [165,167]. Among these, the most studied component of *C. longa* is curcumin, a lipophilic polyphenol extracted from the rhizomes of *C. longa*, which represent 2–5% of turmeric [164,165].

Due to the antioxidant, anti-inflammatory, and anticancer effects of curcumin, it has an important role in the prevention of depression, cancer, and pro-inflammatory, neurodegenerative, diabetic, autoimmune, and cardiovascular diseases [168–172]. Furthermore, curcumin has antimicrobial, insecticidal, larvicidal, and radioprotective activities [165].

Curcumin mediates its effects through direct or indirect interactions with growth factors, kinases, enzymes, transcription factors, receptors, and proteins that regulate cell proliferation and apoptosis [168,173–175]. In the case of OM, the beneficial effects of curcumin may be related with the upregulation of TGF-β-1, which promotes re-epithelialization through the stimulation of fibronectin and collagen production by fibroblasts, while increasing the rate of granulation [168,176,177]. TGF-β-1also promotes the removal of dead tissue by enhancing the recruitment of macrophages [177]. Aside from that, curcumin potently inhibits the activation of nuclear factor-κB (NF-κB), but activates others, such as the nuclear factor erythroid 2-related factor 2 (Nrf2) [168,176,177]. COX-2, the inducible form of COX, can be selectively induced by mitogenic and inflammatory stimuli, resulting in enhanced synthesis of prostaglandins, such as IL-6. The activation of NF-κB significantly upregulates superoxide dismutase (SOD) expression [168,176,177]. Curcumin also enhanced the expression of antioxidant enzymes such as SOD, catalase (CAT), glutathione (GSH), and glutathione peroxidase (GSH-px) through the regulation of Nrf2 [168,176,177].

The wound-healing ability of curcumin is accelerated by its antioxidant activity, as it decreases the levels of lipid peroxides (LPs) and increases the activity levels of superoxide dismutase (SOD), catalase (CAT), and glutathione peroxidase (GPx) [178].

In a placebo-controlled study, an assessment of the tolerability of a curcumin mouthwash for the prevention of OM in pediatric patients undergoing CT was performed in a group of seven pediatric and young-adult patients [179]. In this study, which was developed without a control group for ethical reasons, in addition to the standard preventive oral care consisting of 0.2% chlorhexidine mouthwash for 30 s twice per day, the patients also used a mouthwash with 10 drops of Curcumall® (a dietary supplement containing turmeric, curcumin and ginger) twice per day during the CT treatment. The researcher concluded that curcumin mouthwash was safe and well tolerated by the patients [179]. The efficiency of curcumin mouthwash in cancer patients undergoing RT and suffering from OM was evaluated in a randomized trial involving 20 patients [180]. The study group used 0.004% curcumin mouthwash diluted at a ratio of 1:5 for 1 min three times per day for 20 days, while the control group was treated with standard preventive oral care using a commercially available 0.2% chlorhexidine mouthwash to be used in a 1:1 dilution for 1 min three times per day for 20 days. Curcumin promoted faster wound healing and better patient compliance in the management of RT-induced OM ($p < 0.001$) [180]. In another double-blind randomized clinical trial, the effects of curcumin encapsulated in nanomicelles on OM in 32 head and neck cancer patients receiving RT were evaluated [181]. During the RT, patients in the treatment group received daily one capsule of SinaCurcumin® (Exir Nano Sina Company, Tehran, Iran), which contained 80 mg of curcumin-loaded nanomicelles. The control group received placebo tablets containing lactose. There were statistically significant differences ($p < 0.05$) between the two groups in the severity of OM, as all of the patients in the placebo group developed OM versus the 32% of the case group [181].

3.7. *Olea Europaea*

Olive leaf extract is a natural product extracted from *Olea europaea*, which is traditionally used to treat and prevent hypertension and diabetes due to its antioxidant, anti-inflammatory, anticancer, antiapoptotic, antimicrobial, hypoglycemic, and diuretic properties [182–187]. The leaves of *O. europaea* contain a high concentration of phenolic compounds (1450 mg/100 g of fresh leaf), with secoiridoid oleuropein, verbascoside, rutin, luteolin-7-glucoside, and hydroxytyrosol as the main phenolic constituents [188]. Oleuropein is possibly the main active compound promoting the wound-healing activity of olive leaf extract, as it increases collagen fiber deposition and advanced re-epithelialization [189,190]. Furthermore, it has been demonstrated that oleuropein decreases oxidative stress and inflammation through the modulation of the COX-2, AMPF, eNOS, MAPK, and apoptosis cell signaling pathways in in vivo studies on mice [187]. In addition, olive leaf extract also inhibited the aggregation platelets in in vitro studies [186].

In 2013, the effect of a mouth rinse containing olive leaf extract on the prevention of severe OM in CT-receiving patients, as well as an estimation of its effect on the salivary levels of pro-inflammatory cytokines, was assessed in a prospective, randomized, double-blind, placebo-controlled cross-over study design involving 25 cancer patients [182]. The studied drugs (olive leaf extract at 333 mg/mL, benzydamine hydrochloride at 0.15 g/100 mL, or normal saline) were self-administered 3–4 times daily for 14 days, starting on the first day of chemotherapy. The patients were evaluated weekly until 15 days after CT for each cycle. The findings indicated that the olive leaf extract could effectively reduce the OM rates ($p < 0.001$) by decreasing the salivary levels of IL-1β and TNF-α [182]. Briefly, Ahmed et al. performed an experimental animal study and a prospective, randomized, double-blind, placebo-controlled cross-over study to evaluate the management of OM with mouthwashes containing olive leaf extract [191]. In the animal study, 45 male albino rats received two intraperitoneal injections of 5-fluorouracil (60 mg/kg) on day 0 and day 2. The first group received normal saline, the second group received olive leaf extract (333 mg/mL), and the third group received benzydamine hydrochloride (0.15 g/100 mL). By the end of the study (day 14), the control group presented ulcerated connective tissue that was not completely covered by epithelium, and there was evidence of necrosis and degeneration. The animals with the olive leaf extract and benzydamine hydrochloride presented a totally re-epithelialized mucosal surface with hyperkeratinization and hyperplasia, while the sub-epithelia were more organized, with decreased cellularity of fibrous tissue [191]. In a clinical study, 62 CT-receiving patients were divided to receive olive leaf extract, benzydamine hydrochloride, or a placebo in the form of a mouth rinse, and the treatment was changed in the next chemotherapy cycle for each patient (cross-over design) [191]. Mouth rinses were self-administered 3–4 times per day for 14 days from the start of the CT. When compared to the benzydamine hydrochloride and the control, the olive leaf extract more efficiently reduced the oral pain, dysphagia, and functional impairment of eating ($p < 0.001$) [191].

3.8. *Glycyrrhiza glabra*

Glycyrrhiza glabra, commonly known as licorice, is one of the most important herbal medicines for traditional Chinese medicine and Japanese Kampo medicine [192]. It is traditionally used to relieve inflammation, gastric and peptic ulcers, arthritis, eye and liver disorders, hyperacidity, and sex-hormone imbalance [193–201]. This plant has attracted the attention of the pharmacological field due to its antimicrobial, antiviral, and anti-inflammatory properties [202–205]. The roots of *G. glabra* have been found to possess many secondary metabolites, with numerous pharmacological properties that contribute to their medicinal use, including flavonoids (such as liquirtin, rhamnoliquirilin, liquiritigenin, and prenyllicoflavone A) and volatile components (including pentanol, hexanol, tetramethyl pyrazine, linalool, and terpinen-4-ol) [206]. The essential oil extracted from the roots of *G. glabra* contains propionic acid, 1-methyl-2-formylpyrrole, benzoic acid, 2,3-butanediol, and ethyl linoleate, among other compounds. The roots of *G. glabra* are also composed of 20% moisture, 3–16% sugars, 30% starch, and 6% ash [207].

The main biologically active components of *G. glabra* are dipotassium glycyrrhizinate, glycyrrhizin, also known as glycyrrhizic acid, and its aglycone, glycyrrhetinic acid [206]. Dipotassium glycyrrhizinate has similar properties to those of corticosteroids, namely, anti-inflammatory, antiallergic, and antibiotic activities, without the side effects of allergic reactions on the skin [208]. This property is due to dipotassium glycyrrhizinate's ability to efficiently inhibit the activity of phospholipase A2 enzyme, which is necessary for several inflammatory processes [209–211]. Moreover, it is able to avoid damage to the extracellular matrix by inhibiting the activity of hyaluronidase enzyme, histamine release, inflammatory chemical mediators, leukotrienes, and prostaglandins [212]. Glycyrrhizic acid inhibits prostaglandin E2 synthesis by suppressing the activity of COX-2, resulting in the augmentation of NO production through the enhancement of iNOS mRNA secretion and indirectly preventing platelet aggregation [211,213,214]. The anti-inflammatory activity of glycyrrhizic and glycyrrhetinic acids is realized through cytokines such as 1β, IL-4, IL-5, IL-6, IL-8, IL-10, IL-12, and IL-17, IFN-γ, and TNF-α [215–217]. Moreover, these compounds also present immunomodulatory activity through their interaction with different transcription factors, such as NF-κB, as well as signal transducers and activator of transcriptions (STAT- 3 and STAT-6) [215].

Najafi et al. conducted a double-blind clinical trial to evaluate the potential effect of *G. glabra* extract on cancer patients under head and neck radiotherapy [207]. The experimental group received a 50% extract of *Glycyrrhiza* (hydroalcoholic extract) and the placebo group received a brown-colored water. The patients were asked to use 20 cc twice per day for 14 days after the beginning of RT. According to the results obtained, *Glycyrrhiza* extract efficiently decreased the OM, wound size, and irritation ($p < 0.001$) [207]. The effect of *G. glabra* on head and neck cancer patients receiving RT was also evaluated in a small randomized study with six patients who were assigned to receive a licorice mucoadhesive film or a placebo mucoadhesive film [218]. The level of pain and the mucositis severity were significantly lower in the licorice-mucoadhesive-film-receiving patients in the last 2 weeks of the clinical trial (weeks 3 and 4) ($p < 0.05$) [218].

The efficiency of a *G. glabra* root extract in preventing CT-induced OM in colon cancer patients was evaluated in a double-blind randomized clinical trial that involved 72 patients [219]. The treatment group received 5% licorice root extract, and the control group received a combined mouthwash composed of aluminum, magnesium, diphenhydramine, nystatin powder, and 2% lidocaine. For one week, from the first day of CT, both mouthwashes were used daily, every 8 h, at a dose of 10 cc. The researchers did not observe differences between the two groups in terms of the incidence and severity of OM ($p > 0.05$) [219].

3.9. *Matricaria Recutita*

The chamomile plant, *Chamomilla recutita* or *Matricaria recutita*, one of the most common medicinal plants, is characterized by flowers with anti-inflammatory, antibacterial, and antifungal properties [220]. It is mainly used to treat different inflammatory conditions of the skin and mucosa, as it promotes faster a wound-healing process in comparison to corticosteroids [220,221]. *M. recutita* owes its therapeutic activity to chamazulene, α-bisabolol, bisabolol oxides, spiroethers, and flavonoids [220]. Flavonoids—in particular, apigenin-7-glucoside—have been found to be responsible for the anti-inflammatory activity that may be involved in recuperation from OM [222]. Pre-clinical studies showed evidence of the anti-inflammatory action of *M. recutita* through the inhibition of COX-2 and IL-6 production [223,224].

In a small comparative study with random assignment, dos Reis et al. evaluated the efficacy of *M. recutita* infusion cryotherapy for the prevention and reduction of the intensity of OM in gastric and colorectal cancer patients [225]. The study was performed during the first course (5 days) of CT. The patients in the *M. recutita* group received a cup of ice chips made with an *M. recutita* infusion at 2.5%, while the control group received a cup of ice chips made with pure water. The patients in both groups were instructed to swish the ice chips around in their mouths for at least 30 min, starting 5 min before the CT infusion. The *M. recutita* group presented less pain and had no ulcerations when compared to the control group [225]. The effects and the percentage of extract necessary to reduce the incidence and intensity of OM in patients undergoing hematopoietic stem cell transplantation were assessed in a randomized, controlled, phase II clinical trial [221]. All 40 patients received standard oral care, while the treatment group received an additional mouthwash containing a liquid extract of *M. recutita* at 0.5%, 1%, or 2%. When compared with the control group, the *M. recutita* group at 1% (equivalent to 0.108 mg of apigenin-7-glucoside/mL) demonstrated have reduced incidence, intensity, and duration of OM in patients undergoing hematopoietic stem cell transplantation ($p < 0.01$) [221]. Shabanloei et al. performed a randomized, double-blind clinical trial between alloporinol and *M. recutita* extract in the prevention of OM in CT-receiving patients [226]. Group 1 received 5 mg/mL of allopurinol, group 2 received a solution of 8 g of *M. recutita* in

50 cc, and the control group received a normal saline solution as a mouthwash. All patients gargled daily, four times per day, for the 16 days following the beginning of CT. The researchers concluded that both the allopurinol and *M. recutita* mouthwashes were effective in reducing post-CT OM, with no significant differences in the mean stomatitis ($p = 0.59$) and stomatitis pain ($p = 0.071$) [226].

3.10. *Calendula officinalis*

Calendula officinalis, commonly known as marigold, has been used for centuries as a topical and oral herbal remedy due to its bactericidal, antioxidant, anti-inflammatory, antiseptic, hepatoprotective, and anti-metastatic effects, with applications in blood purification and treatment of herpes, keratolytic radiation dermatitis, wounds, and scars, and as an antispasmodic [227–229]. The main compounds that contribute to its medicinal use are triterpenoids, flavonoids, oleanolic acid, faradiol, glycosides, quinones, tannins, coumarins, carotenoids, saponins, alkaloids, phenolic acids, and amino acids [227]. Triterpenoids provide anti-inflammatory and anti-edematous effects, in addition to stimulating the proliferation of fibroblasts, possibly through the inhibition of COX-2, C3-convertase, and 5-lipoxygenase [230–232]. Flavonoids are reported to have anti-inflammatory, antioxidant, and anti-edematous properties, in addition to their inhibition of lipoxygenase enzymes and mast cells [233].

The potential of *C. officinalis* extract for the healing of 5-fluorouracil-induced OM was studied in hamsters [229]. OM was induced in 60 male hamsters on days 0, 5, and 10 through the intraperitoneal administration of 5-fluorouracil (60 mg/kg). The cheek pouch was scratched with a needle once per day, from day 1 until day 12, when erythematous changes were noted. The treatment of OM started on days 12–17 with the topical application of a gel once a day. The animals were divided into four groups: 12 without treatment as control animals, 15 treated with 5% *C. officinalis* gel, 15 treated with 10% *C. officinalis* gel, and 15 treated with the gel base. The *C. officinalis* gel (5% and 10%) significantly reduced the microscopic and macroscopic scores of OM when compared with the gel base and the control group. Moreover, the animals of the treatment groups gained more weight than those in the gel base and the control groups [229]. In humans, the effect of *C. officinalis* on OM was evaluated in a placebo-controlled clinical trial with 40 patients with neck and head cancers under RT or concurrent CT [234]. Patients were given 5 mL of either placebo or a 2% *C. officinalis* extract gel mouthwash to be held for at least 1 min in the oral cavity two times per day. Compared to the placebo group, the intensity of OM was significantly lower in the *C. officinalis* mouthwash group at weeks 2, 3, and 6 ($p < 0.048$). According to the same study, the high content of flavonoids and phenolic compounds and the antioxidant activity may be responsible for the protective effect of *C. officinalis* in RT-induced OM [234].

3.11. *Other Compounds*

In addition to the compounds mentioned above, different experiments were also performed to evaluate the potential of other natural compounds for preventing/treating OM. However, due to the low number of studies published, not only regarding the OM application, but also with respect to the molecular mechanism of action enrolled, a section was not dedicated to them in this review. Table 4 summarizes the different natural products in these circumstances.

Table 4. Summary of studies with natural products for prevention/treatment of oral mucositis.

Name	Properties/Mechanisms	Application	Experimental Setting/Model	References
Manuka (*Leptospermum scoparium*) essential oil	Anti-inflammatory, analgesic, antimycotic, and antibacterial	Mouthwash	Randomized placebo-controlled trial	[235]
Kanuka (*Kunzea ericoides*) essential oil	Anti-inflammatory, analgesic, antimycotic, and antibacterial	Mouthwash	Randomized placebo-controlled trial	[235]
Indigo root (*Isatis indigotica*)	Anti-inflammatory and antiviral	Mouthwash	Randomized clinical trial	[236]
Rhodiola algida	Immunomodulatory effects	Mouthwash	Randomized clinical trial	[237]
Thymus spp. L	Antiseptic, anti-inflammatory, antimicrobial, and antimycotic	Mouthwash	Randomized pilot study	[238]
Eucalyptus	Antibacterial, antiviral, antifungal, anti-inflammatory, analgesic, and antioxidant	Topical gel	Hamsters	[239]
Zizyphus jujuba	Anti-inflammatory, analgesic, and wound healing	Topical gel and dietary	Hamsters	[240]
Zataria multiflora	Carminative, stimulant, diaphoretic, diuretic, antiseptic, anesthetic, antispasmodic, anti-hermitic, antidiarrheal, and analgesic	Mouthwash	Randomized clinical trial	[241]
Carapa guianensis oil	Anti-inflammatory, analgesic, and antimicrobial	Topical gel/swab	Controlled and randomized clinical trial/ hamsters	[242,243]
Plantago ovata	Antioxidant, anti-inflammatory, and antibacterial	Mouthwash	Randomized cross-over clinical trial	[244]
Achillea millefolium	Antimicrobial and anti-inflammatory	Mouthwash	Double-blind, randomized, controlled trial	[82]
Vaccinium myrtillus	Antioxidant, cardioprotective, neuroprotective, anti-inflammatory, and anticarcinogenic	Topical application, gavage administration, mouthwash	Clinical trials, Hamsters	[245–247]
Carum carvi	Antioxidant, antidiabetic, antifungal, and antimicrobial	Topical gel	Hamsters	[248]
Pistacia atlantica	Antioxidant and anti-inflammatory	Topical gel	Hamsters	[249,250]
Hypericum perforatum	Antioxidant and anti-inflammatory	Topical gel	Hamsters	[251]
Elaeagnus angustifolia	Anti-inflammatory, analgesic, and wound healing	Topical gel	Hamster	[252]
Trachyspermum ammi	Anti-inflammatory, antiviral, antifungal, antioxidant, and analgesic	Topical gel	Hamsters	[250]
Hippophae rhamnoides	Antioxidant, anti-inflammatory, antimicrobial, and anti-ulcerogenic	Gavage administration	Rats	[253,254]

4. Conclusions and Future Perspectives

OM is a common and incapacitating side effect of antineoplastic therapies. The increased knowledge of its pathogenesis allows a better prediction of a patient's risk with the aim of adapting the management protocols and improving the development of new therapies. Nevertheless, standard guidelines for preventing and treating OM do not display significant effectiveness. The interest in natural products as potential therapeutic drugs has increased in recent years, as they have the advantage of being accessible and generating minimal side effects, with potential properties that include anti-inflammatory, antioxidant, antimicrobial, antiulcerative, and wound-healing capacities. In addition, over recent years, there have been multiple efforts to develop naturally based therapies, with natural compounds being tested in model organisms and clinical trials that are currently ongoing. However, the environment of the oral cavity is a complex system that is divided into two functional layers—the epithelium (thick and avascular) and the underlying tissue (vascular)—that are anatomically different, which affects their permeability to drugs and the capacity for maintaining a system for a certain period [255]. The buccal mucosa, which is composed of epithelial cells, provides a large surface area of almost 100 cm^2 [256]. This area is ideal for attaching a drug delivery system, providing a permeability that is 4 to 4000 times higher than that of skin [256–259]. Oral administration provides the advantage of a simple administration that does not suffer from the first-pass metabolism and that is safe and increases the drug availability. In addition, this route has a rapid action, reduced side effects, easy access to the local condition, and great patient compliance [255,260]. These characteristics make the buccal mucosa an optimal solution for the systemic and local treatment of OM [261]. However, it also has limitations that are associated with a functionalized protective barrier. The presence of saliva and its enzymatic action, as well as the constant mechanical pressure caused by eating and speaking movements, may compromise the penetration of the drug present in the delivery system; as such, the application of mucoadhesive components may be required to solve this issue, but this can compromise the therapeutic effectiveness [255,257–259].

Due to the characteristics of the oral cavity, it is necessary to develop novel strategies for overcoming topical delivery, such as mucoadhesive dosage forms (e.g., films, tablets). For the treatment of oral diseases, the most suitable formulations investigated were in the form of tablets, films, sprays, mouthwashes, gels, and pastes [24,259,262]. Gel and film formulations were evaluated in hamsters with CT-induced mucositis. By the 28th day, the hamsters' mucosa appeared to be healed, as no erythema or edema was visible. These results proved their efficiency, as the animals' survival was higher than in the control group, and these treatments showed promising potential for a function as an occlusive patch and for delivering therapeutic compounds [261]. Films containing ethanolic propolis extract also presented optimal mucoadhesion capacity, ensuring the release of propolis compounds, a good stability, a high swelling capacity, and antimicrobial effects against *S. aureus* [263]. In addition, the incorporation of nanoparticles in the forms of dosage for buccal drug delivery has recently been encouraged [24,256,264]. Furthermore, nanoparticles could transport many therapeutic agents [24,256]. Functional and biocompatible carriers that display chemical stability are sought in the innovation of buccal drug delivery systems [264–267]. Chitosan is an example of a biopolymer that is biologically safe and bioadhesive, and it has been used in several studies for the development of drug delivery systems, as it has longer retention periods in the oral mucosa [261,268,269]. In addition, it inhibits the attachment of *C. albicans* to human oral mucosal cells [261,268,269].

A SWOT diagram (Figure 2) was constructed with the aim of summarizing the previously described strengths, weaknesses, opportunities, and threats of employing natural products for the prevention/treatment of OM.

Despite the significant advances made in this area, more investigations are needed to ensure that these formulations reach the pharmaceutical market, and few have been published regarding this topic with natural products.

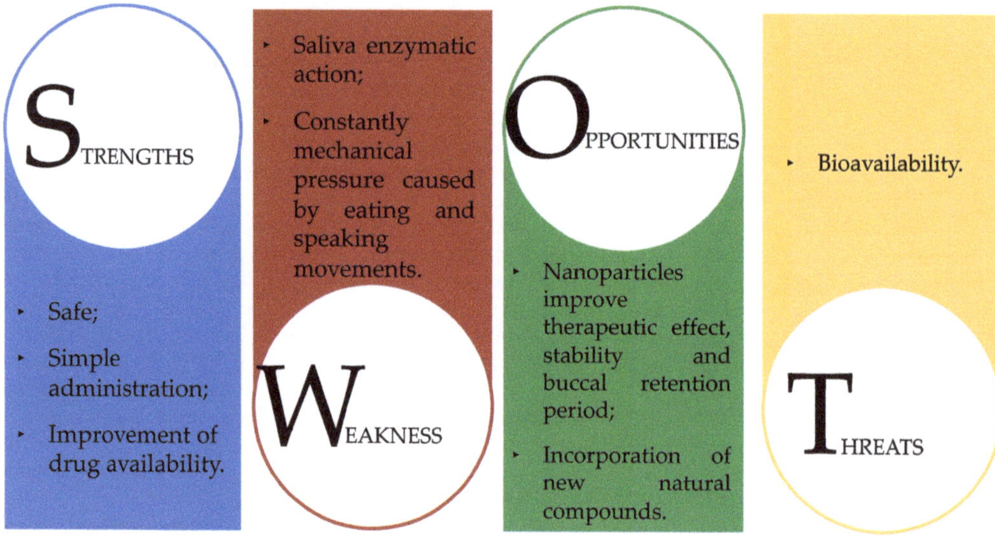

Figure 2. SWOT analysis for the possible use of natural products to prevent/treat OM.

Author Contributions: Conceptualization, F.R.; methodology, A.S.F., C.M., A.M.S. and F.R.; validation, F.R.; investigation, A.S.F. and C.M.; resources, F.R.; writing—original draft preparation, A.S.F., C.M. and A.M.S.; writing—review and editing, F.R., P.C. and C.D.-M.; supervision, F.R., C.D.-M. and P.C.; project administration, F.R.; funding acquisition, F.R. All authors have read and agreed to the published version of the manuscript.

Funding: This research was funded by the project EXPL/BAA-GR/0663/2021—Kiwi4Health—Exploring the Eco-Innovative Re-Use of Kiwiberry, supported by national funds from Fundação para a Ciência e a Tecnologia (Portugal), as well as the project MTS/SAS/0077/2020-Honey+-New reasons to care honey from the Natural Park of Montesinho: A bioindicator of environmental quality & its therapeutic potential.

Institutional Review Board Statement: Not applicable.

Informed Consent Statement: Not applicable.

Data Availability Statement: Data are available on request due to restrictions, e.g., privacy or ethical restrictions.

Acknowledgments: Ana Sofia Ferreira (SFRH/BD/7519/2020) and Ana Margarida Silva (SFRH/BD/144994/2019) are thankful for their Ph.D. grants financed by POPH-QREN and subsidized by the European Science Foundation and Ministério da Ciência, Tecnologia e Ensino Superior. Catarina Macedo is thankful for her scholarship from the project EXPL/BAA-GR/0663/2021. Francisca Rodrigues (CEECIND/01886/2020) is thankful for her contract financed by FCT/MCTES—CEEC Individual Program Contract. This work was financially supported by Portuguese national funds through projects UIDB/50006/2020, UIDP/50006/2020, and LA/P/0008/2020, from the Fundação para a Ciência e a Tecnologia (FCT)/Ministério da Ciência, Tecnologia e Ensino Superior (MCTES). This work was also financed by national funds from FCT—Fundação para a Ciência e a Tecnologia, I.P., in the scope of the project UIDP/04378/2020 and UIDB/04378/2020 of the Research Unit on Applied Molecular Biosciences—UCIBIO and the project LA/P/0140/2020 of the Associate Laboratory Institute for Health and Bioeconomy—i4HB.

Conflicts of Interest: The authors declare no conflict of interest.

References

1. Sung, H.; Ferlay, J.; Siegel, R.L.; Laversanne, M.; Soerjomataram, I.; Jemal, A.; Bray, F. Global Cancer Statistics 2020: GLOBOCAN Estimates of Incidence and Mortality Worldwide for 36 Cancers in 185 Countries. *CA Cancer J. Clin.* **2021**, *71*, 209–249. [CrossRef] [PubMed]
2. Cinausero, M.; Aprile, G.; Ermacora, P.; Basile, D.; Vitale, M.G.; Fanotto, V.; Parisi, G.; Calvetti, L.; Sonis, S.T. New Frontiers in the Pathobiology and Treatment of Cancer Regimen-Related Mucosal Injury. *Front. Pharmacol.* **2017**, *8*, 354. [CrossRef] [PubMed]
3. Sonis, S.T. Mucositis: The impact, biology and therapeutic opportunities of oral mucositis. *Oral Oncol.* **2009**, *45*, 1015–1020. [CrossRef]
4. Oronsky, B.; Goyal, S.; Kim, M.M.; Cabrales, P.; Lybeck, M.; Caroen, S.; Oronsky, N.; Burbano, E.; Carter, C.; Oronsky, A. A Review of Clinical Radioprotection and Chemoprotection for Oral Mucositis. *Transl. Oncol.* **2018**, *11*, 771–778. [CrossRef] [PubMed]
5. Shetty, S.S.; Maruthi, M.; Dhara, V.; de Arruda, J.A.A.; Abreu, L.G.; Mesquita, R.A.; Teixeira, A.L.; Silva, T.A.; Merchant, Y. Oral mucositis: Current knowledge and future directions. *Dis. Mon.* 2021, 101300, *in press*. [CrossRef] [PubMed]
6. Sonis, S.T. A hypothesis for the pathogenesis of radiation-induced oral mucositis: When biological challenges exceed physiologic protective mechanisms. Implications for pharmacological prevention and treatment. *Support. Care Cancer* **2021**, *29*, 4939–4947. [CrossRef]
7. Kawashita, Y.; Soutome, S.; Umeda, M.; Saito, T. Oral management strategies for radiotherapy of head and neck cancer. *Jpn. Dent. Sci. Rev.* **2020**, *56*, 62–67. [CrossRef]
8. Lalla, R.V.; Saunders, D.P.; Peterson, D.E. Chemotherapy or radiation-induced oral mucositis. *Dent. Clin. N. Am.* **2014**, *58*, 341–349. [CrossRef]
9. Singh, V.; Singh, A.K. Oral mucositis. *Natl. J. Maxillofac. Surg.* **2020**, *11*, 159–168. [CrossRef]
10. Pulito, C.; Cristaudo, A.; Porta, C.; Zapperi, S.; Blandino, G.; Morrone, A.; Strano, S. Oral mucositis: The hidden side of cancer therapy. *J. Exp. Clin. Cancer Res.* **2020**, *39*, 210. [CrossRef]
11. Daugelaite, G.; Uzkuraityte, K.; Jagelaviciene, E.; Filipauskas, A. Prevention and Treatment of Chemotherapy and Radiotherapy Induced Oral Mucositis. *Medicina* **2019**, *55*, 25. [CrossRef] [PubMed]
12. Chen, S.C.; Lai, Y.H.; Huang, B.S.; Lin, C.Y.; Fan, K.H.; Chang, J.T. Changes and predictors of radiation-induced oral mucositis in patients with oral cavity cancer during active treatment. *Eur. J. Oncol. Nurs.* **2015**, *19*, 214–219. [CrossRef] [PubMed]
13. De Sanctis, V.; Bossi, P.; Sanguineti, G.; Trippa, F.; Ferrari, D.; Bacigalupo, A.; Ripamonti, C.I.; Buglione, M.; Pergolizzi, S.; Langendjik, J.A.; et al. Mucositis in head and neck cancer patients treated with radiotherapy and systemic therapies: Literature review and consensus statements. *Crit. Rev. Oncol. Hematol.* **2016**, *100*, 147–166. [CrossRef] [PubMed]
14. Fidan, O.; Arslan, S. Development and Validation of the Oral Mucositis Risk Assessment Scale in Hematology Patients. *Semin. Oncol. Nurs.* **2021**, *37*, 151159. [CrossRef]
15. Hearnden, V.; Sankar, V.; Hull, K.; Juras, D.V.; Greenberg, M.; Kerr, A.R.; Lockhart, P.B.; Patton, L.L.; Porter, S.; Thornhill, M.H. New developments and opportunities in oral mucosal drug delivery for local and systemic disease. *Adv. Drug Deliv. Rev.* **2012**, *64*, 16–28. [CrossRef]
16. Elad, S.; Yarom, N.; Zadik, Y.; Kuten-Shorrer, M.; Sonis, S.T. The broadening scope of oral mucositis and oral ulcerative mucosal toxicities of anticancer therapies. *CA Cancer J. Clin.* **2022**, *72*, 57–77. [CrossRef]
17. Kawashita, Y.; Koyama, Y.; Kurita, H.; Otsuru, M.; Ota, Y.; Okura, M.; Horie, A.; Sekiya, H.; Umeda, M. Effectiveness of a comprehensive oral management protocol for the prevention of severe oral mucositis in patients receiving radiotherapy with or without chemotherapy for oral cancer: A multicentre, phase II, randomized controlled trial. *Int. J. Oral Maxillofac. Surg.* **2019**, *48*, 857–864. [CrossRef]
18. Basile, D.; Di Nardo, P.; Corvaja, C.; Garattini, S.K.; Pelizzari, G.; Lisanti, C.; Bortot, L.; Da Ros, L.; Bartoletti, M.; Borghi, M.; et al. Mucosal Injury during Anti-Cancer Treatment: From Pathobiology to Bedside. *Cancers* **2019**, *11*, 857. [CrossRef]
19. Elting, L.S.; Chang, Y.C. Costs of Oral Complications of Cancer Therapies: Estimates and a Blueprint for Future Study. *J. Natl. Cancer Inst. Monogr.* **2019**, *2019*, lgz010. [CrossRef]
20. Rodrigues-Oliveira, L.; Kowalski, L.P.; Santos, M.; Marta, G.N.; Bensadoun, R.J.; Martins, M.D.; Lopes, M.A.; Castro, G., Jr.; William, W.N., Jr.; Chaves, A.L.F.; et al. Direct costs associated with the management of mucositis: A systematic review. *Oral Oncol.* **2021**, *118*, 105296. [CrossRef]
21. Sampson, M.M.; Nanjappa, S.; Greene, J.N. Mucositis and oral infections secondary to gram negative rods in patients with prolonged neutropenia. *IDCases* **2017**, *9*, 101–103. [CrossRef] [PubMed]
22. Alkhouli, M.; Laflouf, M.; Comisi, J.C. Assessing the topical application efficiency of two biological agents in managing chemotherapy-induced oral mucositis in children: A randomized clinical trial. *J. Oral Biol. Craniofac. Res.* **2021**, *11*, 373–378. [CrossRef] [PubMed]
23. Kusiak, A.; Jereczek-Fossa, B.A.; Cichonska, D.; Alterio, D. Oncological-Therapy Related Oral Mucositis as an Interdisciplinary Problem-Literature Review. *Int. J. Environ. Res. Public Health* **2020**, *17*, 2464. [CrossRef] [PubMed]
24. Tran, P.H.L.; Duan, W.; Tran, T.T.D. Recent developments of nanoparticle-delivered dosage forms for buccal delivery. *Int. J. Pharm.* **2019**, *571*, 118697. [CrossRef] [PubMed]
25. Moslemi, D.; Nokhandani, A.M.; Otaghsaraei, M.T.; Moghadamnia, Y.; Kazemi, S.; Moghadamnia, A.A. Management of chemo/radiation-induced oral mucositis in patients with head and neck cancer: A review of the current literature. *Radiother. Oncol.* **2016**, *120*, 13–20. [CrossRef]

26. Patel, P.; Robinson, P.D.; Baggott, C.; Gibson, P.; Ljungman, G.; Massey, N.; Ottaviani, G.; Phillips, R.; Revon-Riviere, G.; Treister, N.; et al. Clinical practice guideline for the prevention of oral and oropharyngeal mucositis in pediatric cancer and hematopoietic stem cell transplant patients: 2021 update. *Eur. J. Cancer* **2021**, *154*, 92–101. [CrossRef]
27. Ayago Flores, D.; Ferriols Lisart, R. Effectiveness of palifermin in the prevention of oral mucositis in patients with haematological cancers. *Farm. Hosp.* **2010**, *34*, 163–169. [CrossRef]
28. de Sousa Melo, A.; de Lima Dantas, J.B.; Medrado, A.; Lima, H.R.; Martins, G.B.; Carrera, M. Nutritional supplements in the management of oral mucositis in patients with head and neck cancer: Narrative literary review. *Clin. Nutr. ESPEN* **2021**, *43*, 31–38. [CrossRef]
29. Yarom, N.; Ariyawardana, A.; Hovan, A.; Barasch, A.; Jarvis, V.; Jensen, S.B.; Zadik, Y.; Elad, S.; Bowen, J.; Lalla, R.V. Systematic review of natural agents for the management of oral mucositis in cancer patients. *Support. Care Cancer* **2013**, *21*, 3209–3221. [CrossRef]
30. Yarom, N.; Hovan, A.; Bossi, P.; Ariyawardana, A.; Jensen, S.B.; Gobbo, M.; Saca-Hazboun, H.; Kandwal, A.; Majorana, A.; Ottaviani, G.; et al. Correction to: Systematic review of natural and miscellaneous agents, for the management of oral mucositis in cancer patients and Clinical Practice Guidelines—Part 1: Vitamins, minerals and nutritional supplements. *Support. Care Cancer* **2021**, *29*, 4175–4176. [CrossRef]
31. Konuk Sener, D.; Aydin, M.; Cangur, S.; Guven, E. The Effect of Oral Care with Chlorhexidine, Vitamin E and Honey on Mucositis in Pediatric Intensive Care Patients: A Randomized Controlled Trial. *J. Pediatric Nurs.* **2019**, *45*, e95–e101. [CrossRef] [PubMed]
32. Lima, I.; de Fatima Souto Maior, L.; Gueiros, L.A.M.; Leao, J.C.; Higino, J.S.; Carvalho, A.A.T. Clinical applicability of natural products for prevention and treatment of oral mucositis: A systematic review and meta-analysis. *Clin. Oral Investig.* **2021**, *25*, 4115–4124. [CrossRef] [PubMed]
33. Yang, C.; Tang, H.; Wang, L.; Peng, R.; Bai, F.; Shan, Y.; Yu, Z.; Zhou, P.; Cong, Y. Dimethyl Sulfoxide Prevents Radiation-Induced Oral Mucositis through Facilitating DNA Double-Strand Break Repair in Epithelial Stem Cells. *Int. J. Radiat. Oncol. Biol. Phys.* **2018**, *102*, 1577–1589. [CrossRef] [PubMed]
34. Anderson, P.M.; Lalla, R.V. Glutamine for Amelioration of Radiation and Chemotherapy Associated Mucositis during Cancer Therapy. *Nutrients* **2020**, *12*, 1675. [CrossRef]
35. Charalambous, M.; Raftopoulos, V.; Paikousis, L.; Katodritis, N.; Lambrinou, E.; Vomvas, D.; Georgiou, M.; Charalambous, A. The effect of the use of thyme honey in minimizing radiation-induced oral mucositis in head and neck cancer patients: A randomized controlled trial. *Eur. J. Oncol. Nurs.* **2018**, *34*, 89–97, (clinicaltrials.gov identifier NCT01465308). [CrossRef]
36. Charalambous, M.; Raftopoulos, V.; Lambrinou, E.; Charalambous, A. The effectiveness of honey for the management of radiotherapy-induced oral mucositis in head and neck cancer patients: A systematic review of clinical trials. *Eur. J. Integr. Med.* **2013**, *5*, 217–225. [CrossRef]
37. Ramsay, E.I.; Rao, S.; Madathil, L.; Hegde, S.K.; Baliga-Rao, M.P.; George, T.; Baliga, M.S. Honey in oral health and care: A mini review. *J. Oral Biosci.* **2019**, *61*, 32–36. [CrossRef]
38. Munstedt, K.; Momm, F.; Hubner, J. Honey in the management of side effects of radiotherapy- or radio/chemotherapy-induced oral mucositis. A systematic review. *Complement. Ther. Clin. Pract.* **2019**, *34*, 145–152. [CrossRef]
39. Pires Marques, E.C.; Piccolo Lopes, F.; Nascimento, I.C.; Morelli, J.; Pereira, M.V.; Machado Meiken, V.M.; Pinheiro, S.L. Photobiomodulation and photodynamic therapy for the treatment of oral mucositis in patients with cancer. *Photodiagn. Photodyn. Ther.* **2020**, *29*, 101621. [CrossRef]
40. de Carvalho, P.A.G.; Lessa, R.C.; Carraro, D.M.; Assis Pellizzon, A.C.; Jaguar, G.C.; Alves, F.A. Three photobiomodulation protocols in the prevention/treatment of radiotherapy-induced oral mucositis. *Photodiagn. Photodyn. Ther.* **2020**, *31*, 101906. [CrossRef]
41. Cotomacio, C.C.; Calarga, C.C.; Yshikawa, B.K.; Arana-Chavez, V.E.; Simoes, A. Wound healing process with different photobiomodulation therapy protocols to treat 5-FU-induced oral mucositis in hamsters. *Arch. Oral Biol.* **2021**, *131*, 105250. [CrossRef] [PubMed]
42. Blakaj, A.; Bonomi, M.; Gamez, M.E.; Blakaj, D.M. Oral mucositis in head and neck cancer: Evidence-based management and review of clinical trial data. *Oral Oncol.* **2019**, *95*, 29–34. [CrossRef] [PubMed]
43. Campos, M.I.; Campos, C.N.; Aarestrup, F.M.; Aarestrup, B.J. Oral mucositis in cancer treatment: Natural history, prevention and treatment. *Mol. Clin. Oncol.* **2014**, *2*, 337–340. [CrossRef] [PubMed]
44. Abt, E. Probiotics May Lower the Risk of Oral Mucositis in Cancer Patients. *J. Evid. Based Dent. Pract.* **2021**, *21*, 101639. [CrossRef]
45. Wardill, H.R.; Sonis, S.T.; Blijlevens, N.M.A.; Van Sebille, Y.Z.A.; Ciorba, M.A.; Loeffen, E.A.H.; Cheng, K.K.F.; Bossi, P.; Porcello, L.; Castillo, D.A.; et al. Prediction of mucositis risk secondary to cancer therapy: A systematic review of current evidence and call to action. *Support. Care Cancer* **2020**, *28*, 5059–5073. [CrossRef]
46. Elad, S.; Cheng, K.K.F.; Lalla, R.V.; Yarom, N.; Hong, C.; Logan, R.M.; Bowen, J.; Gibson, R.; Saunders, D.P.; Zadik, Y.; et al. MASCC/ISOO clinical practice guidelines for the management of mucositis secondary to cancer therapy. *Cancer* **2020**, *126*, 4423–4431. [CrossRef]
47. Logan, R.M.; Stringer, A.M.; Bowen, J.M.; Yeoh, A.S.; Gibson, R.J.; Sonis, S.T.; Keefe, D.M. The role of pro-inflammatory cytokines in cancer treatment-induced alimentary tract mucositis: Pathobiology, animal models and cytotoxic drugs. *Cancer Treat. Rev.* **2007**, *33*, 448–460. [CrossRef]

48. Bertolini, M.; Sobue, T.; Thompson, A.; Dongari-Bagtzoglou, A. Chemotherapy Induces Oral Mucositis in Mice without Additional Noxious Stimuli. *Transl. Oncol.* **2017**, *10*, 612–620. [CrossRef]
49. Raber-Durlacher, J.E.; Elad, S.; Barasch, A. Oral mucositis. *Oral Oncol.* **2010**, *46*, 452–456. [CrossRef]
50. Bailly, C. Potential use of edaravone to reduce specific side effects of chemo-, radio- and immuno-therapy of cancers. *Int. Immunopharmacol.* **2019**, *77*, 105967. [CrossRef]
51. Vigarios, E.; Epstein, J.B.; Sibaud, V. Oral mucosal changes induced by anticancer targeted therapies and immune checkpoint inhibitors. *Support. Care Cancer* **2017**, *25*, 1713–1739. [CrossRef] [PubMed]
52. Kudrimoti, M.; Curtis, A.; Azawi, S.; Worden, F.; Katz, S.; Adkins, D.; Bonomi, M.; Elder, J.; Sonis, S.T.; Straube, R.; et al. Dusquetide: A novel innate defense regulator demonstrating a significant and consistent reduction in the duration of oral mucositis in preclinical data and a randomized, placebo-controlled phase 2a clinical study. *J. Biotechnol.* **2016**, *239*, 115–125. [CrossRef] [PubMed]
53. Ariyawardana, A.; Cheng, K.K.F.; Kandwal, A.; Tilly, V.; Al-Azri, A.R.; Galiti, D.; Chiang, K.; Vaddi, A.; Ranna, V.; Nicolatou-Galitis, O.; et al. Systematic review of anti-inflammatory agents for the management of oral mucositis in cancer patients and clinical practice guidelines. *Support. Care Cancer* **2019**, *27*, 3985–3995. [CrossRef] [PubMed]
54. Davy, C.; Heathcote, S. A systematic review of interventions to mitigate radiotherapy-induced oral mucositis in head and neck cancer patients. *Support. Care Cancer* **2021**, *29*, 2187–2202. [CrossRef]
55. Peterson, D.E.; Lalla, R.V. Oral mucositis: The new paradigms. *Curr. Opin. Oncol.* **2010**, *22*, 318–322. [CrossRef]
56. Satheeshkumar, P.S.; El-Dallal, M.; Mohan, M.P. Feature selection and predicting chemotherapy-induced ulcerative mucositis using machine learning methods. *Int. J. Med. Inform.* **2021**, *154*, 104563. [CrossRef]
57. Pratiwi, E.S.; Ismawati, N.D.S.; Ruslin, M. Prevalence and risk factors of oral mucositis in children with acute lymphoblastic leukemia in Dr. Soetomo Hospital Surabaya Indonesia. *Enfermería Clínica* **2020**, *30*, 289–292. [CrossRef]
58. Soutome, S.; Yanamoto, S.; Nishii, M.; Kojima, Y.; Hasegawa, T.; Funahara, M.; Akashi, M.; Saito, T.; Umeda, M. Risk factors for severe radiation-induced oral mucositis in patients with oral cancer. *J. Dent. Sci.* **2021**, *16*, 1241–1246. [CrossRef]
59. Parkhideh, S.; Zeraatkar, M.; Moradi, O.; Hajifathali, A.; Mehdizadeh, M.; Tavakoli-Ardakani, M. Azithromycin oral suspension in prevention and management of oral mucositis in patients undergoing hematopoietic stem cell transplantation: A randomized controlled trial. *Support. Care Cancer* **2022**, *30*, 251–257. [CrossRef]
60. Aspinall, S.R.; Parker, J.K.; Khutoryanskiy, V.V. Oral care product formulations, properties and challenges. *Colloids Surf. B* **2021**, *200*, 111567. [CrossRef]
61. Yamaguchi, T.; Makiguchi, T.; Nakamura, H.; Yamatsu, Y.; Hirai, Y.; Shoda, K.; Suzuki, K.; Kim, M.; Kurozumi, S.; Motegi, S.I.; et al. Impact of muscle volume loss on acute oral mucositis in patients undergoing concurrent chemoradiotherapy after oral cancer resection. *Int. J. Oral Maxillofac. Surg.* **2021**, *50*, 1195–1202. [CrossRef] [PubMed]
62. Orlandi, E.; Iacovelli, N.A.; Rancati, T.; Cicchetti, A.; Bossi, P.; Pignoli, E.; Bergamini, C.; Licitra, L.; Fallai, C.; Valdagni, R.; et al. Multivariable model for predicting acute oral mucositis during combined IMRT and chemotherapy for locally advanced nasopharyngeal cancer patients. *Oral Oncol.* **2018**, *86*, 266–272. [CrossRef] [PubMed]
63. Hong, C.H.L.; Gueiros, L.A.; Fulton, J.S.; Cheng, K.K.F.; Kandwal, A.; Galiti, D.; Fall-Dickson, J.M.; Johansen, J.; Ameringer, S.; Kataoka, T.; et al. Systematic review of basic oral care for the management of oral mucositis in cancer patients and clinical practice guidelines. *Support. Care Cancer* **2019**, *27*, 3949–3967. [CrossRef] [PubMed]
64. Inoue, Y.; Yamagata, K.; Nakamura, M.; Ohnishi, K.; Tabuchi, K.; Bukawa, H. Are Intraoral Stents Effective for Reducing the Severity of Oral Mucositis during Radiotherapy for Maxillary and Nasal Cavity Cancer? *J. Oral Maxillofac. Surg.* **2020**, *78*, 1214.e1–1214.e8. [CrossRef]
65. Duzkaya, D.S.; Uysal, G.; Bozkurt, G.; Yakut, T. The Effect of Oral Care Using an Oral Health Care Guide on Preventing Mucositis in Pediatric Intensive Care. *J. Pediatr. Nurs.* **2017**, *36*, 98–102. [CrossRef]
66. Edmans, J.G.; Murdoch, C.; Santocildes-Romero, M.E.; Hatton, P.V.; Colley, H.E.; Spain, S.G. Incorporation of lysozyme into a mucoadhesive electrospun patch for rapid protein delivery to the oral mucosa. *Mater. Sci. Eng. C Mater. Biol. Appl.* **2020**, *112*, 110917. [CrossRef]
67. Liao, Y.C.; Hsu, L.F.; Hsieh, L.Y.; Luo, Y.Y. Effectiveness of green tea mouthwash for improving oral health status in oral cancer patients: A single-blind randomized controlled trial. *Int. J. Nurs. Stud.* **2021**, *121*, 103985, (clinicaltrials.gov identifier NCT04615780). [CrossRef]
68. Freires, I.A.; Rosalen, P.L. How Natural Product Research has Contributed to Oral Care Product Development? A Critical View. *Pharm. Res.* **2016**, *33*, 1311–1317. [CrossRef]
69. Yen, S.H.; Wang, L.W.; Lin, Y.H.; Jen, Y.M.; Chung, Y.L. Phenylbutyrate mouthwash mitigates oral mucositis during radiotherapy or chemoradiotherapy in patients with head-and-neck cancer. *Int. J. Radiat. Oncol. Biol. Phys.* **2012**, *82*, 1463–1470. [CrossRef]
70. Nigro, O.; Tuzi, A.; Tartaro, T.; Giaquinto, A.; Vallini, I.; Pinotti, G. Biological effects of verbascoside and its anti-inflammatory activity on oral mucositis: A review of the literature. *Anticancer Drugs* **2020**, *31*, 1–5. [CrossRef]
71. Baysal, E.; Sari, D.; Vural, F.; Cagirgan, S.; Saydam, G.; Tobu, M.; Sahin, F.; Soyer, N.; Gediz, F.; Acarlar, C.; et al. The Effect of Cryotherapy on the Prevention of Oral Mucositis and on the Oral pH Value in Multiple Myeloma Patients Undergoing Autologous Stem Cell Transplantation. *Semin. Oncol. Nurs.* **2021**, *37*, 151146. [CrossRef] [PubMed]

72. Askarifar, M.; Lakdizaji, S.; Ramzi, M.; Rahmani, A.; Jabbarzadeh, F. The Effects of Oral Cryotherapy on Chemotherapy-Induced Oral Mucositis in Patients Undergoing Autologous Transplantation of Blood Stem Cells: A Clinical Trial. *Iran. Red Crescent Med. J.* **2016**, *18*, e24775. [CrossRef] [PubMed]
73. Migliorati, C.A.; Oberle-Edwards, L.; Schubert, M. The role of alternative and natural agents, cryotherapy, and/or laser for management of alimentary mucositis. *Support. Care Cancer* **2006**, *14*, 533–540. [CrossRef] [PubMed]
74. Zhang, X.; Sun, D.; Qin, N.; Liu, M.; Zhang, J.; Li, X. Comparative prevention potential of 10 mouthwashes on intolerable oral mucositis in cancer patients: A Bayesian network analysis. *Oral Oncol.* **2020**, *107*, 104751. [CrossRef]
75. Karavana Hizarcioglu, S.Y.; Sezer, B.; Guneri, P.; Veral, A.; Boyacioglu, H.; Ertan, G.; Epstein, J.B. Efficacy of topical benzydamine hydrochloride gel on oral mucosal ulcers: An in vivo animal study. *Int. J. Oral Maxillofac. Surg.* **2011**, *40*, 973–978. [CrossRef]
76. El-Salamouni, N.S.; Hanafy, A.S. Hyaluronic-benzydamine oromucosal films outperform conventional mouth rinse in ulcer healing. *J. Drug Deliv. Sci. Technol.* **2021**, *65*, 102690. [CrossRef]
77. Yang, C.; Gong, G.; Jin, E.; Han, X.; Zhuo, Y.; Yang, S.; Song, B.; Zhang, Y.; Piao, C. Topical application of honey in the management of chemo/radiotherapy-induced oral mucositis: A systematic review and network meta-analysis. *Int. J. Nurs. Stud.* **2019**, *89*, 80–87. [CrossRef]
78. Saunders, D.P.; Epstein, J.B.; Elad, S.; Allemano, J.; Bossi, P.; van de Wetering, M.D.; Rao, N.G.; Potting, C.; Cheng, K.K.; Freidank, A.; et al. Systematic review of antimicrobials, mucosal coating agents, anesthetics, and analgesics for the management of oral mucositis in cancer patients. *Support. Care Cancer* **2013**, *21*, 3191–3207. [CrossRef]
79. Nielsen, B.N.; Aagaard, G.; Henneberg, S.W.; Schmiegelow, K.; Hansen, S.H.; Romsing, J. Topical morphine for oral mucositis in children: Dose finding and absorption. *J. Pain Symptom Manag.* **2012**, *44*, 117–123. [CrossRef]
80. Sanz, R.; Calpena, A.C.; Mallandrich, M.; Gimeno, A.; Halbaut, L.; Clares, B. Development of a buccal doxepin platform for pain in oral mucositis derived from head and neck cancer treatment. *Eur. J. Pharm. Biopharm.* **2017**, *117*, 203–211. [CrossRef]
81. Chung, M.K.; Wang, S.; Oh, S.L.; Kim, Y.S. Acute and Chronic Pain from Facial Skin and Oral Mucosa: Unique Neurobiology and Challenging Treatment. *Int. J. Mol. Sci.* **2021**, *22*, 5810. [CrossRef] [PubMed]
82. Miranzadeh, S.; Adib-Hajbaghery, M.; Soleymanpoor, L.; Ehsani, M. Effect of adding the herb *Achillea millefolium* on mouthwash on chemotherapy induced oral mucositis in cancer patients: A double-blind randomized controlled trial. *Eur. J. Oncol. Nurs.* **2015**, *19*, 207–213. [CrossRef] [PubMed]
83. Santos, H.T.C.; Coimbra, M.C.; Meri Junior, A.E.; Gomes, A.J.P.S. Effectiveness of topically applied chamomile in the treatment of oral mucositis: A literature review. *Res. Soc. Dev.* **2021**, *10*, e433101422081. [CrossRef]
84. Honda, H.; Onda, T.; Hayashi, K.; Shibahara, T.; Takano, M. Comparison of topical agents that are effective against oral mucositis associated with chemotherapy using a rat anticancer agent-induced oral mucositis model. *J. Oral Maxillofac. Surg. Med. Pathol.* 2021, *in press*. [CrossRef]
85. da Silva, V.C.R.; da Motta Silveira, F.M.; Barbosa Monteiro, M.G.; da Cruz, M.M.D.; Caldas Junior, A.F.; Pina Godoy, G. Photodynamic therapy for treatment of oral mucositis: Pilot study with pediatric patients undergoing chemotherapy. *Photodiagn. Photodyn. Ther.* **2018**, *21*, 115–120. [CrossRef] [PubMed]
86. Simoes, A.; Benites, B.M.; Benassi, C.; Torres-Schroter, G.; de Castro, J.R.; Campos, L. Antimicrobial photodynamic therapy on treatment of infected radiation-induced oral mucositis: Report of two cases. *Photodiagn. Photodyn. Ther.* **2017**, *20*, 18–20. [CrossRef] [PubMed]
87. de Oliveira, A.B.; Ferrisse, T.M.; Basso, F.G.; Fontana, C.R.; Giro, E.M.A.; Brighenti, F.L. A systematic review and meta-analysis of the effect of photodynamic therapy for the treatment of oral mucositis. *Photodiagn. Photodyn. Ther.* **2021**, *34*, 102316. [CrossRef]
88. Yarom, N.; Hovan, A.; Bossi, P.; Ariyawardana, A.; Jensen, S.B.; Gobbo, M.; Saca-Hazboun, H.; Kandwal, A.; Majorana, A.; Ottaviani, G.; et al. Systematic review of natural and miscellaneous agents, for the management of oral mucositis in cancer patients and clinical practice guidelines—Part 2: Honey, herbal compounds, saliva stimulants, probiotics, and miscellaneous agents. *Support. Care Cancer* **2020**, *28*, 2457–2472. [CrossRef]
89. Nagi, R.; Patil, D.J.; Rakesh, N.; Jain, S.; Sahu, S. Natural agents in the management of oral mucositis in cancer patients-systematic review. *J. Oral Biol. Craniofac. Res.* **2018**, *8*, 245–254. [CrossRef]
90. Aghamohamamdi, A.; Hosseinimehr, S.J. Natural Products for Management of Oral Mucositis Induced by Radiotherapy and Chemotherapy. *Integr. Cancer Ther.* **2016**, *15*, 60–68. [CrossRef]
91. Khanal, B.; Baliga, M.; Uppal, N. Effect of topical honey on limitation of radiation-induced oral mucositis: An intervention study. *Int. J. Oral Maxillofac. Surg.* **2010**, *39*, 1181–1185. [CrossRef] [PubMed]
92. Amanat, A.; Ahmed, A.; Kazmi, A.; Aziz, B. The Effect of Honey on Radiation-induced Oral Mucositis in Head and Neck Cancer Patients. *Indian J. Palliat. Care* **2017**, *23*, 317–320. [CrossRef] [PubMed]
93. Lusby, P.E.; Coombes, A.; Wilkinson, J.M. Honey: A potent agent for wound healing? *J. Wound Ostomy Cont. Nurs.* **2002**, *29*, 295–300. [CrossRef]
94. Molan, P.C. Re-introducing honey in the management of wounds and ulcers-theory and practice. *Ostomy Wound Manag.* **2002**, *48*, 28–40.
95. Motallebnejad, M.; Akram, S.; Moghadamnia, A.; Moulana, Z.; Omidi, S. The effect of topical application of pure honey on radiation-induced mucositis: A randomized clinical trial. *J. Contemp. Dent. Pract.* **2008**, *9*, 40–47. [CrossRef]
96. Al-Waili, N.S. An alternative treatment for pityriasis versicolor, tinea cruris, tinea corporis and tinea faciei with topical application of honey, olive oil and beeswax mixture: An open pilot study. *Complement. Ther. Med.* **2004**, *12*, 45–47. [CrossRef]

97. Khanjani Pour-Fard-Pachekenari, A.; Rahmani, A.; Ghahramanian, A.; Asghari Jafarabadi, M.; Onyeka, T.C.; Davoodi, A. The effect of an oral care protocol and honey mouthwash on mucositis in acute myeloid leukemia patients undergoing chemotherapy: A single-blind clinical trial. *Clin. Oral Investig.* **2019**, *23*, 1811–1821, (WHO Trial registration IRCT2015121419919N7). [CrossRef]
98. Raeessi, M.A.; Raeessi, N.; Panahi, Y.; Gharaie, H.; Davoudi, S.M.; Saadat, A.; Karimi Zarchi, A.A.; Raeessi, F.; Ahmadi, S.M.; Jalalian, H. "Coffee plus honey" versus "topical steroid" in the treatment of chemotherapy-induced oral mucositis: A randomised controlled trial. *BMC Complement. Altern. Med.* **2014**, *14*, 293, (Iranian Registry of Clinical Trials IRCT: 201104074737N3). [CrossRef]
99. Davis, J.K.; Green, J.M. Caffeine and anaerobic performance: Ergogenic value and mechanisms of action. *Sports Med.* **2009**, *39*, 813–832. [CrossRef]
100. Miłek, M.; Młodecki, Ł.; Dżugan, M. Caffeine content and antioxidant activity of various brews of specialty grade coffee. *Acta Sci. Pol. Technol. Aliment.* **2021**, *20*, 179–188.
101. Barcelos, R.P.; Lima, F.D.; Carvalho, N.R.; Bresciani, G.; Royes, L.F. Caffeine effects on systemic metabolism, oxidative-inflammatory pathways, and exercise performance. *Nutr. Res.* **2020**, *80*, 1–17. [CrossRef] [PubMed]
102. Sforcin, J.M. Biological Properties and Therapeutic Applications of Propolis. *Phytother. Res.* **2016**, *30*, 894–905. [CrossRef] [PubMed]
103. Melliou, E.; Chinou, I. Chemical analysis and antimicrobial activity of Greek propolis. *Planta Med.* **2004**, *70*, 515–519. [CrossRef] [PubMed]
104. Huang, X.Y.; Guo, X.L.; Luo, H.L.; Fang, X.W.; Zhu, T.G.; Zhang, X.L.; Chen, H.W.; Luo, L.P. Fast Differential Analysis of Propolis Using Surface Desorption Atmospheric Pressure Chemical Ionization Mass Spectrometry. *Int. J. Anal. Chem.* **2015**, *2015*, 176475. [CrossRef]
105. Franco, T. Chemical composition of propolis: Vitamins and aminoacids. *Rev. Bras. Farmacogn.* **1985**, *1*, 12–19.
106. Kamburoğlu, K.; Özen, T. Analgesic effect of Anatolian propolis in mice. *Agri* **2011**, *23*, 47–50.
107. Montero, J.C.; Mori, G.G. Assessment of ion diffusion from a calcium hydroxide-propolis paste through dentin. *Braz. Oral Res.* **2012**, *26*, 318–322. [CrossRef]
108. Ozan, F.; Sümer, Z.; Polat, Z.A.; Er, K.; Ozan, U.; Deger, O. Effect of mouthrinse containing propolis on oral microorganisms and human gingival fibroblasts. *Eur. J. Dent.* **2007**, *1*, 195–201.
109. Cavalcante, D.R.; Oliveira, P.S.; Góis, S.M.; Soares, A.F.; Cardoso, J.C.; Padilha, F.F.; Albuquerque, R.L., Jr. Effect of green propolis on oral epithelial dysplasia in rats. *Braz. J. Otorhinolaryngol.* **2011**, *77*, 278–284. [CrossRef]
110. Mirzoeva, O.K.; Calder, P.C. The effect of propolis and its components on eicosanoid production during the inflammatory response. *Prostaglandins Leukot. Essent. Fat. Acids* **1996**, *55*, 441–449. [CrossRef]
111. AkhavanKarbassi, M.H.; Yazdi, M.F.; Ahadian, H.; SadrAbad, M.J. Randomized DoubleBlind Placebo Controlled Trial of Propolis for Oral Mucositis in Patients Receiving Chemotherapy for Head and Neck Cancer. *Asian Pac. J. Cancer Prev.* **2016**, *17*, 3611–3614. [PubMed]
112. Pavel, C.; Mărghitaş, A.L.; Bobis, O.; Dezmirean, D.; Şapcaliu, A.; Radoi, I.; Mădaş, M. Biological Activities of Royal Jelly-Review. *Lucr. Stiintifice* **2011**, *44*, 108–118.
113. Nagai, T.; Inoue, R. Preparation and the functional properties of water extract and alkaline extract of royal jelly. *Food Chem.* **2004**, *84*, 181–186. [CrossRef]
114. Nagai, T.; Inoue, R.; Suzuki, N.; Nagashima, T. Antioxidant properties of enzymatic hydrolysates from royal jelly. *J. Med. Food* **2006**, *9*, 363–367. [CrossRef]
115. Izuta, H.; Chikaraishi, Y.; Shimazawa, M.; Mishima, S.; Hara, H. 10-Hydroxy-2-decenoic acid, a major fatty acid from royal jelly, inhibits VEGF-induced angiogenesis in human umbilical vein endothelial cells. *Evid. Based Complement. Altern. Med.* **2009**, *6*, 489–494. [CrossRef]
116. Šimúth, J.; Bíliková, K.; Kováčová, E.; Kuzmová, Z.; Schroder, W. Immunochemical Approach to Detection of Adulteration in Honey: Physiologically Active Royal Jelly Protein Stimulating TNF-α Release is a Regular Component of Honey. *J. Agric. Food Chem.* **2004**, *52*, 2154–2158. [CrossRef]
117. Matsui, T.; Yukiyoshi, A.; Doi, S.; Sugimoto, H.; Yamada, H.; Matsumoto, K. Gastrointestinal enzyme production of bioactive peptides from royal jelly protein and their antihypertensive ability in SHR. *J. Nutr. Biochem.* **2002**, *13*, 80–86. [CrossRef]
118. Fujiwara, S.; Imai, J.; Fujiwara, M.; Yaeshima, T.; Kawashima, T.; Kobayashi, K. A potent antibacterial protein in royal jelly. Purification and determination of the primary structure of royalisin. *J. Biol. Chem.* **1990**, *265*, 11333–11337. [CrossRef]
119. Kohno, K.; Okamoto, I.; Sano, O.; Arai, N.; Iwaki, K.; Ikeda, M.; Kurimoto, M. Royal jelly inhibits the production of proinflammatory cytokines by activated macrophages. *Biosci. Biotechnol. Biochem.* **2004**, *68*, 138–145. [CrossRef]
120. Fujii, A.; Kobayashi, S.; Kuboyama, N.; Furukawa, Y.; Kaneko, Y.; Ishihama, S.; Yamamoto, H.; Tamura, T. Augmentation of wound healing by royal jelly (RJ) in streptozotocin-diabetic rats. *Jpn. J. Pharmacol.* **1990**, *53*, 331–337. [CrossRef]
121. Xue, X.; Wu, L.; Wang, K. Chemical Composition of Royal Jelly. In *Bee Products-Chemical and Biological Properties*; Alvarez-Suarez, J.M., Ed.; Springer International Publishing: Cham, Switzerland, 2017; pp. 181–190.
122. Lin, Y.; Shao, Q.; Zhang, M.; Lu, C.; Fleming, J.; Su, S. Royal jelly-derived proteins enhance proliferation and migration of human epidermal keratinocytes in an in vitro scratch wound model. *BMC Complement. Altern. Med.* **2019**, *19*, 175. [CrossRef] [PubMed]
123. Inoue, S.; Koya-Miyata, S.; Ushio, S.; Iwaki, K.; Ikeda, M.; Kurimoto, M. Royal Jelly prolongs the life span of C3H/HeJ mice: Correlation with reduced DNA damage. *Exp. Gerontol.* **2003**, *38*, 965–969. [CrossRef]

124. Suemaru, K.; Cui, R.; Li, B.; Watanabe, S.; Okihara, K.; Hashimoto, K.; Yamada, H.; Araki, H. Topical application of royal jelly has a healing effect for 5-fluorouracil-induced experimental oral mucositis in hamsters. *Methods Find. Exp. Clin. Pharmacol.* **2008**, *30*, 103–106. [CrossRef] [PubMed]
125. Watanabe, S.; Suemaru, K.; Takechi, K.; Kaji, H.; Imai, K.; Araki, H. Oral mucosal adhesive films containing royal jelly accelerate recovery from 5-fluorouracil-induced oral mucositis. *J. Pharmacol. Sci.* **2013**, *121*, 110–118. [CrossRef]
126. Erdem, O.; Güngörmüş, Z. The effect of royal jelly on oral mucositis in patients undergoing radiotherapy and chemotherapy. *Holist. Nurs. Pract.* **2014**, *28*, 242–246. [CrossRef]
127. Yamauchi, K.; Kogashiwa, Y.; Moro, Y.; Kohno, N. The effect of topical application of royal jelly on chemoradiotherapy-induced mucositis in head and neck cancer: A preliminary study. *Int. J. Otolaryngol.* **2014**, *2014*, 974967. [CrossRef]
128. Gomes, M.S.; Lins, R.D.A.U.; Langassner, S.M.Z.; da Silveira, E.J.D.; de Carvalho, T.G.; de Sousa Lopes, M.L.D.; de Souza Araujo, L.; de Medeiros, C.A.C.X.; de Carvalho Leitão, R.F.; Guerra, G.C.B.; et al. Anti-inflammatory and antioxidant activity of hydroethanolic extract of *Spondias mombin* leaf in an oral mucositis experimental model. *Arch. Oral Biol.* **2020**, *111*, 104664. [CrossRef]
129. Cabral, B.; Siqueira, E.M.S.; Bitencourt, M.A.O.; Lima, M.C.J.S.; Lima, A.K.; Ortmann, C.F.; Chaves, V.C.; Fernandes-Pedrosa, M.F.; Rocha, H.A.O.; Scortecci, K.C.; et al. Phytochemical study and anti-inflammatory and antioxidant potential of *Spondias mombin* leaves. *Rev. Bras. Farmacogn.* **2016**, *26*, 304–311. [CrossRef]
130. Nworu, C.S.; Akah, P.A.; Okoye, F.B.C.; Toukam, D.K.; Udeh, J.; Esimone, C.O. The leaf extract of *Spondias mombin* L. displays an anti-inflammatory effect and suppresses inducible formation of tumor necrosis factor-α and nitric oxide (NO). *J. Immunotoxicol.* **2011**, *8*, 10–16. [CrossRef]
131. Khan, N.; Mukhtar, H. Tea polyphenols for health promotion. *Life Sci.* **2007**, *81*, 519–533. [CrossRef]
132. Lin, J.K.; Liang, Y.C.; Lin-Shiau, S.Y. Cancer chemoprevention by tea polyphenols through mitotic signal transduction blockade. *Biochem. Pharmacol.* **1999**, *58*, 911–915. [CrossRef]
133. Moyers, S.B.; Kumar, N.B. Green Tea Polyphenols and Cancer Chemoprevention: Multiple Mechanisms and Endpoints for Phase II Trials. *Nutr. Rev.* **2004**, *62*, 204–211. [CrossRef] [PubMed]
134. Cardoso, R.R.; Neto, R.O.; dos Santos D'Almeida, C.T.; do Nascimento, T.P.; Pressete, C.G.; Azevedo, L.; Martino, H.S.D.; Cameron, L.C.; Ferreira, M.S.L.; de Barros, F.A.R. Kombuchas from green and black teas have different phenolic profile, which impacts their antioxidant capacities, antibacterial and antiproliferative activities. *Food Res. Int.* **2020**, *128*, 108782. [CrossRef] [PubMed]
135. Steinmann, J.; Buer, J.; Pietschmann, T.; Steinmann, E. Anti-infective properties of epigallocatechin-3-gallate (EGCG), a component of green tea. *Br. J. Pharmacol.* **2013**, *168*, 1059–1073. [CrossRef] [PubMed]
136. Musial, C.; Kuban-Jankowska, A.; Gorska-Ponikowska, M. Beneficial Properties of Green Tea Catechins. *Int. J. Mol. Sci.* **2020**, *21*, 1744. [CrossRef] [PubMed]
137. Okabe, S.; Ochiai, Y.; Aida, M.; Park, K.; Kim, S.J.; Nomura, T.; Suganuma, M.; Fujiki, H. Mechanistic aspects of green tea as a cancer preventive: Effect of components on human stomach cancer cell lines. *Jpn. J. Cancer Res.* **1999**, *90*, 733–739. [CrossRef]
138. Pervin, M.; Unno, K.; Takagaki, A.; Isemura, M.; Nakamura, Y. Function of green tea catechins in the brain: Epigallocatechin gallate and its metabolites. *Int. J. Mol. Sci.* **2019**, *20*, 3630. [CrossRef]
139. Carulli, G.; Rocco, M.; Panichi, A.; Chios, C.F.; Ciurli, E.; Mannucci, C.; Sordi, E.; Caracciolo, F.; Papineschi, F.; Benedetti, E.; et al. Treatment of oral mucositis in hematologic patients undergoing autologous or allogeneic transplantation of peripheral blood stem cells: A prospective, randomized study with a mouthwash containing *camelia sinensis* leaf extract. *Hematol. Rep.* **2013**, *5*, 21–25. [CrossRef]
140. Cabrera-Jaime, S.; Martinez, C.; Ferro-Garcia, T.; Giner-Boya, P.; Icart-Isern, T.; Estrada-Masllorens, J.M.; Fernandez-Ortega, P. Efficacy of *Plantago major*, chlorhexidine 0.12% and sodium bicarbonate 5% solution in the treatment of oral mucositis in cancer patients with solid tumour: A feasibility randomised triple-blind phase III clinical trial. *Eur. J. Oncol. Nurs.* **2018**, *32*, 40–47. [CrossRef]
141. Soltani, G.M.; Hemati, S.; Sarvizadeh, M.; Kamalinejad, M.; Tafazoli, V.; Latifi, S.A. Efficacy of the *Plantago major* L. syrup on radiation induced oral mucositis in head and neck cancer patients: A randomized, double blind, placebo-controlled clinical trial. *Complement. Ther. Med.* **2020**, *51*, 102397, (Iranian Registry of Clinical Trials IRCT: 20190312043027N1). [CrossRef]
142. Yazdian, M.A.; Khodadoost, M.; Gheisari, M.; Kamalinejad, M.; Ehsani, A.; Barikbin, B. A Hypothesis on the Possible Potential of *Plantago major* in the Treatment of Urticaria. *Galen Med. J.* **2014**, *3*, 123–126. [CrossRef]
143. Ringbom, T.; Segura, L.; Noreen, Y.; Perera, P.; Bohlin, L. Ursolic acid from *Plantago major*, a selective inhibitor of cyclooxygenase-2 catalyzed prostaglandin biosynthesis. *J. Nat. Prod.* **1998**, *61*, 1212–1215. [CrossRef] [PubMed]
144. Stenholm, A.; Göransson, U.; Bohlin, L. Bioassay-guided supercritical fluid extraction of cyclooxygenase-2 inhibiting substances in *Plantago major* L. *Phytochem. Anal.* **2013**, *24*, 176–183. [CrossRef] [PubMed]
145. Pourmorad, F.; Hosseinimehr, S.J.; Shahabimajd, N. Antioxidant activity of phenol and flavonoid contents of some Iranian medicinal plants. *Afr. J. Adv. Biotechnol.* **2005**, *5*, 1684–5315.
146. Samuelsen, A.B.; Paulsen, B.S.; Wold, J.K.; Otsuka, H.; Yamada, H.; Espevik, T. Isolation and partial characterization of biologically active polysaccharides from *Plantago major* L. *Phytother. Res.* **1995**, *9*, 211–218. [CrossRef]
147. Baechler, B.J.; Nita, F.; Jones, L.; Frestedt, J.L. A novel liquid multi-phytonutrient supplement demonstrates DNA-protective effects. *Plant Foods Hum. Nutr.* **2009**, *64*, 81–85. [CrossRef]

148. Langmead, L.; Makins, R.J.; Rampton, D.S. Anti-inflammatory effects of *Aloe vera* gel in human colorectal mucosa in vitro. *Aliment. Pharmacol. Ther.* **2004**, *19*, 521–527. [CrossRef]
149. Heggers, J.; Pineless, G.; Robson, M. Dermaide *Aloe vera* gel-comparison of the anti-microbial effects. *J. Am. Med. Inform. Assoc.* **1979**, *41*, 293–294.
150. Yagi, A.; Kabash, A.; Okamura, N.; Haraguchi, H.; Moustafa, S.M.; Khalifa, T.I. Antioxidant, free radical scavenging and anti-inflammatory effects of aloesin derivatives in *Aloe vera*. *Planta Med.* **2002**, *68*, 957–960. [CrossRef]
151. Hu, Y.; Xu, J.; Hu, Q. Evaluation of antioxidant potential of *Aloe vera* (Aloe barbadensis miller) extracts. *J. Agric. Food Chem.* **2003**, *51*, 7788–7791. [CrossRef]
152. Heggie, S.; Bryant, G.P.; Tripcony, L.; Keller, J.; Rose, P.; Glendenning, M.; Heath, J. A Phase III study on the efficacy of topical *Aloe vera* gel on irradiated breast tissue. *Cancer Nurs.* **2002**, *25*, 442–451. [CrossRef] [PubMed]
153. Davis, R.H.; Leitner, M.G.; Russo, J.M.; Byrne, M.E. Wound healing. Oral and topical activity of *Aloe vera*. *J. Am. Podiatr. Med. Assoc.* **1989**, *79*, 559–562. [PubMed]
154. Davis, R.H.; Leitner, M.G.; Russo, J.M.; Byrne, M.E. Anti-inflammatory activity of *Aloe vera* against a spectrum of irritants. *J. Am. Podiatr. Med. Assoc.* **1989**, *79*, 263–276. [CrossRef] [PubMed]
155. Meadows, T.P. Aloe as a humectant in new skin preparations. *Cosmet. Toilet.* **1980**, *95*, 51–56.
156. Davis, R.H.; Donato, J.J.; Hartman, G.M.; Haas, R.C. Anti-inflammatory and wound healing activity of a growth substance in *Aloe vera*. *J. Am. Podiatr. Med. Assoc.* **1994**, *84*, 77–81. [PubMed]
157. Vázquez, B.; Avila, G.; Segura, D.; Escalante, B. Anti-inflammatory activity of extracts from *Aloe vera* gel. *J. Ethnopharmacol.* **1996**, *55*, 69–75. [CrossRef]
158. Wei, A.; Shibamoto, T. Antioxidant/lipoxygenase inhibitory activities and chemical compositions of selected essential oils. *J. Agric. Food Chem.* **2010**, *58*, 7218–7225. [CrossRef]
159. Su, C.K.; Mehta, V.; Ravikumar, L.; Shah, R.; Pinto, H.; Halpern, J.; Koong, A.; Goffinet, D.; Le, Q.-T. Phase II double-blind randomized study comparing oral *Aloe vera* versus placebo to prevent radiation-related mucositis in patients with head-and-neck neoplasms. *Int. J. Radiat. Oncol. Biol. Phys.* **2004**, *60*, 171–177. [CrossRef]
160. Mansouri, P.; Haghighi, M.; Beheshtipour, N.; Ramzi, M. The effect of *Aloe vera* solution on Chemotherapy-induced stomatitis in clients with lymphoma and leukemia: A randomized controlled Clinical Trial. *Int. J. Community Based Nurs. Midwifery* **2016**, *4*, 119–126, (Iranian Registry of Clinical Trials: 2014092819318N1).
161. Karbasizade, S.; Ghorbani, F.; Ghasemi Darestani, N.; Mansouri-Tehrani, M.M.; Kazemi, A.H. Comparison of therapeutic effects of statins and *Aloe vera* mouthwash on chemotherapy induced oral mucositis. *Int. J. Physiol. Pathophysiol. Pharmacol.* **2021**, *13*, 110–116.
162. Alkhouli, M.; Laflouf, M.; Alhaddad, M. Efficacy of *Aloe-vera* use for prevention of chemotherapy-induced oral mucositis in children with acute lymphoblastic leukemia: A randomized controlled clinical trial. *Compr. Child Adolesc. Nurs.* **2021**, *44*, 49–62, (Australian New Zealand Clinical Trials Registry: 12618001931268). [CrossRef] [PubMed]
163. Sahebjamee, M.; Mansourian, A.; Hajimirzamohammad, M.; Zadeh, M.T.; Bekhradi, R.; Kazemian, A.; Manifar, S.; Ashnagar, S.; Doroudgar, K. Comparative efficacy of *Aloe vera* and benzydamine mouthwashes on radiation-induced oral mucositis: A triple-blind, randomised, controlled clinical trial. *Oral Health Prev. Dent.* **2015**, *13*, 309–315, (Iranian Registry of Clinical Trials: 2012072410377N1). [PubMed]
164. Kocaadam, B.; Şanlier, N. Curcumin, an active component of turmeric (*Curcuma longa*), and its effects on health. *Crit. Rev. Food Sci. Nutr.* **2017**, *57*, 2889–2895. [CrossRef] [PubMed]
165. Gupta, S.C.; Sung, B.; Kim, J.H.; Prasad, S.; Li, S.; Aggarwal, B.B. Multitargeting by turmeric, the golden spice: From kitchen to clinic. *Mol. Nutr. Food Res.* **2013**, *57*, 1510–1528. [CrossRef]
166. Deogade, S.C.; Ghate, S. Curcumin: Therapeutic applications in systemic and oral health. *Int. J. Biol. Pharm. Res.* **2015**, *6*, 281–290.
167. Li, S. Chemical composition and product quality control of turmeric (*Curcuma longa* L.). *Pharm. Crops* **2011**, *5*, 28–54. [CrossRef]
168. Shehzad, A.; Lee, Y. Curcumin: Multiple molecular targets mediate multiple pharmacological actions: A review. *Drugs Future* **2010**, *35*, 113. [CrossRef]
169. Srinivasan, M. Effect of curcumin on blood sugar as seen in a diabetic subject. *Indian J. Med. Sci.* **1972**, *26*, 269–270.
170. Srimal, R.C.; Dhawan, B.N. Pharmacology of diferuloyl methane (curcumin), a non-steroidal anti-inflammatory agent. *J. Pharm. Pharmacol.* **1973**, *25*, 447–452. [CrossRef]
171. Sharma, O.P. Antioxidant activity of curcumin and related compounds. *Biochem. Pharmacol.* **1976**, *25*, 1811–1812. [CrossRef]
172. Aggarwal, B.B.; Harikumar, K.B. Potential therapeutic effects of curcumin, the anti-inflammatory agent, against neurodegenerative, cardiovascular, pulmonary, metabolic, autoimmune and neoplastic diseases. *Int. J. Biochem. Cell Biol.* **2009**, *41*, 40–59. [CrossRef] [PubMed]
173. Gandhy, S.U.; Kim, K.; Larsen, L.; Rosengren, R.J.; Safe, S. Curcumin and synthetic analogs induce reactive oxygen species and decreases specificity protein (Sp) transcription factors by targeting microRNAs. *BMC Cancer* **2012**, *12*, 564. [CrossRef] [PubMed]
174. Kim, S.G.; Veena, M.S.; Basak, S.K.; Han, E.; Tajima, T.; Gjertson, D.W.; Starr, J.; Eidelman, O.; Pollard, H.B.; Srivastava, M.; et al. Curcumin treatment suppresses IKKβ kinase activity of salivary cells of patients with head and neck cancer: A pilot study. *Clin. Cancer Res.* **2011**, *17*, 5953–5961. [CrossRef] [PubMed]

175. Das, L.; Vinayak, M. Curcumin attenuates carcinogenesis by down regulating proinflammatory cytokine interleukin-1 (IL-1α and IL-1β) via modulation of AP-1 and NF-IL6 in lymphoma bearing mice. *Int. Immunopharmacol.* **2014**, *20*, 141–147. [CrossRef] [PubMed]
176. Rujirachotiwat, A.; Suttamanatwong, S. Curcumin upregulates transforming growth factor-β1, its receptors, and vascular endothelial growth factor expressions in an in vitro human gingival fibroblast wound healing model. *BMC Oral Health* **2021**, *21*, 535. [CrossRef]
177. Mani, H.; Sidhu, G.S.; Kumari, R.; Gaddipati, J.P.; Seth, P.; Maheshwari, R.K. Curcumin differentially regulates TGF-beta1, its receptors and nitric oxide synthase during impaired wound healing. *Biofactors* **2002**, *16*, 29–43. [CrossRef]
178. Panchatcharam, M.; Miriyala, S.; Gayathri, V.S.; Suguna, L. Curcumin improves wound healing by modulating collagen and decreasing reactive oxygen species. *Mol. Cell. Biochem.* **2006**, *290*, 87–96. [CrossRef]
179. Elad, S.; Meidan, I.; Sellam, G.; Simaan, S.; Zeevi, I.; Waldman, E.; Weintraub, M.; Revel-Vilk, S. Topical curcumin for the prevention of oral mucositis in pediatric patients: Case series. *Altern. Ther. Health Med.* **2013**, *19*, 21–24.
180. Patil, K.; Guledgud, M.V.; Kulkarni, P.K.; Keshari, D.; Tayal, S. Use of Curcumin mouthrinse in radio-chemotherapy induced Oral Mucositis patients: A pilot study. *J. Clin. Diagn. Res.* **2015**, *9*, ZC59–ZC62. [CrossRef]
181. Delavarian, Z.; Pakfetrat, A.; Ghazi, A.; Jaafari, M.R.; Homaei Shandiz, F.; Dalirsani, Z.; Mohammadpour, A.H.; Rahimi, H.R. Oral administration of nanomicelle curcumin in the prevention of radiotherapy-induced mucositis in head and neck cancers. *Spec. Care Dent.* **2019**, *39*, 166–172. [CrossRef]
182. Ahmed, K.M. The effect of olive leaf extract in decreasing the expression of two pro-inflammatory cytokines in patients receiving chemotherapy for cancer. A randomized clinical trial. *Saudi Dent. J.* **2013**, *25*, 141–147. [CrossRef] [PubMed]
183. Lee-Huang, S.; Huang, P.L.; Zhang, D.; Lee, J.W.; Bao, J.; Sun, Y.; Chang, Y.T.; Zhang, J.; Huang, P.L. Discovery of small-molecule HIV-1 fusion and integrase inhibitors oleuropein and hydroxytyrosol: Part I. fusion inhibition. *Biochem. Biophys. Res. Commun.* **2007**, *354*, 872–878. [CrossRef] [PubMed]
184. Markin, D.; Duek, L.; Berdicevsky, I. In vitro antimicrobial activity of olive leaves. *Mycoses* **2003**, *46*, 132–136. [CrossRef] [PubMed]
185. Coni, E.; Di Benedetto, R.; Di Pasquale, M.; Masella, R.; Modesti, D.; Mattei, R.; Carlini, E.A. Protective effect of oleuropein, an olive oil biophenol, on low density lipoprotein oxidizability in rabbits. *Lipids* **2000**, *35*, 45–54. [CrossRef] [PubMed]
186. Singh, I.; Mok, M.; Christensen, A.M.; Turner, A.H.; Hawley, J.A. The effects of polyphenols in olive leaves on platelet function. *Nutr. Metab. Cardiovasc. Dis.* **2008**, *18*, 127–132. [CrossRef] [PubMed]
187. Campolo, M.; Di Paola, R.; Impellizzeri, D.; Crupi, R.; Morittu, V.M.; Procopio, A.; Perri, E.; Britti, D.; Peli, A.; Esposito, E.; et al. Effects of a polyphenol present in olive oil, oleuropein aglycone, in a murine model of intestinal ischemia/reperfusion injury. *J. Leukoc. Biol.* **2013**, *93*, 277–287. [CrossRef] [PubMed]
188. Silva, S.; Gomes, L.; Leitão, F.; Coelho, A.; Boas, L.V. Phenolic compounds and antioxidant activity of *Olea europaea* L. fruits and leaves. *Food Sci. Technol. Int.* **2006**, *12*, 385–395. [CrossRef]
189. Koca, U.; Süntar, I.; Akkol, E.K.; Yilmazer, D.; Alper, M. Wound repair potential of *Olea europaea* L. leaf extracts revealed by in vivo experimental models and comparative evaluation of the extracts' antioxidant activity. *J. Med. Food* **2011**, *14*, 140–146. [CrossRef]
190. Mehraein, F.; Sarbishegi, M.; Aslani, A. Evaluation of effect of oleuropein on skin wound healing in aged male BALB/c mice. *Cell J.* **2014**, *16*, 25–30.
191. Ahmed, K. Olive leaf extract as a new topical management for oral mucositis following chemotherapy: A microbiological examination, experimental animal study and clinical trial. *Pharm. Anal. Acta* **2013**, *4*, 4–9.
192. Kondo, K.; Shiba, M.; Nakamura, R.; Morota, T.; Shoyama, Y. Constituent properties of licorices derived from *Glycyrrhiza uralensis*, *G. glabra*, or *G. inflata* identified by genetic information. *Biol. Pharm. Bull.* **2007**, *30*, 1271–1277. [CrossRef] [PubMed]
193. Gupta, V.K.; Fatima, A.; Faridi, U.; Negi, A.S.; Shanker, K.; Kumar, J.K.; Rahuja, N.; Luqman, S.; Sisodia, B.S.; Saikia, D.; et al. Antimicrobial potential of *Glycyrrhiza glabra* roots. *J. Ethnopharmacol.* **2008**, *116*, 377–380. [CrossRef] [PubMed]
194. Saxena, R.C.; Garg, K.C.; Bhargava, K.P.; Gupta, G.P. A clinical trial of glycyrrhitinic acid in allergic conditions of the eye. *J. Indian Med. Prof.* **1965**, *12*, 5487–5490. [PubMed]
195. Lv, H.; Yang, H.; Wang, Z.; Feng, H.; Deng, X.; Cheng, G.; Ci, X. Nrf2 signaling and autophagy are complementary in protecting lipopolysaccharide/d-galactosamine-induced acute liver injury by licochalcone A. *Cell Death Dis.* **2019**, *10*, 313. [CrossRef] [PubMed]
196. Feng, Y.; Mei, L.; Wang, M.; Huang, Q.; Huang, R. Anti-inflammatory and pro-apoptotic effects of 18beta-glycyrrhetinic acid in vitro and in vivo models of rheumatoid arthritis. *Front. Pharmacol.* **2021**, *12*, 681525. [CrossRef] [PubMed]
197. Meng, X.; Zhang, X.; Su, X.; Liu, X.; Ren, K.; Ning, C.; Zhang, Q.; Zhang, S. Daphnes Cortex and its licorice-processed products suppress inflammation via the TLR4/NF-κB/NLRP3 signaling pathway and regulation of the metabolic profile in the treatment of rheumatoid arthritis. *J. Ethnopharmacol.* **2022**, *283*, 114657. [CrossRef]
198. Sadinpour, A.; Seyedi, Z.S.; Arabdolatabadi, A.; Razavi, Y.; Ajdary, M. The synergistic effect of Paeonia spp and *Glycyrrhiza glabra* on polycystic ovary induced in mice. *Pak. J. Pharm. Sci.* **2020**, *33*, 1665–1670.
199. Hsu, Y.W.; Chen, H.Y.; Chiang, Y.F.; Chang, L.C.; Lin, P.H.; Hsia, S.M. The effects of isoliquiritigenin on *endometriosis* in vivo and in vitro study. *Phytomedicine* **2020**, *77*, 153214. [CrossRef]
200. Zadeh, J.B.; Kor, Z.M.; Goftar, M.K. Licorice (*Glycyrrhiza glabra* Linn) as a valuable medicinal plant. *Int. J. Adv. Biol. Biomed. Res.* **2013**, *1*, 1281–1288.

201. Pastorino, G.; Cornara, L.; Soares, S.; Rodrigues, F.; Oliveira, M. Liquorice (*Glycyrrhiza glabra*): A phytochemical and pharmacological review. *Phytother. Res.* **2018**, *32*, 2323–2339. [CrossRef]
202. Vispute, S.; Khopade, A. *Glycyrrhiza glabra* Linn.-"Klitaka": A Review. *Int. J. Pharma Bio Sci.* **2011**, *2*, 42–51.
203. Damle, M. *Glycyrrhiza glabra* (Liquorice)-a potent medicinal herbInt. *J. Herb. Med.* **2014**, *2*, 132–136.
204. Kaur, R.; Kaur, H.; Dhindsa, A.S. *Glycyrrhiza glabra*: A phytopharmacological review. *Int. J. Pharm. Sci. Res.* **2013**, *4*, 2470.
205. Tewari, D.; Mocan, A.; Parvanov, E.D.; Sah, A.N.; Nabavi, S.M.; Huminiecki, L.; Ma, Z.F.; Lee, Y.Y.; Horbańczuk, J.O.; Atanasov, A.G. Ethnopharmacological approaches for therapy of jaundice: Part II. Highly used plant species from Acanthaceae, Euphorbiaceae, Asteraceae, Combretaceae, and Fabaceae families. *Front. Pharmacol.* **2017**, *8*, 519. [CrossRef]
206. El-Saber Batiha, G.; Magdy Beshbishy, A.; El-Mleeh, A.; Abdel-Daim, M.M.; Prasad Devkota, H. Traditional uses, bioactive chemical constituents, and pharmacological and toxicological activities of *Glycyrrhiza glabra* L. (Fabaceae). *Biomolecules* **2020**, *10*, 352. [CrossRef]
207. Najafi, S.; Koujan, S.E.; Manifar, S.; Kharazifard, M.J.; Kidi, S.; Hajheidary, S. Preventive effect of *Glycyrrhiza glabra* extract on oral mucositis in patients under head and neck radiotherapy: A randomized clinical trial. *J. Dent.* **2017**, *14*, 267–274.
208. Lee, S.H.; Bae, I.H.; Choi, H.; Choi, H.W.; Oh, S.; Marinho, P.A.; Min, D.J.; Kim, D.Y.; Lee, T.R.; Lee, C.S.; et al. Ameliorating effect of dipotassium glycyrrhizinate on an IL-4- and IL-13-induced atopic dermatitis-like skin-equivalent model. *Arch. Dermatol. Res.* **2019**, *311*, 131–140. [CrossRef]
209. Shim, J.Y.; Yim, S.B.; Chung, J.H.; Hong, K.S. Antiplaque and antigingivitis effects of a mouthrinse containing cetylpyridinium chloride, triclosan and dipotassium glycyrrhizinate. *J. Periodontal Implant Sci.* **2012**, *42*, 33–38. [CrossRef]
210. Vitali, R.; Palone, F.; Cucchiara, S.; Negroni, A.; Cavone, L.; Costanzo, M.; Aloi, M.; Dilillo, A.; Stronati, L. Dipotassium glycyrrhizate inhibits hmgb1-dependent inflammation and ameliorates colitis in mice. *PLoS ONE* **2013**, *8*, e66527. [CrossRef]
211. Okimasu, E.; Moromizato, Y.; Watanabe, S.; Sasaki, J.; Shiraishi, N.; Morimoto, Y.M.; Miyahara, M.; Utsumi, K. Inhibition of phospholipase A2 and platelet aggregation by glycyrrhizin, an antiinflammation drug. *Acta Med. Okayama* **1983**, *37*, 385–391.
212. Leite, C.D.S.; Pires, O.C.; Tenis, D.G.; Ziegler, J.V.N.; Priolli, D.G.; Rocha, T. Effects of dipotassium glycyrrhizinate on wound healing. *Acta Cir. Bras.* **2021**, *36*, e360801. [CrossRef] [PubMed]
213. Harwansh, R.; Patra, K. Pharmacological studies on *Glycyrrhiza glabra*. *Pharmacologyonline* **2011**, *2*, 1032–1038.
214. Bhattacharjee, S.; Bhattacharjee, A.; Majumder, S.; Majumdar, S.B.; Majumdar, S. Glycyrrhizic acid suppresses Cox-2-mediated anti-inflammatory responses during *Leishmania donovani* infection. *J. Antimicrob. Chemother.* **2012**, *67*, 1905–1914. [CrossRef] [PubMed]
215. Richard, S.A. Exploring the pivotal immunomodulatory and anti-inflammatory potentials of glycyrrhizic and glycyrrhetinic acids. *Mediat. Inflamm.* **2021**, *2021*, 6699560. [CrossRef] [PubMed]
216. Okamoto, T. The protective effect of glycyrrhizin on anti-Fas antibody-induced hepatitis in mice. *Eur. J. Pharmacol.* **2000**, *387*, 229–232. [CrossRef]
217. Abe, N.; Ebina, T.; Ishida, N. Interferon induction by glycyrrhizin and glycyrrhetinic acid in mice. *Microbiol. Immunol.* **1982**, *26*, 535–539. [CrossRef]
218. Pakravan, F.; Salehabad, N.H.; Karimi, F.; Isfahani, M. N, Comparative study of the effect of licorice muco-adhesive film on radiotherapy induced oral mucositis, a randomized controlled clinical trial. *Gulf J. Oncol.* **2021**, *1*, 42–47.
219. Sattari, A.; Shariati, A.; Maram, N.; Ehsanpour, A.; Maraghi, E. Comparative study of the effect of licorice root extract mouthwash and combined mouthwash on the incidence and severity of chemotherapy-induced mucositis symptoms in colon cancer patients admitted to intensive care units. *Jundishapur J. Health Sci.* 2019, *in press*. [CrossRef]
220. Martins, M.D.; Marques, M.M.; Bussadori, S.K.; Martins, M.A.; Pavesi, V.C.; Mesquita-Ferrari, R.A.; Fernandes, K.P. Comparative analysis between *Chamomilla recutita* and corticosteroids on wound healing. An in vitro and in vivo study. *Phytother. Res.* **2009**, *23*, 274–278. [CrossRef]
221. Braga, F.T.M.M.; Santos, A.C.F.; Bueno, P.C.P.; Silveira, R.C.C.P.; Santos, C.B.; Bastos, J.K.; Carvalho, E.C. Use of *Chamomilla recutita* in the prevention and treatment of oral mucositis in patients undergoing hematopoietic stem cell transplantation: A randomized, controlled, phase ii clinical trial. *Cancer Nurs.* **2015**, *38*, 322–329. [CrossRef]
222. Srivastava, J.K.; Shankar, E.; Gupta, S. Chamomile: A herbal medicine of the past with bright future. *Mol. Med. Rep.* **2010**, *3*, 895–901. [PubMed]
223. Srivastava, J.K.; Pandey, M.; Gupta, S. Chamomile, a novel and selective COX-2 inhibitor with anti-inflammatory activity. *Life Sci.* **2009**, *85*, 663–669. [CrossRef] [PubMed]
224. Smolinski, A.T.; Pestka, J.J. Modulation of lipopolysaccharide-induced proinflammatory cytokine production in vitro and in vivo by the herbal constituents apigenin (chamomile), ginsenoside Rb(1) (ginseng) and parthenolide (feverfew). *Food Chem. Toxicol.* **2003**, *41*, 1381–1390. [CrossRef]
225. dos Reis, P.E.D.; Ciol, M.A.; de Melo, N.S.; de Souza Figueiredo, P.T.; Leite, A.F.; de Melo Manzi, N. Chamomile infusion cryotherapy to prevent oral mucositis induced by chemotherapy: A pilot study. *Support. Care Cancer* **2016**, *24*, 4393–4398. [CrossRef] [PubMed]
226. Shabanloei, R.; Ahmadi, F.; Vaez, J.; Ansarin, K.; Hajizadeh, E.; Javadzadeh, Y.; Dolathkhah, R.; Gholchin, M. Alloporinol, chamomile and normal saline mouthwashes for the prevention of chemotherapy-induced stomatitis. *J. Clin. Diagn. Res.* **2009**, *3*, 1537–1542.

227. Muley, B.; Khadabadi, S.S.; Banarase, N. Phytochemical constituents and pharmacological activities of *Calendula officinalis* Linn (Asteraceae): A review. *Trop. J. Pharm. Res.* **2009**, *8*, 455–465. [CrossRef]
228. Hadfield, R.A.; Vlahovic, T.C.; Khan, M.T. The use of marigold therapy for podiatric skin conditions. *Foot Ankle J.* **2008**, *1*, 1–8. [CrossRef]
229. Tanideh, N.; Tavakoli, P.; Saghiri, M.A.; Garcia-Godoy, F.; Amanat, D.; Tadbir, A.A.; Samani, S.M.; Tamadon, A. Healing acceleration in hamsters of oral mucositis induced by 5-fluorouracil with topical *Calendula officinalis*. *Oral Surg. Oral Med. Oral Pathol. Oral Radiol.* **2013**, *115*, 332–338. [CrossRef]
230. Fronza, M.; Heinzmann, B.; Hamburger, M.; Laufer, S.; Merfort, I. Determination of the wound healing effect of Calendula extracts using the scratch assay with 3T3 fibroblasts. *J. Ethnopharmacol.* **2009**, *126*, 463–467. [CrossRef]
231. Dzubak, P.; Hajduch, M.; Vydra, D.; Hustova, A.; Kvasnica, M.; Biedermann, D.; Markova, L.; Urban, M.; Sarek, J. Pharmacological activities of natural triterpenoids and their therapeutic implications. *Nat. Prod. Rep.* **2006**, *23*, 394–411. [CrossRef]
232. Givol, O.; Kornhaber, R.; Visentin, D.; Cleary, M.; Haik, J.; Harats, M. A systematic review of *Calendula officinalis* extract for wound healing. *Wound Repair Regen.* **2019**, *27*, 548–561. [CrossRef] [PubMed]
233. Middleton, E., Jr.; Kandaswami, C.; Theoharides, T.C. The effects of plant flavonoids on mammalian cells: Implications for inflammation, heart disease, and cancer. *Pharmacol. Rev.* **2000**, *52*, 673–751. [PubMed]
234. Babaee, N.; Moslemi, D.; Khalilpour, M.; Vejdani, F.; Moghadamnia, Y.; Bijani, A.; Baradaran, M.; Kazemi, M.T.; Khalilpour, A.; Pouramir, M.; et al. Antioxidant capacity of *Calendula officinalis* flowers extract and prevention of radiation induced oropharyngeal mucositis in patients with head and neck cancers: A randomized controlled clinical study. *Daru* **2013**, *21*, 18, (Iranian Registry of Clinical Trials: 201106076734N1). [CrossRef] [PubMed]
235. Maddocks-Jennings, W.; Wilkinson, J.M.; Cavanagh, H.M.; Shillington, D. Evaluating the effects of the essential oils *Leptospermum scoparium* (manuka) and *Kunzea ericoides* (kanuka) on radiotherapy induced mucositis: A randomized, placebo controlled feasibility study. *Eur. J. Oncol. Nurs.* **2009**, *13*, 87–93. [CrossRef]
236. You, W.C.; Hsieh, C.C.; Huang, J.T. Effect of extracts from indigowood root (*Isatis indigotica* Fort.) on immune responses in radiation-induced mucositis. *J. Altern. Complement. Med.* **2009**, *15*, 771–778. [CrossRef]
237. Loo, W.T.; Jin, L.J.; Chow, L.W.; Cheung, M.N.; Wang, M. *Rhodiola algida* improves chemotherapy-induced oral mucositis in breast cancer patients. *Expert Opin. Investig. Drugs* **2010**, *19*, S91–S100. [CrossRef] [PubMed]
238. Mutluay Yayla, E.; Izgu, N.; Ozdemir, L.; Aslan Erdem, S.; Kartal, M. Sage tea-thyme-peppermint hydrosol oral rinse reduces chemotherapy-induced oral mucositis: A randomized controlled pilot study. *Complement. Ther. Med.* **2016**, *27*, 58–64. [CrossRef] [PubMed]
239. Tanideh, N.; Badie, A.D.; Habibagahi, R.; Koohi-Hosseinabadi, O.; Haghnegahdar, S.; Andisheh-tadbir, A. Effect of topical 2% eucalyptus extract on 5-fu-induced oral mucositis in male golden hamsters. *Braz. Dent. J.* **2020**, *31*, 310–318. [CrossRef]
240. Koohi-Hosseinabadi, O.; Andisheh-Tadbir, A.; Bahadori, P.; Sepehrimanesh, M.; Mardani, M.; Tanideh, N. Comparison of the therapeutic effects of the dietary and topical forms of *Zizyphus jujuba* extract on oral mucositis induced by 5-fluorouracil: A golden hamster model. *J. Clin. Exp. Dent.* **2015**, *7*, e304–e309. [CrossRef]
241. Aghamohammadi, A.; Moslemi, D.; Akbari, J.; Ghasemi, A.; Azadbakht, M.; Asgharpour, A.; Hosseinimehr, S.J. The effectiveness of Zataria extract mouthwash for the management of radiation-induced oral mucositis in patients: A randomized placebo-controlled double-blind study. *Clin. Oral Investig.* **2018**, *22*, 2263–2272. [CrossRef]
242. Soares, A.D.S.; Wanzeler, A.M.V.; Cavalcante, G.H.S.; Barros, E.; Carneiro, R.C.M.; Tuji, F.M. Therapeutic effects of andiroba (*Carapa guianensis* Aubl) oil, compared to low power laser, on oral mucositis in children underwent chemotherapy: A clinical study. *J. Ethnopharmacol.* **2021**, *264*, 113365. [CrossRef] [PubMed]
243. Wanzeler, A.M.V.; Júnior, S.M.A.; Gomes, J.T.; Gouveia, E.H.H.; Henriques, H.Y.B.; Chaves, R.H.; Soares, B.M.; Salgado, H.L.C.; Santos, A.S.; Tuji, F.M. Therapeutic effect of andiroba oil (*Carapa guianensis* Aubl.) against oral mucositis: An experimental study in golden Syrian hamsters. *Clin. Oral Investig.* **2018**, *22*, 2069–2079. [CrossRef] [PubMed]
244. Hasheminasab, F.S.; Hashemi, S.M.; Dehghan, A.; Sharififar, F.; Setayesh, M.; Sasanpour, P.; Tasbandi, M.; Raeiszadeh, M. Effects of a Plantago ovata-based herbal compound in prevention and treatment of oral mucositis in patients with breast cancer receiving chemotherapy: A double-blind, randomized, controlled crossover trial. *J. Integr. Med.* **2020**, *18*, 214–221, (Iranian Registry of Clinical Trials IRCT: 20180923041093N1). [CrossRef] [PubMed]
245. Bertoglio, J.C.; Folatre, I.; Bombardelli, E.; Riva, A.; Morazzoni, P.; Ronchi, M.; Petrangolini, G. Management of gastrointestinal mucositis due to cancer therapies in pediatric patients: Results of a case series with SAMITAL®. *Future Oncol.* **2012**, *8*, 1481–1486. [CrossRef] [PubMed]
246. Davarmanesh, M.; Miri, R.; Haghnegahdar, S.; Tadbir, A.A.; Tanideh, N.; Saghiri, M.A.; Garcia-Godoy, F.; Asatourian, A. Protective effect of bilberry extract as a pretreatment on induced oral mucositis in hamsters. *Oral Surg. Oral Med. Oral Pathol. Oral Radiol.* **2013**, *116*, 702–708. [CrossRef]
247. Morazzoni, P.; Petrangolini, G.; Bombardelli, E.; Ronchi, M.; Cabri, W.; Riva, A. SAMITAL®: A new botanical drug for the treatment of mucositis induced by oncological therapies. *Future Oncol.* **2013**, *9*, 1717–1725. [CrossRef]
248. Mardani, M.; Afra, S.M.; Tanideh, N.; Tadbir, A.A.; Modarresi, F.; Koohi-Hosseinabadi, O.; Iraji, A.; Sepehrimanesh, M. Hydroalcoholic extract of *Carum carvi* L. in oral mucositis: A clinical trial in male golden hamsters. *Oral Dis.* **2016**, *22*, 39–45. [CrossRef]

249. Tanideh, N.; Davarmanesh, M.; Andisheh-Tadbir, A.; Ranjbar, Z.; Mehriar, P.; Koohi-Hosseinabadi, O. Healing acceleration of oral mucositis induced by 5-fluorouracil with *Pistacia atlantica* (bene) essential oil in hamsters. *J. Oral Pathol. Med.* **2017**, *46*, 725–730. [CrossRef]
250. Tanideh, N.; Zareh, A.A.; Fani, M.M.; Mardani, M.; Farrokhi, F.; Talati, A.; Koohi Hosseinabadi, O.; Kamali, M. Evaluation of the effect of a topical gel form of *Pistacia atlantica* and *Trachyspermum ammi* on induced oral mucositis in male golden hamsters by bio-marker indices and stereological assessment. *J. Dent.* **2019**, *20*, 240–248.
251. Tanideh, N.; Namazi, F.; Andisheh Tadbir, A.; Ebrahimi, H.; Koohi-Hosseinabadi, O. Comparative assessment of the therapeutic effects of the topical and systemic forms of *Hypericum perforatum* extract on induced oral mucositis in golden hamsters. *Int. J. Oral Maxillofac. Surg.* **2014**, *43*, 1286–1292. [CrossRef]
252. Koohi-Hosseinabadi, O.; Ranjbar, Z.; Sepehrimanesh, M.; AndisheTadbir, A.; Poorbaghi, S.L.; Bahranifard, H.; Tanideh, N.; Koohi-Hosseinabadi, M.; Iraji, A. Biochemical, hematological, and pathological related healing effects of *Elaeagnus angustifolia* hydroalcoholic extract in 5-fluorouracil-induced oral mucositis in male golden hamster. *Environ. Sci. Pollut. Res. Int.* **2017**, *24*, 24447–24453. [CrossRef] [PubMed]
253. Kuduban, O.; Mazlumoglu, M.R.; Kuduban, S.D.; Erhan, E.; Cetin, N.; Kukula, O.; Yarali, O.; Cimen, F.K.; Cankaya, M. The effect of *Hippophae rhamnoides* extract on oral mucositis induced in rats with methotrexate. *J. Appl. Oral Sci.* **2016**, *24*, 423–430. [CrossRef] [PubMed]
254. Erhan, E.; Terzi, S.; Celiker, M.; Yarali, O.; Cankaya, M.; Cimen, F.K.; Malkoc, I.; Suleyman, B. Effect of *Hippophae rhamnoides* extract on oxidative oropharyngeal mucosal damage induced in rats using methotrexate. *Clin. Exp. Otorhinolaryngol.* **2017**, *10*, 181–187. [CrossRef] [PubMed]
255. Shahiwala, A. Applications of Polymers in Buccal Drug Delivery. In *Applications of Polymers in Drug Delivery*; Elsevier: Amsterdam, The Netherlands, 2021; pp. 43–76.
256. Sandri, G.; Ruggeri, M.; Rossi, S.; Bonferoni, M.C.; Vigani, B.; Ferrari, F. (Trans) buccal drug delivery. In *Nanotechnology for Oral Drug Delivery*; Academic Press: London, UK, 2020; pp. 225–250.
257. Moroz, E.; Matoori, S.; Leroux, J.C. Oral delivery of macromolecular drugs: Where we are after almost 100 years of attempts. *Adv. Drug Deliv. Rev.* **2016**, *101*, 108–121. [CrossRef]
258. Li, T.; Lalla, R.V.; Burgess, D.J. Enhanced drug loading of in situ forming gels for oral mucositis pain control. *Int. J. Pharm.* **2021**, *595*, 120225. [CrossRef]
259. Paderni, C.; Compilato, D.; Giannola, L.I.; Campisi, G. Oral local drug delivery and new perspectives in oral drug formulation. *Oral Surg. Oral Med. Oral Pathol. Oral Radiol.* **2012**, *114*, e25–e34. [CrossRef]
260. Tedesco, M.P.; Monaco-Lourenco, C.A.; Carvalho, R.A. Characterization of oral disintegrating film of peanut skin extract-Potential route for buccal delivery of phenolic compounds. *Int. J. Biol. Macromol.* **2017**, *97*, 418–425. [CrossRef]
261. Aksungur, P.; Sungur, A.; Unal, S.; Iskit, A.B.; Squier, C.A.; Senel, S. Chitosan delivery systems for the treatment of oral mucositis: In Vitro and in vivo studies. *J. Control. Release* **2004**, *98*, 269–279. [CrossRef]
262. Fonseca-Santos, B.; Chorilli, M. An overview of polymeric dosage forms in buccal drug delivery: State of art, design of formulations and their in vivo performance evaluation. *Mater. Sci. Eng. C Mater. Biol. Appl.* **2018**, *86*, 129–143. [CrossRef]
263. Borges, J.G.; De Carvalho, R.A. Orally disintegrating films containing propolis: Properties and release profile. *J. Pharm. Sci.* **2015**, *104*, 1431–1439. [CrossRef]
264. Agrawal, U.; Sharma, R.; Gupta, M.; Vyas, S.P. Is nanotechnology a boon for oral drug delivery? *Drug Discov. Today* **2014**, *19*, 1530–1546. [CrossRef] [PubMed]
265. Morantes, S.J.; Buitrago, D.M.; Ibla, J.F.; García, Y.M.; Lafaurie, G.I.; Parraga, J.E. Composites of hydrogels and nanoparticles. In *Biopolymer-Based Composites*; Woodhead Publishing: Duxford, UK, 2017; pp. 107–138.
266. Allaker, R.P.; Yuan, Z. Nanoparticles and the control of oral biofilms. In *Nanobiomaterials in Clinical Dentistry*; Elsevier: Amsterdam, The Netherlands, 2019; pp. 243–275.
267. Campos, J.C.; Cunha, D.; Ferreira, D.C.; Reis, S.; Costa, P.J. Oromucosal precursors of in loco hydrogels for wound-dressing and drug delivery in oral mucositis: Retain, resist, and release. *Mater. Sci. Eng. C Mater. Biol. Appl.* **2021**, *118*, 111413. [CrossRef] [PubMed]
268. Ryu, J.H.; Choi, J.S.; Park, E.; Eom, M.R.; Jo, S.; Lee, M.S.; Kwon, S.K.; Lee, H. Chitosan oral patches inspired by mussel adhesion. *J. Control. Release* **2020**, *317*, 57–66. [CrossRef] [PubMed]
269. Takeuchi, I.; Kamiki, Y.; Makino, K. Therapeutic efficacy of rebamipide-loaded PLGA nanoparticles coated with chitosan in a mouse model for oral mucositis induced by cancer chemotherapy. *Colloids Surf. B* **2018**, *167*, 468–473. [CrossRef]

MDPI
St. Alban-Anlage 66
4052 Basel
Switzerland
Tel. +41 61 683 77 34
Fax +41 61 302 89 18
www.mdpi.com

International Journal of Molecular Sciences Editorial Office
E-mail: ijms@mdpi.com
www.mdpi.com/journal/ijms

www.ingramcontent.com/pod-product-compliance
Lightning Source LLC
Chambersburg PA
CBHW041945260926
51850CB00018B/845
* 9 7 8 3 0 3 6 5 8 0 1 3 5 *